Deep Energy Retrofit

Alexander Zhivov • Rüdiger Lohse

Deep Energy Retrofit

A Guide to Achieving Significant Energy Use Reduction With Major Renovation Projects

Springer

iea Energy Technology Network

EBC
Energy in Buildings and Communities Programme

Alexander Zhivov
US Army Engineer Research
and Development Center, Construction
Engineering Research Laboratory
Champaign, IL, USA

Rüdiger Lohse
Linkenheim-Hochstetten
Baden-Württemberg, Germany

Participating countries in EBC:
Australia, Austria, Belgium, Canada, P.R. China, Czech Republic, Denmark, France, Germany, Ireland, Italy, Japan, Republic of Korea, the Netherlands, New Zealand, Norway, Portugal, Singapore, Spain, Sweden, Switzerland, United Kingdom and the United States of America.

ISBN 978-3-030-30678-6 ISBN 978-3-030-30679-3 (eBook)
https://doi.org/10.1007/978-3-030-30679-3

This Springer imprint is published by the registered company Springer Nature Switzerland AG.
The registered company address is: Gewerbestrasse 11, 6330 Cham, Switzerland

Preface

The International Energy Agency

The International Energy Agency (IEA) was established in 1974 within the framework of the Organization for Economic Cooperation and Development (OECD) to implement an international energy program. A basic aim of the IEA is to foster international cooperation among the 29 IEA participating countries and to increase energy security through energy research, development, and demonstration in the fields of technologies for energy efficiency and renewable energy sources.

The IEA Energy in Buildings and Communities Programme

The IEA coordinates international energy research and development (R&D) activities through a comprehensive portfolio of Technology Collaboration Programmes. The mission of the IEA Energy in Buildings and Communities (IEA EBC) Programme is to develop and facilitate the integration of technologies and processes for energy efficiency and conservation into healthy, low emission, and sustainable buildings and communities through innovation and research. (Until March 2013, the IEA EBC Programme was known as the IEA Energy in Buildings and Community Systems Programme, ECBCS.)

The R&D strategies of the IEA EBC Programme are derived from research drivers, national programs within IEA countries, and the IEA Future Buildings Forum Think Tank Workshops. These R&D strategies aim to exploit technological opportunities to save energy in the building sector and to remove technical obstacles to market the penetration of new energy-efficient technologies. The R&D strategies apply to residential, commercial, office buildings and community systems and will impact the building industry in five areas of focus for R&D activities:

- Integrated planning and building design
- Building energy systems

- Building envelope
- Community scale methods
- Real building energy use

The Executive Committee

Overall control of the IEA EBC Programme is maintained by an Executive Committee, which not only monitors existing projects but also identifies new strategic areas in which collaborative efforts may be beneficial. As the Programme is based on a contract with the IEA, the projects are legally established as Annexes to the IEA EBC Implementing Agreement. At the present time, the following projects have been initiated by the IEA EBC Executive Committee, with completed projects identified by (∗):

Annex 1: Load Energy Determination of Buildings (∗)
Annex 2: Ekistics and Advanced Community Energy Systems (∗)
Annex 3: Energy Conservation in Residential Buildings (∗)
Annex 4: Glasgow Commercial Building Monitoring (∗)
Annex 5: Air Infiltration and Ventilation Centre
Annex 6: Energy Systems and Design of Communities (∗)
Annex 7: Local Government Energy Planning (∗)
Annex 8: Inhabitants Behaviour with Regard to Ventilation (∗)
Annex 9: Minimum Ventilation Rates (∗)
Annex 10: Building HVAC System Simulation (∗)
Annex 11: Energy Auditing (∗)
Annex 12: Windows and Fenestration (∗)
Annex 13: Energy Management in Hospitals (∗)
Annex 14: Condensation and Energy (∗)
Annex 15: Energy Efficiency in Schools (∗)
Annex 16: BEMS 1 – User Interfaces and System Integration (∗)
Annex 17: BEMS 2 – Evaluation and Emulation Techniques (∗)
Annex 18: Demand Controlled Ventilation Systems (∗)
Annex 19: Low Slope Roof Systems (∗)
Annex 20: Air Flow Patterns Within Buildings (∗)
Annex 21: Thermal Modelling (∗)
Annex 22: Energy Efficient Communities (∗)
Annex 23: Multi Zone Air Flow Modelling (COMIS) (∗)
Annex 24: Heat, Air and Moisture Transfer in Envelopes (∗)
Annex 25: Real Time HVAC Simulation (∗)
Annex 26: Energy Efficient Ventilation of Large Enclosures (∗)
Annex 27: Evaluation and Demonstration of Domestic Ventilation Systems (∗)
Annex 28: Low Energy Cooling Systems (∗)

Annex 29: Daylight in Buildings (*)
Annex 30: Bringing Simulation to Application (*)
Annex 31: Energy-Related Environmental Impact of Buildings (*)
Annex 32: Integral Building Envelope Performance Assessment (*)
Annex 33: Advanced Local Energy Planning (*)
Annex 34: Computer-Aided Evaluation of HVAC System Performance (*)
Annex 35: Design of Energy Efficient Hybrid Ventilation (HYBVENT) (*)
Annex 36: Retrofitting of Educational Buildings (*)
Annex 37: Low Exergy Systems for Heating and Cooling of Buildings (LowEx) (*)
Annex 38: Solar Sustainable Housing (*)
Annex 39: High Performance Insulation Systems (*)
Annex 40: Building Commissioning to Improve Energy Performance (*)
Annex 41: Whole Building Heat, Air and Moisture Response (MOIST-ENG) (*)
Annex 42: The Simulation of Building-Integrated Fuel Cell and Other Cogeneration Systems (FC+COGEN-SIM) (*)
Annex 43: Testing and Validation of Building Energy Simulation Tools (*)
Annex 44: Integrating Environmentally Responsive Elements in Buildings (*)
Annex 45: Energy Efficient Electric Lighting for Buildings (*)
Annex 46: Holistic Assessment Tool-Kit on Energy Efficient Retrofit Measures for Government Buildings (EnERGo) (*)
Annex 47: Cost-Effective Commissioning for Existing and Low Energy Buildings (*)
Annex 48: Heat Pumping and Reversible Air Conditioning (*)
Annex 49: Low Exergy Systems for High Performance Buildings and Communities (*)
Annex 50: Prefabricated Systems for Low Energy Renovation of Residential Buildings (*)
Annex 51: Energy Efficient Communities (*)
Annex 52: Towards Net Zero Energy Solar Buildings (*)
Annex 53: Total Energy Use in Buildings: Analysis and Evaluation Methods (*)
Annex 54: Integration of Micro-Generation and Related Energy Technologies in Buildings (*)
Annex 55: Reliability of Energy Efficient Building Retrofitting – Probability Assessment of Performance and Cost (RAP-RETRO) (*)
Annex 56: Cost Effective Energy and CO2 Emissions Optimization in Building Renovation
Annex 57: Evaluation of Embodied Energy and CO2 Equivalent Emissions for Building Construction (*)
Annex 58: Reliable Building Energy Performance Characterization Based on Full Scale Dynamic Measurements (*)
Annex 59: High Temperature Cooling and Low Temperature Heating in Buildings (*)

Annex 60: New Generation Computational Tools for Building and Community Energy Systems
Annex 61: Business and Technical Concepts for Deep Energy Retrofit of Public Buildings
Annex 62: Ventilative Cooling
Annex 63: Implementation of Energy Strategies in Communities
Annex 64: LowEx Communities – Optimized Performance of Energy Supply Systems with Exergy Principles
Annex 65: Long-Term Performance of Super-Insulating Materials in Building Components and Systems
Annex 66: Definition and Simulation of Occupant Behavior in Buildings
Annex 67: Energy Flexible Buildings
Annex 68: Indoor Air Quality Design and Control in Low Energy Residential Buildings
Annex 69: Strategy and Practice of Adaptive Thermal Comfort in Low Energy Buildings
Annex 70: Energy Epidemiology: Analysis of Real Building Energy Use at Scale
Annex 71: Building Energy Performance Assessment Based on In-Situ Measurements
Annex 72: Assessing Life Cycle Related Environmental Impacts Caused by Buildings
Annex 73: Towards Net Zero Energy Public Communities
Annex 74: Energy Endeavour
Annex 75: Cost-Effective Building Renovation at District Level Combining Energy Efficiency and Renewables

Working Group – Energy Efficiency in Educational Buildings (*)
Working Group – Indicators of Energy Efficiency in Cold Climate Buildings (*)
Working Group – Annex 36 Extension: The Energy Concept Adviser (*)
Working Group – HVAC Energy Calculation Methodologies for Non-residential Buildings

Champaign, IL, USA
Linkenheim-Hochstetten
Baden-Württemberg, Germany

Alexander Zhivov
Rüdiger Lohse

Acknowledgements

The "Deep Energy Retrofit – A Guide to Achieving Significant Energy Use Reduction with Major Renovation Projects" was developed within the IEA EBC Program Annex 61, "Business and technical Concepts for Deep Energy Retrofit of Public Buildings," as the result of a joint effort by researchers and practitioners from Austria, Canada, China, Denmark, Estonia, Germany, Latvia, the UK, and the United States. The authors express their appreciation to the many international contributors and organizations whose volunteer efforts allowed to develop this practical Guide. Special gratitude to the ASHRAE and its Technical Committees TC 7.6 "Building Energy Performance," TC 5.2 "Duct Design," and a Standing Standard Project Committee 90.1 for providing a platform for public discussion of the project progress and a review of the document by its members and to CIBSE for sharing its technical information and providing valuable input into the document. The authors would like to personally thank the members of the EBC Program Executive Committee for their directions, guidance, and support to the project. Special appreciation to the following reviewers, who provided their valuable comments and suggested improvements to this Guide: Isabella Zwerger (BMVIT, Austria), Ala Hasan (VTT, Finland), Peter Spafford (ABAA, USA), Sean Garrett and Ann Kosmal (GSA, USA), Brian Nohr, Roderick Bridgeman and Billy Tindell (USACE, USA), James Works (USAF, USA), and Brian Cooper and John Miller (CH2M, USA). The authors gratefully acknowledge William J. Wolfe, writer-editor, ERDC-ITL, for his help in coordinating the preparation of this document. The authors would like to acknowledge the financial and other support of the following: the office of the Assistant Secretary of the US Army for Installations, Energy, and Environment; the US Army Corps of Engineers; the US Department of Energy (Federal Energy Management Program); Austria; the EUDP program of the Danish Energy Agency; and Bundesministerium für Wirtschaft und Energie, Germany.

Contents

About the Authors and Contributors

Authors

Dr. Alexander Zhivov US Army Engineer Research and Development Center, Construction Engineering Research Laboratory, Champaign, IL, USA

Rüdiger Lohse, Linkenheim-Hochstetten, Baden-Württemberg, Germany

Contributors

George Adams SPIE, UK (Ch. 1); Nickolas Alexander, US Army Corps of Engineers (USACE), USA (Ch. 9, App. F); Wagdy Anis, Anis Building Enclosure Consulting, USA (Ch. 9, App. F); Jeff Boldt, KJWW, USA (Ch. 12); Dr. Anatolis Borodinecs, Riga Technical University, Latvia (Ch. 10); John Arcidiacono, TMI Air Barrier Testing, USA (Ch. 8, App. H); Herman Behls, Behls & Associates, USA (Ch 12, App H); Eric Brandt, Aalborg University Copenhagen, Denmark (Ch. 6); Kim Brandt, Rockwool, Denmark (App. C); Terry Brennan, Camroden Associates, USA (Ch. 10); Brian Cahill, Aspen Aerogels, USA (App. C); Dr. Michael Case, US Army Engineer Research and Development Center (ERDC), USA (App. A); Anthony Cinnamon, Wiss, Janney, Elstner Associates, USA (Ch. 7); Brian Clark, ERDC, USA (App. H); Laverne Dalgleish, ABAA, Canada (App. F); Alexander Geissels, Saint-Gobain, Germany (App B, C); Christopher Glover, Airxchange, USA (Ch. 12), Chad Groshart, Atelier Ten, USA (App. G, H), Paul Johnson, SmithGroup, USA (App. H); Dr. Targo Kalamees, Tallinn University of Technology, Estonia (App. A); Dr. Berthold Kaufmann, Passivhaus Institut (PHI), Germany (Apps. B, E); Dr. Andris Kreslins, Riga Technical University, Latvia (App. A); Kalle Kuusk, Tallinn University of Technology, Estonia (App. A); Mark Lawton, Canada (Ch. 8, Appendix D); Mark Lien, Illuminating Engineering Society, USA (App. G),

Dr. Richard Liesen, ERDC, USA (Ch. 6, App. I); Chris Mathis, Mathis Consulting, USA (Ch. 7); Lars-Ake Mattson, Lindab, Sweden (Ch.12); Dr. Tõnu Mauring, University of Tartu, Estonia (App. A); Dr. Ove Christen Mørck, Cenergia Energy Consultants, Denmark (App. A); Alejandra Nieto, ROXUL, Canada (App. B, C); Tomas O'Leary, Passive House Academy (PHA), Ireland (App. D, E); Axy Pagan-Vazquuez, ERDC, USA (App. D); Vincent Paladino, TREMCO/Canam, USA (App. F); Dr. Christian Pohl, Monier, Germany (Ch. 7, App. C); Martina Riel, KEA, Germany (App. A); Dr. Jørgen Rose, Aalborg University Copenhagen, Denmark (Ch. 8, App. A, D); William Rose, University of Illinois at Urbana-Champaign (UIUC), USA (Ch. 10); Rahul Shira, Philips, USA (App. G), Dr. Martin Spitzner, FIW München, Germany (Ch. 6 App. B, C); Larry Smith, Lindab, USA (Ch. 12); Peter Spafford, ABAA, USA (App. F); Heimo Staller (App A, E); Dr. Galina Stankevica, Riga Technical University, Latvia (App. A); Dr. John Straube, RDH, Canada (App D,E); Kirsten Engelund Thomsen, Aalborg University Copenhagen, Denmark (App. A); Steven Tratt, Canam Building Envelope Specialists, Canada (Ch. 9, App. F); Jay Tulley, Presidio of Monterey Army Garrison (AG), USA (App. H); Simone Weber, Steep, Germany (App. F).; Dr. Runming Yao, Reading University, United Kingdom (UK) (App. A); Justine Yu, US Army Engineer Research and Development Center (ERDC), USA (App. D).

List of Figures

List of Tables

Chapter 1
Introduction

Many governments worldwide are setting more stringent targets for reductions in energy use in government/public buildings. Buildings constructed more than 10 years ago account for a major share of energy used by the building stock. However, the funding and "know-how" (applied knowledge) available for owner-directed energy retrofit projects have not kept pace with new requirements. With typical retrofit projects, reduction of energy use varies between 10% and 20%, while experience from executed projects around the globe shows that energy use reduction can exceed 50%, and renovated buildings can cost-effectively achieve the passive house standard or even approach net zero-energy status (IEA-EBC 2017; Hermelink and Müller 2010; NBI 2014; RICS 2013; GreenBuildingAdvisor.com 2013; Shonder and Nasseri 2015; Miller and Higgins 2015; Emmerich et al. 2011).

Research conducted under the International Energy Agency Energy in Buildings and Communities Program (IEA-EBC) Annex 46 (2009) identified and analyzed more than 400 energy efficiency measures (EEMs) that can be used when buildings are retrofitted. Implementation of some individual measures (such as building envelope insulation, improved airtightness, and cogeneration) can significantly reduce building heating and cooling loads or minimize of energy waste, but require considerable investments with long paybacks.

Research under the IEA EBC Program Annex 61 has been conducted with a goal of providing a framework, selected tools, and guidelines to significantly reduce energy use (by more than 50%) in government and public buildings.

One of the Annex 61 deliverables is the book of *Deep Energy Retrofit – Case Studies*, which contains 26 well-documented case studies from Europe (Austria, Denmark, Estonia, Germany, Ireland, Latvia, Montenegro, The Netherlands, the United Kingdom) and the United States. After these data were collected, the case studies were analyzed with respect to energy use before and after renovation, reasons for undertaking the renovation, co-benefits achieved, resulting cost-effectiveness, and the business models followed. Finally, the lessons learned were compiled and compared. Lessons learned from case studies and experiences of the team led to the conclusion that Deep Energy Retrofit (DER) can be achieved with a limited core

A. Zhivov, R. Lohse, *Deep Energy Retrofit*,
https://doi.org/10.1007/978-3-030-30679-3_1

technologies bundle readily available on the market. Characteristics of some of these core technology measures depend on the technologies available on an individual nation's market, on the minimum requirements of national standards, and on economics (as determined by a life cycle cost [LCC] analysis). Also, requirements to building envelope-related technologies (e.g., insulation levels, windows, vapor and water barriers, and requirements for building airtightness) depend on specific climate conditions.

When a limited number of core technologies are implemented together (i.e., "bundled"), they can often significantly reduce energy use for a smaller investment and thereby provide a faster payback. The synergistic effect of bundled technologies strategy proposed for the Deep Energy Retrofit distinguishes it from the more commonly used energy-saving approach based on an analysis of the financial payback of individual technologies and measures. For more details on economics of DER, business models, project structuring and project financing options, direct cost savings benefits and monetizing of indirect benefits, macro- and microeconomic barriers for implementation of DER projects in the public building sector, see Annex 61 "Deep Energy Retrofit – Business Guide."

Deep Energy Retrofit – A Guide to Achieving Significant Energy User Reduction with Major Renovation Projects contains **recommendations for characteristics** of some **of the core technologies and measures** that are based on studies conducted by national teams associated with the International Energy Agency Energy Conservation in Buildings and Communities Program (IEA-EBC) Annex 61 (Lohse et al. 2016; Case et al. 2016a, b; Rose et al. 2016; Yao et al. 2016a, b, c; Dake 2014; Stankevica et al. 2016; Kiatreungwattana 2014). Results of these studies provided a base for setting minimum requirements to the building envelope-related technologies to make Deep Energy Retrofit feasible and, in many situations, cost-effective. Use of energy efficiency measures (EEMs) in addition to core technologies bundle and high-efficiency appliances will foster further energy use reduction. This Guide also provides **best practice examples of how to apply these technologies in different construction situations.**

The Guide is focused on public buildings constructed before the 1980s with low internal loads (e.g., office buildings, dormitories, barracks, public housing, and educational buildings) undergoing major renovation. This bundle of technologies allows site (end) energy use reduction of 50% or better (defined as Deep Energy Retrofit) in most climate conditions and results in significant reduction of carbon footprint. This task is more difficult in hot climate conditions with significant cooling needs and may require additional EEMs to be applied (e.g., reduction of plug loads, water conservation measures, advanced heating, ventilating, and air-conditioning [HVAC] systems). Deep Energy Retrofit is easier to achieve in heating-dominated climates and in cases when either by cultural or normative reasons, cooling is not desired and building users can tolerate temporary increases in indoor air temperature (e.g., up to 77 °F [25 °C]).

Additional energy efficiency technologies, especially those that result in process load reduction, are necessary to achieve Deep Energy Retrofit in buildings with significant ventilation requirements and high internal loads such as dining facilities, hospitals, or data centers.

High levels of energy use reduction using core technology bundles along with improvements in indoor climate and thermal comfort can be only achieved when a Deep Energy Retrofit (DER) adopts a quality assurance (QA) process. In addition to design, construction, commissioning, and postoccupancy phases of the quality assurance process, the Guide emphasizes the importance of clearly and concisely formulating and documenting the owner's goals, expectations, and requirements for the renovated building during development of the statement of work (SOW). Another important component of the quality assurance (QA) process is a procurement phase, during which bidders' qualifications, their understanding of the SOW and its requirements, and their previous experience are analyzed.

The target audience for this Guide are technical experts: Energy Service Companies (ESCOs) and general contractors, design, architectural and engineering professionals, and manufacturers of energy-efficient products and systems. It can be also of interest to building owners and to executive decision-makers and energy managers of public, governmental, and military organizations.

Chapter 2
What Is Deep Energy Retrofit?

Though the DER concept is currently widely used all over the world, there is no established global definition of this term. Since the energy crisis of the 1970s, energy requirements for new construction and building renovation worldwide have significantly improved. Tables 2.1 and 2.2 list standards and requirements used to design and construct buildings pre-1980s and today.

Since the 1980s, building energy use requirements in the United States (Table 2.1) have been reduced by more than 50% (calculated without consideration of plug loads). Also, buildings and building systems degrade over time. They develop cracks in the building envelope, and dirty and leaky ducts; HVAC systems are not regularly commissioned, etc. This results in a reduction of their energy performance of at least 10%. Therefore, it is theoretically possible to reduce building energy use by more than 50% simply by using technologies readily available on the market and by adapting current requirements for new buildings to the refurbishment of existing building stock.

The recently rewritten European Energy Performance Building Directive (EPBD 2010), which requires buildings to "be refurbished to a nearly zero-energy condition," states that "member states should not be required to set minimum energy performance requirements that are not cost effective over a building's estimated economic lifecycle." By the EPBD definition, a nearly zero-energy building (NZEB) is "a building that has a very high energy performance." In the United States, the "Massachusetts Save Energy Retrofit Builder Guide" refers to DER as the "the retrofit of the building enclosure and other building systems in a way that results in a high performance building." Not many national and international bodies take their definition beyond this level of specificity except for Austria, Germany, and the Czech Republic, which decided that a high performance or "nearly zero-energy" building is a building meeting approximately the Passive House Institute Standard. Denmark has decided to use a new standard defined in the Danish Building Regulations 2010 referred to as the "2020 definition of NZEB."

Analysis conducted by the Annex 61 team (Subtask A report) shows that a significant number of commercial and public buildings have reduced their energy

A. Zhivov, R. Lohse, *Deep Energy Retrofit*,
https://doi.org/10.1007/978-3-030-30679-3_2

Table 2.1 Historical improvement of the ASHRAE Standard 90.1 (data are compared without plug loads)

ASHRAE Standard 90.1 version	Energy use index
1975	100
1980	100
1989	86
1999	81.5
2001	82
2004	69.7
2007	65.2
2010	46.7
2013	43.4

Source: Tillou (2014)

Table 2.2 Historical improvement in European National Energy Requirements for Buildings

Country	National Standard/Code	Energy Use Intensity (EUI), kWh/m2	
		Pre-1980	Current
Denmark	BR10 (2010)	Dwellings:	Dwellings: 52.5 kWh/m^{2}y + 1650 kWh/GFA[*] (16.7 MBtu/sq. ft-y + 5631 MBtu/GFA)
		167.1 kWh/m^{2}y (53 MBtu/sq. ft-y)	Office: 71.3 kWh/m^2 y + 1650 kWh/GFA (22.6 MBtu/sq. ft-y + 5631 MBtu/GFA)
Germany	Pre-1980:	Dwellings:	Dwellings (new): 50–60 kWh/m^{2}y (15.9–19 MBtu/sq. ft-y) schools new:
	WSVO[†] (1977)	150–250 kWh/m^{2}y (47.6–79.3 MBtu/sq. ft-y) (IWU 2011)	Heating demand: 80–105 kWh/m^{2}y (25.4–33.3 MBtu/sq. ft-y),
			Electricity: 10—20 kWh/m^{2}y (3.2–6.3 MBtu/sq. ft)
	Current: Energy-Saving Ordinance (EnEV 2014) for new buildings	Schools: 210 kWh/m^{2}y (66.7 MBtu/sq. ft-y) (IWU 2011)	Examples of renovation of other building types show that much lower EUIs are achievable office buildings as well (e.g., <40 kWh/m^{2}y [12,687.6 Btu/sq. ft-y]) (annex 612,015)
	Refurbishment: EnEV 2009 + <40%		
Austria	OIB[‡] (2011) RL 6	Maximum U-values	Heating energy demand:
			Residential buildings: max. 87.5 kWh/m^{2}y (27.8 MBtu/sq. ft-y)
			Nonresidential buildings: max. 30 kWh/m3 (2.9 MBtu/cu ft) y

Source: ASHRAE 2013a, b, c; WSVO 1977; OIB 2011
[*]Gross Floor Area (GFA)
[†]Wärmeschutzverordnung (German: Ordinance on Thermal Insulation)
[‡]Österreichisches Institut für Bautechnik (Austrian Institute for Structural Engineering)

consumption by more than 50% after renovation and that some have met the Passive House Institute energy efficiency standard or even net zero-energy state. According to the Global Building Performance Network prognosis (RICS 2013), deep retrofit that follows the most recent and proposed European Union (EU) guidance can improve the building energy performance by at least 80%.

Based on experiences described above, the IEA-EBC Annex 61 (Zhivov et al. 2015) team has proposed the following definition of the DER:

Deep Energy Retrofit (DER) is a major building renovation project in which site energy use intensity (including plug loads) has been reduced by at least 50% from the pre-renovation baseline with a corresponding improvement in indoor environmental quality and comfort.

Deep Energy Retrofit requires a whole-building analysis approach along with an integrative design process. A "whole-building analysis" means that the building is considered as a single, integrated system rather than as a collection of standalone systems, such as building envelope, HVAC system, renewable energy system, building operations, etc. The whole-building approach facilitates the identification of synergistic relationships between the component systems. Analyzing systems in isolation does not effectively identify synergies between systems. For example, improving the building envelope, providing solar heat gain control, and improving lighting systems could substantially reduce a building's heating and cooling energy demand. This would in turn reduce the size of duct systems, smaller air-handling units, boilers, and chillers. Also, replacing an aging air-handling unit with a smaller, more efficient unit could improve indoor air quality and further reduce energy demand. Such cascading benefits would not be achievable if the building were not analyzed as an integrated whole.

The key to whole-building analysis is the use of an integrated design process. This approach differs from a traditional design process in that it brings all relevant disciplines together for an initial charrette-based study of the problem as a whole, based on collaboration and shared information, whereas a more traditional process is based on a linear flow of information passing from one discipline to another. Appendix H of ASHRAE 189.1 gives more information on integrative design and the charrette process.

Chapter 3
Deep vs. Shallow Energy Retrofit

A literature analysis by Bettgenhäuser et al. (2014) shows that an alternative to deep renovation allowing for fossil fuel use reduction is a shallow renovation track that focuses less on energy efficiency and more on renewable energy supply. However, such "shallow renovation" with very high shares of renewable energy often turns out to be more expensive.

According to Nock and Wheelock (2010), shallow renovation projects, such as retro-commissioning, provide savings of 10–20% with an average payback slightly over 1 year. These projects are low-risk investment opportunities with an average cost savings over 3–5 years. Energy savings using Energy Performance Companies focusing primarily on HVAC and lighting systems usually result in 20–40% with a simple payback in public buildings between 3 and 12 years. Energy audits conducted by the Pacific Northwest National Laboratory (Baechler et al. 2011) have identified numerous low-cost and no-cost energy efficiency measures with a simple payback below 5 years.

Dynamic simulations conducted by Hermelink and Müller (2011) for building stock scenarios have shown that using available annual budget for many cheaper shallow renovations saving instead of using fewer, more expansive deep energy renovations saving may lead to unwanted, irreversible long-term consequences (Fig. 3.1). Although the return on investment (ROI) ratio of a "shallow renovation" may look better than that of a DER for meeting short-term goals, the shallow renovation will fail to achieve long-term energy goals.

However, when funds for building retrofits are not expected in the foreseeable future, different cost-effective combinations of EEMs that maximize the energy savings at a building can be implemented with energy projects using operation and maintenance funds or through energy performance contracting. Real data for US Federal Facilities collected by the US Department of Energy (USDOE) Federal Energy Management Program (FEMP) (Fig. 3.2) shows that it is difficult to achieve long-term energy goals without implementing an aggressive strategy for energy reduction in existing building stock.

A. Zhivov, R. Lohse, *Deep Energy Retrofit*,
https://doi.org/10.1007/978-3-030-30679-3_3

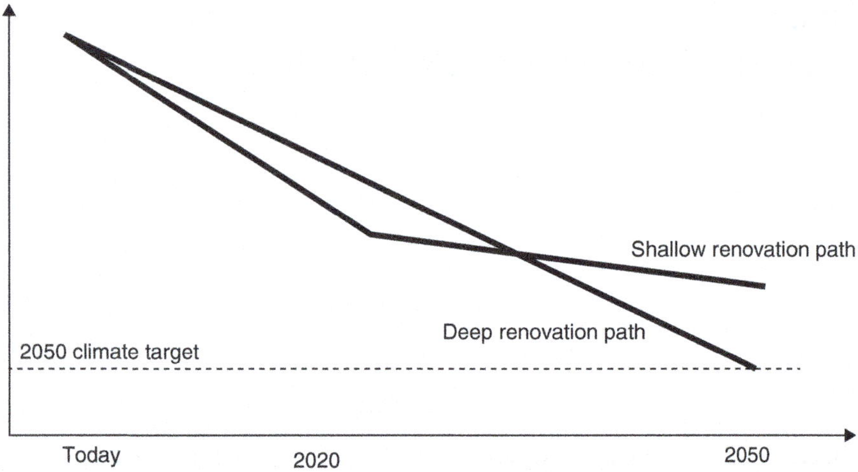

Fig. 3.1 The speed trap of shallow renovation (from "Economics of Deep Renovation"). (Source: Hermelink and Müller 2011)

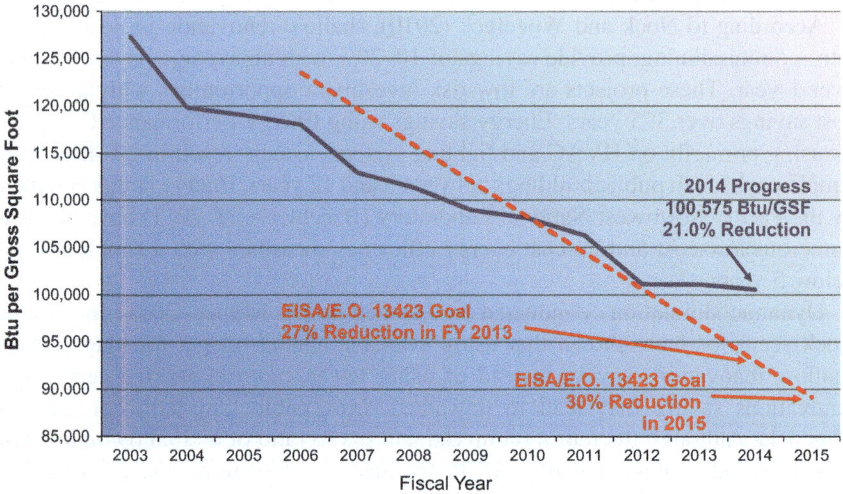

Fig. 3.2 Overall government progress toward facility energy efficiency goals, fiscal year (FY) 2003–2014

Chapter 4
Major Renovation and Deep Energy Retrofit

Typical Energy Efficiency Improvement Projects are planned as:

- A part of major building renovation[1]
- A part of minor building renovation
- Utilities modernization projects
- Mechanical and electrical equipment/systems replacement
- System retro-commissioning
- Dedicated energy projects using an Energy Savings Performance Contract (ESPC) or Utility Energy Service Contract (UESC).

Buildings usually undergo major renovations for many reasons, including the need to reduce energy consumption. Among some of the most common reasons are:

- Extension of the useful life requiring overhaul of its structure, internal partitions, and systems
- Repurposing of the building (e.g., renovation of old warehouses into luxurious apartments)
- Bringing the building to new or updated codes
- Remediation of environmental problems (mold and mildew), improvement of the visual or thermal comfort, or indoor air quality
- Adding to the value with improvements to increase investment (increasing useful space and/or space attractiveness/quality) resulting in a higher sale or lease price.

The DOE (2010) and EPBD (EU 2010) define a major building renovation as any renovation with the cost that exceeds 25% of the replacement value of the building. EPBD also defines building renovation as a major renovation if more than 25% of the surface of building envelope undergoes renovation. US DOD policy (2013)

---

[1]The US Department of Energy (DOE) (DOE 2010) and EPBD (EU 2010) define a major building renovation as any renovation where the cost exceeds 25% of the replacement value of the building. EPBD also defines building renovation as a major renovation if more than 25% of the surface of building envelope undergoes renovation.

A. Zhivov, R. Lohse, *Deep Energy Retrofit*,
https://doi.org/10.1007/978-3-030-30679-3_4

defines major renovation project with renovation cost exceeding 50% of estimated replacement cost (ERC).

Timing a DER to coincide with a major renovation is best since the building is typically evacuated and gutted; scaffolding is installed; single pane and damaged windows are scheduled for replacement; building envelope insulation is replaced and/or upgraded; and most of mechanical, electrical lighting, and energy conversion systems (e.g., boiler and chillers) along with connecting ducts, pipes, and wires will be replaced. A significant sum of money covering the cost of energy-related scope of the renovation designed to meet minimum energy code is already budgeted anyway.

Chapter 5
Core Bundles of Technologies to Achieve Deep Energy Retrofit with Major Building Renovation Projects

IEA ECBCS Annex 46 (2014) has identified and analyzed numerous energy efficiency measures (EEMs) that can be used when buildings are retrofitted. Measures include those related to the building envelope, mechanical and lighting systems, energy generation and distribution, internal processes, etc. Implementation of some individual measures (such as building envelope insulation, improved airtightness, cogeneration, etc.) can significantly reduce building heating and cooling loads or minimize energy waste but require significant investments with long paybacks. However, when a limited number of "core technologies" are implemented together ("bundled"), they can significantly reduce energy use for a smaller investment, thereby providing a faster payback.

A list of core energy efficiency technologies (Table 5.1) was generated from the results of case studies of DERs conducted in Europe and North America, from surveys and discussions conducted at the ASHRAE Technical Committee (TC) 7.6 "Public Buildings" working group meetings in 2013 and 2014, and from previous experience and research conducted by the Annex 61 team members. These technologies, when applied together (as a bundle), will reduce the total building site energy use by about 50% (including plug loads). Technical characteristics of these building envelope-related technologies grouped into a "core technologies bundle" have been studied through modeling and LCC analysis for representative national climate conditions. Other characteristics of these technology bundles are based on the requirements of national standards (Tables A-3 and A-4) or on best international practices, which have been collected and summarized and presented in this Guide.

When buildings are retrofitted, additional EEMs can be used to gain greater energy savings than can be achieved by using a "core technologies bundle" alone. The use of some of these measures may depend on the end-user, rather than on the contractor (e.g., when purchasing and installing more energy-efficient appliances and other plug loads or installing separate power lines and timers to turn off some of the electrical appliances). Other measures might include those that are specific to a

A. Zhivov, R. Lohse, *Deep Energy Retrofit*,
https://doi.org/10.1007/978-3-030-30679-3_5

Table 5.1 Core technologies bundles for DER

Category	Name	Source for characteristics
Building envelope	Roof insulation	Modeling results
	Wall insulation	Modeling results
	Slab insulation	National requirements (Table A-5)
	Windows	Modeling results
	Doors	National requirements (Table A-5)
	Thermal bridges remediation	DER guide based on best practices
	Airtightness	The most stringent national requirements
	Vapor barrier	DER guide based on best practices
	Building envelope QA	DER guide based on best practices
Lighting and electrical systems	Lighting design, technologies, and controls	DER guide based on best practices
HVAC	High-performance motors, fans, furnaces, chillers, boilers, etc.	The most stringent national requirements
	Dedicated outdoor air system (DOAS)	DER guide based on best practices
	Heat recovery (HR) (dry and wet)	The most stringent national requirements
	Duct insulation	The most stringent national requirements
	Duct airtightness	The most stringent national requirements
	Pipe insulation	The most stringent national requirements

particular building type (e.g., water-saving showerheads and clothes washers, which can significantly reduce domestic hot water usage) or measures specific to the project (e.g., use of such low exergy heating and cooling systems as indirect evaporative cooling, radiant heating and cooling systems, heating and cooling return water energy, and other waste streams).

Chapter 6
Building Envelope Technologies

Core technologies comprising the DER bundle include insulation of walls, roofs (attic), and slabs; existing windows and doors replacement with high performance products (including efficient sun shading systems in climates where the air-conditioning (A/C) will be turned on in warm periods); mitigation of thermal bridges; improvement of building airtightness and vapor control through the building envelope; and implementation of a QA process for the building envelope retrofit. This section describes these technologies and their characteristics required for DER.

6.1 Thermal Insulation

Many older buildings undergoing major renovation may have inadequate thermal insulation in the exterior envelope. Existing building insulation types and thermal performance should be evaluated as part of the initial energy audit. The information generated during the energy audit will inform the decisions the design and construction team will need to make in developing the DER strategies and the final DER plan. Insulation type and location and the options for its installation depend on whether the building is considered to be historic/listed, local climate, building structure and its existing fabric, building height, fire code requirements, and a few other factors.

6.2 Thermal Improvements to Existing Building

The Annex 61 modeling team has optimized building envelope minimum heat transmission (U-values) for wall and roof assemblies and windows thermal requirements through computational modeling (described in Appendix A). Based on the results of these studies, the total U-values for wall and roof assemblies

A. Zhivov, R. Lohse, *Deep Energy Retrofit*,
https://doi.org/10.1007/978-3-030-30679-3_6

Table 6.1 Total wall U-value ("c.z." = climate zone)

Country	U-value W/(m^2*K) (Btu/[hr*ft^2*°F])	R-value (m^2*K)/W ([hr*ft^2*°F]/Btu)
Austria (c.z. 5A)	0.135 (0.024)	7.4 (42)
c.z. 6A	0.24 (0.043)	4.17 (23)
China c.z. 2A	0.96(0.169)	1.0(6)
c.z. 3A	0.96(0.169)	1.0(6)
c.z. 3C	0.60(0.106)	1.7(9)
c.z. 4A	0.48(0.084)	2.1(12)
c.z. 7	0.31(0.054)	3.2(19)
Denmark (c.z. 5A)	0.15 (0.026)	6.7 (38)
Estonia (c.z. 6A)	0.17 (0.03)	5.9 (33)
Germany (c.z. 5A)	0.17 (0.03)	5.9 (33)
Latvia (c.z. 6A)	0.19 (0.033)	5.3 (30)
UK (c.z. 4A)	0.22(0.039)	4.5(26)
5A	0.22(0.039)	4.5(26)
USA c.z. 1	0.76 (0.133)	1.3 (8)
c.z. 2	0.38 (0.067)	2.6. (15)
c.z. 3	0.28 (0.050)	3.6 (20)
c.z. 4	0.23 (0.040)	4.3 (25)
c.z. 5	0.19 (0.033)	5.3. (30)
c.z. 6	0.14 (0.025)	7.1. (40)
c.z. 7	0.11 (0.020)	9.1 (50)
c.z. 8	0.11 (0.020)	9.1 (50)

required to achieve DER in different climate conditions have been identified and are summarized in Tables 6.1 and 6.2. These values should be used as starting points for the energy modeling analysis. The final thermal performance values will be determined by results of the detailed modeling.

Building envelope insulation requirements depend on the types of walls and roofs that comprise the building envelope. For example, steel-framed wall constructions have larger U-values for the same insulation combinations than wood-framed wall constructions. This is due to the large thermal heat loss through structural elements in the cavity.

The total thermal resistance R_T and the thermal transmittance, U_T, of a flat building assembly component (wall or roof) comprised of parallel layers between the environments at both sides are given by (ASHRAE Handbook 2013a, b, c):

$$R_T = R_i + R_{assembly} + R_o, U_T = 1/R_T \tag{6.1}$$

where:

R_i = combined inner surface film resistance, m^2K/W or h*ft^2*F/Btu
R_s = resistance of building assembly surface to surface, including thermal resistance of possible air layers in component, m^2K/W or h*ft^2*F/Btu.
R_o = combined outer surface film resistance, m^2K/W or h*ft^2*F/Btu.

Table 6.2 Total roof U-value

Country	Climate zone	U-value W/(m^2*K) (Btu/[hr*ft^2*°F])	R-value (m^2*K)/W ([hr*ft^2*°F]/Btu)
Austria	5A	0.159 (0.028)	6.3 (36)
	6A	0.23 (0.041)	4.4 (25)
China	2A	0.53 (0.093)	1.9(11)
	3A	0.53 (0.093)	1.9(11)
	3A	0.53 (0.093)	1.9(11)
	4A	0.38(0.067)	2.6(15)
	7	0.30 (0.053)	3.3(19)
Denmark	5A	0.10 (0.018)	1 (57)
Estonia	6A	0.11 (0.02)	9.1 (52)
Germany	5A	0.14 (0.025)	7.1 (40)
Latvia	6A	0.16 (0.029)	6.3 (35)
UK	4A	0.13(0.023)	7.7 (44)
	5A	0.13(0.023)	7.7 (44)
USA	1	0.16 (0.029)	6.3 (35)
	2	0.14 (0.025)	7.1 (40)
	3	0.12 (0.022)	8.3 (45)
	4	0.12 (0.022)	8.3 (45)
	5	0.11 (0.020)	9.1 (50)
	6	0.09 (0.0167)	11.1 (60)
	7	0.09 (0.0154)	11.1 (65)
	8	0.08 (0.0133)	12.5 (75)

Note that R_s can be comprised of external sheathing, one or two layers of continuous insulation, solid wall layer (concrete or masonry) and/or the layer with cavities, and the internal layer of the gypsum board (Fig. 6.1) and can be calculated by:

$$R_{assembly} = R_{ext.sheathing} + R_{cont\ insulation,1} + R_{cont\ insulation,2} + R_{effective\ cavity} + R_{int\ gypsum} \quad (6.2)$$

If the thermal conductivity λ of the material of a homogeneous flat layer is known (e.g., from the insulation layer, the gypsum boards, etc.), the thermal resistance R_{layer} of that layer may be calculated by dividing the thickness of the layer by the thermal conductivity of the material of that layer:

$$R_{layer} = thickness_{layer}/\lambda \quad (6.3)$$

and used in Eq. 6.2 as summed on the right-hand side.

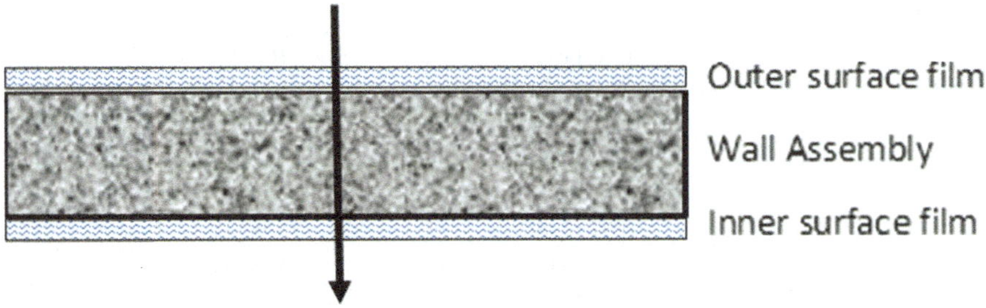

Fig. 6.1 Simplified schematic of heat transfer through wall (roof) assembly

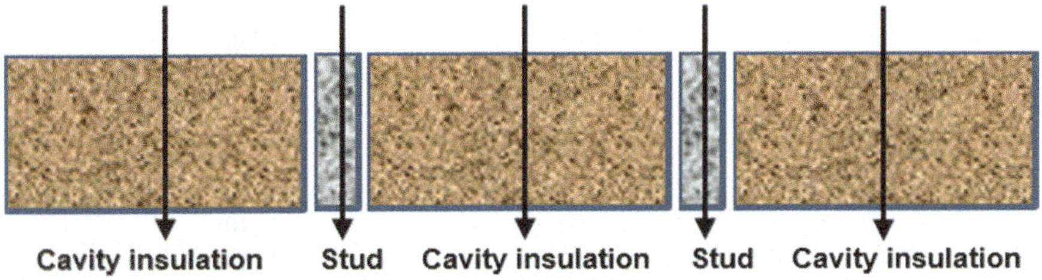

Fig. 6.2 Influence of studs on U-value

Therefore, using the total U-values for walls and roofs listed in Tables 6.1 and 6.2 and the information on existing wall and roof assemblies and their thermal properties, one can evaluate the required U-value(s) of contiguous insulation.

In many building assemblies (e.g., wood frame, metal frame, concrete frame construction), components are arranged so that heat flows in parallel paths of different conductance (Fig. 6.2). If no heat flows through lateral paths, the average thermal transmittance U_{av}-value of the wall assembly can be calculated using a parallel heat flow path method given by:

$$U_{av} = aU_a + bU_b + \ldots. + nU_n \tag{6.4}$$

where a, b,…n are surface-weighted path fractions for a typical basic assembly composed of several different paths with transmittances $U_{a,}$ U_b …,U_n.

If heat flow through lateral paths is to be taken into account, the average thermal transmittance U_{av}-value of the wall assembly can be calculated using a combined parallel and lateral heat flow path method. The method first calculates a U-value only with parallel heat flows as above (which is slightly lower than the precise result), then calculates a U-value with lateral heat flow within the nonhomogeneous layer (which is slightly higher than the precise result), and then takes the reciprocal average between the two as average thermal transmittance U_{av} for the assembly:

$$U_{tot,lower} = aU_a + bU_b + \ldots. + nU_n \tag{6.5}$$

$$U_{tot,upper} = 1/\left(R_i + R_{element,nonhomog\ layers\ averaged} + R_o\right) \tag{6.6}$$

$$1/U_{av} = \left(1/U_{tot,lower} + 1/U_{tot,upper}\right)/2 \tag{6.7}$$

The thermal resistance R $_{element,\ nonhomog\ layers\ averaged}$ can then be calculated according to Eq. 6.2, but for all nonhomogeneous layers (i.e., a layer with insulation between studs), its thermal resistance is calculated from its thickness divided by an area-weighted average (areas *a, b, . . . n*) of the thermal conductivities $\lambda_a, \lambda_b \ldots \lambda_n$ of the *n* neighboring materials within that layer. For example, if insulation layer 1 is continuous (e.g., mineral wool matt over the studs) and insulation layer 2 consists of insulation between the studs:

$$\begin{aligned} R_{assembly,nonhomog\ layers\ averaged} = {} & R_{ext.sheathing} + R_{cont\ insulation,1} \\ & + \left[d_{nonhomog\ insulation\ 2}/\left(a_{(studs)}\,\lambda_{studs} + b_{(insulation)}\lambda_{insulation}\right)\right] \\ & + R_{effective\ cavity} + R_{int\ gypsum} \end{aligned} \tag{6.8}$$

Note that wooden studs or beams in a thermal insulation layer may be calculated with these formulas (6.3 or 6.7), which give an average U-value for walls or roofs with nonhomogeneous material composition. Use of these equations to calculate the influence of metal studs or the metal-framed assembly is not recommended because thermal conductivity of steel is too high. The effect of metal penetrations through insulation layers has to be calculated using 2D or 3D thermal flux calculations resulting in explicit psi- or chi-values. Please note that in countries where ISO 6946 has been introduced, lateral heat flows are to be taken into account in U_{av}; therefore, Eqs. 6.5, 6.6, 6.7 and 6.8 should be used (according to ISO 6946).

Parallel heat flows can occur in wall assemblies with wooden, metal, or concrete frames where the insulation is installed in a cavity. Another example of parallel heat flows is the situation when highly conductive materials like steel or concrete are passing through insulation layers, creating "thermal bridges," resulting in building envelopes with higher overall thermal transmittance (Section 8, "Thermal Bridges," p 51).

When the existing wall or roof assembly composition is known, the level of continuous (internal and/or external) insulation and/or cavity insulation can be calculated based on the total U-values listed in Tables 6.1 and 6.2 and on the thermal properties of the existing assembly, for example:

$$R_{cont.insulation} = R_T - R_i - R_{exiting\ assembly} - R_o \tag{6.9}$$

In this equation, R $_{existing\ assembly}$ represents the effectiveness of the wall assembly before renovation, which accounts for existing insulation and framing with associated thermal bridges.

Table 6.3 U-values for the wood and steel wall type calculation

Element	R-Steel Framing (m^2*K)/W (h*ft^2*°F)/Btu	R-Wood Framing (m^2*K)/W ([h*ft^2*°F]/Btu)
Outdoor air film	0.03 (0.17)	0.03 (0.17)
Stucco	0.01 (0.08)	0.01 (0.08)
Exterior sheathing board	0.1 (0.56)	0.10 (0.56)
R13 effective cavity insulation	1.06 (6.0)	1.62 (9.185)
Interior gypsum board	0.10 (0.56)	0.10 (0.56)
Interior air film	0.12 (0.68)	0.12 (0.68)
Total R-value	1.42 (8.05)	1.98 (11.235)
Total U = 1/R-total	0.02 (0.124)	0.016 (0.089)
Overall U assembly – Base wall	0.02 (0.124)	0.016 (0.089)

Therefore, using the total U-values for walls and roofs listed in Tables 6.1 and 6.2 and the information on existing wall and roof assemblies and their thermal properties, one can evaluate the required U-value(s) of continuous insulation.

This analysis can be done, e.g., using the ASHRAE Standard 90.1, Tables A3.3 and A3.4 for "Assembly U-Factors for Steel Frame Walls" and "Assembly U-Factors for Wood Frame Walls," or through calculation using a series path for heat flow, or a parallel heat flow path method or the "modified zone method." The data listed in Table 6.3 illustrate how the overall U-value resulting in a given amount of continuous insulation with R-13 installed in a cavity of an 88.9-mm (3.5-in.) thick steel-framed wall can differ from the overall U-value in a wood-framed wall with the same amount of insulation.

Adding to the existing wall assembly with R-8.05, as illustrated in the example above, continuous insulation with R-10 (preferably on the outside of the cavity layer) results in the new total R-value = 18.05 and the $U_{assembly} = 1/18.05 = 0.055$. This value can be found in Table A3.3 of the ASHRAE 90.1 Standard.

The following examples show calculations for the above-grade walls using requirements for the U-values from Table 6.4. We are going to take three specifications to explore the process from Table 6.4:

- China c.z. 3A U = 0.106 and R = 9.
- Germany c.z. 5A U = 0.030 and R = 33.
- USA c.z. 7 U = 0.020 and R = 50.

6.3 Masonry or Concrete Walls

For this example, it is assumed that the base masonry wall or concrete wall will have a continuous insulation installed on the interior, exterior, or within the wall. It is assumed that the U-factors for a masonry block wall with a base wall are: R-0.17 for exterior air film; R-0.68 for interior air film and for vertical surfaces; and R-0.45 for 0.5-in. gypsum board; and that the rated R-value is for continuous insulation uninterrupted by framing:

Table 6.4 Overall U-values for the base masonry or concrete wall types

Element	China c.z. 3A R min = 9 (m^2*K)/W ([h*ft^2*°F]/Btu)	Germany c.z. 5A R min = 33 (m^2*K)/W ([h*ft^2*°F]/Btu)	USA c.z. 7 R min = 50 (m^2*K)/W ([h*ft^2*°F]/Btu)
R min continuous insulation	6.36 = 9–2.64	30.36 = 33–2.64	47.36 = 50–2.64
Extruded polystyrene (XPS) foam insulation thickness	R-7.5 = 1 ½ in.	R-30 = 6 in.	R-50 = 10 in.
Total R-value	1.79 (10.14)	5.75 (32.64)	9.27 (52.64)
Overall U assembly = 1/R tot	0.098	0.0306	0.019

$$R_{assembly} = R_{air\ film} + R_{masonry\ block\ wall\ partially\ grouted\ cores} + R_{int\ gypsum} + R_{air\ film}$$
$$R_{base\ assembly} = 0.17 + 1.23 + 0.56 + 0.68 = 2.64$$

The data listed in Table 6.4 indicate that the U-factor would not be affected by the placement of the continuous insulation although the reactivity of the capacitance would be. These examples and insulation specifications only consider the U-factor specification for the energy retrofits. Adding continuous insulation and determining the U-factor is a straightforward 1D calculation.

6.4 Wood-Framed Wall

The wood-framed wall is more complex. This example assumes a wood-framed wall in which the framing cavity is filled first, after which continuous insulation is applied to meet the specified value for each climate zone (Tables 6.5 and 6.6).

Calculations for the wood-framed base wall can be done using the parallel path method. The base assembly is a wall in which the insulation is installed between 2-in. nominal wood framing, cavity insulation is full depth, and headers are double 2-in. nominal wood framing. The U-factors include R-0.17 for exterior air film, R-0.08 for stucco, R-0.56 for 15.88-mm (0.625-in.) exterior sheathing board on the exterior, R-0.56 for 15.87-mm (0.625-in.) gypsum board on the interior, and R-0.68 for interior air film, vertical surfaces. Standard framing: wood framing at 406-mm (16-in.) on center with cavities filled with 368-mm (14.5-in.) wide insulation for both 89-mm (3.5-in.) deep and 140-mm (5.5-in.) deep wall cavities. Double headers leave no cavity. Weighting factors are 75% insulated cavity, 21% studs, plates, and sills, and 4% headers. Based on this input data:

$$U\ assembly = (0.75 * 0.066) + (0.25 * 0.156) = 0.089 \text{ for the base wood-framed wall.}$$
$$R_{effective\ cavity} = \text{Cavity Insulation} + \text{wood framing} = 9.19$$
$$R_{assembly} = R_{air\ film} + R_{stucco} + R_{ext.sheathing} + R_{cont\ insulation} + R_{effective\ cavity} + R_{int\ gypsum} + R_{air\ film}$$
$$R_{assembly} = 0.17 + 0.08 + 0.56 + 0 + 9.19 + 0.56 + 0.68 = 11.24$$

Table 6.5 Overall U-values for the wood-framed wall

Element	R (insulated cavity (m^2*K)/W ([h*ft^2*°F]/Btu at 75%	R (studs, plates, headers) (m^2*K)/W ([h*ft^2*°F]/Btu at 25%
Outdoor air film	0.030 (0.17)	0.030 (0.17)
Stucco	0.014 (0.08)	0.014 (0.08)
Exterior sheathing board	0.099 (0.56)	0.099 (0.56)
Cavity insulation	2.290 (13)	—
Wood stud	–	0.772 (4.38)
Interior gypsum board	0.099 (0.56)	0.099 (0.56)
Interior air film	0.120 (0.68)	0.120 (0.68)
Total	2.652 (15.05)	1.133 (6.43)
Total U	0.012 (0.066)	0.027 (0.156)
Overall U assembly – Base wall	0.016 (0.089)	

Table 6.6 Amount of continuous insulation required for each location and climate zone

Element	China c.z. 3A R min = 9 (m^2*K)/W ([h*ft^2*°F]/Btu)	Germany c.z. 5A R min = 33 (m^2*K)/W ([h*ft^2*°F]/Btu)	USA c.z. 7 R min = 50 (m^2*K)/W ([h*ft^2*°F]/Btu)
R min continuous insulation	Base wall meets specification	**21.76** = 33–11.24	**38.76** = 50–11.24
XPS foam insulation thickness	R-0 = 0-in. ci	R-22.5 = 45-in. ci	R-40 = 8-in. ci
Total R-value	1.980 (11.24)	5.945 (33.74)	9.028 (51.24)
Overall U assembly = 1/R tot	0.016 (0.089)	0.005 (0.0296)	0.003 (0.0195)

This example shows the methodology that is used to calculate the values that can be obtained from ASHRAE Standard 90.1 appendices for low conductance wood framing.

For steel stud framing with insulation, the modified zone method is used to account for the thermal anomalies around a metal stud. All of the methods are described.

6.5 Steel-Framed Wall

The steel-framed wall is even more complex. The calculation of $R_{effective}$ uses the modified zone method (described in detail in the ASHRAE 2013 Fundamentals in Chapter 27 in the Two-Dimensional Assembly U-Factor Calculation section). It is assumed that the steel-framed wall cavity will be insulated first and then continuous insulation will be applied to meet the specified value for each climate zone.

The base wall assembly has the insulation installed within the cavity of the steel stud framing. The steel stud framing is a minimum uncoated thickness of 0.043 in. for 18 gage or 0.054 in. for 16 gage. The U-factors include R-0.17 for exterior air film, R-0.08 for stucco, R-0.56 for 0.625-in. gypsum board on the exterior, R-0.56 for 0.625-in. gypsum board on the interior, and R-0.68 for interior vertical surfaces air film. The performance of the insulation/framing layer is determined using the values from Table A9.2–2 in ASHRAE Standard 90.1.

The R-13 steel wall with the studs at 16 in. on center has a $U_{overall} = 0.124$ or $R_{overall} = 8.05$, which takes into account the framing factor and the metal studs in the cavity. The entire base wall is $R_{overall} = 8.05$, although the effective R for the cavity with the installed insulation is R = 6:

$$R_{effective\ cavity} = \text{Cavity Insulation} + \text{steel framing} = 6.0$$

$$R_{assembly} = R_{air\ film} + R_{stucco} + R_{ext.sheathing} + R_{cont\ insulation} + R_{effective\ cavity} + R_{int\ gypsum} + R_{air\ film}$$

$$R_{assembly} = 0.17 + 0.08 + 0.56 + 0 + 6.0 + 0.56 + 0.68 = 8.05$$

Compared to the wood-framed existing wall, R-values have changed from R-11.24 to R-8.05 with the base wall and the cavity insulated. Table 6.7 lists the requirements to continuous insulation to meet the specifications for each climate zone.

These examples demonstrate how the base wall construction type information provided in the ASHRAE Standard 90.1 can be used to obtain the required continuous insulation levels. More complex scenarios can be evaluated by programs like THERM, a two-dimensional conduction heat transfer analysis program based on the finite-element method, which can model the complicated geometries of building constructions. This program and others like it can provide the overall U-value to see if it complies with a given specification.

6.6 Thermal Insulation Materials

An insulating material layer's resistance to conductive heat flow is rated in terms of its thermal resistance or R-value – the higher the R-value, the greater the insulating effectiveness of the layer. The R-value depends on the type of insulation, its thickness, and its density. When calculating the R-value of a multilayered installation, add the R-values of the individual layers. The effectiveness of an insulation material's resistance to heat flow also depends on how and where the insulation is installed. For example, insulation that is compressed will not provide its full rated R-value. The overall R-value of a wall or ceiling will be somewhat different from the R-value of the insulation itself because heat flows more readily through studs, joists, and other building materials, creating "thermal bridging." Thermal insulation that fills building cavities reduces air circulation in these cavities and therefore minimizes convective heat loss.

Table 6.7 Requirements to continuous insulation to meet the specifications for several climate zones

Element	China c.z. 3A R min = 9 $(m^2*K)/W$ $([h*ft^2*°F]/Btu)$	Germany c.z. 5A R min = 33 $(m^2*K)/W$ $([h*ft^2*°F]/Btu)$	USA c.z. 7 R min = 50 $(m^2*K)/W$ $([h*ft^2*°F]/Btu)$
R min continuous insulation	0.95 = 9–8.05	24.95 = 33–8.05	41.95 = 50–8.05
XPS foam insulation thickness	R-2.5 = 0.5 in ci	R-25 = 5 in ci	R-42.5 = 8.5 in ci
Total R-value	1.859 (10.55)	5.823 (33.05)	8.906 (50.55)
Overall U assembly = 1/R tot	0.017 (0.095)	0.005 (0.030)	0.003 (0.0198)

6.6.1 How Much Insulation Is Required?

This section discusses how to calculate the thickness of insulation given the properties and location of existing wall insulation (which will be different for continuous insulation placed outside, cavity insulation with studs creating thermal bridges, internal insulation that has some thermal bridging with uninsulated slabs, etc.). The thickness of continuous insulation layer(s) depends on the insulation strategy selected and on the thermal conductivity of selected insulation materials. Recommendations on insulation strategies and different insulation materials are based on their thermal properties (Appendix B), on required assembly U-values described in the previous section, on space availability, and on costs.

6.6.2 Location of Insulation

The required U-value of the wall can be achieved by adding:

- Internal insulation.
- Cavity insulation.
- External insulation.
- A combination of insulation strategies listed above.

6.6.2.1 Internal Insulation

The major advantage using internal insulation is that it does not change the appearance of the building and that it is relatively cheap to install (compared to external insulation). For historic buildings with protected façades (listed buildings, etc.), internal insulation is the only possibility.

Internal insulation is often added with a stud wall that is placed close to the original wall that does not touch it to avoid forming cold bridges. Insulation behind the studs is necessary to avoid forming a cold bridge if profiled steel studs are used (unless the studs are designed for the purpose). This type of internal insulation was already studied in detail in the late 1970s, e.g., in a project reported in Christensen et al. (1978).

Other solutions for internal insulation are available (Figs. 6.3 and 6.4). Instead of a stud wall, a wall of lightweight concrete, e.g., aerated concrete blocks, may be placed in front of the existing wall with a suitable cavity filled with insulation material between the existing wall and the new one. This type of construction is especially well suited in locations where there is a special need to avoid moisture problems and where internal relative humidity (RH) is anticipated to be high, e.g., in bathrooms.

A tight vapor barrier is used to prevent water vapor from entering the construction. This work is often difficult, and good detailing and high-quality workmanship is crucial to achieve tightness, especially at adjoining building elements.

Sometimes, special insulation material is glued to a gypsum board with a built-in vapor barrier. This material is normally only 20–50-mm thick and is either glued or mechanically attached to the interior surface of the existing wall. This method is especially well suited where the space is limited, for example to reduce cold bridges on solid walls in staircases, where it is not possible to use thicker insulation. However, depending on situation, climate, wall material, room usage, insulation material, etc., larger thicknesses of insulation material and/or omission of the built-in vapor barrier may be possible.

Over the last decade, new types of materials have come on the market, e.g., special calcium silicate boards that have a great capacity to alleviate moisture problems by capillary action. When there is condensation or high RH, the moisture is drawn through capillaries back to the surface where it evaporates. These materials have shown good performance although experience with these relatively new products is still limited to about 10 years of use. It is a prerequisite that the surface is only treated with the prescribed kind of paint (very open to diffusion). The insulation

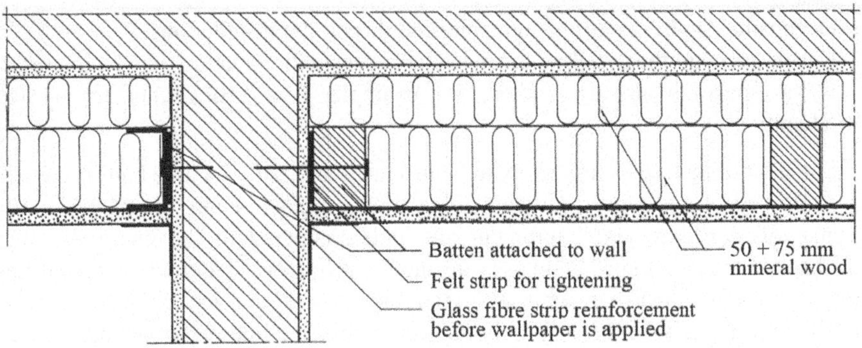

Fig. 6.3 Detail of internally insulated façade around adjoining partition (perpendicular to façade)

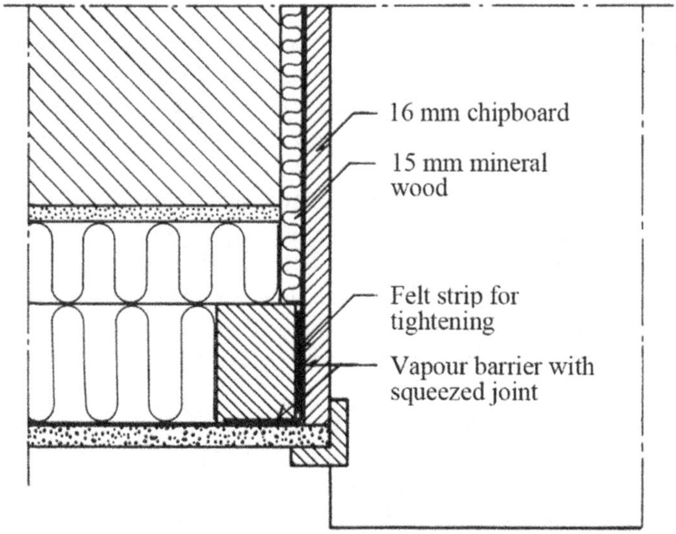

Fig. 6.4 Additional insulation placed behind new opening reveal linings to reduce cold bridges around windows

properties of the materials are not as good as for traditional insulation materials. These materials are especially suited to eliminate cold bridges where the space is limited or to be used on damp walls.

The major drawbacks for internal insulation are:

- Moisture problems/mold growth. As the temperature on the old surface becomes low, there is a considerable risk for high RH or even condensation on the old surface if the vapor barrier is not sufficiently tight. Similar problems occur if the façade is not watertight from the outside. In both cases, the conditions are conducive to mold growth.
- Cold bridges cannot be avoided, i.e., the effect of the insulation is reduced.
- Some rooms, e.g., bathrooms and kitchens, are often not insulated due to high costs.
- Occupants can be greatly inconvenienced during the installation.
- Usable space of the dwelling is reduced; even a small reduction may be important if the space is already limited, for example, if a bed cannot fit in after rehabilitation.

Some building spaces less likely to be insulated internally include bathrooms and kitchens because expenses for new tiling and/or repositioning (or may be even change of) furniture and equipment are high compared to the energy savings (payback time is too long). Staircases are another problem as requirements regarding free space cannot always be maintained if a thick insulation is used. Finally, the size of small rooms may be unacceptably reduced if they are internally insulated.

6.6.2.2 Problems During Construction

Internal insulation can cause great inconvenience to occupants who are often forced to live in a building site for 1–2 months. It is almost impossible to avoid problems with dust, noise, and reduced space during the work.

6.6.2.3 Reduced Area

Internal insulation will always reduce the usable area. For smaller dwellings, the area is reduced by 3–5% dependent on the dwelling – especially dwellings with a gable wall will experience a significant loss in area.

6.6.2.4 Cavity Insulation

Insulating the cavity in brick walls has become popular. For example, in Denmark and in northern regions of Germany it is normally seen as a fast and relatively cheap retrofit that causes no changes to the inside or the outside of the building, and that causes little inconvenience to the occupants. In existing buildings with cavity walls, the only way to add insulation in the cavity is to fill it with a loose material; in these cases, special machines are needed to blow the insulation into place (Fig. 6.5). Of course, the possible insulation thickness is then limited to the thickness of the existing cavity.

The effectiveness of the cavity insulation depends to a large extent on the geometry of the cavity. Normally, for Danish brick walls, about 75-mm space is available, but the available area may be reduced due to headers (rather than wall ties of steel) and solid brickwork around window and door openings. In some buildings, only the upper part of the outer wall has a cavity. The use of headers was the standard until the late 1930s, and cold bridge insulation in cavity walls was first introduced in the 1990s.

6.6.2.5 External Insulation

The advantages of exterior insulation are that it is relatively sensitive to workmanship and that it reduces cold bridges significantly for the entire façade, except where there are cantilevered beams, etc. (which is normally a minor problem). Also, the application of exterior insulation carries almost no risk of moisture problems. The work can be done with little inconvenience to the occupants, and it will protect the structure of the building from temperature fluctuations.

In Denmark, external insulation is most often done using mineral wool insulation mechanically fastened to the existing façade and rendered with a resin mortar (Fig. 6.6). In most cases, this functions well even though some problems have been reported mainly due to bad workmanship, e.g., to thin layers of mortar or poor detail design, e.g., around windows.

Fig. 6.5 An example of a wall insulation, where insulation in the form of loose mineral wool is blown into the cavity. The experience with this kind of additional insulation is very good with low cost and almost no nuisance

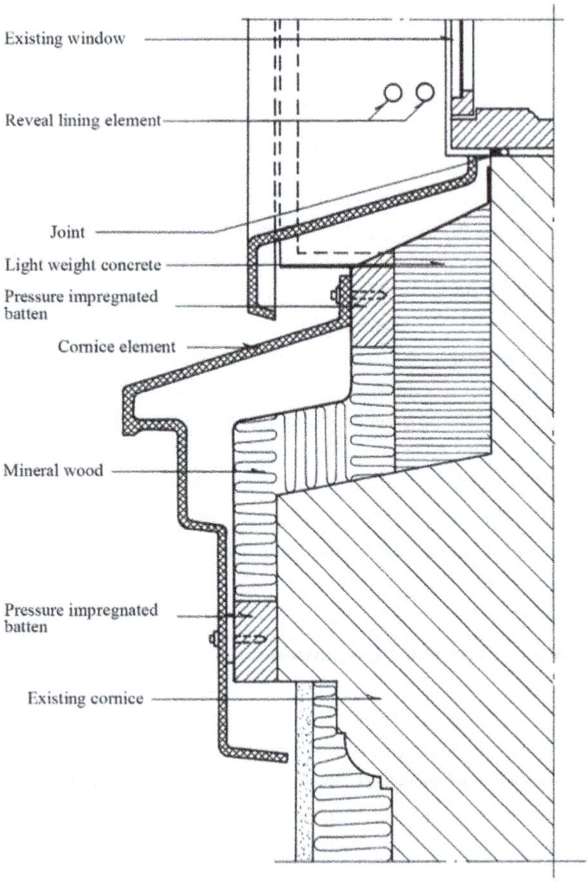

Fig. 6.6 An example of partial replacement of a cornice with prefabricated elements made from glass-fiber reinforced polyester

Also, it is very common to use a ventilated, insulated skeleton structure mounted to the external façade and finished, e.g., with flat boards as external cladding. Such solutions normally perform well.

However, many different systems are used (and many more have been used over the years) including paneling of hard mineral wool, special 10-mm thick brick facing (like tiles) to be glued on the insulation material, prefabricated and painted blocks of mineral wool, and prefabricated elements made especially for gable walls (and other rather large "simple" areas).

The major drawbacks for external insulation are:

- It changes the appearance of the building.
- It requires space around the building.
- It normally requires rather extensive work.

6.6.2.6 Changed Appearance

External insulation will inevitably change the appearance of the building, finer details in old façades especially cannot be reproduced in the new façade. An early (1980s) study by the SBi emphasized preserving the appearance as far as possible (Nørregaard et al. 1981). In this case, the geometry of the façade and the appearance of the windows (even if windows were changed) were especially considered. Some details may be reproduced although somewhat more roughly than the original appearance (Fig. 6.7).

Similar elements may also be made of high-density mineral wool (Rockwool). However, as this example shows, the finest details are normally lost.

In other cases, external insulation is applied quite differently, in ways that change the building's appearance radically. In some cases, a new brick facing is applied to an old concrete building (Figs. 6.7 and 6.8). In such cases, it is normally necessary to

Fig. 6.7 Old concrete walls are used as internal leave in a new construction. Note that the new windows are placed in plane with the additional insulation to reduce cold bridges

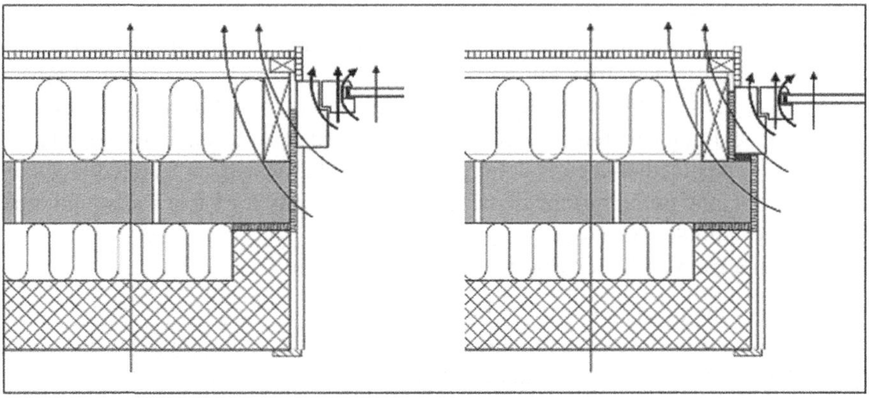

Fig. 6.8 Example on how a window may be positioned in an externally insulated wall. Cold bridges are reduced by use of a thin insulation layer behind reveal linings and positioning the thermal glazing in plane with the additional insulation. To the right, the window is changed to a slightly larger one allowing more daylight. The arrows in the figure indicate the (reduced) energy loss at cold bridges. (Source: Nørregaard et al. 1984)

change the foundation to accommodate the additional load. Experience suggests that these methods are no more costly than other types of external insulation.

For façades made from stud walls, external insulation is very simple to apply as furrings and insulation material can be easily added to the original skeleton construction.

6.6.2.7 Requires Space Around the Building

External insulation requires space around the building. In most cases, this is not a problem for single-family dwellings although it might be a problem for blocks of apartments that face a sidewalk or that have entries placed between buildings. In such cases, the addition of external insulation could cause the sidewalk or the entrance to become too narrow to meet building code requirements, or to allow passage of vehicles.

6.6.2.8 Extensive Works Are Often Needed

To insulate externally, it is normally necessary to change details, e.g., around the roof, doors, and windows, and to move drainpipes, external lamps, etc. For windows, it is recommended to change the position of the window so the thermal glazing is flush with the insulation (Figs. 6.8 and 6.9). The detailing of the façade is crucial to achieve a good technical and aesthetic result. Figure 6.10 shows an example of detailing around roofs from Nørregaard et al. (1984).

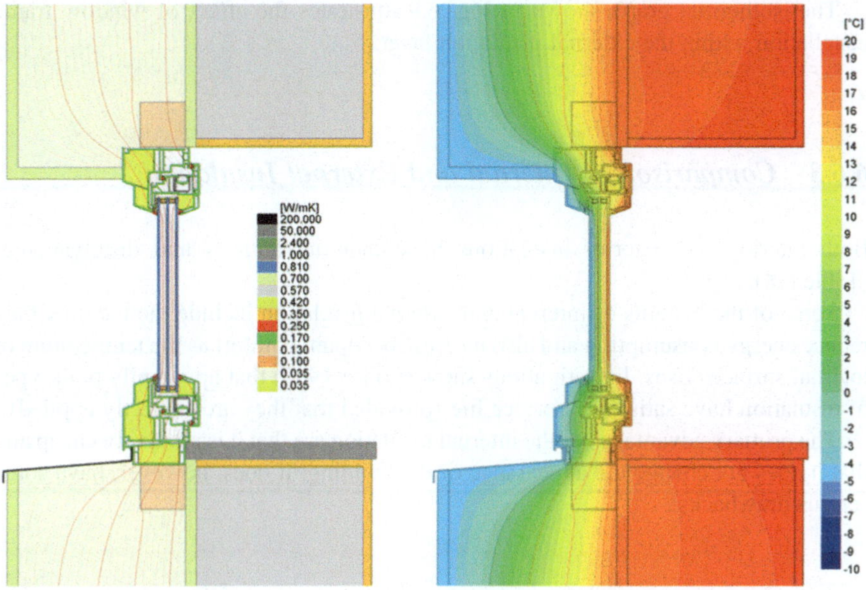

Fig. 6.9 Example of window installation within exterior finish insulation system (EFIS) (reproduced with permission from PHI)

Fig. 6.10 Example on detail at roof for a ventilated wall construction. In other cases, it might be necessary to extend the roof to make room for the additional insulation

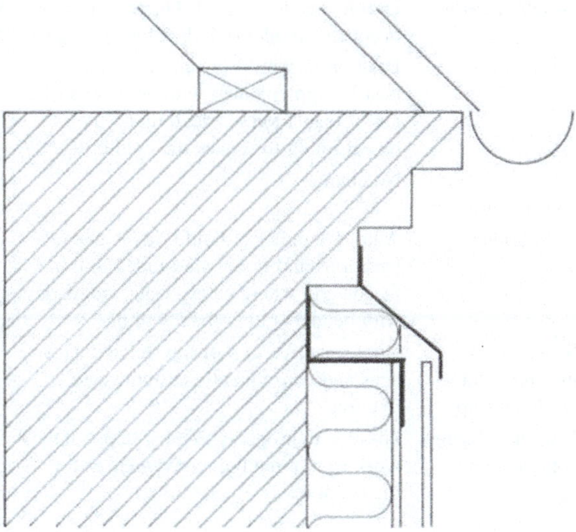

The isothermal graph shown in Fig. 6.9 illustrates the effect of window frame installation within the external insulation layer.

6.6.3 Comparison of Internal and External Insulation

Both interior and exterior insulations have their advantages and disadvantages (Table 6.8).

Some of the benefits of internal and external insulation include the fact that they reduce energy consumption and also increase occupant comfort as the temperature of internal surfaces rises. Investigations show (Lohse 1992) that apparently both types of insulation have satisfying service life (provided that they are correctly applied).

The primary advantages of the internal insulation are that it is relatively cheap and that it does not change the appearance of the building. It does, however, have some serious drawbacks:

Table 6.8 Internal and external insulation methods

Factor	Interior insulation	Exterior insulation
Design exterior	No changes	Normally big changes to appearance
Design interior	Usable area is reduced. The reveal lining becomes broader (window niche deeper)	Window reveal lining becomes deeper
Windows	Position of windows can be maintained best if supplemented with an inner sash and if insulation is added under the reveal lining to reduce the cold bridge	Windows are best moved out in the façade (flush with the additional insulation) to reduce cold bridges
Roof construction	No change	It might be necessary to modify the overhang, flashings, etc.
Cold bridges	A large number of cold bridges where partitions and floors are jointed to the façade and where cavity walls are made with headers rather than wall ties	Cold bridges from the interior are reduced significantly. Cold bridges from cantilevered beams, e.g., in balconies, are increased
Sensitivity to moisture and workmanship	Very sensitive to moisture-related damages and consequently extreme care is required	Sensitivity to failures is modest
Nuisance during construction	Possible to progress step-by-step, but the work causes many inconveniences for the occupiers	Scaffolding and possibly cover is needed. Work may be rather noisy, but rehousing is hardly needed
Other refurbishment works	Can be done together with other interior refurbishment	Especially profitable if combined with necessary external refurbishment of façades
Radiators and other interior installations	Must be repositioned when placed on an outer wall	No change

1. It brings a considerable risk of moisture damages, e.g., mold (depending on the climate zone).
2. It leaves a lot of cold bridges that can only be avoided to a certain extent.
3. It will most likely leave some spaces uninsulated.
4. The work causes an inconvenience to occupants.

The primary advantages of external insulation are that it to a large extent eliminates cold bridges and that it is relatively robust as regards workmanship. The most serious drawbacks are the changed appearance of the building and the price. If a changed appearance can be tolerated – or even is considered a chance to make something new/better – the external insulation is the safest solution with regard to moisture. The price is normally so high that it is hardly ever feasible to make external insulation unless the façade for other reasons needs refurbishment. (A major share of the price goes to scaffolding and surface work.) However, the necessary extent of refurbishment may be reduced as the original façade is being protected by the insulation.

6.7 Roofs: General

Existing roof types covered by the Guide include:

- Flat reinforced concrete slabs.
- Flat roofs in timber with decking and bitumen-based felt.
- Lightweight trusses in timber and steel, supporting a flat concrete slab.
- Pitched roofs of profiled metal sheet panels or concrete or clay interlocking tiles.

While most of the aforementioned (Sect. 6.6) also hold true for flat roofs, pitched roofs require some special considerations. A pitched roof is a complex system composed of several functional layers, each of which performs a specific function. The main functional layers are:

- Tiles.
- Under-tile ventilation.
- Waterproofing underlay.
- Thermal insulation.

The roofing tiles provide the primary waterproofing if you have correct batten gage, correct overlap, etc.

For a good under-tile ventilation, it is recommended to use either:

- Eaves line:
  - Use a first raised row of tiles to allow air flow while preventing leaves and bird intrusion.
  - Batten and counter-batten system (or use preformed insulating boards, which act as substitute for counter battens and allow air flow).

- Ridge line:
  - The use of a ventilated ridge roll effectively protects from infiltration of rainwater and allows air to escape from the ridge.
  - Vent tiles in the roof can increase the under-tile ventilation. Remember: it is not advisable to use cement mortar or polyurethane foam to fix eaves and ridge line as they would prevent air flow.

The waterproofing underlay plays an important role as additional protection element against water infiltration that may cause serious damage to the building. The use of a waterproofing membrane is indicated when the roof pitch is less than 30% and in any geographical area where climatic conditions are adverse. Even under normal conditions, a waterproofing layer is always recommended for better security and airtightness.

The thermal insulation is the main contributor to the overall thermal resistance of the roof and provides resistance to the heat flux to reduce energy loss and/or inflow. The effect of insulation changes with the season:

- Winter: to limit the escape of heat from the heated rooms → energy savings (heating).
- Summer: to limit the entry of heat due to sensible heat and solar radiation → energy savings (air-conditioning), improvement of living comfort.

Depending on the type of thermal insulation, the waterproofing underlay may also be installed underneath the insulation (e.g., with rigid polyurethane (PUR)/ polyisocyanurate (PIR) and some expanded polystyrene [EPS] boards) with or without a second waterproofing layer on top of the insulation.

To further decrease energy inflow in hot and sunny weather (in addition to the use of very good thermal insulation, which is the first important step), there are waterproofing membranes available with an infrared-reflective lining on the outside. When placed on top of the thermal insulation, they reflect the radiative part of the energy inflow back upward to the roofing tiles. Also available are insulation boards with an IR-reflective surface lining on the upper side and/or with steel battens already installed (or with board protrusions) to directly fix the roofing tiles (Fig. 6.11).

6.8 Roofs – Possible Solution to Condensation Risk in Attic Insulation of Pitched Roofs

The renovation of buildings that are 30–40 years old (or even older) usually requires roof renovation including upgrades to their thermal insulation. The building's use, the roof type, and conditions will determine the approach to be used in roof renovation. Two typical situations with pitched roof renovation are:

Fig. 6.11 Under-tile ventilation. Roof layout with (top to bottom) roofing tiles, battens, counter battens for under-tile ventilation space, secondary waterproofing (breather membrane), thermal insulation, waterproofing underlay (vapor control membrane)

- Unused roof attic with horizontal insulation of the attic floor and cold attic roof. Increasing the insulation of the attic floor creates cold bridges and the risk of condensation at the area of the outer wall, where the eaves purlin leave not enough space for sufficient insulation. The installation of a new vapor barrier is also difficult. If the attic space is to be transferred into additional thermally controlled living/working space or a storage, the rafter height is typically not enough to meet modern requirements for insulation thickness. Additional insulation below the rafter reduces the height of usable space. Note that the installation of a vapor control membrane (mainly the airtight connection to the outer walls) is very difficult.
- Roof over the thermally controlled attic space, which can be used as a living/working space or as storage. Such roofs typically have insulation installed between the rafters. The rafter height is usually between 100 and 120 mm, which is not sufficient to meet modern requirements to insulation thickness. Removing the inside plaster board or wooden cladding is a huge cost factor, and as the room height is already fixed, there is no room for insulation below the rafters.

In both situations described above, it would be possible to install all insulation over rafters, which will increase the thickness of the roof system by more than 160 mm. A common currently used approach is to combine between-rafter and over-rafter insulation. Such combination of insulation is essential to avoid condensation risks and thermal bridges.

An important additional benefit resulting from installation of modern pitched roof insulation systems and creating thermally controlled attics is the increase of valuable usable space, which can contribute to cost-effectiveness of the DER project. Examples of such roof renovations are shown in Figs. 6.12, 6.13, and 6.14. Technical details with best practices of roof insulation are presented in Appendix A.

Fig. 6.12 Roof renovation with creating additional working space (Example provided by Braas Monier)

Fig. 6.13 Example of attic renovation at the Klosterhagen Hotel in Bergen. The hotel building has been renovated in 2010. The roof and walls have been insulated and high performance windows been installed in walls and in the roof. That allowed to increase the number of rentable rooms to 15. Windows provide sufficient daylighting during the day and are equipped with shades to be used during nights (when there is almost no dark time throughout nights.)

Fig. 6.14 Example of attic renovation of the building at Rock Island Arsenal. Building has been constructed between 1866 and 1872 and used as a combination of administration and technological shop building. The building had uninsulated roof and walls and single pane windows; the attic has not been used. In 2006, the building has been upgraded to become a landscape office space. As part of the renovation, walls have been insulated using blow-in cellulose insulation, windows have been replaced and the gaps between windows and walls sealed, the roof has been insulated using spray foam insulation and a new reflecting ambient lighting system has been installed, new air heating and ventilation systems have been installed to meet IAQ and thermal comfort needs. Source: Library of Congress (top left, top right); Alexander Zhivov (middle-left, middle right, bottom left, bottom right)

Chapter 7
Windows

7.1 Introduction

Fenestration products have a considerable impact on the total amount of building energy usage since windows typically constitute a large percentage of the area of contemporary building facades. This significant amount of fenestration area combined with much higher U-factors than typical opaque wall areas make them disproportionate contributors to building heating loads. Additionally, fenestration decisions made without due consideration for the management of solar gain often result in these same fenestration systems driving the building air-conditioning load. Therefore, any DER project must pay particular attention to window replacement, including area, U-factors, and Solar Heat Gain Coefficients (SHGCs).

Windows also allow daylight into the building and provide occupants visual contact with their surroundings. As such, the optical properties of the selected window replacements can play a key role in defining interior light loads and visual comfort. The local climate of a project and its existing building design constraints (orientation, overhangs, etc.) may further influence proper fenestration decisions. Some climates might imply benefits from transmitted solar energy to offset heating loads, but only if the building design specifically accommodates such beneficial solar gains. However, most existing buildings are not designed to benefit from simple fenestration solar gain, and improperly selected fenestration can easily result in perimeter space overheating, as well as visual and thermal discomfort.

Whether used in new construction or provided as part of a building's DER rehabilitation, windows can represent a sizeable portion of the project cost while having a large impact on facility energy efficiency, operating costs (energy costs and peak power demand charges), and occupant productivity.

Older windows are commonly single- or double-pane and are likely to have highly conductive metal frames or possibly rotten or damaged wood frames. Older windows can also have cracked glass, locks that do not work, operators and balances

A. Zhivov, R. Lohse, *Deep Energy Retrofit*,
https://doi.org/10.1007/978-3-030-30679-3_7

that no longer function, significant leaks of outside air (both summer and winter) due to missing weatherstripping or badly fitting operable sashes.

When older windows are to be replaced, whether for aesthetic, operational, or for other-than-energy-savings reasons, it is always the proper choice and an economically sound investment to install high-efficiency window systems.

7.2 Diverse Functional Attributes of Fenestration Systems

The energy and daylighting characteristics of windows and fenestration products are a small subset of the diverse functional requirements that must considered for successful product specification, selection, and installation. Each climate and building type may have dramatically different functional requirements for fenestration. For example, some buildings plan for and effectively use their fenestration products to supplement or support natural ventilation systems. For others, particularly those in very severe climates, operable fenestration products that have very durable weather seals may be the priority.

Often, it is a combination of fenestration performance attributes that define the right product for a particular new construction or retrofit project. These combination attributes may include, but are not limited to thermal protection (winter and summer), water shedding, ventilation support, daylighting support, thermal comfort, air sealing, acoustic performance (noise suppression), and other attributes.

Figure 7.1 shows some typical manufacturing and functional characteristics of a window or fenestration systems. In general, these include the frame, glazing, spacer, installation aids, air sealing components, gas fills, coatings, and anchoring.

Figure 7.2 shows schematics of different types of unitized windows commonly used in both residential and commercial applications. Often called "punched windows," these configurations are intended to fill an opening in a wall system.

Note: In general, all sliding windows and doors suffer from poor airtightness if only brush seals are used.

Acceptable airtightness of the closed window can only be achieved, when the movable part of window is pressed actively against rubber seal by the window's metal mounts. Thus, metal mounts must be specially designed. A few other fenestration systems commonly used in commercial applications include storefront, curtain walls, roof windows, and others. These systems may contain both fixed and operable elements.

7.3 Energy Performance

The energy performance of windows and fenestration systems is complex. Windows and fenestration systems have multiple heat transfer mechanisms, each of which impacts different building loads, sometimes simultaneously. To select the proper

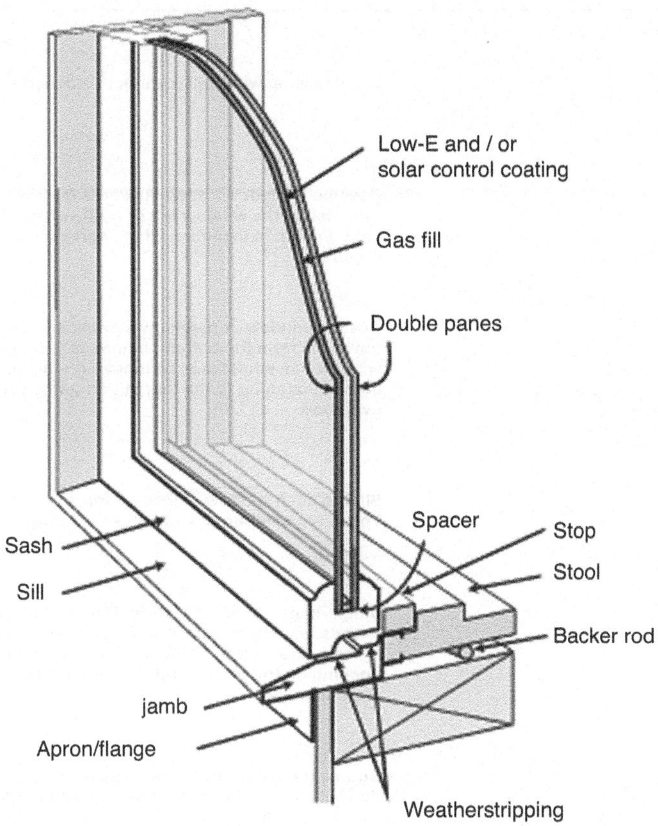

Fig. 7.1 Typical window and fenestration components. (Not shown are anchoring, wall integration, and other common characteristics.) For a more complete listing of components and definitions, see ANSI/AAMA/WDMA I.S.2

fenestration product for any given project, one must have a clear understanding of these different heat transfer mechanisms. The principal energy performance parameters of any fenestration product or system include:

- Thermal transmittance.
- Solar heat gain coefficient.

Additional performance indices that may also impact total building energy savings and long-term operations success include:

- Air leakage.
- Optical properties (spectral properties that define visible transmittance, solar reflectance, and other wavelength-dependent characteristics of the glazing system).

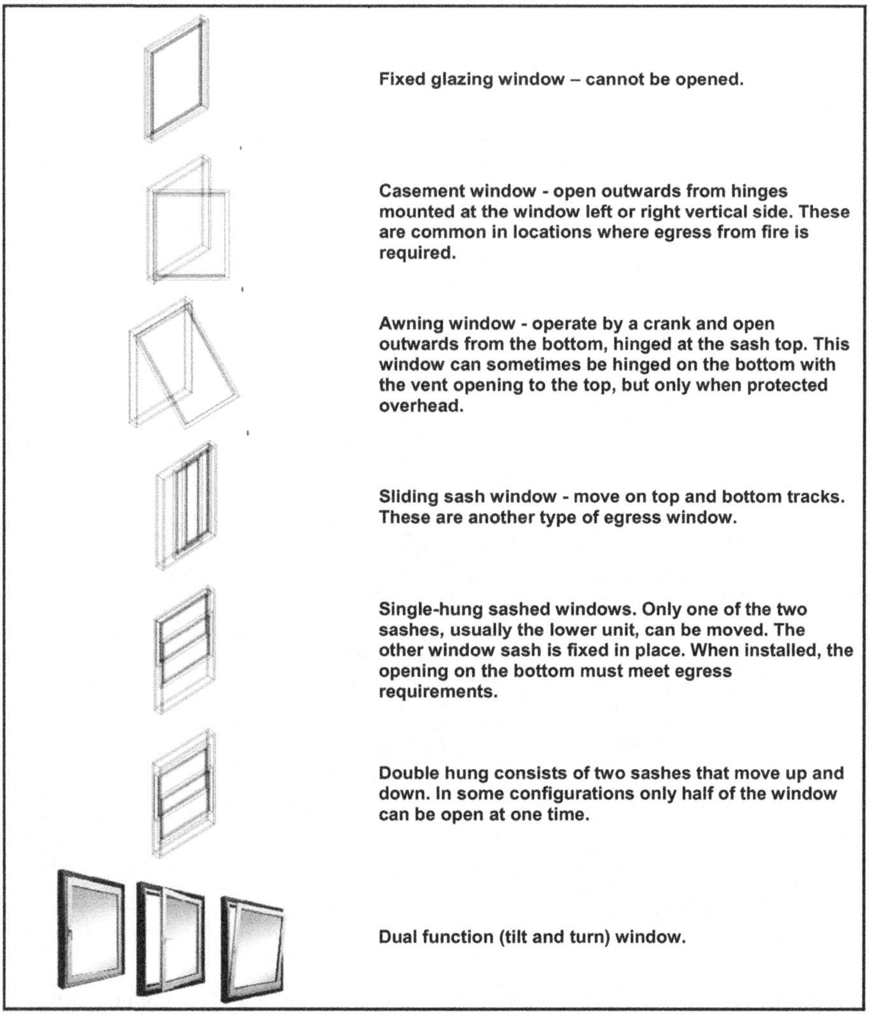

Fig. 7.2 Different types of unitized windows commonly used in both residential and commercial applications

Building design parameters and constraints can also impact proper fenestration product selection. Some of these variables include, but are not limited to:

- Building orientation.
- Window-to-wall ratio (whole building and space-by-space).
- Window-to-floor area ratio (space-specific).
- Quality and integrity of the overall thermal envelope (defining loads, equipment sizing, occupant comfort, and more).

- Depth of perimeter zone (impacting daylight utilization potential, HVAC distribution decisions, and occupant comfort and productivity).
- Fenestration operability (opening force, egress implications, natural ventilation, etc.)

The following sections of this chapter describe each of the fenestration energy performance parameters in more detail and give examples of how each of these indices can impact whole building energy performance.

7.3.1 Thermal Transmittance

U-factor is a measure of a fenestration product's heat transfer due to a temperature difference between inside and outside. U-factor is expressed in units of $W/(m^2{*}K)$ (SI) or $Btu/(hr{*}ft^2{*}F)$ (IP). The U-factor is the primary performance index of consideration when focused solely on heating loads. U-factors range from 1.23 (1.2 $Btu/hr.{*}ft^2{*}F]$) for an aluminum-framed, single-glazed product or system (such as one likely being replaced) to 0.41 (0.25 $Btu/[hr{*}ft^2{*}F]$) for a high performance, double-glazed system in low-conductivity frames common today to 0.28 (0.18 $Btu/[hr{*}ft^2{*}F]$) for a high performance triple- or quad-glazed system that might be selected for climates with extreme heating loads. Maximum U-factors are often prescribed by building and energy codes and vary with climate severity.

7.3.2 Solar Heat Gain Coefficient

SHGC is a measure of a fenestration product or system's ability to transmit or reject incident solar radiation. SHGC is the primary performance index of consideration when the project includes air-conditioning. It is especially important when significant glazing areas are involved. SHGC values range from 0.9 for an aluminum-framed, single-glazed product or system (such as one likely being replaced) to 0.30 for a high performance, double-glazed system in low-conductivity frames common today to 0.20 or less for systems where significant solar control is a priority, such as for projects with significant fenestration area and especially in climates implying extremely high cooling loads. Maximum SHGC values are often prescribed by building and energy codes and vary with climate severity.

7.3.3 Air Leakage

Air leakage is a measure of a window or fenestration system's resistance to air flow from wind and environmental driving forces. Air leakage is typically evaluated by measuring the air flow of a standard-sized window under specific environmental conditions (temperature, wind speed, wind direction, etc.) The pressure difference and air flow through the window or assembly are measured and reported. Standardized air leakage performance of windows has been a key component of window performance certification for decades. Air leakage is usually expressed as $L/(min*m^2)$ in SI units or $ft^3/(min*ft^2)$ in IP units, both under a fixed pressure difference. This is one of the most common window performance indices required by building and energy codes and does not typically vary with climate. Windows that leak a large amount of air can lead to a higher demand on building HVAC systems as well as higher operating costs.

7.3.4 Optical Properties

Today's typical glazing systems can embody an array of different optical properties that can impact many of the other functional objectives of a given project. Different optical properties result from substrates, substrate coatings (or both) that impact particular wavelengths of solar energy. Highly prioritized among optical properties is "visible transmittance" – a measure of how well the fenestration system transmits solar energy in the visible portion of the solar spectrum. If a given project plans or hopes to use daylighting to offset electric lighting expenses, then a high visible transmittance is desired. Conversely, if daylighting and visible clarity are not as important as shedding additional potential solar gain, then products are available that have a very low SHGC and that sacrifice some of the visible clarity. These often have some color or tint in addition to low-e coating designed for reducing heat gain.

Figure 7.3 shows the relationship between energy transmission and optical properties for several different double-pane glazing systems common today. Note that all of the glazing systems illustrated include "low-e" coatings, yet all have dramatically different impacts on heating and cooling loads. Figure 7.4 shows how each of these factors can affect the comfort within in a space, sometimes simultaneously. These simultaneous impacts can exist in both residential and commercial building applications.

7.3.5 Windows and Climate

Proper selection of window and fenestration systems is a function of many variables, but all of the decisions start with climate. The climate conditions (combined with building design, type, and occupancy patterns) define peak heating and cooling loads and also the window energy performance variables to be prioritized.

The window required for subarctic climate (DOE c.z. Zone 8) will certainly be different than that required for very hot and humid climate (DOE c.z. Zone 1)

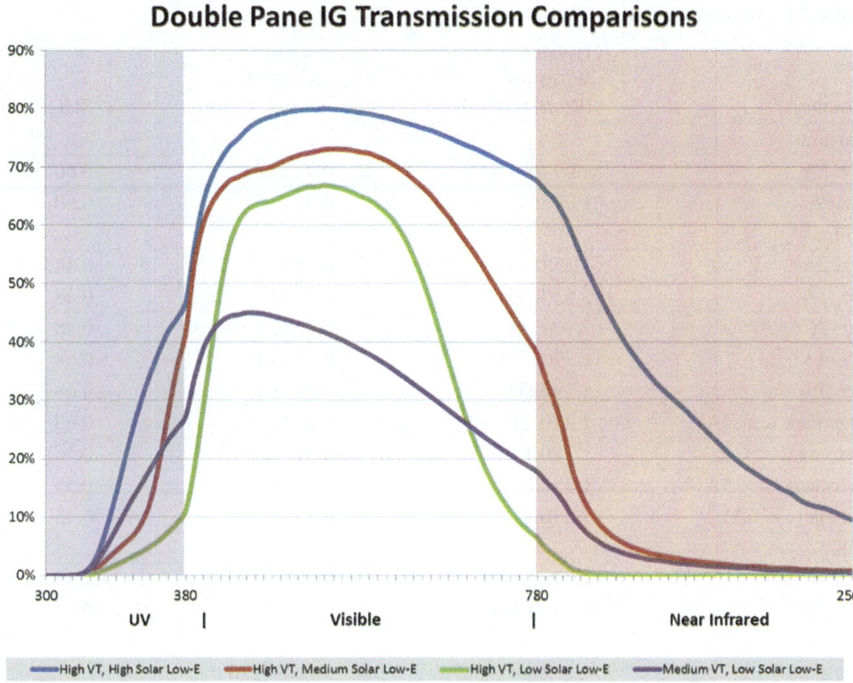

Fig. 7.3 Optical property variations in typical low-e glazing systems

Fig. 7.4 Primary window energy performance elements at work

Table 7.1 Minimum window requirements

Country	U-value W/(m^2*K) (Btu/(hr*ft^2*°F))	R-value (m^2*K)/W (hr*ft^2*°F)/Btu	SHGC
Austria			
c.z. 5A	1.09 (0.19)	0.92 (5.3)	0.60
c.z.6A	1.09 (0.19)	0.92 (5.3)	0.60
China			
c.z. 2A	2.55(0.45)	0.39 (2.2)	0.48
c.z. 3A	2.55(0.45)	0.39 (2.2)	0.48
c.z. 3C	2.70(0.48)	0.37 (2.1)	0.48
c.z. 4A	1.79(0.32)	0.56 (3.1)	0.68
c.z. 7	1.79(0.32)	0.56 (3.1)	0.68
Denmark (c.z. 5A)	1.2 (0.21)	0.83 (4.8)	0.63
Estonia (c.z. 6A)	1.1 (0.19)	0.91 (5.3)	0.56
Germany (c.z. 5A)	1.3 (0.23)	0.77 (4.3)	0.55
Latvia (c.z. 6A)	1.2 (0.21)	0.83 (4.8)	0.43
UK			
(c.z. 4A)	1.32 (0.23)	0.76 (4.3)	0.48
c.z. 5A	1.79 (0.32)	0.56 (3.1)	0.68
USA			
c.z. 1	< 2.15 (< 0.38)	> 0.46 (2.6)	< 0.22
c.z. 2	< 1.98 (< 0.35)	> 0.51 (2.9)	< 0.25
c.z. 3	< 1.81 (< 0.32)	> 0.55 (3.1)	< 0.32
c.z. 4	< 1.70 (< 0.30)	> 0.59 (3.3)	< 0.35
c.z. 5	< 1.53 (< 0.27)	> 0.65 (3.7)	< 0.40
c.z. 6	< 1.36 (< 0.24)	> 0.74 (4.2)	< 0.45
c.z. 7	< 1.25 (< 0.22)	> 0.80 (4.5)	0.40–0.50
c.z. 8	< 1.02 (< 0.18	> 0.98 (5.6)	NR

(Table 7.1). For example, in a Climate Zone 8, a window's ability to retain heat inside the building is most important, so an extremely low U-factor will likely be prioritized. For Climate Zones 1 or 2, the capacity to block heat gain from the sun will likely be a higher priority.

Table 7.1 lists window characteristic determined in modeling studies (Appendix A). These characteristics are *minimum* requirements that are better than current minimum national standards. They are based on the specific national market conditions and climate-specific considerations, and assume that windows are installed without creating thermal bridges between the frame and the wall. U-values represent *average values for the assembly* (frame and glazing) and do not reflect variations due to different window sizes and therefore frame-to-glazing area ratio. Projects seeking a higher level of energy performance should select windows with lower U-values than those listed in Table 7.1.

7.3.6 Window Materials

7.3.6.1 Window Materials and Other Performance Attributes

While window and fenestration performance are the priorities focused on in this Guide, it is helpful to understand some of the various material properties that can influence these performance indices. This material discussion is intended to be informative, with the understanding that it is the ultimate suite of energy performance parameters (U-factor, SHGC, air leakage, etc.) that remain the priority in product selection.

When specifying windows, the energy performance parameters described above are key. However, many other window aspects can affect product specifications. Additional material and product attributes that may figure into specification include, but are not limited to:

- Aesthetics – How does it look or fit with other building envelope elements?
- Type of hardware and operators – What types of handles, levers, balances, etc.
- Operating force – How easy or difficult is it to open and close?
- Compatibility with existing materials – Consider coefficients of expansion, dissimilar metals, etc.
- Maintenance and durability characteristics – Do I have to paint or clean it? How easy is it to clean?
- Ease of Installation.
- Water shedding characteristics – High exposure? Low? Frequency of severe storms, etc.
- Forced entry resistance.
- Cost – First costs, operating costs, maintenance costs, and life cycle costs.

Almost all manufacturers of windows and fenestration system today offer products of many materials. Different combinations of materials can often deliver some or all of the performance attributes described above.

Existing buildings undergoing DERs often present difficult, sometimes conflicting, product specification objectives. For example, a given building in a cold climate might imply specifying a product with low U-factor as a priority. Are these punched openings or curtain wall retrofits? One might imply an operable window solution while the other a fixed system solution. Is this building in a high wind zone? If so, the structural requirements needed for the window may be a higher priority. Are the building occupants going to be able to open and close windows? If so, the operating force for the window may be a priority in specification.

7.3.6.2 Frame Materials

Wood Frames

Wood windows are the earliest and most prevalent window type used in construction in the United States. The primary market for all-wood windows today is in residential construction. A few major manufacturers are also equipped to provide wood windows for some commercial projects. Wood windows, when properly maintained, can have a very long service life. They offer a high degree of thermal comfort and have a warm, natural appearance. However, they require a high level of maintenance, and their operation is affected by climatological changes. Many wood windows today are actually composite windows, incorporating aluminum, vinyl, or other materials into frame elements to reduce maintenance costs.

Aluminum Frames

Aluminum windows were introduced into the construction market in the 1930s and represent the largest share of the commercial window industry. The principal advantage of aluminum windows is the ability to make fairly strong, yet thin frame profiles that have low maintenance requirements. Today's advanced finishes can help provide aluminum systems that can stand up to the exterior environment for a very long time. These advanced finishes come in a wide array of colors as well, making aluminum systems attractive to architects and designers. However, aluminum is highly conductive. Until the advent of good thermal break technologies that break the path of heat conduction, aluminum frames were unable to achieve the low U-factors desired by most markets today. When specifying aluminum-framed window systems, be sure to specify whole product U-factors that more accurately capture the effectiveness of the thermal break technology employed.

Steel Frames

Historically, steel windows were ubiquitous in commercial construction. Today, however, the use of steel windows has almost entirely been replaced by aluminum windows or windows made from modern composites. The inevitable corrosion of the metal is a major concern with steel windows. Manufacturers have addressed this issue with high performance paint and coatings. In some applications, the thin elegant sight lines of steel windows are a requirement to achieve aesthetic goals. Steel windows do not incorporate thermal breaks and therefore have absolute limits in their ability to achieve the desired the thermal performance.

Vinyl Frames

Window frames made of polyvinyl chloride (PVC) have come to dominate the residential window market in the United States over the past 20 years. PVC frames are primarily used in residential construction due to typically small sizes and loads. PVC is a low-conductivity material like wood that can achieve the low U-factors desired in most marketplaces. Among the advantages of PVC window frames are lower initial cost, availability of colors, and low maintenance. Early PVC frames experienced premature failures in extreme hot and extreme cold climates. New chemistry formulations incorporating ultraviolet (UV) stabilizers and other additives have increased the durability of these systems. As window sizes increase, PVC often must be reinforced with aluminum, steel, or other strengtheners, which can negatively impact the overall product thermal performance. Highly thermal conductive materials are allowed to be used as window claddings and surface materials, e.g., to protect against the effects of extreme weather. Still, any penetration through the window frame insulation layer should be avoided as much as possible to prevent severe thermal bridge effects. Figure 7.5 shows examples of window frames made out of different materials.

7.3.6.3 Glazing Technology and Materials

Glass

Glass typically makes up the largest portion of the window or fenestration system. As such, it can be responsible for the biggest energy penalty or the greatest energy savings, depending on proper product specification, selection, and installation.

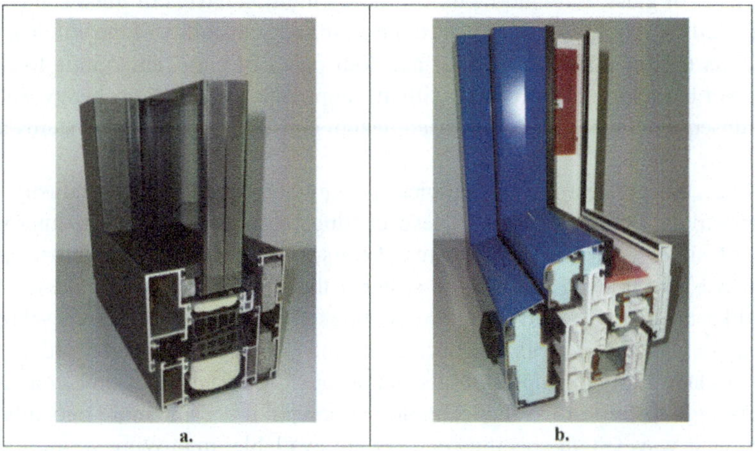

Fig. 7.5 Examples of high performance windows with frames made from different materials: (**a**) aluminum frame with full thermal separation; (**b**) plastic PVC frame with additional cladding including insulation layer

Fortunately, the glass industry has been at the forefront of fenestration energy performance innovation, addressing every aspect of heat transfer with advances in coatings, spacer systems, and gas fills. Additional glazing industry innovations that can influence product specification include: lamination (to increase strength, safety, impact and blast resistance, and acoustic performance); tints and colors (aesthetics and solar control); friable coatings (for ease of cleaning and maintenance); protective packaging (for improved handling); and other innovations. Some glazing innovations include electro-chromic and other "switchable" types of glazing that yield optical properties that can be changed with electrical charge.

Critical specification decisions concerning glass and glazing usually begin with strength characteristics. Specifiers need to know what wind speed, temperature, building-specific pressures, and other local climate forces will be at play that influence the ultimate glass structural properties needed for any given project. While many of these properties are well defined by the local building code, it never hurts the designer or specifier to revisit specification fundamentals regarding proper glass selection such as:

- 3 mm or 6 mm?
- Annealed, heat-treated, or tempered?
- Laminated or monolithic?
- Building corners or midwall?
- High wind zone or protected location?
- Ground floor or tenth story?

These are just a few of the types of fundamental structural questions that are embodied in basic fenestration product specification.

Because lower window U-factor was one of the earliest energy improvement targets, insulating glass (IG) technology was one of the first energy improvement innovations. Until the oil embargo of the early 1970s, most windows contained a single pane of glass. IG window units employ multiple layers of glass sandwiching an enclosed air space, which improved the insulating capability of the window. Early innovators tried windows with three and four panes of glass, attempting to achieve ever lower U-factors. The problem with this approach is that glass is heavy, such that each subsequent layer places increased design constraints on frame systems and on the job site installer.

By far, the greatest energy efficiency innovation in the glass industry is the low-emissivity (low-e) coating. These coatings, usually made from microscopic layers of silver and various protective layers, were initially designed to address long wave heat transfer in window systems that reflect heat back into the space, much like an invisible mirror. These coatings gave new life to double-glazed IG units, making their U-factors approach triple and quad-glazed systems with only two panes of glass. Early examples of this technology commonly had some colored tints and off-angle distortion. As the technology evolved, it became more optically clear. Today, near wavelength-specific coatings are available, improving window energy performance for both heating and cooling savings. Today's advanced low-e coatings can be near optically clear, have U-factors akin to quad-glazing for heating savings, and also block 70% of the unwanted solar gain to reduce building cooling loads.

Since the radiative heat transfer had been so effectively addressed by low-e coatings, the window and glazing industry turned to other areas of performance improvement. Conduction through the edge of the IG units was the next big target. Early IG units had highly conductive aluminum spacers used to keep the glass panes apart, and that were sealed to prevent air flow between the panes. Innovations in low-conductivity spacers – from materials to physical designs and both – resulted in further improvements to system U-factors.

Conduction and convection of the molecules of air between the panes was the next arena for innovation. Numerous gas mixtures were analyzed and investigated to further suppress the heat transfer due to a temperature difference between inside and outside. Gases such as argon, krypton, and others have been investigated and used in the space between the glass panes. The advantage these gases have is that they are heavier than air so they reduce convection and conduction through the air space. Critical to this innovation was the integrity of the seal between the panes of glass to retain whatever gas mixture used. Today, these systems are called "sealed insulation glass units," in which the seal not only holds in the gas mixtures but also serves to protect the invisible metallic low-e coatings. The integrity of this seal is crucial to the longevity of the performance claims. IG warranties can be a good surrogate for seal durability expectations.

Sealed IG units today typically incorporate all of these innovations. Low-E coatings are applied to one or more of the glass surfaces of the IG unit. Gas mixtures fill the space between the panes; low-conductivity spacers maintain the air space; and advanced sealant technologies protect the coatings and fills. These glazing innovations reduce the overall heat transfer of the fenestration system in both summer and winter while maximizing visible light transmission for views and daylighting.

Tables 7.2, 7.3 and 7.4 list examples of U-factors for single-, double-, triple-, and quad-glazed window systems, in a variety of frame materials, and have a variety of low-e coatings and gas fills. Table 7.2 includes U-values and SHGCs for a subset of fenestration products from the ASHRAE Handbook of Fundamentals (ASHRAE 2013a, b, c). Table 7.3 provides a small subset of examples of how various frame materials and glazing combinations can impact whole product U-values. Note that examples listed in Table 7.3 are all for a window of the same size. Table 7.4 lists representative window characteristics in Germany. Table 7.4 lists examples of the Passive House-certified windows for different climates. Figure 7.6 shows example of the high performance window with a low-conductivity closed-cell foam insulated fiberglass frame, lightweight-suspended coated film incorporated into the insulated glazing unit with argon- and krypton-filled cavities and low-e glass coatings with U-value, which, according to NFRC (National Fenestration Rating Council), can be as low as 0.625 W/(m^2∗K), 0.11 (Btu/[hr∗ft^2∗°F]), R = 1.6. (m^2∗K)/W, (9.1 [hr∗ft^2∗°F]/Btu).

Please note that U-values illustrated in Fig. 7.6 and listed in Tables 7.2, 7.3 and 7.4 are intended only to illustrate the effect of window type, number of glazing

Table 7.2 U-factors for fenestration products in Btu/h • ft² • °F (SI)

Glazing type	Wood or vinyl frame	Aluminum frame with thermal break	Aluminum frame without thermal break
Single glazing ⅛ in. Glass	0.91 (5.17)	1.07 (6.08)	1.23 (6.98)
Double glazing ½ in. Air space	0.50 (2.84)	0.58 (3.29)	0.76 (4.32)
Double glazing, e = 0.20 ½ in. Air space	0.41 (2.33)	0.48 (2.73)	0.65 (3.69)
Double glazing, e = 0.10 ½ in. Air space	0.39 (2.21)	0.46 (2.61)	0.63 (3.58)
Double glazing, e = 0.10 ½ in. Argon space	0.36 (2.044)	0.42 (2.385)	0.59 (3.35)
Double glazing, e = 0.05 ½ in. Air space	0.38 (2.16)	0.45 (2.56)	0.61 (3.46)
Double glazing, e = 0.05 ½ in. Argon spaces	0.34 (1.93)	0.41 (2.33)	0.57 (3.24)
Triple glazing, e = 0.10 ½ in. Air space	0.28 (1.59)	0.34 (1.93)	0.50 (2.84)
Triple glazing, e = 0.10 ½ in. Argon spaces	0.26 (1.48)	0.30 (1.70)	0.47 (2.67)
Quadruple glazing, e = 0.10 ½ in. Air space	0.26 (1.48)	0.31 (1.76)	0.48 (2.73)
Quadruple glazing, e = 0.10 ½ in. Argon spaces	0.24 (1.36)	0.29 (1.65)	0.45 (2.56)

Source: ASHRAE Handbook of Fundamentals (2009)

Table 7.3 Representative U-factors for windows in Germany

Construction	Properties	Construction year category		
		1979–1994	EnEV 2014	Passive house
Wooden or synthetic windows	U_w (W/[m²·K])	2.8	1.3	0.85
	Glass	Double	Double	Triple
	g-value	0.75	0.65	0.55

panes, frame materials, and design on the window U-value. When using information from such tables, it is important to remember that the total heat flow is a function of many variables, including product size, that affect the ratio of the glazing area and the frame area and the type of frame (operable or nonoperable windows). Most of these tables and typical product certification programs provide U-factors, SHGC, and Tvis (visible light transmittance) values for a standard size. When selecting the product and conducting whole building energy analyses for systems that are dramatically different from these typical sizes listed, one should always refer to the manufacturer's catalogues to verify and consider the performance values (U, SHGC) for the actual fenestration sizes. This is especially true when conducting peak load analyses to size heating and cooling systems.

Table 7.4 Examples of Passive House-certified frames and glass assembly available on the market for different climate conditions. For more information, see www.passiv.de

Type/Region	Arctic	Cold	Cool,template	Warm, temperature	Hot
Glazing, Glass assembly, Glazing U-value	quadruple 4/12/3/12/3/12/4 0.35 W/(m²k)	triple 6/18/2/18/6 0.52 W/(m²K)	triple 6/16/6/16/6 0.70 w(m²K)	triple 6/16/6/16/6 0.70 w(m²K)	double 6/16/6 1.20 W/(m²K)
Wood/aluminium integral	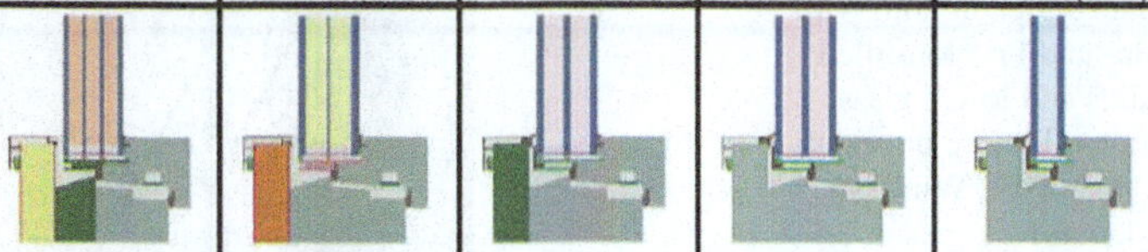				
U_f[W/(m²K)]	0.48	0.62	0.73	0.87	1.03
Wood/aluminium	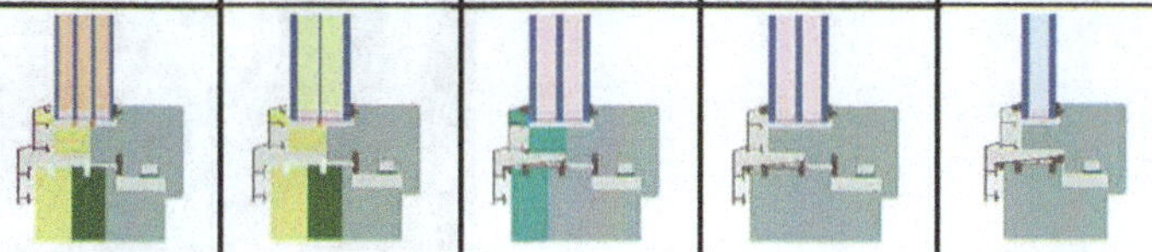				
U_f[W/(m²K)]	0.54	0.57	0.75	0.97	1.19
Wood	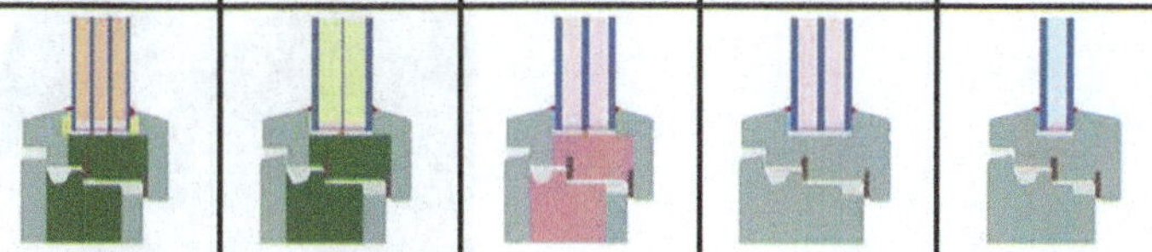				
U_f[W/(m²K)]	0.51	0.53	0.78	0.86	0.99
PVC	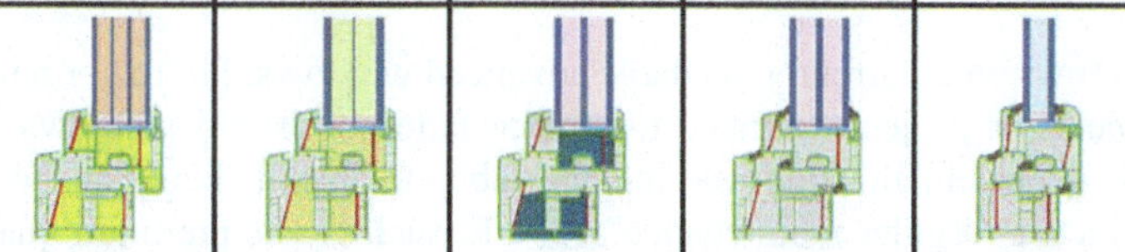				
U_f[W/(m²K)]	0.70	0.75	0.82	1.02	1.16
Aluminium	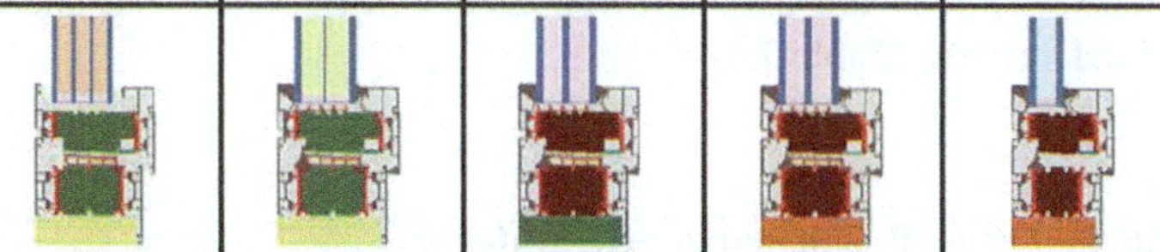				
U_f[W/(m²K)]	0.60	0.61	0.71	0.73	1.17

Current high performing windows can have U-values as low as 0.35 W/(m^2 °C) (0.06 Btu/[hrft^2°F]) (Table 7.2). However, to get to that level, the window must incorporate a quadruple pane IG unit with argon-filled, multiple low-e coatings or vacuum constructed IG units with a warm edge spacer, and multilevel thermal breaks in the framing. For certain projects, this type of window may be desirable if the budget allows, but for Arctic climates, it can be a necessity to ensure comfort and to eliminate condensation, mold, and structural damage.

Modern window technologies are mature and ready for use. Assuming a 10-year payback threshold, it is generally justifiable in all climate zones to replace existing

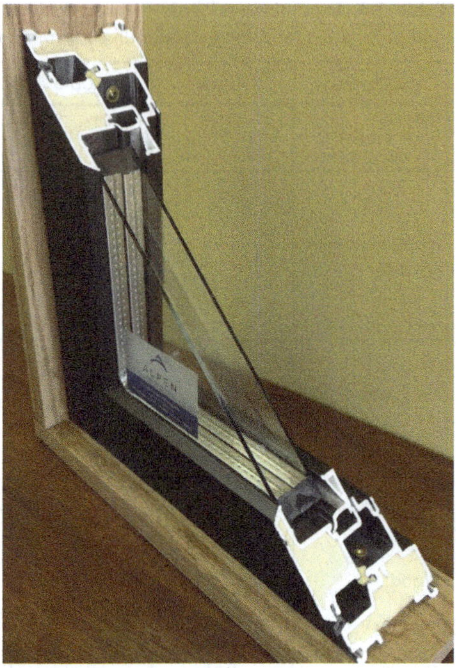

Fig. 7.6 525 Series high performance window with closed-cell foam-insulated fiberglass frame, lightweight suspended coated film incorporated into the insulated glazing unit with argon- and krypton-filled cavities and low-e glass coatings (Picture provided by Alpen HPP Windows and Doors)

windows with currently available advanced windows. For major building renovation projects or projects initiated to replace failed or failing windows, the cost of base case replacement windows and the labor to install them can be considered as a budgeted "regular maintenance" cost. In such cases, premium quality replacement windows options are available for each climate zone that satisfies the 10-year payback criteria (Table 7.4).

7.3.6.4 Other Performance Attributes

As mentioned, many different performance parameters can affect proper window selection. The following sections list and discuss a few of these performance attributes that may be considered for any particular DER project.

Water Penetration Resistance

Resistance to water penetration may be critical for projects in high rainfall areas or in areas where wind-driven rain is common. Most windows are tested and certified to meet certain levels of water penetration resistance. The amount of resistance is usually expressed as a percentage of the design load (the structural performance expectations for the window). For more information on water resistance, see ANSI/

AAMA/WDMA I.S.2. It is important to note that there are two important aspects of resistance to water penetration: the resistance of the window itself (through sash, frames, operators, hardware, etc.) and the resistance of the installed window systems (how well it is attached to, and properly integrated with, the rest of the building's water management plan). Proper window installation is critical to achieving the energy and durability expectation embodied in proper performance specification.

Condensation Resistance

Cold window surfaces can result in condensation on the windows in colder climates. While not an energy performance attribute specifically, unwanted condensation can be an indicator of other performance issues (high interior RH, air leakage at the window, poor performing windows, failed IG units, etc.). Condensation can also result in water damage to interior surfaces and materials.

Operating Force

If the windows are to be a component of building ventilation or fire egress, then specifiers may want to consider windows that have a specific threshold of operating force. Typically measured in Newtons or pounds of force, the seals within the window system and the sealant between the window system and the surrounding substrate are key factors when considering window performance. They have a direct impact on air and water infiltration into the window system itself and into the building.

Impact Resistance

High wind speed areas – such as hurricane-prone, tornado-prone, high mountain locations, or other likely high wind speed areas – may require or benefit from windows specifically designed to resist wind-borne debris. Building codes in the United States have identified certain areas where window resistance to wind-borne debris is now mandatory. Manufacturers may use laminated glazing or other glazing technologies to increase a window system's resistance to such events. Impact-resistant glazing systems are typically designed to hold the glass in place even if it breaks – as is common for car windshields. Some systems are designed for even greater impact protection – such as blast- or bullet-resistant systems.

Acoustic Performance

Often retrofit projects have improved acoustic environment as an objective. Older buildings and windows typically leaked significant amounts of air, resulting in easy transmission of unwanted noise. Many products today are rated for their Sound Transmission Coefficients (STCs) or Sound Transmission Ratings (STRs). Some products are specifically designed for superior resistance to sound transmission – such as might be needed near an airport, school, or busy highway. Specifiers should consider these acoustic ratings when project objectives demand them.

7.3.6.5 Design Parameters for Window Selection

As mentioned above, key building design parameters and constraints play a significant role in proper fenestration product specification and selection. Some DER projects offer extensive opportunities for new product options, sizes, and configurations – such as with an entirely new façade or curtain wall system. Other projects may be more constrained, such as with historic structures or façades that are to remain essentially unchanged in their outward appearance.

While this Guide is not meant to serve as a detailed design guide, the following sections list some of the most critical design parameters that should be considered before window performance specification and selection. A few, more detailed websites and design resources are listed at the end of this section.

Building Orientation

Most retrofit projects do not include reorienting a building. As such, owners, designers, and others seeking to achieve effective DERs must play the "orientation cards" they are dealt. DER constraints such as building orientation need not be an impediment to proper specification and selection of fenestration products. They merely require the specifier to place more emphasis on performance indices that resist heat flows for both heating and cooling. In fact, because most buildings are not constructed with "heat flow specific" orientation in mind, low U-factors and low SHGC are likely to be the norm for almost any DER project. Rarely are buildings designed to effectively capture, store, and distribute "free" solar gains. Usually, the opposite is true. Solar gains are often coincident with peak cooling loads, which demand that fenestration and solar heat gain be minimized where possible (see V. Olgay *Design With Climate*, and www.efficientwindows.org).

Window-to-Wall Ratio

Existing building energy performance is almost always directly proportional to how much of the thermal envelope is opaque, insulated wall area, and to how much is window and glazing. Again, most buildings are not designed to maximize beneficial solar gains nor are they particularly sensitive to occupant comfort. As window-to-wall ratio increases, the challenges of peak power for cooling, occupant discomfort, and high energy bills increase. A balance must be achieved between the heat transfer management objectives (U-factor, SHGC) and daylighting and electric load management objectives (Tvis, kW, etc.). Generally speaking, minimizing U and SHGC while maximizing Tvis is appropriate for most buildings. However, when window-to-wall ratios are greater than 30%, one should expect a significant energy penalty even without greater emphasis on lowering U-factors and SHGC values. Buildings having significant window-to-wall ratios have the added complication of maintaining occupant comfort near that window area. Improper window and glazing selection can easily result in substantially larger and more expensive HVAC systems than anticipated.

Window-to-Floor Area Ratio (Space-Specific)

This is a not-so-subtle variation on the concerns raised regarding window-to-wall area. Perimeter zones of both commercial and residential buildings can vary in depth and in the occupant's proximity to the fenestration system. Additionally, depending on occupancy type, the room aspect ratios have a different set of comfort and energy impacts during the daytime versus at night. Again, designers in new construction are (one would hope) sensitive to these space-specific constraints and design challenges. However, for existing buildings undergoing DER, there may be no option to change these spatial geometries. Specifiers should be careful to consider occupant needs during peak cooling periods as well as peak heating periods. Room aspect ratios combined with how much glazing area the occupant is exposed to in those spaces can dramatically impact comfort and occupant response at the thermostat. Additionally, these spatial geometries (ceiling height, room depth, window-to-floor area ratio) have a direct influence on the effectiveness of planned daylighting strategies.

Access to Daylight and View

Many studies have shown that access to daylight and view yields enormous benefits. In schools, children with access to broad-spectrum, natural light miss fewer days of school and do better on standardized tests. Office workers with access to natural light and views demonstrate dramatically higher levels of productivity and worker happiness. Such benefits can easily exceed the costs of proper window and glazing systems (see Heschong-Mahone, PNNL).

7.4 Window Selection

In addition to the many design and performance considerations described above, there are often specific legal requirements that must be met in fenestration specification and selection. These are often defined by local, state, or national building and energy codes. Specifiers should be very familiar with the minimum performance, certification, and energy requirements of the local codes governing their project.

Caution: Most building codes are in a near constant state of change and update. Specifiers should regularly consult with their governing code agencies to be sure to have the latest updates on fenestration product requirements.

DER teams should have a shared understanding that codes establish minimum performance attributes. DER projects will almost certainly require substantially better performing products and systems than those defined in the local/national codes.

Table 7.5 lists the primary energy codes addressing construction in Europe and North America.

7.5 Performance Certification

Most code bodies require that certain fenestration performance attributes be "certified," i.e., verified by a third party and appropriately labeled and listed. Common performance attributes required to be certified in the United States include structural performance ratings, water resistance, air leakage, U-factor, SHGC, optical properties, and others. Specific projects, such as those for governmental buildings, police stations, and hospitals, may have certification requirements for acoustics, impact resistance, operating force, and many other attributes.

For DER projects, certified energy performance attributes provide a level of protection to the specifier and the project that the energy performance targets for the project might actually be met. In the United States, the energy performance attributes of windows and fenestration systems are required by code to be certified in accordance with the procedures of the National Fenestration Rating Council

Table 7.5 Current energy codes and sections by country

Country	Codes governing building envelope elements
Austria	OIB RL 6, 2011
China	GB 50198-2005
	GB/T 7016-2008
Denmark	Danish Building Regulations 2010
Estonia	EVS-EN ISO 10077, EVS-EN 1026, EVS-EN 12207, EVS-EN 12208
Germany	EnEV 2014, DIN 18361, DIN 18355, DIN V 18599/2, DIN 4102, DIN 4108, DIN EN 13162, DIN EN 13163, DIN EN 13165
UK	Building regulations-conservation of fuel and power in new buildings other than dwelling: Part L2A
Ireland	National standards authority of Ireland (NSAI)
USA	ASHRAE standard 90.1–2013 (commercial and residential over three stories)
	International energy conservation code (IECC) 2015 (low-rise residential)

Fig. 7.7 Example of National Fenestration Rating Council (NFRC) label certifying a window's energy performance attributes

(NFRC). Unitized (punched opening) products typically carry a certification label (Figs. 7.7 and 7.8). Custom and site-assembled systems such as curtain walls must also be certified and their performance provided to the code official and specifier by use of a label certificate showing each fenestration component used (printed or accessed online).

7.6 Blast-Resistant Windows

In certain cases, blast-resistant fenestration may be required when considering a window replacement. Typically, the primary objective of blast-resistant windows is to reduce the risk posed by flying glass and debris produced as window systems fail under explosively induced blast loads. The highest goal is to reduce the risk to building occupants and visitors. A secondary goal is to reduce the damage to and loss of property and facilities. This can still be achieved in a window replacement project by installing the required glazing. This typically entails the use of laminated glass to limit the amount of glass fragments that are projected into the room. Laminated glass can be incorporated in IG units to maintain the thermal performance of the glass package. The laminating layer can also have the added benefit of nearly eliminating ultraviolet light from passing through the glass. However, depending on the blast performance required, the window framing may need to be structurally reinforced with steel elements that can reduce the thermal performance of the frame. This is a trade-off that needs to be considered when specifying blast-resistant systems. Gypsum products with high density are very useful for both fire and blast resistance window frame cases.

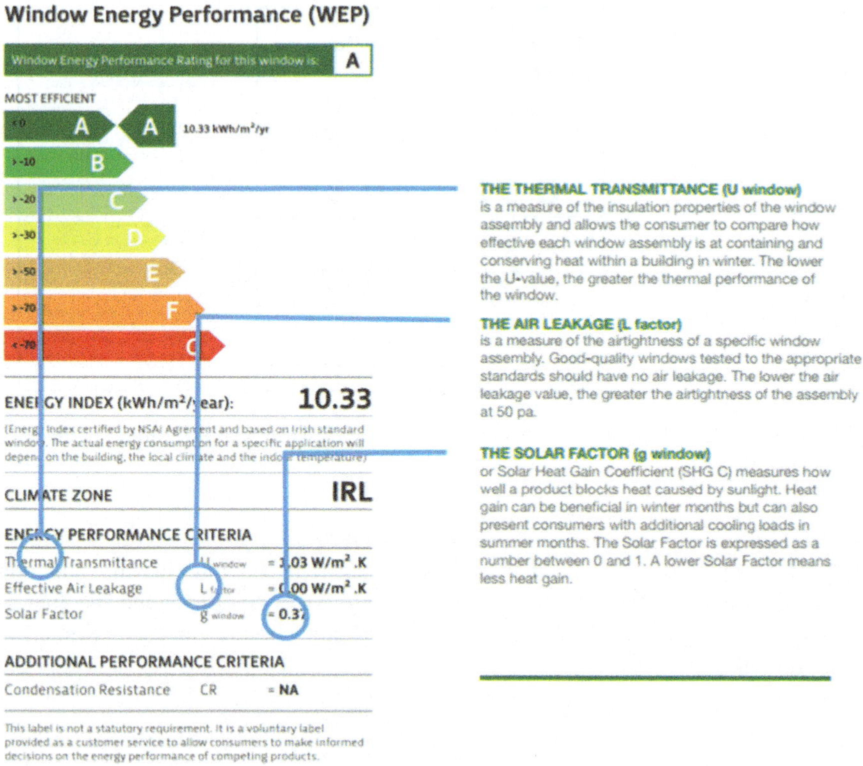

Fig. 7.8 European window energy performance label

7.7 Historic Windows

In some instances, window replacement may not be feasible or possible. This is the case with windows in historic buildings. The windows are considered an integral part of the historic fabric of the façade and therefore need to remain in place. However, these windows can be renovated to improve energy performance. Weatherstripping can be added or replaced to reduce air infiltration and, depending on the window size and configuration, IG units may be installed in place of the existing single pane glass. In projects where it is not possible to perform this type of renovation, either from aesthetic or technical standpoint, an interior storm window may be a viable option (Fig. 7.9). This can have a similar impact on energy performance as a window renovation without the disruption of removing and reinstalling windows. Storm windows can include low-E glazing and can also provide enhanced acoustic performance if outside noise reduction is desired. In some cases, storm windows can even provide some blast resistance if laminated or filmed glass is installed.

Fig. 7.9 View of interior storm window

Other factors to be considered when selecting windows include:

- *Building Code Requirements.*
  The new windows will have to meet the requirements of the local building code including air infiltration, water penetration, and structural capacity.
- *Aesthetics.*
  Exterior finish, sightlines, and profiles should be considered when choosing a window. Will the new windows match the existing, or will they provide a new look for the building?
- *Operability.*
  Determine if the new windows will be operable or fixed. This may depend on code requirements as well as occupancy. If the windows will be operable, will they be hung, sliders, casements, or a combination of various window types?
- *Maintenance.*
  Exterior perimeter sealants will still need to be maintained on a regular basis. Also, depending on the exterior finish, regular painting may be necessary as well.
- *Initial Costs and Life Cycle Costs.*

The initial cost of the window replacement, including all labor and materials, should be evaluated not only as a separate item but also in the light of life cycle costs including maintenance, cleaning, etc.

7.8 Window Installation

When installing retrofit/replacement window into an existing building, the installer should clearly understand the existing wall and roof conditions where the replacement fenestration is to be installed. The installer must be well aware of the existing and planned bulk water management and air leakage management of the wall system, and of the planned thermal barrier system to eliminate/reduce thermal bridges. The installer must exercise caution to ensure that the retrofit/replacement installation does not impair the existing/planned building wall system from working properly, or that it does not destroy the weather resistance of the existing/planned system. Anchoring of window or door products should always be done according to the manufacturer's instructions, the approved construction documents, and the recommendations of a qualified structural engineer.

For more information on windows installation, see Appendix D.

Chapter 8
Thermal Bridges

A thermal bridge is an area in an insulated construction that has a significantly poorer degree of insulation than the construction as an average. Thermal bridges are characterized by multidimensional heat flows. Figure 8.1 shows pictures and thermographic images that reveal thermal bridges.

The magnitude of thermal bridges depends on many aspects. Typically, the more insulation a construction has, the greater the relative importance of the thermal bridges. Previous studies show that thermal bridges can increase the total transmission loss from buildings by 14–50%, depending on the extent of the thermal bridges and insulation level of the building envelope in general. This means that thermal bridges are most important to address in new buildings with high insulation levels or existing buildings undergoing DER where insulation levels are increased significantly.

In addition to increasing the transmission of heat loss, thermal bridges also lead to lower internal surface temperatures for constructions, which can lead to:

- Poor indoor climate (drafts)
- Contamination of surfaces (dust condensation)
- Moisture damage (mold growth, fungi, etc.).

Thermal bridges typically occur in building joints where different constructions are joined together, i.e., windows or doors in walls. Figure 8.2 shows some typical examples of locations where thermal bridges occur in buildings.

Thermal bridges can be divided into four subgroups:

1. Geometrical thermal bridges
2. Convective thermal bridges
3. Constructive thermal bridges
4. Systematic thermal bridges

The following sections briefly describe each thermal bridge subgroup.

A. Zhivov, R. Lohse, *Deep Energy Retrofit*,
https://doi.org/10.1007/978-3-030-30679-3_8

Fig. 8.1 Examples of thermal bridges: parapets, doors, slabs

8.1 Geometrical Thermal Bridges

Geometrical thermal bridges are thermal bridges that occur due to the geometry of the construction. All constructions where there is a difference between internal and external dimensions will contain geometric thermal bridges. Building corners are one typical example of this type of thermal bridge. Figure 8.3 shows typical geometrical thermal bridges.

8.2 Convective Thermal Bridges

As the name suggests, convective thermal bridges are thermal bridges caused by unintentional air flows in a construction. For example, convective thermal bridges can occur where there is a possibility of air flowing through the insulation layer from the warm to the cold side of the construction either because the insulation does not fill the gap entirely or because air can flow through the insulation material itself. Convective thermal bridges will also occur where the room air is allowed to flow out

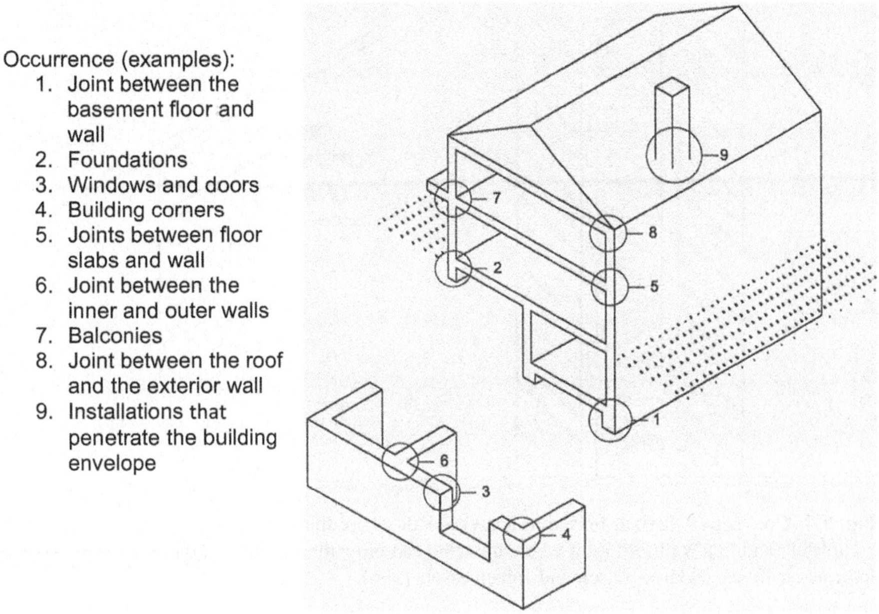

Fig. 8.2 Typical occurrence of thermal bridges in a building. (Source: Olsen and Johannesson 1996)

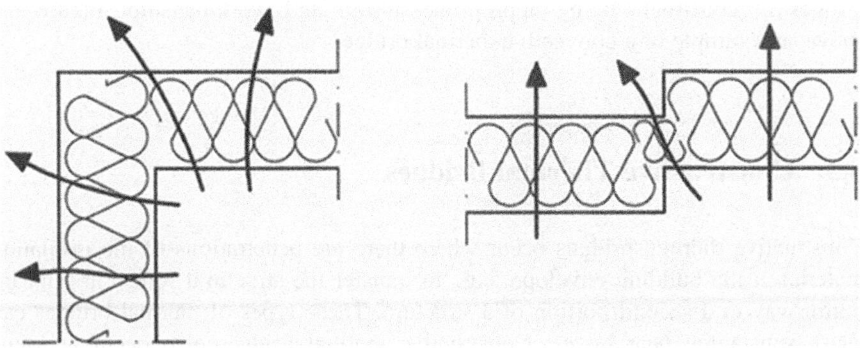

Fig. 8.3 Geometrical thermal bridges. Arrows indicate heat flows. (Source: Olsen and Johannesson 1996)

into the construction (i.e., where the construction is not airtight). Note that, per International Standards Organization (ISO) building energy code, air gaps within the building envelope (room air flowing throughout the building component to the outside) are not considered to be thermal bridges; they are addressed in the context of building envelope airtightness. ISO building energy codes only address geometrical, constructive, and systematic thermal bridges.

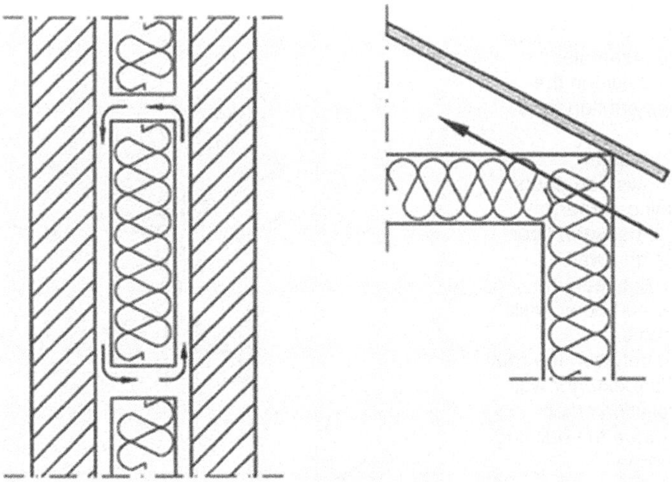

Fig. 8.4 Convective thermal bridge in a cavity wall where the insulation does not fill the cavity completely and in a wall/roof joint where the wind can blow through part of the insulation. Arrows indicate air flows. (Source: Olsen and Johannesson 1996)

Convective thermal bridges and convection in insulation material are to some extent taken into account in the thermal conductivity of the insulation materials. Furthermore, there are good opportunities for minimizing this type of thermal bridges in constructions by appropriate design and workmanship. Figure 8.4 shows an example of a convective thermal bridge.

8.3 Constructive Thermal Bridges

Constructive thermal bridges occur where there are penetrations of the insulation material in the building envelope, e.g., to transfer the structural forces in a meaningful way to a second portion of a structure. These types of thermal bridges can cause substantial heat losses. Constructive thermal bridges also occur at joints between exterior walls and doors/windows and penetrations of the envelope by electrical installations, water pipes, etc. Penetrations of the building envelope can occur both as full penetrations and partial penetrations of the insulation material. Full penetrations will naturally cause the greatest thermal bridges. Figure 8.5 shows examples of constructive thermal bridges.

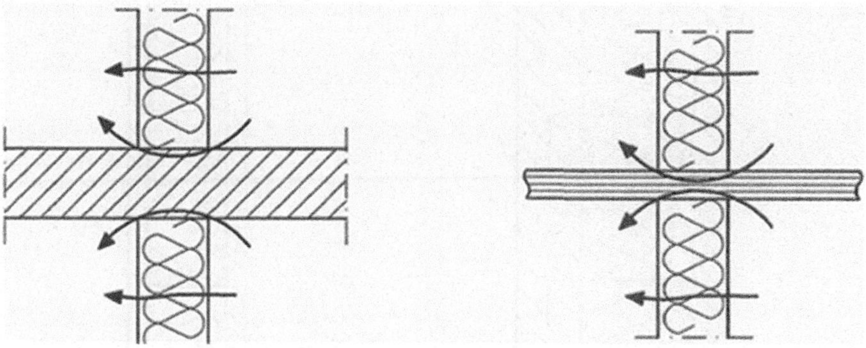

Fig. 8.5 Constructive thermal bridges. Arrows indicate heat flows. (Source: Olsen and Johannesson 1996)

8.4 Systematic Thermal Bridges

Systematic thermal bridges could actually be attributed to the group of constructive thermal bridges, but are typically attributed to a separate group. These thermal bridges often represent a relatively small additional contribution to heat loss. For example, systematic thermal bridges could be wall binders or wood studs in walls, i.e., thermal bridges that occur repetitively. Typically, the heat loss of systematic thermal bridges is included in the overall U-value of the construction in which they are present. Figure 8.6 shows examples of typical systematic thermal bridges.

The most significant thermal bridges tend to occur at interfaces between building enclosure assemblies (Fig. 8.2). The significance of the heat loss depends on the transmission rates and the quantity of thermal bridges. Some of the most significant heat losses occur at:

- Places where slabs or structure penetrate the insulation
- Roof wall intersections such as parapets or roof truss extensions
- Window/wall connections
- At grade assemblies
- Interior wall/exterior connections that penetrate insulation
- Curbs or pedestals that penetrate roof insulation

In many buildings, the heat loss through these thermal bridges can exceed the heat loss through the opaque walls and roofs. Figure 8.7 shows examples of thermal bridges.

There are, of course, many other smaller thermal bridges at penetrations (e.g., corners, door thresholds, ducts, wire, and pipes), which individually add only small heat losses, but which together contribute a significant heat loss to the building as a whole (Figs. 8.8 and 8.9).

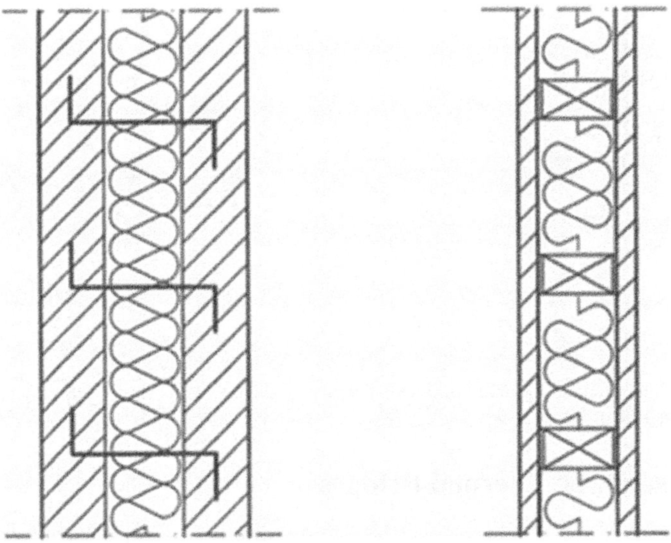

Fig. 8.6 Typical systematic thermal bridges. (Source: Olsen and Johannesson 1996)

8.5 National Requirements to Thermal Bridges with Renovation Projects

Issues related to thermal bridges in European countries are discussed in (Citterio et al. 2008). Most of the Northern and Central European countries are dealing with the problem as far as new construction is concerned. This is not the case for renovation projects. Specific attention has been given to collecting information on simplified approaches: a simplified approach is most used in Northern and Southern Europe. Only Finland applies special assessment methods (dependent on the λ ratio).

Different countries use many different methods to deal with the maximum value for thermal bridges in regulations: in Germany, the dimensionless temperature factor fRsi is used; in Denmark and the Czech Republic, a ψ_{max} value is set depending on the type of join; in France, the ψ_{max} depends on the type of building; in Austria, there are no specific requirements on linear thermal transmittance Ψ and point thermal transmittance χ, but there are requirements related to thermal bridges in the Austrian standard ÖNORm^2 B 8110-2 to internal surface temperature factor concerning moisture safety and condensation prevention.

In Germany and the Netherlands, every effort is made to minimize the influence of thermal bridges with economically justifiable measures. In building modeling, they are accounted for using default values provided by NEN 1068 (2001), DIN 4108 (2003), and DIN V 18599 (2007) standards or by accurate analysis of thermal bridges done in accordance with calculation standards (e.g., DIN EN ISO 10211-1/2 [1995, 2002]). The default value is the increase in building envelope element overall

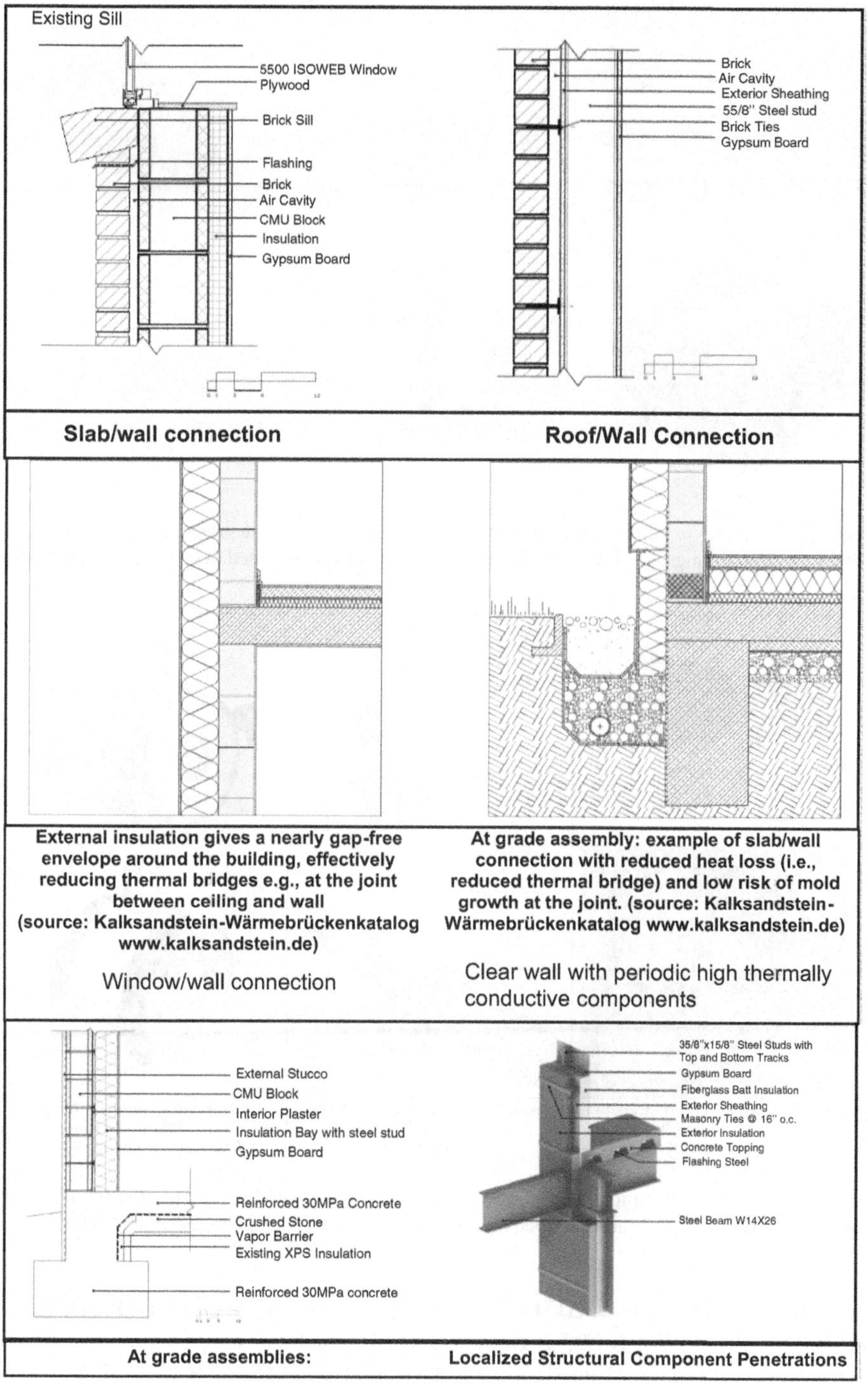

Fig. 8.7 Examples of thermal bridges

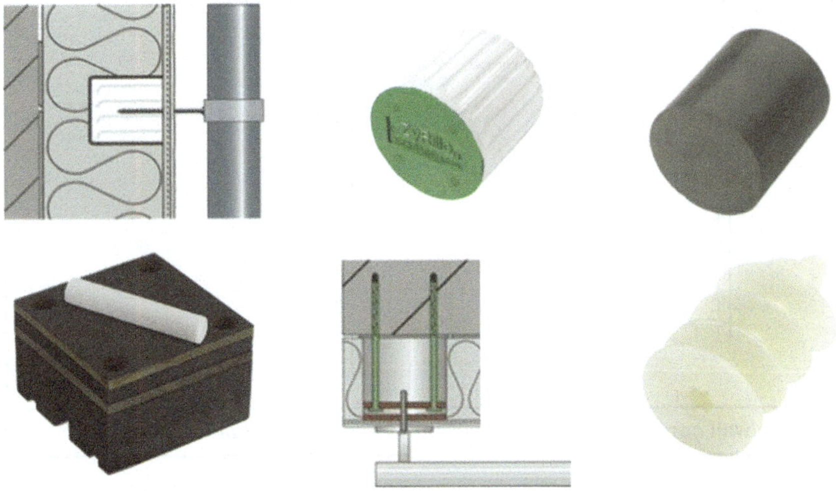

Fig. 8.8 Examples of smaller penetrations: mounting devices to be integrated in insulation layer to give support to railings, shading devices, small roofs, etc., to be mounted outside of insulation layer

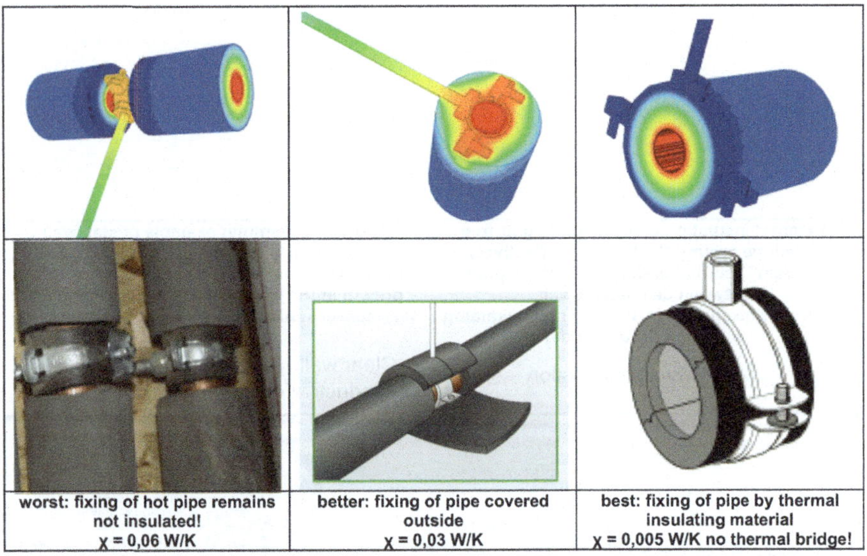

Fig. 8.9 Fixings of hot pipes are worthwhile to be well insulated (example: 20-mm pipe with insulation 1∗DN): large heat losses because of high temperatures of pipes. Source (AKKP 49)

heat transfer coefficient ΔUTB due to the effect of the thermal bridge by ΔUTB = 0.10 W/m^2K (0.57 Btu/hr-sq ft °F) for new construction and major renovation with external insulation and by ΔUTB = 0.15 W/m^2K [0.85 Btu/hr-sq ft °F]) for existing buildings with internal insulation (DIN V 18599-2 [2007]).

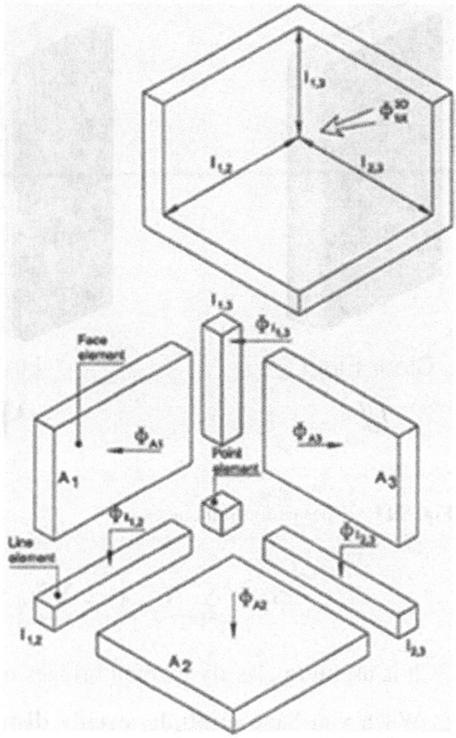

Fig. 8.10 Principle in the division of transmission heat losses in one-, two-, and three-dimensional losses. (Source: Olsen and Johannesson 1996)

Germany provides a catalogue of principles of thermal bridge avoidance measures for various building element connections, which can be applied both to new and existing (retrofit) buildings.

In North America, there are no legal requirements governing thermal bridges. However, several guides based on recent research have been produced that address the impact of thermal bridges on the building envelope performance and that provide guidance for mitigation of thermal bridges (Pagan-Vazquez et al. 2015a, b; Lawton et al. 2014; Schild 2010).

8.6 Calculation Methods Accounting for Thermal Bridges

Calculation of heat losses including thermal bridge effects is done by dividing the transmission heat losses into one-, two-, and three-dimensional losses. Figure 8.10 shows the principle in the division of heat losses.

The transmission heat losses are divided into one-dimensional losses (face elements, U-value, W/m^2K), two-dimensional losses (line elements, ψ-values, W/mK), and three-dimensional losses (point elements, χ-values, W/K). The total heat loss can be determined by:

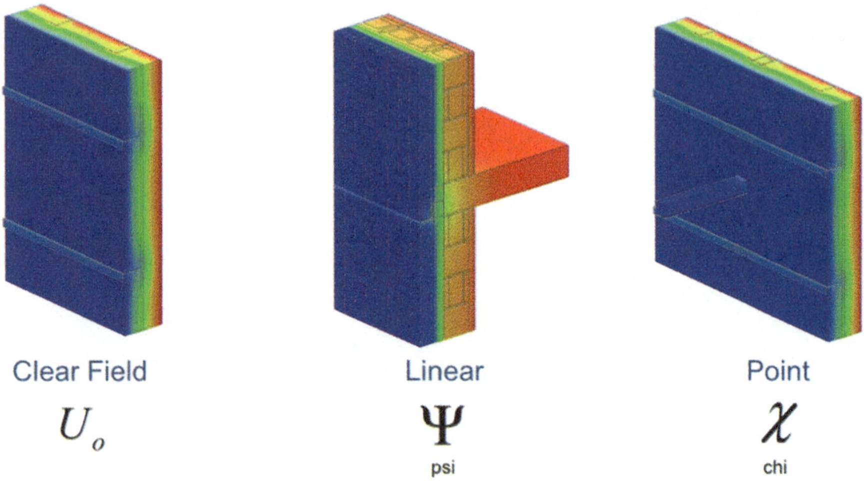

Fig. 8.11 Types of transmittances

$$\Phi = \left(\sum(U \cdot A) + \sum(\psi \cdot l) + \sum(\chi)\right) \cdot (\theta_i - \theta_u) \tag{8.1}$$

It is useful to classify thermal bridges into three types (Fig. 8.11):

- When you have **multiple, evenly distributed thermal bridges** such as metal studs, brick ties, or cladding attachments, it is usual to calculate their impact and add it to the thermal transmission or **U-value** of the wall or roof type.
- **Linear thermal bridges** are details that occur along a length. For these, one can determine that the linear thermal transmission or Psi value (Ψ) would be multiplied by the length over which it occurs to get the heat loss per degree Fahrenheit and then add this to the building heat losses.
- **Point thermal bridges** are those that occur, for example, when a steel beam penetrates a wall/roof/floor.

Another way of looking at the basic concept of linear transmittance is by superimposing the heat flows from the full assembly (Fig. 8.12) with an interface detail, over the clear field assembly without the interface detail. Figure 8.12 shows the lateral heat flows to the path of least resistance through the interface detail assembly (i.e., through the slab). This results in a higher heat flow through the slab compared to heat flow through the clear field. At a certain distance from the slab, the heat flow through the slab reaches the same level as through the clear field. Subtracting the heat flow through clear field from the total interface detail assembly leaves the additional heat flow from just the slab, from which one can derive the linear transmittance.

Figure 8.13 shows linear thermal bridge using an example of a balcony or concrete overhang in an old building with a massive concrete slab penetrating the wall. The slab and balcony are directly connected and form a severe thermal bridge. Without any thermal insulation (graphs on left), inside surface temperatures at

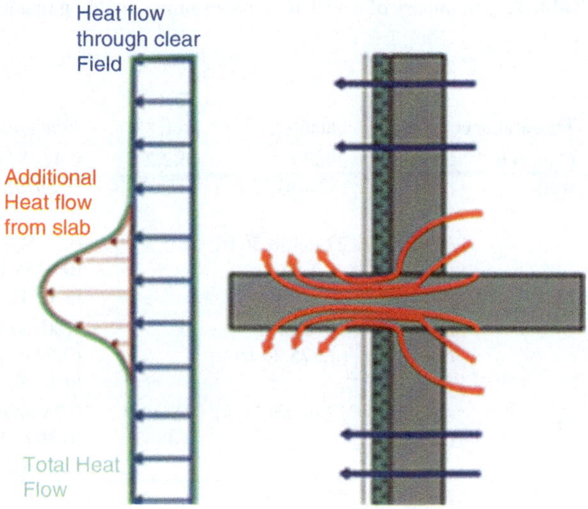

Fig. 8.12 Superimposing heat flows

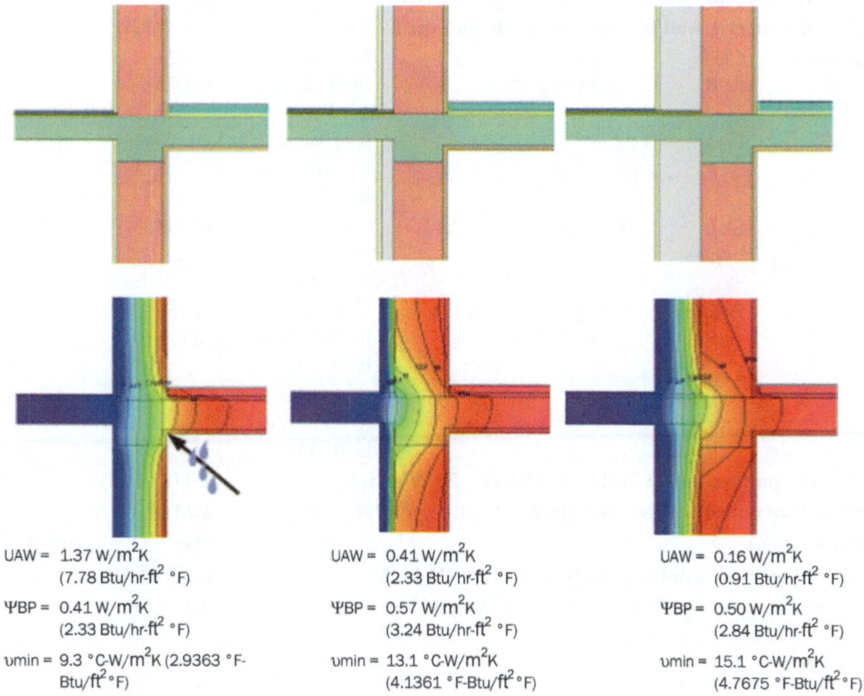

Fig. 8.13 Example of thermal bridge created by a balcony. Left: prerenovation situation; middle: conventional insulation thickness; right: well-insulated wall with ≥20 cm of insulation

Table 8.1 Summary of calculation, for example, building opaque wall

Transmittance type		Quantity	Detail ref.	Transmittance	Heat flow W/K (Btu/hr °F)	% of total heat flow
Concrete wall	Clear field	2987 m^2 (32,140.12 ft^2)	6.2.2	0.42 W/m^2K (2.39 Btu/hr-ft^2 °F)	1254 (20.4)	16%
	Parapet	27 m (88.56 ft)	6.5.3	0.78 W/mK (0.4131 Btu/hr-ft °F)	21 (0.34)	<1%
	Exposed floor slab	1090 m (3,575.20 ft)	6.2.5	1.00 W/mK (0.5296 Btu/hr-ft °F)	1085 (17.68)	14%
	At grade transition	27 m (88.56 ft)	ISO/IEC 14863	0.75 W/mK (0.3972 Btu/hr-ft °F)	20 (0.33)	<1%
	Partition wall	1315 m (4,313.20 ft)	6.2.2	0.67 W/mK (0.3548 Btu/hr-ft °F)	876 (14.27)	11%
Overall concrete wall U-value, Btu/hr ft^2 °F (W/m^2K)					**0.192 (1.09)**	
Overall concrete wall R-value, hr ft^2 °F/ Btu (m^2K/W)					**5.2 (0.92)**	
Window-wall spandrel	Clear field	1792 m^2 (19,281.92 ft^2)	1.1.1	1.07 W/m^2K (6.08 Btu/hr-ft^2 °F)	1917 (31.25)	24%
	Parapet	82 m (268.96 ft)	1.3.2	0.72 W/mK (0.3813 Btu/hr-ft °F)	59 (0.96)	<1%
	Slab bypass	1635 m (5,362.80 ft)	1.2.1	0.58 W/mK (0.3072 Btu/hr-ft °F)	945 (15.4)	12%
	Balcony slab	1635 m (5,362.80 ft)	8.1.9	1.11 W/mK (0.5879 Btu/hr-ft °F)	1815 (29.58)	23%
	At grade transition	82 m (268.96 ft)	2.5.1 (est.)	0.86 W/mK (0.4555 Btu/hr-ft °F)	70 (1.14)	<1%
Overall spandrel wall U-value, Btu/hr ft^2 °F (W/m^2K)					**0.472 (2.68)**	
Overall spandrel wall R-value, hr ft^2 °F/ Btu (m^2K/W)					**2.11 (0.37)**	
Total (W/K)					**8063**	**100%**
Overall opaque wall U-value, Btu/hr ft^2 °F (W/m^2K)					**0.297 (1.68)**	
Overall opaque wall R-value, hr ft^2 °F/ Btu (m^2K/W)					**3.4 (0.59)**	

Table 8.2 Summary of calculation, for example, building opaque roof

Transmittance type		Quantity	Detail ref.	Transmittance	Heat flow (W/K)	% of total heat flow
Roof	Clear field	743 m^2 (7,994.68 ft^2)	9.2.2	0.27 W/m^2K (1.53 Btu/hr-ft^2 °F)	200 W/K (3.2600 Btu/ hr °F)	82%
	Curbs	20 m (65.60 ft)	9.2.2	0.93 W/mK (0.4925 Btu/hr-ft °F)	19 W/K (0.3097 Btu/ hr °F)	8%
	Beam penetrations	#20	Modeled	1.2 W/K (0.0196 Btu/hr °F)	24 W/K (0.3912 Btu/ hr °F)	10%
Overall roof U-value, W/m²K (Btu/hr ft² °F)					**0.33 (0.058)**	
Overall roof R-value, (m²*K)/W ([hr*ft²*°F]/Btu)					**3.05 (17.31)**	

critical points are quite low. When the wall is covered from outside with a 20-cm (7.86-in.) thermal insulation layer, the U-value of wall will be significantly reduced. In addition, the inside surface temperatures at concrete ceiling increase will increase to a comfortable level and the issues causing condensation on cold surfaces and mold will be solved (graphs in the center and right). Note that the thermal bridge effect and thus ψ-value at this point will remain high, so this is not recommended for new construction. However, for old house renovation, this is a good compromise; it is not necessary to cover and "wrap" all the concrete overhangs with insulation.

There are now a number of sources of data of linear and point heat transmission rates (Tables 8.1 and 8.2) (ISO 2007, ASHRAE 2011, Pagan-Vazquez et al. 2015a, b, BC Hydro 2016) that allow the impact of thermal bridges to be calculated by superimposing the heat loss from the thermal bridges onto the heat loss over the clear field area of the enclosure assemblies. Calculating the overall heat transmission rate of any specific part of the enclosure is simply a matter of summing the heat transmission from each element. This is easy to do in a spreadsheet. Figure 8.14 shows an example.

Appendix D gives more information on thermal bridges and their mitigation with DER projects, including examples of how a DER can mitigate thermal bridges. Appendix F includes a catalogue of typical architectural details. The QA process will require IR thermography testing to identify whether the DER has mitigated project thermal bridges (see Appendix H, Annex H-4).

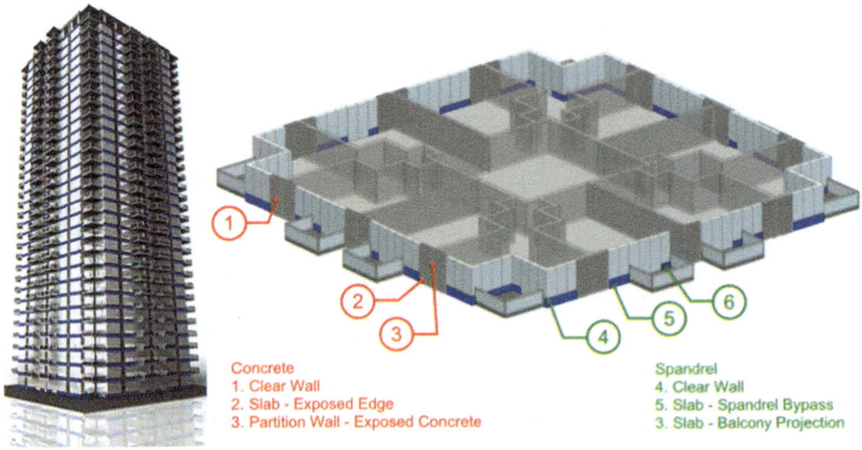

Fig. 8.14 Example of calculation of the overall heat transmission rate of any specific part of the enclosure by summing the heat transmission from each element

Chapter 9
Improved Building Airtightness

Uncontrolled air transfer through enclosures markedly increases the energy required to heat, cool, control humidity, and regulate indoor climate conditions in buildings. Investigations into building enclosure problems indicate that air leakage is a leading cause of moisture problems (Anis 2001; Zhivov et al. 2014). These include problems of mold, moisture penetration, and durability, especially in intersections between exterior walls, roofs, and windows, excessive rain penetration into wall cavities, unstable indoor temperature, and humidity profiles. In many cases, buildings with insufficient airtightness may suffer from moisture-related construction failures and losses of equity values. In colder climates, air leakage problems can cause such problems as ice damming in roof eaves, icicles on exterior facades, spalling of masonry, efflorescence, premature corrosion of metal parts in exterior walls, high wood moisture content, and rot. In hot humid climates, infiltrating air in combination with insufficient construction thermal bridges causes mold due to condensation on cold air-conditioned surfaces. Some of nonenergy-related and building sustainability issues related to uncontrolled air transfer through the building envelope include migration of bugs, insects, rodents, and outside pollutants entering into the conditioned space. Sealing penetrations and reducing the chimney effect of interior ventilation can address these concerns. Application of air barrier theory in a building design requires the selection of a component or layer in an assembly to serve as the airtight layer. It is important to clearly identify all air barrier components of each envelope assembly on construction documents and detail the joints, interconnections, and penetrations of the air barrier components (Anis 2001).

In exiting building airtightness is often poor, mainly due to imperfect joints between the panels, blocks, building components (especially around windows), and services penetrating the wall. It is important to address building airtightness with all major and even some minor building renovation projects. If the scope of work of the major renovation projects includes gutting the building, the air barrier concept can be similar to the one used with new construction. Increasing building airtightness can easily account for 10–40% of the total energy savings, depending on climate.

A. Zhivov, R. Lohse, *Deep Energy Retrofit*,
https://doi.org/10.1007/978-3-030-30679-3_9

Table 9.1 Airtightness best practice requirements

Country	Source	Requirement[a]	cfm/ft² @ 75Pa[a]
Estonia	Ordinance No. 58. RT I, 09.06.2015, 21, 2015	$\leq 6\ m^3/(h \cdot m^2)$ @ 50 Pa for renovation $\leq 3\ m^3/(h \cdot m^2)$ @ 50 Pa for new construction	0.42 cfm/ft² 0.21 cfm/ft²
Austria	OIB RL 6, 2011 for buildings with mechanical ventilation	1.5 1/h at 50 Pa	0.28 cfm/ft²
Denmark	Danish Building Regulations BR10 (2010)	1.5 1/h at 50 Pa	0.28 cfm/ft²
Germany	DIN 4108-2	1.5 1/h at 50 Pa	0.28 cfm/ft²
USA	USACE ECB for all buildings (HQUSACE 2010), ASHRAE Standard 189.1-2011, 2013 Supplement, ASHRAE Standard 189.1.–2013 Supplement, ASHRAE Standard 90.1 – 2013		0.25 cfm/ft²
	USACE HP Buildings and DER proposed requirement		0.15 cfm/ft²
Latvia	Latvian Construction Standard LBN 002-01 for buildings with mechanical ventilation	$2\ m^3/(m^2h)$ at 50 Pa	0.14 cfm/ft²
UK	ATTMA-TSL2	$2\ m^3/h/m^2$ at 50 Pa	0.14 cfm/ft²
CAN	R-2000	1 sq in. EqLA @10 Pa/100 ft²	0.13 cfm/ft²
Germany	Passive House Std	0.6 1/h at 50 Pa	0.11 cfm/ft²
Sweden	FEBY 12 Std	$1.08\ m^3/h/m^2$ at 50 Pa	0.08 cfm/ft²

[a]Based on example for four-story building, 120 × 110 ft, n = 0.65. Note that values are expressed in the units used in the subject country's national standards

Table 9.1 lists requirements for building airtightness, which differ in different countries (IEA ECBCS 2014) and which are included into the DER core technologies bundle. Existing buildings undergoing major renovations, especially those located in cold or hot and humid climates, should be sealed to the same standard as new construction if construction details allow for this. The QA of that process will require a "blower door" test.

9.1 Minor Renovations, Energy-Focused Projects, and Major Renovation Projects

The level of airtightness that can be achieved during the building renovation and approaches to be used depend on the overall scope of work of the project. Buildings undergoing major renovations are typically evacuated and gutted; scaffolding is installed; windows are scheduled for replacement; building envelope insulation is replaced and/or upgraded; and in many cases, roofs are replaced or repaired; most of mechanical, electrical lighting, and energy conversion systems along with connecting ducts, pipes, and wires will be replaced, which requires opening of vertical chases. Therefore, with regard to the air barrier, many of major renovation projects can be treated similarly to new construction.

In projects with a limited scope of work that does not include gutting of the whole building, there may be only a limited opportunity to improve the building envelope and vertical chases.

Before starting a minor renovation or energy-focused project, a building enclosure assessment may be performed to determine the existing paths of air leakage. This is performed in the predesign or planning phase to establish if the scope and overall goals of the project would benefit from envelope improvements to mitigate airflow and reduce energy loss through the envelope. The results must be documented on drawings of the building and by photographs of existing conditions. Open louvers, conduit penetrations, and elevator or lift machine room vents service shafts or like features should be identified and inspected, or tested to establish an overall condition of the envelope. A full envelope airtightness test and infrared or smoke tests are often a good starting point to locate major air leakages in the envelope structures and to establish a scope for a renovation project.

This test can be performed in accordance with national standards, e.g., ASTM E 779, *Standard Method of Determining Air Leakage Rate by Fan Pressurization*, and ASTM C 1060, *Standard Practice for Thermographic Inspection of Insulation Installations in Envelope Cavities of Frame Buildings in the USA*. Other diagnostics such as the use of a smoke tracer, with either handheld or theatrical smoke fog machine, can help to locate and determine the magnitude of leakage in a particular location. Air leakage testing around openings can also be done using ASTM E 1886, *Standard Practices for Air Leakage Site Detection in Building Envelopes and Air Barrier Systems*.

The strategy to increase airtightness depends on different factors, including building type, building system of walls and roof, window assembly, facade penetrations and their locations, condition and the level of existing insulation, and air and vapor barriers. Appendix F summarizes measures to improve airtightness in renovation projects.

The following aspects of building renovation with regard to the air barrier should be considered:

- What materials will best meet the needs or scope of the project?
- How will they be best applied or installed to achieve design expectations?
- Are windows and doors in the scope of the project?
- What is the intended use for the building after the project is complete?
- Compartmentalization of vented rooms, i.e., high-voltage rooms, shipping docks, etc.
- Incorporating an air barrier and additional insulation into an older building may have an effect on how the building behaves hygrothermally.

- Moisture analysis and possibly an investigation should be performed to assess and establish the expected moisture conditions.
- What are current mechanical systems and their components, and which new systems are included in the scope of the project?

Historical building renovations will also include these same considerations, along with a few more unique aspects. Historical preservation requirements and limitations may significantly restrict the approaches used to incorporate an air barrier into a wall system. It is possible that the walls of the building or their construction may have specific historical significance; in such cases, the proper authorities should be approached to make exception to the air barrier requirements. It is also possible that the wall and roof construction may be such that only minor areas of the building need to be addressed for significant envelope improvements.

9.2 What Is an Air Barrier?

Air barriers are systems of materials designed and constructed to control airflow between the conditioned and unconditioned spaces. Air barrier design is a 3D task that requires one airtight layer all around the whole conditioned (heated, cooled, and humidity controlled) volume. This concept may be illustrated using the "rule of drawing pencil" (Fig. 9.1). Figure 9.2 shows an example of the air barrier concept implementation on the building drawings.

The air barrier material, which must be structurally supported to withstand the maximum positive and negative air pressures to which it will be exposed, may have only a limited air permeance. An air barrier material is specifically defined as a material that has an air permeance less than or equal to 0.004 cfm/ft^2 at 1.57 lb/ft^2 (0.02 [L/s]/m^2 at 75 Pa) per ASTM E 2178. Air barriers can use vapor retarders and/or water-resistant barriers, or other appropriate materials, e.g., a liquid-applied material applied to the exterior wall sheathing.

9.3 Design Documents and the Air Barrier System

Air barrier drawing details are critical to a successful project. In the design-build project, the architect/engineer (AE) firm and the contractor shall be engaged as soon as possible in the design phase. For design-bid-build, the designing entity should be fully aware of the scope of the project and whether an air barrier is to be incorporated into the building envelope. At the point of submittal to the request for proposal (RFP) for a design-build project, it is important to ensure that a building science/air barrier narrative and, at a minimum, that the air barrier are called out in the drawings. If these two items are not present, then the submittal should be considered nonresponsive in this aspect. Although this may seem like a "hard line" approach,

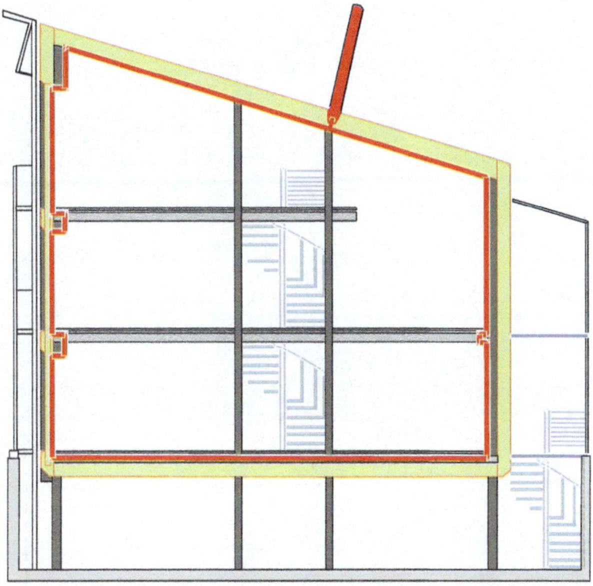

Fig. 9.1 Air barrier concept illustrated using the "rule of the drawing pencil" that ensures a continuous air barrier boundary in the contract drawings for a project

Used courtesy of PHI.

these elements must be addressed at this point to avoid significant problems down the road. Drawings, specifications, and design analysis should be reviewed fully through the various phases of the design process. If the 65 or 95% submittal design drawings of the air barrier and specifications of its materials are still not present, the fix for a poorly designed air barrier will be costly and may possibly delay the project. Appendix F provides examples of typical drawing details for major renovation projects.

9.4 Air Barrier Material Selection

Air barrier material selection depends on the wall type, building use, building shape, or even on the range of materials that are locally available. Other factors that might influence the selection may include material compatibility, climactic conditions, or local contractor experience with specific materials. Designers commonly exercise a good deal of flexibility in choosing materials for the air barrier system. Among the mostly used materials are:

- Mechanically Attached Flexible Sheets (Fig. 9.3):
  - Come from different manufacturers and are widely available
  - Are flexible and can be cut to fit odd or difficult spaces
  - May be applied in most weather conditions

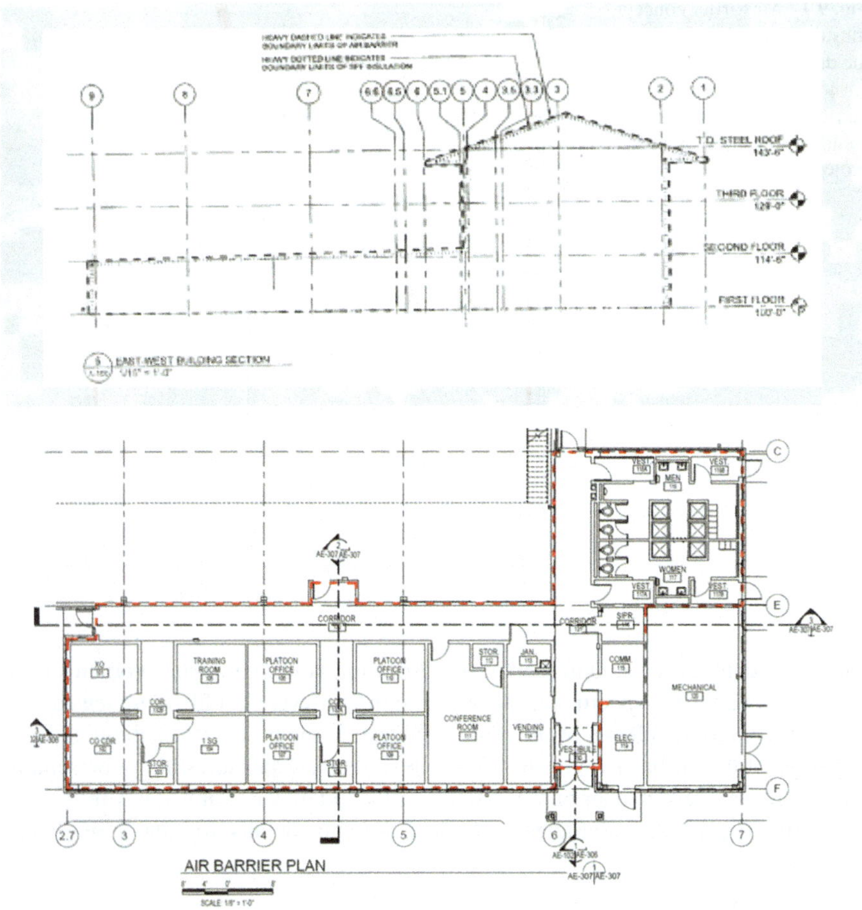

Fig. 9.2 Air barrier boundaries shown on a building section (**a**) and a plan (**b**)

- Have ancillary products that form a full system such as tapes, flashings, window and door treatments, fasteners, etc.
- Are simple to work with and are easy to repair
- Are usually familiar to contractors and installers who can be manufacturer-trained in their installation

• Spray-Applied Polyurethane Foams (Fig. 9.4):

- Come in different densities and are either open or closed cell
- When applied, create a continuous barrier with proper application and can fill or cover difficult to address areas
- Provide added thermal value to a wall system
- Are widely available with many manufactures to choose from
- Tend to need specialty contractors with experience to properly apply

Fig. 9.3 Mechanically attached flexible sheets

Fig. 9.4 Spray-applied polyurethane foam

- Are restricted to specific application temperatures, which are a factor for both interior and exterior applications
- Are restricted to application thickness, which is a factor, particularly for air barrier purposes
- Are messy and do off-gas compounds after application
- Are susceptible to sunlight exposure

• Self-Adhered Sheet Materials (Fig. 9.5):

- Come from a variety of manufactures and are widely available
- Come in various sizes
- Can be cut to fit and are flexible
- Use adhesives that generally stick to most surfaces

Fig. 9.5 Self-adhered sheet materials

Fig. 9.6 Fluid-applied membrane

- Are somewhat temperature sensitive when applied in cold temperatures
- Will self-heal to a degree when a fastener is put through it
- Can be used in a variety of assemblies including wall and roofs
- Are used as flashings and for other purposes
- Are used with other materials to form a complete system
- Are easy to apply and do not typically need any specialty training for application

• Fluid-Applied Membranes (Fig. 9.6):

- Come from a variety of manufactures and are widely available
- Are generally restricted by application temperatures, which are a concern for this material. Some manufacturers make low-temperature application products
- Need to be stored properly

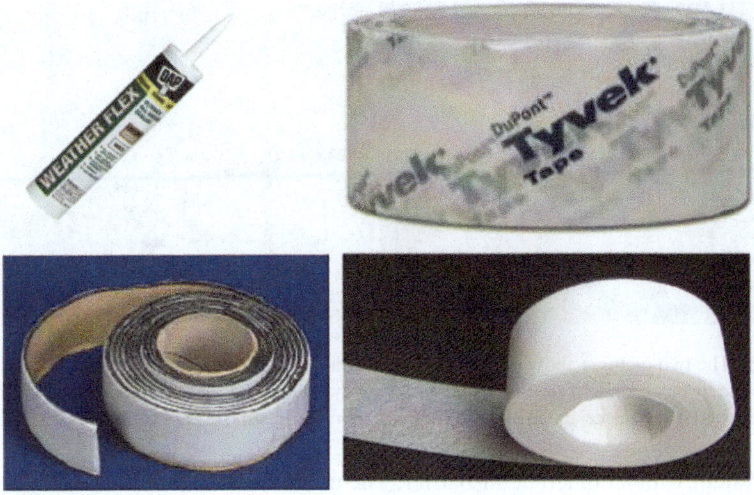

Fig. 9.7 Examples of other materials used to complete the air barrier system

- Are generally restricted by application thickness, which is also a concern for this material. Must be applied per manufacturer's recommendations for air permeance
- Must be applied by an experienced individual or company
- Are typically applied by specialty contractors
- Typically come as a system with other components and materials for windows doors and penetrations
- Can be applied to most substrates
- Create a high-quality continuous barrier when applied properly

9.5 Other Materials

Materials that are used to complete the air barrier system: tapes, gaskets, flashings, sill seal materials, specialty gasketed fasteners, brick ties furring systems, a variety of caulks, and sealants are available (Fig. 9.7).

9.6 Precast or Tilt-Up Concrete Panels for an Air Barrier

Other materials and wall types can be used as an air barrier system, e.g., tilt-up or precast concrete (Fig. 9.8). Because the concrete in both of these systems acts to block the air from entering the envelope, these materials are an excellent option as a

Fig. 9.8 The two pictures above are of new construction; however, if a select project was found to be feasible to incorporate a precast or tilt-up panel system as the new envelope for a DER, these same aspects of the material use and areas of concern would still be applicable

wall system with an inherent air barrier. The use of these wall types greatly reduces the complexity of the interface between air barrier components in which the primary interfaces are between panels, penetrations, and fenestrations and roof-to-wall and wall-to-foundation spaces. In a renovation or retrofit situation, panels may be reinstalled after a building envelope is fully deconstructed and secured to the original structure as the new building envelope.

In precast construction, the typical areas of concern for leakage are:

- Panel joints between the interior and exterior
- Joints between interior and exterior panels
- Door and window openings
- Top of wall to roof or ceiling connection
- Mechanical or plumbing or other through-wall penetrations

9.7 Wood Sheathing, CMU, or Gypsum Board Air Barriers

In addition to concrete, other materials such as wood, concrete masonry units (CMU), gypsum board, and fire-rated assemblies can be used as an effective air barrier. Though these materials are not tested to a specific material standard, they effectively perform by creating air barriers when applied and constructed properly.

- Gypsum board can be used as an air barrier when applied throughout the interior of a building. This is sometimes necessary when the exterior cladding cannot be removed, such as in a historic renovation.
- Joints need to be treated properly and penetrations need to be sealed, including electrical outlets, switches, wall-mounted equipment, lights, fire alarm devices, and controls. Anything that penetrates through the wall plane must be sealed.

Fig. 9.9 Gypsum air barrier partition wall used to compartmentalize an interior space: Left, see rear wall; Right, CMU and OSB sheathing is used as an air barrier assembly. (Photo courtesy of Nicholas Alexander, USACE)

- Slip tracks, roof deflection (if the wall goes to the roof), and control joints need to be considered when using gypsum board as an air barrier.
- Ceiling types may need to be considered when using gypsum board as an air barrier.
- Air barrier continuity will require a continuous wall/ceiling/floor barrier.

Figure 9.9 shows an example of a gypsum air barrier partition wall used to compartmentalize an interior space (rear wall in picture on the left), an example both CMU and OSB sheathing used as an air barrier assembly (picture on the right). The application of joint treatments between sheets and the transition locations is critical for the OSB sheathing.

CMU can be used as an effective air barrier assembly provided that the walls are fully grouted and the joints are complete, without any voids or deficiencies. In some instances, it may be necessary to treat joints and corners in CMU assemblies to ensure that the air barrier plane is continuous as required. Wood sheathing can also be used as an air barrier. If wood sheathing is used, the joints must be treated to properly ensure there is no leakage around or between the sheets. Fire-rated assemblies are also often used as interior air barrier planes due to the fact that all of these assemblies must be sealed to meet a required fire rating. In many situations, interior walls of an air barrier system can employ features such as fire-rated slip tracks or assembly configurations to achieve airtightness.

Overall, all of these materials can be used to improve envelope airtightness whether in a DER, in which, effectively, a new envelope is constructed, or in a minor or major renovation in which these materials are selectively used to improve the envelope in targeted locations. The overall QA process involved for each material is the same with regard to the materials, to the specific attention directed toward the applications, and to the scope that is specific to each project.

9.8 Construction Quality Assurance, Quality Control for the Air Barrier System

In a DER or any project that includes a renovation or new construction, high-quality construction is vital to ensure the building's proper function and performance.

The responsibility for QA and quality control (QC) falls to the Government or Owners Quality Representative and the contractor, respectively. Ultimately, the contractor is responsible for quality construction of the project; however, both the QA and QC representative(s) must be thoroughly knowledgeable in quality installation and application of these aspects of the construction for the constructed product to be successful.

General points for inspection:

- Several critical points for inspection of the air barrier system or the control layers of a building envelope are:
  - Roof-to-wall transition
  - Wall-to-foundation transition
  - Penetrations
  - Fenestrations
  - Material transitions
  - Substrate transitions
  - Access hatches
  - Soffits
  - Building corners
  - Expansion joints
  - Structural penetrations
  - Wall type and control or air barrier installation
- Also, particular inspection points and aspects of material storage or application for each type of material, in brief, are:
  - Material storage temperatures
  - Application temperatures
  - Substrate temperatures
  - Material compatibility
  - Fasteners
  - Substrate conditions
  - Internal or external environments

Appendix H provides more extensive detail and information on the QA aspects throughout the construction process for control layers, air barrier systems, and air barrier testing (Annex H-5).

Chapter 10
Water Vapor Control

10.1 Introduction

There are flows of heat, air, and water (liquid, capillary, and vapor) through building envelope assemblies. We seek to reduce heat flow and airflow through assemblies for energy savings. These are discussed in Sects. 7, 8, and 9 of this document. Liquid water flow depends on proper detailing for the exposure. Capillary water flow occurs in all sorptive materials, and its impact is minimal except in materials with ground contact.

Building materials are installed in assemblies for all of the reasons above, as well as for structure and aesthetics. They all have an impact on vapor flow, and that impact depends on their permeance or their openness to the flow of water vapor under diffusion or the flow driven by vapor pressure differences. Water vapor control consists of ensuring that the permeances of the materials or components in an assembly are such that they do not lead to an unsafe accumulation of water within the building assembly.

10.2 Above-Grade Walls and Roofs

The discussion in this subsection is limited to above-grade walls and roofs.

- Windows are treated in Sect. 8 of this document.
- Water vapor control pertains to foundations only when those foundations can be shown to exclude both liquid and capillary water available to the foundation from ground contact. This requirement is rarely met in actual foundations. Liquid and capillary controls are beyond the scope of this section. In addition:
  - Water leakage through the envelope cannot be assumed never to occur if the foundation space is below grade.

A. Zhivov, R. Lohse, *Deep Energy Retrofit*,
https://doi.org/10.1007/978-3-030-30679-3_10

- Two- and three-dimensional effects of water leakage and capillary transport are very important.
- Liquid water loads are difficult to quantify. They are changing with changes to climate and sea level rise.
- Conditions surrounding foundation assemblies are nonuniform.

It is assumed that:

- The walls and roofs comply with the thermal requirements found in the document.
- The walls and roofs are designed and constructed to be sufficiently airtight that the building can meet the airtightness requirements found elsewhere in this document. For that reason, water vapor transport due to air movement through the building envelope assembly is ignored.
  - If water vapor transport due to air movement is to be included, the flows shall be calculated using the parallel permeance methods described in ASHRAE Handbook (ASHRAE 2009).
- There is no leakage of liquid water into the assembly.
- The only damage to components in the building assembly that is considered in this section is mold growth on surfaces other than the exposed interior and exterior surfaces of the assembly.
  - Table 10.1 below lists all common forms of damage to wall and roof assemblies and explains why and how the form of damage is included or excluded from consideration in this section.
  - With the adoption of Addendum E, ASHRAE 160 has become a satisfactory method for estimating damage to assemblies due to mold growth on interior surfaces.
- The analysis used is one-dimensional. Two- and three-dimensional effects are ignored.

Given the scope and the assumptions, the requirements for vapor control can be stated simply: "Above-grade walls and roofs shall comply with ASHRAE Standard 160."

ASHRAE Standard 160 (and the comparable ISO document 13,788) was developed precisely to estimate the wetness within building envelope assemblies given the various inputs: components, material properties, boundary conditions, surface films, etc. With the adoption of Addendum E, Standard 160 is capable of estimating the extent and severity of mold growth, which is the focus of this section.

It is not necessary for all building assemblies to run a compliance check against Standard 160. In fact, almost all common assemblies in nonextreme conditions and climates will be expected to comply. We may develop tables for various assemblies under different conditions of temperature difference and vapor pressure difference and show the safe ranges for these assemblies, in "deemed to comply" tables. By

Table 10.1 Wall and roof performance problems

	Addressed?	Explain
Mold on inside of wall sheathing	Yes. Standard 160	Standard 160 was designed specifically for this purpose
Mold on interior surface	No, not vapor control	Will be associated with thermal bridges, diffuser throw, HVAC design/operation, or a combination, not with arrangement of permeances
Freeze-thaw in masonry	No, not vapor control	Freeze-thaw effects are associated with outdoor exposure, as possibly worsened by temperature of exterior materials. Insulation requirements that exterior elements will be cold in cold weather. No arrangement of permeances at the interior will impact freeze-thaw
Exterior or interior coating failure	Vapor control plays minor and undefinable role	Industry must specify coatings, substrates, and appropriate substrate preparation using baseline (mothball) conditions, i.e., conditions with no indoor climate conditioning
Mold on back of vinyl wallcovering	No	Occurs under conditions of airflow in the assembly, precluded under the airtightness requirement
Foundation walls (leaking spotting, mold growth, etc.)	No. Outside of scope	Usually damaged by water leakage, thus outside this specification
Corrosion	No	Corrosion resistance of metal elements must be specified for design conditions
Construction moisture effects	Yes	Accounted for in Standard 160 and WUFI
Loss of shingle service life	No	Product quality issue
Full attic sheathing darkening, winter	No, with qualifications	With relatively airtight ceiling, the full range of ventilation ratios is permissible (to discuss: ratio of ceiling airtightness to roof plane airtightness)
Full attic sheathing darkening, summer	No. Not a vapor control issue	Associated with powered ventilation and cooling loss into the attic
Water spotting at flashing and details	No. Not a vapor control issue	Water penetration exclusion
Mold in "vented" cathedral ceiling	Yes	It is impossible to estimate or anticipate airflow in cathedral vents. Therefore, the assembly should be modeled using the roof assumptions in the specification
Spalling or coating failure in protruding elements	No. 2- or 3-D effects	Protruding elements have highest exposure and are least likely to be affected by flows from the interior

using highly at-risk assemblies in the tables, it can be shown that less risky assemblies will comply within the more at-risk assembly "deemed to comply" range.

Further work is necessary to develop those tables. A recent check was conducted on three wood frame OSB-sheathed assemblies, R-30, R-40, and R-50, in which one third of the total insulation resistance was provided by highly impermeable insulation (XPS). The assembly was tested against Standard 160 for four climates, Anchorage, Minneapolis, Chicago, and Seattle, and for four different indoor climate classes (defined in ISO 13788) 1, 2, 3, and 4 (driest to wettest). The assembly was shown to comply for all outdoor climates and all indoor climate classes. A second assembly, using only fiberglass insulation, delivered results in which the coldest and wettest conditions led to failure and the mildest and driest conditions led to success, with the opportunity to define a limited range for such an at-risk assembly.

Can the use of Standard 160, which is shown to be capable only of determining the likelihood (or not) of mold growth within an assembly, be used as the basis for this standard? It can be shown by listing the types of damage to envelope assemblies, and how they are or are not addressed in this section.

"Condensation" is not listed as form of moisture damage to walls and roofs. This is because:

1. The building assemblies shown here are opaque assemblies. The formation of water droplets on a surface is of concern in transparent (glazed) assemblies, which are out of the scope of this document.
2. The building materials considered here (almost entirely) are sorptive. That is, the building materials here do not show the formation of droplets condensed from adjacent air cavities containing water vapor.
   (a) Instead, sorptive materials will absorb or adsorb water, they will become heavier or lighter with daily and other cycling, and this weight change or moisture uptake is natural, not a cause for concern.
   (b) Moisture uptake in sorptive materials becomes a concern when the wetness of the materials, together with the temperature of the surface, permits mold to grow. Mold growth is precisely the focus of ASHRAE 160, which serves as the basis for this section.

10.3 Attic Ventilation

In sloped roof construction, the insulation may be placed at the ceiling plane or at the roof plane. In either case, the air barrier should be located contingent to or integral with the thermal barrier.

1. For ceiling-insulated thermal barriers, with high levels of insulation and with low-airflow ceilings, the presence or absence of attic ventilation makes little difference in water vapor control. Attic ventilation carries an energy penalty and a resilience (wind and firestorm resistance) penalty.

2. Low slope roofs that comply with Standard 160 typically use high moisture-resistance materials in the assembly. They are not vented.
3. Sloped roofs with the thermal barrier placed within the roof assembly should be designed to comply with Standard 160 without reliance on ventilation. Achieving useful ventilation is difficult in simple roofs and is practically impossible in roofs with other than simple geometry: hips, valleys, sloped and low slope adjacencies, and long lengths.

10.4 Foundations

Water management in building envelope assemblies consists of managing liquid water flow, capillary flow, and vapor flow (diffusion). In above-grade assemblies, liquid water flow control is a matter of good design and detailing, for loads that are small, and can be easily anticipated. Capillary control is a minor matter where there are no liquid sources.

Water management in foundations consists almost entirely of liquid water management and capillary water management. Vapor management plays a minor role. Liquid water management rarely relies on a single element, but instead it includes several elements in series, each of which may be expected to be less than perfect. Liquid water management may occur at several levels:

1. Gutter and downspout configuration to keep water away from the foundation.
2. Sloping of soil surface away from the building.
3. "Flashing" the building into the soil so that surface water close to the building is directed away from the soil in contact with the foundation
4. Drainage of surface water downward so that it cannot apply a head of liquid water to the foundation walls.
5. Collection tiles at the base of the building, leading to daylight, to storm drains or to sump pumps.
6. Waterproofing (membrane, coating, expansive clays) applied to walls.
7. Coatings to resist capillary flow applied to the footing/wall joint, or to the insides of foundation materials.
8. Collection methods for rising ground water, together with discharge of collected water.
9. Ground covers (ideally sealed against water leakage and evaporation) in crawl spaces.
10. Low permeance membranes beneath slabs.

Chapter 11
Lighting Systems

Lighting accounts for almost 32% of the energy used in commercial buildings. Related energy codes are becoming more rigorous as the need to reduce energy consumption increases. Since reduction in lighting energy consumption can significantly affect a building's energy performance, lighting is a practical target. Many lighting solutions are simple and easy-to-implement, others more complex; many can yield substantial results. Advanced lighting systems should be considered in all renovation projects of federal and public facilities. Table 11.1 lists current minimum national standards for lighting systems.

A number of lighting technologies have been available for decades, but were not often implemented in federal and public facilities due to either budgetary constraints, lack of guidance, undocumented results, or other application issues. Other technologies in the lighting field are emerging with potential for even greater energy savings if used in the right applications.

When considering energy retrofits, the following basic principles should be considered:

- Cornerstone design strategies:
  - Provide appropriate illuminance levels without overlighting.
  - Use efficient lamps, ballasts, and luminaires.
  - Reduce electric lighting usage with controls.
- Energy-saving lighting design tactics that help create visually comfortable, effective, and efficient lighted environments:
  - Optimize architecture to provide daylight in frequently occupied spaces.
  - Apply light-colored (high-reflectance) surface finishes.
  - Cluster similar tasks to improve lighting system energy efficiency.

A. Zhivov, R. Lohse, *Deep Energy Retrofit*,
https://doi.org/10.1007/978-3-030-30679-3_11

Table 11.1 International requirements and standards for lighting systems

Country	Standards and requirements
Austria	ÖNORM EN 12464–1 and 12,464–2 for working spaces (1 = indoor, 2 outdoor spaces)
	ÖNORM EN 15193 – energy performance of buildings – energy requirements for lighting
China	GB 50034–2013, GB 50189–2015
Denmark	DS/EN ISO 12464-1 (Tables 5.1–5.53)
Estonia	Requirements for building service systems. Ministry of Economic Affairs and Communications' Ordinance No. 70. RT I, 09.11.2012, 12
Germany	DIN 18599–4, DIN 5035 T 1–14
Latvia	LVS-EN 12464–1:2011
	LVS-EN 12464–2:2014
	LVS-EN 15193:2009
UK	British Standard (BS) EN 12464–1:2011, Non-Domestic Building Services Compliance Guide:2013, Non-Domestic Building Services Compliance Guide for Scotland:2015
USA	ASHRAE Std 90.1 + Illuminating Engineering Society of North America (IESNA) recommended practices, tenth edition 2010

- Locate luminaires close to tasks that require higher illuminance.
- Use linear fluorescent and light-emitting diode (LED) luminaires predominately.
- Use high-efficiency ballasts with appropriate ballast factors.
- Use high-efficacy versions of lamps (e.g., 3100 lumen fluorescent T8).
- Illuminate walls and ceilings to increase perception of brightness.
- Use task lighting in general office areas and locations where additional illumination is needed for detailed tasks.
- Use daylight responsive lighting controls in frequently occupied spaces with daylight access.
- Use occupancy sensors in spaces without daylight access.
- Control lighting with astronomic time clocks for building-wide energy conservation.

There are numerous international guidelines for lighting systems retrofits (e.g., CIBSE 1993; USACE 2013; DS/EN 12464–1:2011; ZVEI 2005). The USACE Lighting Design Guide (USACE 2013) that has been adopted by the Annex 61 provides best practice guidance for lighting strategies in different building types and spaces along with illumination levels and maximum lighting power density (LPD) (Table 11.2).

Table 11.2 Illuminance and LPD targets for some building spaces

Space type	Target luminance fc	Lux	Target LPD W/ft^2	W/m^2
Common spaces				
Conference room	40	430	0.80	8.90
Corridor	10	108	0.50	5.56
Living quarters	5–30	54–323	0.60	6.67
Mechanical/electrical	30	323	0.70	7.78
Reception/waiting	15–30	161–323	0.50	5.56
Restroom/shower	20	215	0.80	8.90
Stair	10	108	0.50	5.56
Storage (general)	10	108	0.50	5.56
Telecom/SIPRnet	50	538	1.20	13.33
Vault	40	430	0.70	7.78
Restaurants/cafeteria dining area	20	215	0.60	6.67
Dishwashing/tray return	50	538	0.65	7.22
Kitchen/food prep/drive thru	50	538	0.65	7.22
Storage (dry food)	10	108	0.70	7.78
Training				
Readiness bay	40	430	0.75	8.33
Educational/training room (small)	15–30	161–323	0.70 W	7.78
Education and sport facilities				
Kindergarten classroom	15–30	161–323	0.70	7.78
High school classroom	15–30	161–323	0.70	7.78
Active play room	30–50	323	0.50	5.56
Staff lounge	15–30	161–323	0.50	5.56
Athletic – gym	40–50	430–538	0.70 W/ft^2	7.78
Gym with jogging track	50–60	538	0.80 W/ft^2	8.90
Racquetball	50	538	1.10 W/ft^2	12.22
Combative	50	538	0.90 W/ft^2	10.00
Health				
Resident work area	30–45	323	0.40 W/ft^2	4.44
Nurses station	15–30	161–323	0.80 W/ft^2	8.90
Exam room	30–50	323–538	0.60 W/ft^2	6.67

Source: USACE (2013)

Chapter 12
HVAC Equipment and Systems

12.1 Introduction

Mechanical HVAC systems are designed to provide thermal comfort and indoor air quality. The type of system, its efficiency, and its components and control strategies should be selected and designed to minimize energy consumption and to maximize comfort throughout the range of operating conditions.

When building heating, cooling, and electrical loads are significantly reduced, the importance of selecting one type of heating and cooling system over another diminishes.

However, a few aspects must addressed to achieve DER:

- When replacing HVAC systems with new ones, use high-performance motors, fans, furnaces, chillers, boilers, etc. according to the most stringent current national standard and requirements to energy systems.
- Separating systems for ventilation, make-up air, humidity control, and building pressurization from systems providing temperature control can be an effective means of reducing energy use, downsizing mechanical equipment, and improving systems controllability. Use DOASs.
- Use well-sealed and insulated ducts and insulated hot water and chilled water pipes.
- Use energy recovery from exhaust air to preheat and precool outdoor air supplied by the DOAS.

A. Zhivov, R. Lohse, *Deep Energy Retrofit*,
https://doi.org/10.1007/978-3-030-30679-3_12

Table 12.1 National standards for HVAC systems

Country	HVAC
Austria	EN 1507, EN 12237, ÖNORM H 5057, OIB RL 6, 2011
China	GB 50736-2012, GB 50189-2015
Denmark	Standard 447, Standard 452
Estonia	EVS-EN 13779, EN 12237, Ordinance No. 70. RT I, 09.11.2012, 12
Germany	EnEV 2014, DIN V 18599, DIN 1946-6, DIN EN 13779
	DIN 24192 II/III/IV, DIN 4108-6, DIN 4701-10
Latvia	Latvian Construction Standard LBN 231-03
	Latvian Construction Standard LBN 003-01
UK	Non-Domestic Building Services Compliance Guide: 2013
	Non-Domestic Building Services Compliance Guide for Scotland: 2015
	BS EN 15727:2010, BS 5422:2009
USA	ASHRAE Std 90.1, ASHRAE Std 189

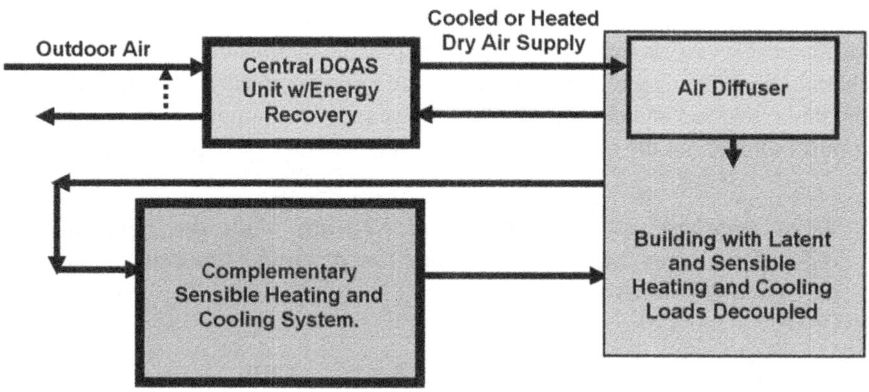

Fig. 12.1 DOAS schematic

12.2 Requirements for Mechanical Equipment

Mechanical equipment should be properly sized and meet the efficiency requirements specified in national standards listed in Table 12.1 or better. Equipment efficiency should be sized based on optimal performance at part load versus peak its performance. The most important efficiency aspect of HVAC performance is the overall efficiency of the whole system, not just the efficiencies of its components.

One common misconception is that making a building's airtight will result in inferior indoor air quality. Including a DOAS (Fig. 12.1) into a "core technologies bundle" ensures that building spaces receive sufficient outside air in amounts that are not affected by the building temperature control strategies. A DOAS delivers 100% outside air (OA) to each individual space in the building via its own duct system (PIER 2009) for ventilation, make-up air, humidity control, and building pressurization. Heating and cooling of building spaces is provided by a separately controlled system (Ministry of Economics of the Republic of Latvia 2013).

As a general rule, a DOAS operates at constant volume. For most applications, the DOAS cannot meet all of the thermal loads in the space by itself; it requires a parallel system to accommodate any sensible and latent loads the DOAS cannot accommodate.

DOAS airflow rates generally are dictated by:

- Indoor air quality needs (based on national standards)
- Latent load (humidity control needs)
- Make-up air for bathroom and kitchen exhausts (when needed)
- Building pressurization to prevent infiltration, which helps to reduce heating/ cooling and moisture loads

12.3 Energy Recovery

Energy recovery is an important component of the DOAS that recycles energy contained in the air normally exhausted from the building to preheat or precool the incoming ventilation air. In many existing buildings, heating and cooling the ventilation air accounts for up to 50% of the total HVAC system load. In such older designs, the ventilation air is treated with conventional components such as an oil or gas burner, hot and chilled water coils, and DX coils. A typical energy recovery process is three to six times more efficient at treating ventilation air compared to a conventional component. Figure 12.2 shows schematics of typical heat recovery components.

To characterize component efficiency, a value called "recovery efficiency ratio" (RER) has been developed. RER is the rate of energy recovered relative to the power supplied to perform that work. Similar to "energy efficiency ratio" used in chillers and unitary equipment, RER allows designers to model efficiency gains in whole systems.

The performance of the energy recovery component is typically measured by its effectiveness and pressure drop. Energy transfer effectiveness is defined as the actual energy or moisture recovered to the maximum possible between the exhaust and supply air streams. The energy transfer effectiveness can vary between 15% and 85% with a typical value of 75% for cross-flow energy wheel arrangement (Harvey 2006). Energy recovery effectiveness using the synthetic counterflow heat exchanger (Brink International 2016) can be up to 95%. Table 12.2 lists minimum legal standards in several countries for energy recovery equipment, and Table 12.3 lists typical total effectiveness and limitations for energy recovery components.

Figure 12.3 shows an example of a cross-counterflow heat exchanger with an HR effectiveness of 85%.

It is important to consider the reduced ventilation load and its impact on the primary heating and cooling requirements of the DOAS. Oversizing the system may

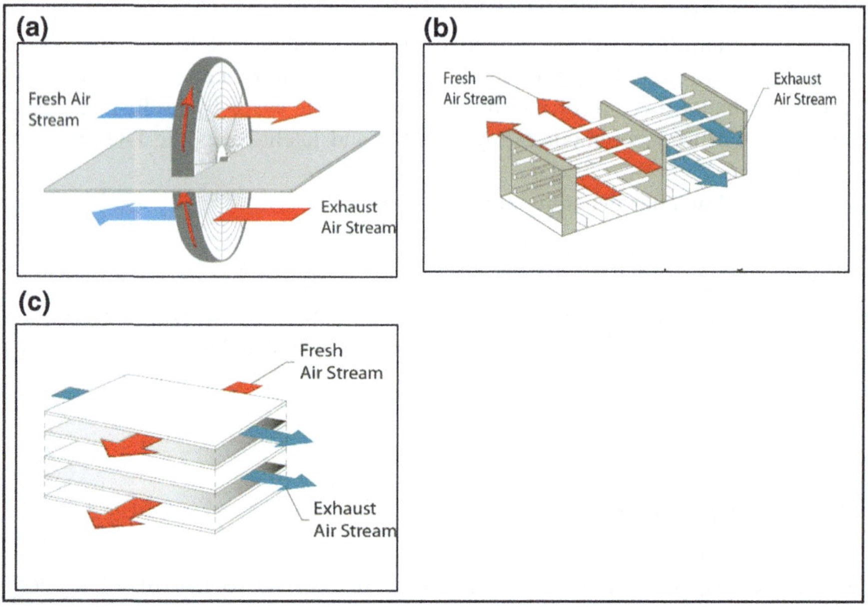

Fig. 12.2 Typical heat recovery components: (**a**) rotary air-to-air energy exchanger, (**b**) heat pipe heat exchanger, and (**c**) fixed plate cross-flow heat exchanger

Table 12.2 Minimum legal national standards and requirements for heat recovery from return air

Country	Standard for HR equipment	Energy type recovered (total, sensible, latent)	Efficiency, %
Austria	ÖNORM EN 13141-7	Total	70
China	GB 50189-2015	Total and sensible	60
Denmark	Danish Building Regulations BR10 (2010)	Sensible	70
Estonia	Ministry of Economic Affairs and Communications Ordinance No. 70. Rt I, 09.11.2012, 12	Sensible	70
Germany	DIN 4108-6, DIN 4701-10, DIN EN 13053, EnEV 2014	Total	50% in average (depending on m^3/h and h/a the range is 0.4–0.65)
Latvia	Latvian Construction Standard LBN 231-15 of 16 June 2015	Not defined	Not defined
UK	Non-Domestic Building Services Compliance Guide: 2013	Sensible	Heat pipe – 60
	Non-Domestic Building Services Compliance Guide for Scotland: 2015		Heat wheel – 65
			Runaround coil – 45
USA	ASHRAE Standard 90.1-2013.	Total	>50%

Table 12.3 Energy effectiveness and limitation of typical energy recovery components

	Energy wheels	Fixed plate	Heat pipe
Typical total effectiveness	65–85%	35–70%	15–35%
Advantages	Highest total effectiveness	No moving parts	No moving parts
	Low pressure drop	Low pressure drop	No air leakage
	Scales in size easily		Low pressure drop
	Widely available in standard ventilation equipment		
Limitations	Requires a motor and belt to operate	Large sizes at higher airflows	Minimal effectiveness
	Some air leakage	Some air leakage	Requires seasonal reset
			Few suppliers

Fig. 12.3 State-of-the-art cross-counterflow heat exchanger with HR effectiveness of 85%. (Source: www.passipedia.org)

result in excess energy consumption and may compromise the device's ability to maintain space conditions. Proper system sizing will result in optimized comfort control, energy savings, and minimized capital expense. For ventilated-only (not air-conditioned) buildings, energy recovery is generally more cost-effective in cold and moderate climates with a payback ~2 years. For air-conditioned buildings, this technology is applicable to all climates.

Life cycle energy savings and operational performance will vary by component. Furthermore, equipment availability, capital cost, maintenance cost, and technology advantages and disadvantages should be considered when selecting a component. Certified performance ratings such as effectiveness, leakage parameters, and pressure drop should be used when comparing energy savings. To maximize energy savings, it is important to select a device that can be adequately maintained.

Components that are easy to access and to properly clean will provide greater energy savings over the life cycle of the device.

Heat recovery can be combined with any air-conditioning system and will then support both heating in winter and cooling in summer. Ventilation heat recovery reduces energy losses resulting from exchange of cold air in winter and hot and humid air in summer for ventilation purpose.

Heat recovery combined with humidity recovery in a heat exchanger with a water-permeable membrane can additionally support dehumidification of air in hot and humid climates.

In general, heat recovery can be combined with split heat pump units for air-conditioning (AC) systems and also with any other heat pump system.

12.4 Heat Recovery Combined with Indirect Evaporative Cooling

In some instances, when an old building with ventilation and heating systems is renovated, cooling and dehumidification capabilities are added, which results in a significant increase in the building's energy use. Mechanical cooling provided by direct expansion (DX) units or chilled water systems is one of the major electric loads in buildings, particularly in southern climates. In many climate zones, the cooling load for many buildings could be satisfied most of the year without mechanical refrigeration by cooling the buildings' ventilation air using indirect and direct evaporative cooling (IDEC). This would result in much lower operating costs, especially during peak electrical demand periods. In buildings with significant cooling loads, indirect evaporative cooling can be used in combination with DX units or chilled water coils to reduce the size (first cost) and the load (operating costs) of mechanical cooling and humidity control systems by up to 65% (USACE 2016). The primary advantage of the IDEC in relation to the other evaporative cooling units is that it cools without adding any moisture to the conditioned space. This advantage comes at a price, though, and not just in terms of initial cost. The heat and mass exchange modules have a very high resistance to flow, which results in decreased delivered airflow and higher power consumption than other evaporative cooler systems (Dean et al. 2012). However, this system still uses considerably less power than a conventional air conditioner and will likely keep a space more comfortable than other evaporative coolers for most of the cooling season.

To implement this concept, a DOAS should be equipped with the IDEC heat exchanger installed in the return air duct to provide energy recovery ventilation. In most situations, the return air in the cooling season has a temperature around 22 °C (72 °F) and RH around 50% in most climate conditions and applications. In cooling mode, the return air is used on the working side of the IDEC heat exchanger and provides indirect cooling of the outdoor air passing on the other side of the heat exchanger. During the heating mode, there is no water supply to the IDEC unit.

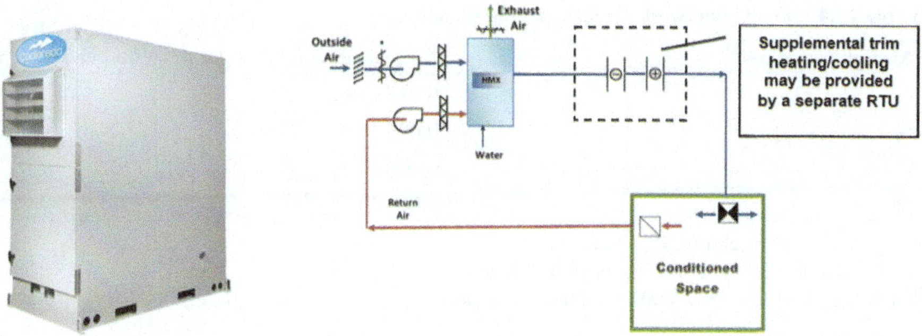

Fig. 12.4 Hybrid DOAS using indirect evaporative cooling and heating with energy recovery ventilation (ERV) (USACE 2016)

Return air has a temperature higher than the OA flow such that the IDEC heat exchanger acts like a (dry) heat recovery device. During the entire year, an IDEC is used for OA preheating and precooling/cooling. Figure 12.4 shows the configuration of an IDEC unit to be connected to supply and return air ducts.

12.5 HVAC System Air Leakage Requirements

12.5.1 Introduction

When a building undergoes deep energy retrofits (DERs) that are conducted as part of a major renovation project, the duct system is usually replaced. Replacing it with a well-insulated and airtight system is one of the most cost-effective energy efficiency measures available. One reason is that the airflows through air-handling units (AHUs) connected to ducts must be sufficient to compensate for air leakage involving ducts and duct-mounted components (HVAC system air leakage). These AHU airflows account for a large fraction of the energy use in buildings. For example, in the United States, HVAC system air leakage is ranked as the primary source of energy inefficiency in commercial buildings; it wasted an estimated $2.9 billion in 2005 (Mills 2009). Based on research results and experience from the United States and around the world (especially in Scandinavia), this section of the Guide provides recommendations on cost-effective targets for ductwork air tightness and a quality assurance process for achieving these targets.

For the purposes of this Guide, *ductwork* includes straight duct, flexible duct, sheet metal and rigid fiberglass plenums, and fittings (e.g., elbows, transitions, tees, wyes) for distribution and extraction of air. It does not include duct-mounted components (e.g., terminal units, access doors/panels, attenuators, coils, fire/smoke dampers, balancing, and control dampers). A *system* consists of all ductwork that connects the air handler to the conditioned space and includes duct-mounted components where leaks through joints or penetrations can also occur (ASHRAE 2016b).

Table 12.4 HVAC ductwork air leakage classifications

Air tightness class[a,b]	Air leakage limit, L/s per m^2 (cfm per 100 ft^2)
A	0.027 (19.2) $p_{test}^{0.65}$
B	0.009 (6.4) $p_{test}^{0.65}$
C	0.003 (2.1) $p_{test}^{0.65}$
D	0.001 (0.7) $p_{test}^{0.65}$

Note: p_{test} = test pressure, Pa (in. of water)
[a]CEN Standard EN 12237-2003 (circular ducts)
[b]CEN Standard EN 1507-2006 (rectangular ducts)

Duct air leakage classifications. In Europe, air leakage limits for round and rectangular ducts, respectively, are defined in CEN Standards EN 12237 (CEN 2003) and EN 1507 (CEN 2006). The corresponding air tightness classes (Table 12.4), in order of increasing air tightness, are A, B, C, and D. In countries with leakage-related regulations such as Sweden, these classes are generally applied to newly constructed ductwork as follows (Guyot and Carrié 2010; Andersson 2013):

- Class A (greatest leakage allowed) applies to *visible* ductwork in the space being served. Although supply duct leakage and supply air are entering the same space, air leakage will negatively affect the ventilation and heating/cooling effectiveness, which depend on the location of air supply and type of diffusers used. Also, balancing the system so that the correct amount of air is supplied at the diffusers will result in increased airflow and energy use at the supply fan.
- Class B (three times tighter than A) applies to ductwork with surface area $\leq$20 m^2 ($\leq$215 ft^2). This class generally applies to small houses.
- Class C (three times tighter than B) applies to ductwork with surface areas >20 m^2 (>215 ft^2). This class applies to most other buildings. Class C ductwork typically results in about 30% less fan power than Class A ductwork.
- Class D (three times tighter than C) is not a standard requirement, but can optionally be specified for ductwork for which air tightness is essential (e.g., in hospitals).

In the United States, ASHRAE originally developed "Leakage Class (C_L)" to *predict* air leakage from round, rectangular, flexible, and fibrous glass ducts, based on air leakage versus seam, joint, and sealant type test data from ASHRAE research project RP-308 (ASHRAE 1985). Today, energy standards such as ANSI/ASHRAE/IES Standard 90.1 (ASHRAE 2016a) use air leakage classes as a function of duct surface area in a manner similar to the way standards in European countries use those classes. Figure 12.5 (SI) and Fig. 12.6 (IP) show the relationship of European air tightness classes and ASHRAE leakage classes. The equivalency between European and ASHRAE leakage classes is noted in these figures as A ($C_L = 19$), B ($C_L = 6.4$), C ($C_L = 2.1$), and D ($C_L = 0.7$).

How to achieve quality ductwork? Apart from Scandinavia (Denmark, Finland, Norway, and Sweden), many countries in Europe generally have very leaky duct systems. Air leakage is typically three times higher than Class A, which is up to 30 times higher than the leakage observed in Sweden (Schild and Railio 2009).

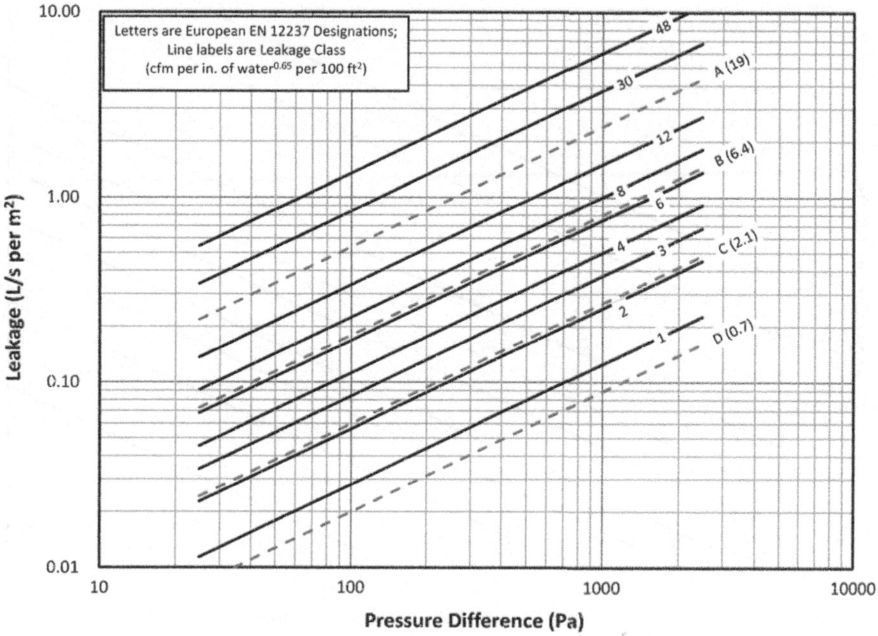

Fig. 12.5 Relationship between European air tightness classes (dashed lines) and ASHRAE leakage classes (solid lines) in SI units

Nearly all Swedish buildings and their installations meet the voluntary AMA[1] series of quality guidelines (e.g., Svensk Byggtjänst 2015) that are referenced in building contracts between the owner and contractors (Andersson 2015). These guidelines in part address ductwork air tightness. According to Schild and Railio, the three ingredients for success in achieving airtight duct systems in Scandinavia are:

- Increased awareness of the benefits for quality ductwork and duct-mounted components
- Guidelines and requirements with incentives provided in building contracts
- Verification of guidelines and requirements provided in each project with predefined penalties

High-performance airtight duct systems can be achieved in many ways. One example is through the installation of circular spiral ductwork and low-leakage system components. In particular, some manufacturers of spiral ducts offer duct and duct fittings with factory-fitted integrated rubber gaskets (Figs. 12.7 and 12.8) that enhance the air tightness of HVAC systems. The ducts are made of continuous strips of sheet metal formed into continuous spiral tubes with spiral lockseams that also serve as continuous reinforcing. These construction and installation features

[1] 1AMA: Allmän Material och Arbetsbeskrivning; Swedish for "General Material and Workmanship Specifications"

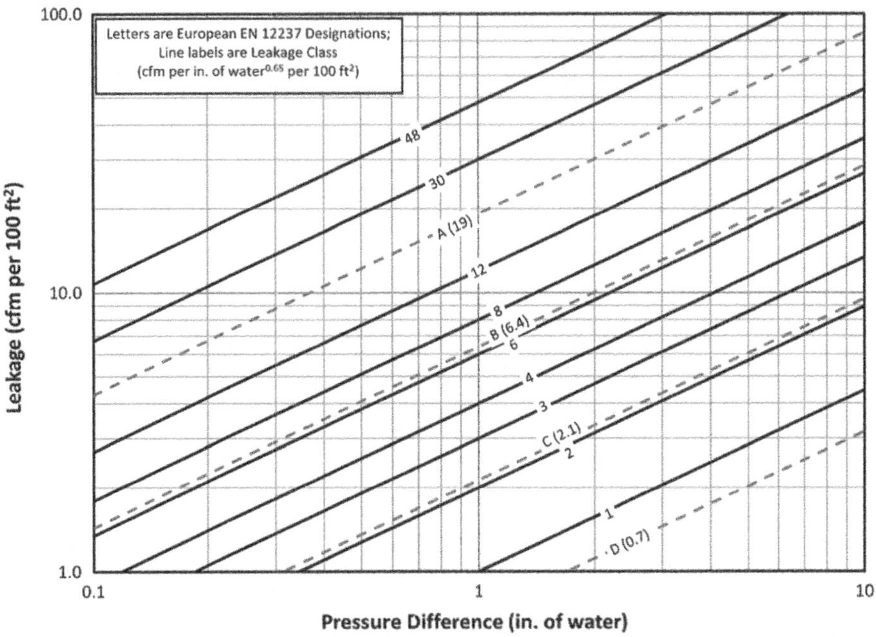

Fig. 12.6 Relationship between ASHRAE leakage classes (solid lines) and European air tightness classes (dashed lines) in IP units

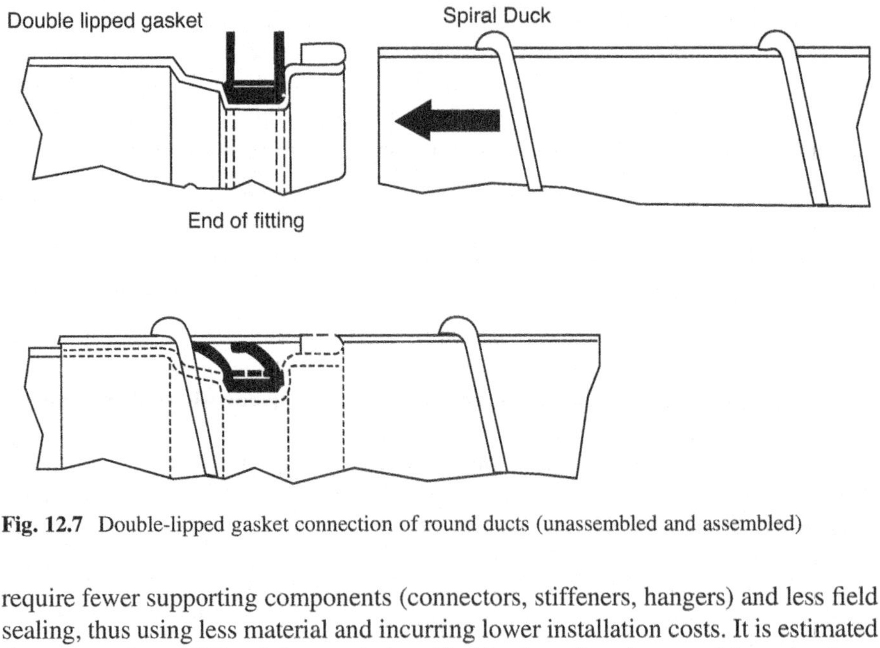

Fig. 12.7 Double-lipped gasket connection of round ducts (unassembled and assembled)

require fewer supporting components (connectors, stiffeners, hangers) and less field sealing, thus using less material and incurring lower installation costs. It is estimated that as much as 95% of ductwork installed in Scandinavia uses this technology (Schild and Railio 2009).

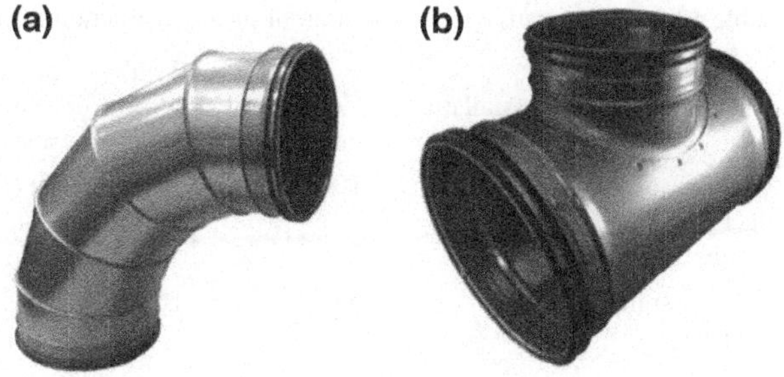

Fig. 12.8 Circular duct fittings with double-lipped gasket connections: (**a**) elbow and (**b**) tee

Although some buildings with limited ceiling space may indeed require that rectangular or flat oval ducts be used instead of circular ducts, it is important to recognize that round ducts often need less installation space than rectangular ducts with the same pressure drop. Furthermore, using tighter duct systems can also result in smaller ducts, because system airflows are reduced. Another alternative when space is limited is to use multiple parallel round ducts.

Other technologies can also achieve tight ducts, but they involve more labor-intensive field sealing instead of using gasketed joints. They should always be used when joints do not have gaskets. In these cases, clamps, tapes, brushed- or spray-on coatings, or aerosol-based sealants as may be appropriate to the application can be used during or after the system is constructed. An example of an "airtight" rectangular duct system with brushed-on sealants is described by Diamond et al. (2003). Supply air leakage from this system while it was operating at "maximum" flow[2] was about 4% of the supply fan flow. Equivalent leakage classes were not determined for ductwork sections in this system.

12.5.2 *Economics*

To better understand the economics of the various construction technologies, consider the following. Tests carried out by BSRIA (2002) of circular spiral ductwork that meets air tightness Class C or tighter requirements have demonstrated a 39% saving on installation time compared to traditional ungasketed circular spiral ductwork, which requires the application of an external tape, mastic, or rolled sealant to seal each joint; the total installation costs (materials and labor) can be reduced as much as 12%. Results of TÜV's[3] (2008) comparative test evaluation (Table 12.5)

[2]The "maximum" flow in this case corresponds to the reference operating conditions described in ANSI/ASHRAE Standard 215 (2018) for a variable-air-volume (VAV) system with direct digital controls (DDC).

[3]TÜV is an abbreviation of the German Technischer Überwachungsverein or Technical Inspection Association, which is an umbrella organization for German businesses that provide inspection and product certification services.

Table 12.5 Comparative test evaluation of duct construction techniques

Duct type /seal type	Installation time index	Component price index	Total price index	Pressure, Pa (in. of water)	Measured air leakage, L/s (cfm)	Air tightness class
Circular spiral ducts/double-sealing gasket	100	100	100	400 (1.6)	0.71 (1.5)	D
				−750 (−3.0)	0.85 (1.8)	D
Circular spiral ducts/tape	155	95	108	400 (1.6)	7.63 (16.2)	B
				−750 (−3.0)	11.46 (24.3)	B
Circular spiral ducts/mastic	164	96	110	400 (1.6)	7.18 (15.2)	B
				−750 (−3.0)	8.42 (17.8)	B
Rectangular ducts/ tape + sealing compound	228	115	139	400 (1.6)	7.25 (15.4)	B
				−750 (−3.0)	9.0 (19.1)	C

show that circular spiral ducts with double-sealing gaskets require a shorter installation time, have lower costs, and achieve better air tightness compared to other duct construction techniques.

Airtight systems also support other energy efficiency measures (EEMs), including demand-control ventilation and heat recovery that affect energy use for both heating and cooling. For example, Guyot and Carrié (2010) calculated that air leaking out of duct systems can reduce the global efficiency of a heat recovery system from 85% (nominal value) to less than 60% (ducts three times leakier than Class A ducts). This equates to approximately 5 kWh/m^2/year (15.8 kBtu/ft^2/year) of space heating. Guyot and Carrié did not provide associated cooling energy impacts.

What is acceptable air leakage for high-performance buildings? To enable a proper accounting of leakage-related impacts on fan energy and space-conditioning loads, the allowable air leakage for each fan system should be established as a *percentage of fan airflow* at the maximum system operating conditions.

For the ductwork, the maximum acceptable *air leakage flow* for a test section corresponding to a specified air tightness class (air leakage class) can be expressed by Eq. 12.1 (ASHRAE 2017):

$$\begin{aligned} Q_{leak,section,i} &= C_{L,section,i}\, A_{section,i}\, p_{section,i}^{0.65} \quad \text{SI} \\ Q_{leak,section,i} &= C_{L,section,i} \left(\frac{A_{section,i}}{100}\right) p_{section,i}^{0.65} \quad \text{IP} \end{aligned} \tag{12.1}$$

where

$Q_{leak,section,I}$ =air leakage flow for test section i, L/s (cfm)
$C_{L,section,i}$=air tightness class (air leakage class) for test section i, L/s per Pa$^{0.65}$ per m^2 (cfm per [in. of water]$^{0.65}$ per 100 ft^2) of duct surface area
$A_{section,i}$ =surface area of test section i, m^2 (ft^2)

$p_{section,i}$=static pressure difference across the duct wall corresponding to maximum fan airflow that would occur during operation for test section i, Pa (in. of water)

Equation 12.1 shows that the air leakage flow depends on section air tightness class, section surface area, and the section static pressure raised to the 0.65 power. For example, a test section with $C_{Lsection,i} = 0.003$ L/s per $\text{Pa}^{0.65}$ per m^2 (2.1 cfm per $[\text{in. of water}]^{0.65}$ per 100 ft^2) of duct surface area, $A_{section,i} = 45\ \text{m}^2$ (484 ft^2), and $p_{section,i} = 750$ Pa (3.0 in. of water), the air leakage for that section is 10.0 L/s (20.8 cfm). With $p_{section,i} = 125$ Pa (0.5 in. of water) instead (six times less), the air leakage is about three times less (3.1 L/s [6.5 cfm]).

To obtain the system air leakage fraction *due only to ductwork leakage* as a function of fan airflow, one needs to add the leakage flows from each section (Eq. 12.1) together as follows:

$$Q_{leak,frac} = 100 \left[\frac{\sum_{i=1}^{N} \left(C_{L,section,i} A_{section,i} p_{section,i}^{0.65} \right)}{\left(\frac{Q_{fan}}{A_{system}} \right) \left(\sum_{i=1}^{N} A_{section,i} \right)} \right] \quad \text{SI}$$

$$Q_{leak,frac} = 100 \left[\frac{\sum_{i=1}^{N} \left(C_{L,section,i} \left(\frac{A_{section,i}}{100} \right) p_{section,i}^{0.65} \right)}{\left(\frac{Q_{fan}}{A_{system}} \right) \left(\sum_{i=1}^{N} A_{section,i} \right)} \right] \quad \text{IP} \qquad (12.2)$$

where

$Q_{leak,frac}$=system air leakage fraction due only to ductwork leakage and corresponding to maximum fan airflow that would occur during operation, %
Q_{fan}=maximum fan airflow that would occur during operation, L/s (cfm)
A_{system}=total system duct surface area, m^2 (ft^2)
N=total number of ductwork sections included in A_{system}, dimensionless

Equation 12.2 can be simplified to:

$$Q_{leak,frac} = 100 \left[\frac{\sum_{i=1}^{N} \left(C_{L,section,i} p_{section,i}^{0.65} \right)}{\left(\frac{Q_{fan}}{A_{system}} \right)} \right] \quad \text{SI}$$

$$Q_{leak,frac} = \left[\frac{\sum_{i=1}^{N} \left(C_{L,section,i} p_{section,i}^{0.65} \right)}{\left(\frac{Q_{fan}}{A_{system}} \right)} \right] \quad \text{IP} \qquad (12.3)$$

For "typical" normalized airflows (Q_{fan}/A_{system}), such as those described by SMACNA (2012, Appendix A), Tables 12.6 and 12.7 lists a tabulation of the relationship in Eq. 12.3 for a system with one test section (or a combination of sections each with the same airtightness class and static pressure). For example, the data in Table 12.6 (SI) show that for an HVAC system with $Q_{fan}/A_{system} = 10$ L/s per m^2 and $p_{section} = 1000$ Pa, the corresponding air leakage fraction for European air tightness classes A, B, C, and D is 24.1%, 8.0%, 2.7%, and 0.9%, respectively. In equivalent IP units, for an HVAC system with $Q_{fan}/A_{system} = 2$ cfm per 100 ft^2 and $p_{section} = 4$ in. of water, the data in Table 12.7 (IP) show that the air leakage fraction for ASHRAE air leakage classes 20, 6, 2.1, and 1 is 24.6%, 7.4%, 2.6%, and 1.2%, respectively.

Many European countries have successfully attained air tightness Class C and Scandinavia uses Class D ductwork as a standard practice for buildings such as hospitals. Based on this success and on the many benefits of Class C as the paper has discussed earlier, and assuming that the first cost of a Class C system is less than that of a Class D system, the proposed air tightness goal for DER project duct systems is Class C (EBC 2017b) for all ducts. The equivalent for European air tightness Class C is ASHRAE air leakage Class 2.1. The data listed in Tables 12.8 and 12.9, abstracted from Tables 12.6 and 12.7, show that the maximum air leakage for Class C (Class 2.1) ductwork with 1500 Pa (6 in. of water) maximum static pressure or less in each duct section (from Tables 12.6 and 12.7) is about 4.5%. It is important to note that actual systems usually do not have uniform static pressures when operating, so the leakage fraction for duct sections will be less than this maximum for these systems.

In addition to the guidance offered by many European countries, the "Duct Construction" chapter of the *ASHRAE Handbook – HVAC Systems and Equipment* (ASHRAE 2016b) provides preliminary guidance for the maximum total *system* air leakage (and economic rationale) as follows:

- Five percent of the design airflow for ductwork located in the conditioned space, except 2% for exhaust ductwork
- Two percent of the design airflow for supply ductwork outside the conditioned space
- One percent for air-handling units

The "Duct Design" chapter of the *Handbook – Fundamentals* (ASHRAE 2017) also provides guidance for leakage characteristics related to duct-mounted components. The maximum allowable leakage characteristics of the duct-mounted equipment to be installed should be specified so that the combined leakage of the ductwork and all duct-mounted components does not exceed the specified maximum allowable system leakage fraction at actual operating conditions.

Table 12.6 Leakage as percentage of flow (SI units)

Air tightness class[a], L/s per m² per Pa$^{0.65}$		Q_{fan}/A_{system}, L/s per m²	Static pressure, Pa								
			125	250	500	750	1000	1250	1500	2000	2500
A	0.027	8	7.8	12.2	19.2	24.9	30.1	34.8	39.2	47.2	54.6
		10	6.2	9.8	15.3	20.0	24.1	27.8	31.3	37.8	43.7
		12.5	5.0	7.8	12.3	16.0	19.3	22.3	25.1	30.2	34.9
		15	4.2	6.5	10.2	13.3	16.0	18.5	20.9	25.2	29.1
		20	3.1	4.9	7.7	10.0	12.0	13.9	15.7	18.9	21.8
		25	2.5	3.9	6.1	8.0	9.6	11.1	12.5	15.1	17.5
B	0.009	8	2.6	4.1	6.4	8.3	10.0	11.6	13.1	15.7	18.2
		10	2.1	3.3	5.1	6.7	8.0	9.3	10.4	12.6	14.6
		12.5	1.7	2.6	4.1	5.3	6.4	7.4	8.4	10.1	11.6
		15	1.4	2.2	3.4	4.4	5.3	6.2	7.0	8.4	9.7
		20	1.0	1.6	2.6	3.3	4.0	4.6	5.2	6.3	7.3
		25	0.8	1.3	2.0	2.7	3.2	3.7	4.2	5.0	5.8
C	0.003	8	0.9	1.4	2.1	2.8	3.3	3.9	4.4	5.2	6.1
		10	0.7	1.1	1.7	2.2	2.7	3.1	3.5	4.2	4.9
		12.5	0.6	0.9	1.4	1.8	2.1	2.5	2.8	3.4	3.9
		15	0.5	0.7	1.1	1.5	1.8	2.1	2.3	2.8	3.2
		20	0.3	0.5	0.9	1.1	1.3	1.5	1.7	2.1	2.4
		25	0.3	0.4	0.7	0.9	1.1	1.2	1.4	1.7	1.9
D	0.001	8	0.3	0.5	0.7	0.9	1.1	1.3	1.5	1.7	2.0
		10	0.2	0.4	0.6	0.7	0.9	1.0	1.2	1.4	1.6
		12.5	0.2	0.3	0.5	0.6	0.7	0.8	0.9	1.1	1.3
		15	0.2	0.2	0.4	0.5	0.6	0.7	0.8	0.9	1.1
		20	0.1	0.2	0.3	0.4	0.4	0.5	0.6	0.7	0.8
		25	0.1	0.1	0.2	0.3	0.4	0.4	0.5	0.6	0.6

[a]Air tightness classes are per EN 12237 and EN 1507 (CEN 2003, 2006)

12.5.3 Leakage Fraction Goal for DER Projects

Many European countries have successfully attained air tightness Class C (the proposed goal for DER projects is Class C duct systems); Sweden uses Class D ductwork as a standard practice. The equivalent for European air tightness Class C in the United States is ASHRAE air leakage Class 2.1. The data in Tables 12.8 and 12.9, abstracted from Table 12.6, shows that the maximum air leakage for systems 1500 Pa (6 in. water) static pressure maximum (from Table 12.6) is 4.5%.

Table 12.7 Leakage as percentage of flow (SI units)

Air leakage class, cfm per 100 ft^2 per (in. of water)$^{0.65}$	Q_{fan}/ A_{system}, cfm per ft^2	Static pressure, in. of water								
		0.5	1	2	3	4	5	6	7	10
20	1.5	8.5	13.3	20.9	27.2	32.8	38.0	42.7	51.5	59.6
	2	6.4	10.0	15.7	20.4	24.6	28.5	32.0	38.6	44.7
	2.5	5.1	8.0	12.6	16.3	19.7	22.8	25.6	30.9	35.7
	3	4.2	6.7	10.5	13.6	16.4	19.0	21.4	25.8	29.8
	4	3.2	5.0	7.8	10.2	12.3	14.2	16.0	19.3	22.3
	5	2.5	4.0	6.3	8.2	9.8	11.4	12.8	15.5	17.9
8	1.5	3.4	5.3	8.4	10.9	13.1	15.2	17.1	20.6	23.8
	2	2.5	4.0	6.3	8.2	9.8	11.4	12.8	15.5	17.9
	2.5	2.0	3.2	5.0	6.5	7.9	9.1	10.3	12.4	14.3
	3	1.7	2.7	4.2	5.4	6.6	7.6	8.5	10.3	11.9
	4	1.3	2.0	3.1	4.1	4.9	5.7	6.4	7.7	8.9
	5	1.0	1.6	2.5	3.3	3.9	4.6	5.1	6.2	7.1
6	1.5	2.5	4.0	6.3	8.2	9.8	11.4	12.8	15.5	17.9
	2	1.9	3.0	4.7	6.1	7.4	8.5	9.6	11.6	13.4
	2.5	1.5	2.4	3.8	4.9	5.9	6.8	7.7	9.3	10.7
	3	1.3	2.0	3.1	4.1	4.9	5.7	6.4	7.7	8.9
	4	1.0	1.5	2.4	3.1	3.7	4.3	4.8	5.8	6.7
	5	0.8	1.2	1.9	2.5	3.0	3.4	3.8	4.6	5.4
5	1.5	2.1	3.3	5.2	6.8	8.2	9.5	10.7	12.9	14.9
	2	1.6	2.5	3.9	5.1	6.2	7.1	8.0	9.7	11.2
	2.5	1.3	2.0	3.1	4.1	4.9	5.7	6.4	7.7	8.9
	3	1.1	1.7	2.6	3.4	4.1	4.7	5.3	6.4	7.4
	4	0.8	1.3	2.0	2.6	3.1	3.6	4.0	4.8	5.6
	5	0.6	1.0	1.6	2.0	2.5	2.8	3.2	3.9	4.5
4	1.5	1.7	2.7	4.2	5.4	6.6	7.6	8.5	10.3	11.9
	2	1.3	2.0	3.1	4.1	4.9	5.7	6.4	7.7	8.9
	2.5	1.0	1.6	2.5	3.3	3.9	4.6	5.1	6.2	7.1
	3	0.8	1.3	2.1	2.7	3.3	3.8	4.3	5.2	6.0
	4	0.6	1.0	1.6	2.0	2.5	2.8	3.2	3.9	4.5
	5	0.5	0.8	1.3	1.6	2.0	2.3	2.6	3.1	3.6
3	1.5	1.3	2.0	3.1	4.1	4.9	5.7	6.4	7.7	8.9
	2	1.0	1.5	2.4	3.1	3.7	4.3	4.8	5.8	6.7
	2.5	0.8	1.2	1.9	2.5	3.0	3.4	3.8	4.6	5.4
	3	0.6	1.0	1.6	2.0	2.5	2.8	3.2	3.9	4.5
	4	0.5	0.8	1.2	1.5	1.8	2.1	2.4	2.9	3.4
	5	0.4	0.6	0.9	1.2	1.5	1.7	1.9	2.3	2.7

(continued)

Table 12.7 (continued)

Air leakage class, cfm per 100 ft² per (in. of water)$^{0.65}$	Q_{fan}/ A_{system}, cfm per ft²	Static pressure, in. of water								
		0.5	1	2	3	4	5	6	7	10
2.1	1.5	0.9	1.4	2.2	2.9	3.4	4.0	4.5	5.4	6.3
	2	0.7	1.1	1.6	2.1	2.6	3.0	3.4	4.1	4.7
	2.5	0.5	0.8	1.3	1.7	2.1	2.4	2.7	3.2	3.8
	3	0.4	0.7	1.1	1.4	1.7	2.0	2.2	2.7	3.1
	4	0.3	0.5	0.8	1.1	1.3	1.5	1.7	2.0	2.3
	5	0.3	0.4	0.7	0.9	1.0	1.2	1.3	1.6	1.9
1	1.5	0.4	0.7	1.0	1.4	1.6	1.9	2.1	2.6	3.0
	2	0.3	0.5	0.8	1.0	1.2	1.4	1.6	1.9	2.2
	2.5	0.3	0.4	0.6	0.8	1.0	1.1	1.3	1.5	1.8
	3	0.2	0.3	0.5	0.7	0.8	0.9	1.1	1.3	1.5
	4	0.2	0.3	0.4	0.5	0.6	0.7	0.8	1.0	1.1
	5	0.1	0.2	0.3	0.4	0.5	0.6	0.6	0.8	0.9

Table 12.8 Air leakage fraction for EU air tightness Class C systems (SI units)

Air tightness class, L/s per m² per Pa$^{0.65}$		Q_{fan}/A_{system}, L/s per m²	Static pressure, Pa						
			125	250	500	750	1000	1250	1500
C	0.003	8	0.9	1.4	2.1	2.8	3.3	3.9	4.4
		10	0.7	1.1	1.7	2.2	2.7	3.1	3.5
		12.5	0.6	0.9	1.4	1.8	2.1	2.5	2.8
		15	0.5	0.7	1.1	1.5	1.8	2.1	2.3
		20	0.3	0.5	0.9	1.1	1.3	1.5	1.7
		25	0.3	0.4	0.7	0.9	1.1	1.2	1.4

Table 12.9 Air leakage fraction for ASHRAE air leakage Class 2.1 systems (I-P units)

Air leakage class, cfm per 100 ft² per (in. of water)$^{0.65}$	Q_{fan}/A_{system}, cfm per ft²	Static pressure, in. of water						
		0.5	1	2	3	4	5	6
2.1	1.5	0.9	1.4	2.2	2.9	3.4	4.0	4.5
	2	0.7	1.1	1.6	2.1	2.6	3.0	3.4
	2.5	0.5	0.8	1.3	1.7	2.1	2.4	2.7
	3	0.4	0.7	1.1	1.4	1.7	2.0	2.2
	4	0.3	0.5	0.8	1.1	1.3	1.5	1.7
	5	0.3	0.4	0.7	0.9	1.0	1.2	1.3

12.5.4 Testing for Air Leakage

It is critical that supply and exhaust air systems be tested for air leakage during construction to verify the quality of the workmanship and the use of low-leakage components, as required to achieve the air tightness specified by the owner's representative in the contract. Leakage tests should be conducted by an independent party responsible to the owner's representative.

During construction, ductwork sections should be pressure tested using EN 12599 (CEN 2012) or SMACNA's HVAC Air Duct Leakage Test Manual (2012: Sections 4, 6, and 7). The resulting air leakage of each test section should not exceed the air leakage calculated using Eq. 12.1 for European air tightness Class C (0.003 L/s per m^2 per $Pa^{0.65}$), which is the same as ASHRAE air leakage Class 2.1 (cfm per 100 ft^2 per $[\text{in. of water}]^{0.65}$).

As described in the "Duct Design" chapter of the *ASHRAE Handbook – Fundamentals* (ASHRAE 2017), at a minimum, 25%[4] of the ductwork based on duct surface area (including 100% of the ducts to be enclosed in chases and other concealed space and ducts installed outdoors) should be pressure tested and another 25% should be tested if any of the initial sections fail. If any section of the second 25% fails, the entire ductwork should be air leakage tested. Sections to be tested should be selected by the owner's representative. The selected sections should include a representative variety of duct dimensions and fittings (CEN 2003, 2006). Duct surface area should be calculated as described in EN 14239 (CEN 2004).

12.5.5 Example

Consider a duct system as depicted in Fig. 12.9 with characteristics as listed in Tables 12.10 and 12.11. This example illustrates how to use Eq. 12.1 to calculate the maximum allowable air leakage for the duct system sections when they are pressure tested using two different test setups (Sections 1 through 5 together and Sections 4 and 5 together). It also illustrates how to use Eq. 12.3 to calculate the leakage fraction for the system's ductwork to meet the Class C (2.1) requirement, based on the combined leakage of all five of the tested sections. Tables 12.10 and 12.11 shows the results of these calculations.

It is important to recognize that the pressurization-based leakage test procedures currently used by industry focus on determining component air tightness (e.g., for ductwork located upstream of terminal box inlet dampers). Air tightness alone, however, is insufficient to determine leakage airflows. One must also then estimate system pressures *during system operation* to determine these flows. Determining the location of every leak and the pressure difference across each leak is practically impossible for most systems and can cause significant uncertainty in leakage airflow estimates using this approach.

In particular, comparisons of leakage fractions based on "measured" leakage flows with fractions estimated using the industry method (Wray et al. 2005) show

[4]The minimum of 25% is based on the related requirement in Section 6.4.4.2.2 of ANSI/ASHRAE/IES Standard 90.1 (2016b).

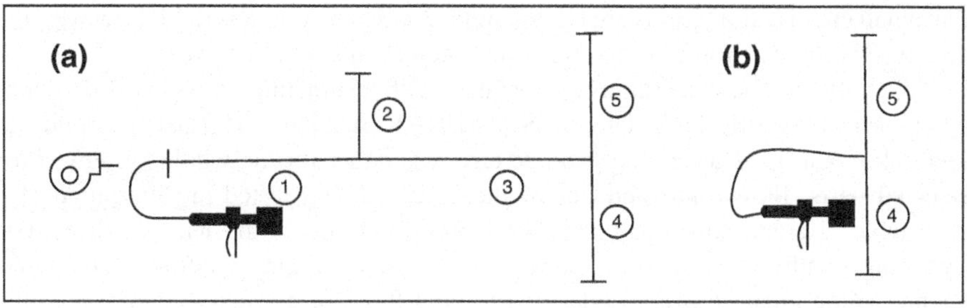

Fig. 12.9 Ductwork layout and instrumentation, for example: (**a**) pressurizing entire system; (**b**) pressurizing sections 4 and 5 combined (note: [. . . . . .] denotes a temporary seal of a duct opening)

Table 12.10 Allowable air leakage for the example system test (SI)

	Inlet flow, L/s	$A_{section,i}$, Section surface area, m^2	$p_{section,i}$, section static pressure, Pa	$C_{L,section,i}$, air tightness class, L/s per m^2 per Pa$^{0.65}$	$Q_{leak,section,i}$, allowable leakage, L/s (Eq. 121)	$Q_{leak,frac}$, ductwork leakage fraction, % (Eq. 12.3)
Section	Characteristics				Calculated	
1	2360	46.5	750	0.003	10.3	
2	944	62.0	250	0.003	6.7	
3	1416	69.7	500	0.003	11.9	
4	472	18.6	250	0.003	2.0	
5	944	92.9	250	0.003	10.1	
System	**2360**	**289.7**			**41.0**	**1.7**
4 and 5	1416	111.5	250	0.003	12.1	

Table 12.11 Allowable air leakage for the example system test (IP)

	Inlet flow, cfm	$A_{section,i}$, section surface area, ft^2	$p_{section,i}$, section static pressure, in. of water	$C_{L,section,i}$, air leakage class, cfm per ft^2 per (in. of water)$^{0.65}$	$Q_{leak,section,i}$, allowable leakage, cfm (Eq. 12.1)	$Q_{leak,frac}$, ductwork leakage fraction, % (Eq. 12.3)
Section	Characteristics				Calculated	
1	5000	500	3.0	2.1	21.4	
2	2000	667	1.0	2.1	14.0	
3	3000	750	2.0	2.1	24.7	
4	1000	200	1.0	2.1	4.2	
5	2000	1000	1.0	2.1	21.0	
System	**5000**	**3117**			**85.4**	**1.7**
4 and 5	3000	1200	1.0	2.1	25.2	

that small estimated fractions are a good indicator that a system is tight; however, the converse is not true for large fractions that indicate a system is leaky.

To eliminate the uncertainty associated with estimating pressure differences across leaks, recently published ANSI/ASHRAE Standard 215 (2018) provides a method of test for determining leakage airflows, either for the whole system or for selected parts. Flows into and out of the section being tested are measured at a repeatable reference operating condition: the difference is the leakage flow. The operating condition is not necessarily the design operating condition, but corresponds to the greatest system inlet flow (outlet flow for exhaust systems) possible without being detrimental to the occupants of the building, the building structure, or the HVAC mechanical components while maintaining the duct static pressure set point (where applicable) specified in design documents. Once the system is constructed and operable, it is recommended that this standard be used when commissioning DER buildings.

12.6 Duct Insulation

Duct insulation saves energy and keeps warmed or cooled air as close to a desired temperature as possible while it is being moved to spaces where it is needed. If reduced heat transfer through insulated ducts is accounted for in the HVAC load calculations, it may even be possible to reduce the size of HVAC equipment. Requirements for duct insulation are stated in national codes (listed in Table 12.12); and levels of insulation depend on the building climate zone and duct location (attic, crawlspace, basement, inside conditioned space, or other). Return ducts connected to heat recovery equipment should also be insulated. In addition to energy savings, there are several other reasons to insulate ductwork: prevent condensation on the duct surface from causing either a safety or mold health hazard; and achieve sound attenuation if a perforated liner is used (US constructions methods). For obvious reasons, the integrity of the thermal and vapor barrier must be maintained. For construction guidelines in the United States, refer to the latest edition of SMACNA HVAC Duct Construction Standards (ASHRAE 2009). Insulation must be protected from damage, including that due to sunlight, moisture, equipment maintenance, etc. Duct insulation can be placed inside or outside of the air duct. While in Europe the internal duct is typically used as a pressure shell (Fig. 12.10), in the United States, it is a common practice to provide double-wall insulated ducts with the perforated internal wall acting as a sound attenuator and/or preventing water condensation on the duct's exterior shell. The latter technology is usually applied where the duct is exposed to the occupants or to the outdoor environment.

Table 12.12 Duct insulation requirements

Country	Standard for duct insulation	Insulation requirements[a]
Austria	ÖNORM H 5057	U-value 0.6 W/(m^2·K), 0.106 Btu/(hr∗ft^2∗°F)
China	GB50736-2012	U-value 1.234 W/(m^2·K), 0.217 Btu/(hr∗ft^2∗°F) – Normal air-conditioning duct
		0.877 W/(m^2·K), 0.154 Btu/(hr∗ft^2∗°F) – Low-temperature air-conditioning duct
Denmark	Danish Standard 452	0.36–1.40 W/(m^2·K), 0.063–0.246 Btu/(hr∗ft^2∗°F)
Estonia	EVS-EN 13779:2007; EN 12097	0.36–1.4 W/(m^2·K), 0.063–0.246 Btu/(hr∗ft^2∗°F)
Germany	EnEV 2014	20–200 mm (0.79–7.87 in.) dependent on pipe diameter and surrounding air temperature
Latvia	LVS-EN 13779:2007	0.36–1.4 W/(m^2·K), 0.063–0.246 Btu/(hr∗ft^2∗°F)
UK	BS5422:2009	5–120 mm (0.2–4.7 in.) ∗ (condensation control)
		6.45 W/m^2 (chilled or dual-purpose duct heat gain control)
		16.34 W/m^2 (ductworks carry warm air heat loss control)
USA	ASHRAE Std 90.1-2013 (6.8.2B)	U-value: 0.7–1.62 W/(m^2·K), 0.125–0.2860 Btu/(hr∗ft^2∗°F) – supply and return for heating or cooling
		U-value: 0.95–1.62 W/(m^2·K), 0.167–0.286 Btu/(hr∗ft^2∗°F) – return ducts

[a]Depending on pipe radius and temperature difference

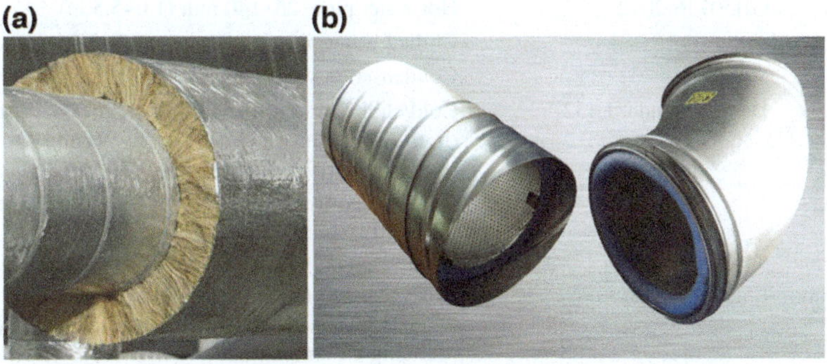

Fig. 12.10 Examples of supply duct insulation: (**a**) external duct insulation (courtesy of Isover-Saint-Gobain), (**b**) internal duct insulation (courtesy of Linx Industries, Inc., Division of DMI)

12.7 Insulation of Hot and Cold Water Pipes

Piping that serves as part of a heating, domestic hot water, or cooling systems should be thermally insulated to reduce energy loss and eliminate moisture condensation on cold piping surfaces. This also applies to chilled water, liquid refrigerant, cooling coil condensate, storm water roof drainage, and domestic cold water piping systems with a temperature below 60 °F (15 °C) or below the adjacent space DPT. All piping, fittings, and devices should be insulated and be provided with a well-sealed vapor barrier applied to the entire insulation system to prevent diffusion of water vapor into the insulation system. Exceptions can be made for piping that conveys fluids that have a design temperature range between 60 °F (15 °C) and 105 °F (41 °C). The minimum insulation thickness depends on the pipe size and fluid temperature (Table 12.13).

For cold pipe and duct insulation, it is more advantageous to use closed-cell caoutchouc foam (Fig. 12.11) insulation to prevent condensing water entering the insulating foam.

Table 12.13 Hot water and cold water pipe insulation requirements

Country	Minimum standard	U-value or other requirement
Austria	OIB RL 6, 2011	Insulation = pipe diameter with thermal conductivity of the insulation material of 0.035 W/(m*K) (0.0019 Btu/[h*ft*°F])
		Cold/chilled water: 0.6–0.7 W/(m^2·K), 0.106–0.124 Btu/(hr*ft^2*°F), hot water: 0.7 W/(m^2·K), 0.124 Btu/(hr*ft^2*°F)
China	GB50736-2012	Hot water pipe: 25–140 mm (1.0–5.5 in)
		Chilled water pipe: 19–60 mm (0.7–2.4 in)
		Condensation water pipe: 9–15 mm (0.4–0.6 in)
Denmark	Danish Standard 452	Cold/chilled water: 0.33 W/(m^2*K) (0.0587 Btu[/(h*ft^2*°F)]), hot water: 0.11–0.24 W/(m^2*K) (0.0196–0.0427 Btu/(h*ft^2*°F))*
Estonia		Cold/chilled water: ≈1.5 W/(m^2*K) (0.267 Btu/[h*ft^2*°F]), hot water: ≈ 0.5 . . . 0.8 W/(m^2*K) (0.0889 . . . 0.1422 Btu/[h*ft^2*°F])*
Germany	EnEV 2014	Hot, cold, and chilled water pipe: 20–200 mm (0.79–7.87 in.) dependent on pipe diameter and surrounding air temperature
Latvia	Construction Standard LBN 221-15 of 30 June 2015	Requirement for pipe insulation defined, while U-values are not provided
UK	BS5422:2009	Insulation requirement
		Cooled and chilled water systems Cooled water temperature >10 °C (>50 °F), 2.48–14.74 W/m (2.48–14.7400 Btu/hr*ft)
		Cooled and chilled water systems Chilled water temperatures >4.9 °C (41 °F) to <10 °C (50 °F), 2.97–16.28 W/m (2.97–16.28 Btu/hr*ft)

(continued)

Table 12.13 (continued)

Country	Minimum standard	U-value or other requirement
		Cooled and chilled water systems Chilled water temperatures of 0 °C (32 °F) to <4.9 °C (41 °F), 3.47–17.48 W/m (3.47–17.4800 Btu/hr∗ft)
		Nondomestic hot water service: 6.6–32.4 W/m (6.6–32.4 Btu/hr∗ft)
		Domestic heating and hot water systems: 7.06–14.12 W/m (7.06–14.12 Btu/hr∗ft)
		Nondomestic heating services Low-temperature heating services (≤95 °C [203 °F]) 8.9–38.83 W/m (8.9–38.83 Btu/hr∗ft)
		Nondomestic heating services Medium-temperature heating services (96–120 °C [205–248 °F]) 13.34–43.72 W/m (13.34–43.72 Btu/hr∗ft)
		Nondomestic heating services High-temperature heating services (121–50 °C [250–302 °F]) 17.92–48.48 W/m (17.92–48.48 Btu/hr∗ft)
USA	ASHRAE Std 90.1 2013 s 6.8.3 A and B	Heating and hot water pipes[a]
		40 °C–60 °C (105 °F–140 °F): (1.25–2.38) W/(m^2·K), (0.22–0.42) Btu/(hr∗ft^2∗°F)
		61 °C–93 °C (141 °F–200 °F): (2.12–3.28) W/(m^2·K), (0.375–0.58) Btu/(hr∗ft^2∗°F)
		Cooling systems pipes∗:
		4 °C–16 °C (40 °F–60 °F): 0.6–1.53 W/(m^2·K), (0.105–0.27) Btu/(hr∗ft^2∗°F)
		<4 °C (<40 °F): 0.57–2.2 W/(m^2·K), (0.1–0.39) Btu/(hr∗ft^2∗°F)

[a]Depending on pipe inner diameter

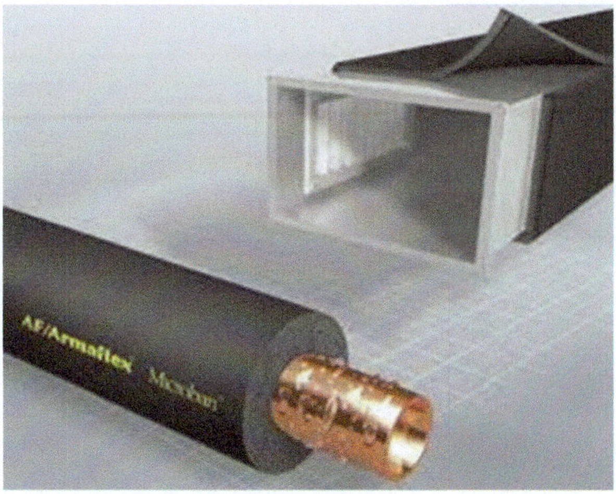

Fig. 12.11 Example of pipe and duct insulation using closed-cell caoutchouc foam (Armaflex)

Chapter 13
Quality Assurance

To increase a building's value and improve its indoor climate and thermal comfort, DER must adopt a Product Delivery Quality Assurance (PDQA) process that in addition to conventional understanding of QA includes:

- Formulation of detailed technical specification (e.g., SOW or Owner's Project Requirements Document), against which tenders (i.e., bids) will be made, and verification of understanding of these specifications by potential contractors
- Specification in SOW/Office of Primary Responsibility (OPR) of areas of major concern to be addressed and checked during the bid selection, design, construction, commissioning, and postoccupancy phases
- Clear delineation of the responsibilities and qualifications of stakeholders in this process

When established and well understood, this process requires minimum or no additional cost. It is critical for achieving desired energy performance of the building, durability of its systems and their elements, and reduction in maintenance and repair costs. The PDQA process described in Appendix H can be applied to major renovation projects using both D-B-B and D-B procurement methods. Some elements of this process can be used also with minor renovation and system replacement projects.

A. Zhivov, R. Lohse, *Deep Energy Retrofit*,
https://doi.org/10.1007/978-3-030-30679-3_13

Chapter 14
Economic Analysis

The scope of deep energy retrofit project and its attractiveness to investors depends on the project's cost-effectiveness. The standard method to analyze a project's cost-effectiveness is by performing a life-cycle cost analysis (LCCA), which accounts for present and future costs of the project. Life-cycle costs typically include the following two categories: **investment costs and operational costs. Investment-related costs** include costs related to planning, design, purchase, and construction as well as capital replacement costs, which are usually incurred when replacing major systems or components. An LCCA also typically includes energy use cost and the cost of operation and maintenance.

An important consideration in an LCCA is the selection of the **Base Case Scenario**, against which the cost-effectiveness of the DER will be evaluated. Note that the Base Case Scenario should not be confused with the **Baseline**, which is used for benchmarking energy use in the building before renovation. Most of major renovation projects include a scope of work, which can be either **nonenergy-related** or **energy-related**. A **nonenergy-related scope of work** may include such elements as different construction jobs related to changing floor layouts (e.g., moving/removing internal partitions), adding bathrooms, removing asbestos, adding sprinkler system, etc.

An **energy-related scope of work** of a major renovation project typically includes replacement of existing mechanical, lighting, and electrical systems, replacement of some or all windows, replacement of existing ductwork and plumbing systems, etc. A **major renovation** with the energy-related scope of work undertaken to **meet current minimum standard** requirements will be considered to be a **Base Case** for the LCCA.

Energy-related investment costs will usually be higher in a DER compared to the Base Case. Some energy-related improvements included into DER, e.g., building envelope insulation and mitigation of thermal bridges, installation of high-performance windows, and air tightening the building envelope, are expensive and are rarely included in the scope of major renovation or are performed following lower energy performance requirements. However, reduction of heating, cooling, or

A. Zhivov, R. Lohse, *Deep Energy Retrofit*,
https://doi.org/10.1007/978-3-030-30679-3_14

humidity loads resulting from implementation of these measures will result in the need for a **smaller and sometimes simpler HVAC system, which will, in turn, reduce both initial investment and capital replacement costs related to these systems.**

Also, combining DER with major renovation will result in smaller investment costs compared to performing a DER on its own. In any case, costs for building evacuation and gutting, installation of scaffolding, replacement of single pane and damaged windows, or even replacement and/or upgrade of building envelope insulation to meet minimum standards, are already a part of the major renovation project and do not have to be accounted as an additional cost related to DER.

Compared to the Base Case, DER may or may not result in the following operating cost savings:

- Energy use and cost reduction due to improved efficiency of the building and its systems
- Energy cost reduction due to shifting energy peaks, switching to different fuels (e.g., using cogeneration or trigeneration), or replacing fossil fuel-based thermal or electrical systems to systems from renewable energy sources
- Maintenance cost reduction with replacement of worn equipment at the end of its life cycle
- Maintenance cost reduction due to downsizing of mechanical systems with reduced heating and cooling loads
- Operation cost reduction using advanced building automation systems

In some scenarios, energy use may increase compared to the Base Case due to new indoor air quality or thermal comfort requirements. For example, adding cooling or humidity control requirement will result in additional energy use for cooling systems. Maintenance costs of some replacement systems may increase due to the complexity of their controls system, but they may also be offset by reduced energy use resulting from more efficient operation of the HVAC system.

Early studies and pilot projects performed around the world indicate that, in addition to the traditional areas of operating cost reduction listed above, there are other **bankable cost reduction and income-generating opportunities related to DER** that shall be considered in LCCA (see Chap. 14 for more details):

- **Improved building durability** due to better temperature and humidity control (e.g., reduced cost of building envelope repair and mold mitigation)
- **Grants, rebates, and other financial subsidies** for energy-efficient and sustainable design (one-time payment)
- **Reduced costs and time of accommodating "churn" of employees** in flexible and sustainable work spaces (single or multiple time cost reduction)
- **Increased usable space** due to downsized and consolidated mechanical equipment (additional cost reduction or income-generating cash flows)
- **Increased usable space** due to improved thermal comfort in areas close to external walls (additional cost reduction or income-generating cash flows)

- **Increased usable space** due to thermal insulation and ventilation of the attic space (additional cost reduction or income-generating cash flows)
- **Reduced short-term absenteeism** due to improved indoor air quality and comfort (additional cost reduction)
- **Improved workers' productivity** due to improved indoor air quality and comfort (additional income-generating cash flows)
- **Recruiting and retention cost** savings through employees satisfaction (additional cost reduction that can be spread over time)
- **Additional revenues from the enhanced demand for deep retrofit properties** from potential tenants (additional income-generating cash flow)
- **Reduced insurance premiums** with building components' replacement and improved protection against losses (additional cost reduction)

Based on the above discussion, the cost-effectiveness of a DER project can be evaluated by conducting an LCC analysis using incremental investment cost increase (ΔC) required to achieve a DER compared to the Base Case scenario. For a DER to be cost-effective, this delta investment cost increase shall be smaller than the Net Present Value of Operating and Maintenance Costs saving combined with the NPV of Replacement Cost Savings and the NPV of other **bankable cost reduction and income-generating opportunities related to DER**. Since most of parameters required for the LCC analysis differ not only by the individual country but also within the country (first costs and labor rates, energy rates, life of the project, inflation and discount rates, etc.), the following methodology has been proposed (Zhivov et al. 2015) to evaluate the effectiveness of an LCC analysis of an integrated energy technology bundle to be used for a DER:

- Step 1: Calculate annual operational costs and income-generating opportunities per the DER scenario.
- Step 2: Calculate annual operational costs per the Base Case scenario.
- Step 3: Subtract costs calculated in Step 1 from those calculated in Step 2 and calculate NPV of cost savings over the project life using scalar ratio, as described in Appendix I.
- Step 4: The NPV of operational savings and income-generating opportunities can be used to estimate the extent of the budget increase compared to the Base Case that can be used for energy enhancements with DER compared to building renovation based on minimum energy requirements.

The budget increase allowance will depend on the difference in operational costs savings resulting from reduced energy costs, reduced maintenance and system insurance costs, and, when applicable, increased revenues from renting the space. In buildings having undergone DER, reduced thermal loads reduce HVAC system's size and complexity, which in turn results in reduced Operations and Maintenance (O&M) costs. According to the Rocky Mountain Institute (RMI) (Bendewald et al. 2014), high-performance buildings have 9–14% smaller maintenance costs compared to the business as usual baseline. German Standard VDI 2067(2010) and

Danish Green Building Council (DK-DGNB [2014] provide estimates for maintenance costs as a percentage of the building system cost.

A DER that results in a more energy-efficient, more sustainable building will accelerate lease-up time vs. average market downtime and provide additional value, i.e., the property can be leased for a rent 9–14% higher than that of comparable unimproved properties in the local market. Also, many DER projects result in adding rentable/usable space, e.g., due to reduced size of mechanical rooms, adding thermally controlled areas (mansards, basements, repurposing storage spaces, etc.), which can be accounted for in the estimation of the budget increase allowance. For more information on Economics of DER, see Annex 61 "Deep Energy Retrofit – Business Guide."

Chapter 15
Conclusions

The core bundle of technologies described in this Guide make it possible to achieve DER in major renovations of buildings with low internal loads (e.g., office buildings, dormitories, barracks, and educational buildings). This task is more difficult in hot climate zones (DOE c.z. 1–3) with significant cooling needs and may require the application of additional EEMs (e.g., reduction of plug loads, water conservation measures, advanced HVAC systems). A DER is easier to achieve in heating-dominated climates and in cases where, either by cultural or normative reasons, cooling is not desired and building users can tolerate temporarily increases in indoor air temperature (e.g., up to 77 °F [25 °C]). Additional energy efficiency technologies, especially those that result in process load reduction, are necessary to achieve Deep Energy Retrofit in buildings with significant ventilation requirements and high internal loads (e.g., dining facilities, hospitals, data centers, etc.).

In building simulations conducted for locations in China, the pre-renovation baseline (based on pre-1980s design) was developed for a naturally ventilated office building with inferior insulation levels and poorer building airtightness as compared to European countries and the United States. This resulted in lower insulation levels of the building envelope and window characteristics required for the DER scenario (50% energy use reduction compared to the baseline) presented in Tables 6.1, 6.2, and 7.3. If parameters similar to those adopted by western courtiers in similar climate conditions were used for DER in China, this will result in a greater energy use reduction in all climates.

The energy efficiencies and cost-effectiveness of a Deep Energy Retrofit depend on more than the simple characteristics of the core bundle of technologies, how they are implemented, or how they are used. **For example, it is important to pay attention to the continuous thermal barrier (no thermal bridges) when building envelope insulation is designed and installed, to the continuous air barrier (to achieve required building airtightness), to proper installation of windows in walls, etc.**

In addition to building energy use reduction, the proper selection, design, and installation of technologies selected for DER result in improvements in

A. Zhivov, R. Lohse, *Deep Energy Retrofit*,
https://doi.org/10.1007/978-3-030-30679-3_15

indoor air quality and thermal and visual comfort. A DER usually reduces cold and hot radiation from external walls, prevents drafts created either by air diffusers or by air infiltrating through cracks in the building envelope, improves illumination levels, and eliminates glare through windows.

For a DER project to be successful, it is critical to implement a quality assurance process, which in addition to design, construction, commissioning, and postoccupancy phases includes formulation of clear and concise documentation of the owner's goals, expectations, and requirements for the renovated building during development of the SOW. Another important component of the QA process is a procurement phase, during which bidders' qualifications, their understanding of the SOW and its requirements, and their previous experience are analyzed.

The key to making a DER cost-effective is to time the retrofit as part of a major building renovation that already has allocated funds, including those required to meet minimum energy requirements. Since there is an overlap between the funds allocated for the retrofit and those required for the DER, achieving the DER requires only an incremental cost because the DER is evaluated based on a bundle of core technologies, not on individual EEMs. Some "core" technologies (e.g., those related to building envelope insulation, replacement of windows, etc.), which may not be cost-effective when implemented individually, become economically attractive when implemented in a technology bundle. Implementation of these technologies can significantly reduce building heating and cooling loads and, consequently, reduce the size and cost of HVAC mechanical equipment, which subsequently results in reduced annual maintenance and insurance costs of these systems.

For a DER to be cost-effective, this incremental increase in the investment cost will be smaller (in comparison with a major building renovation undertaken to meet minimum code requirements) than the NPV of Operating and Maintenance Costs Savings combined with the NPV of Replacement Cost Savings, the NPV of other bankable cost reduction, and income-generating opportunities related to the DER.

The real budget required for a DER may depend on many factors including the state of the national or local economy, local labor costs, the local and national system of energy-related incentives available at the time of the project, the availability and prices of specific energy efficient technologies, familiarity and experience of contractors with DER projects, contracting mechanisms used, etc. When a DER is cost-effective, additional funding can become available either from the government or public funds or from the private funding sources (using ESPC or UESC models) (Zhivov et al. 2015). For more details on implementation models, economics, and financing of DER, see the "Deep Energy Retrofit – Business Guide."

Appendixes

Appendix A: Building Envelope Optimization Through Modeling

The "core bundle of technologies" include building envelope insulation levels and window characteristics optimized by the Annex 61 modeling team by computational simulation of representative buildings for different climate zones (c.z.) of participating countries (Table A.1).

ASHRAE has developed a standard for climate zone classification that is derived from accumulated weather data from all over the world. ASHRAE Standard 169, *Weather Data for Building Design Standards*, includes dry-bulb, dewpoint, and wet-bulb temperatures, enthalpy, HR, wind conditions, solar irradiation, latitude, longitude, and elevation for locations worldwide. These data have been compiled into broad climate zones to generally characterize climates for building codes and energy analysis. This standard also identifies representative cities for each broad climate zone to aid in building modeling and energy analyses.

Figure A.1 shows these broad climate zones for North America, and Figure A.2 shows these broad climate zones for the world. More detail for each country and representative city can be found in ASHRAE Standard 169-2013.

Modeling was conducted for 17 US climate zones (c.z.) and for representative climates in Austria, China, Denmark, Estonia, Germany, Sweden, and the UK (Table A.2).

The following scenarios were modeled:

- *Scenario 1*. This baseline scenario uses a pre-1980 standard to describe the building envelope and systems. Building use, systems operation schedules, and appliances and their use (expressed in W/m^2) used in Scenario 1 were fixed for all scenarios even though, in actual conditions, it is likely that such scenario elements would be improved/reduced over time.
- *Scenario 2*. This "business as usual" (base case) scenario describes a major renovation with energy-related measures included in the scope of work that

A. Zhivov, R. Lohse, *Deep Energy Retrofit*,
https://doi.org/10.1007/978-3-030-30679-3

Table A.1 Modeled buildings

Team	Buildings modeled		Climate zone (DOE)	Results
ERDC, USA	Barracks	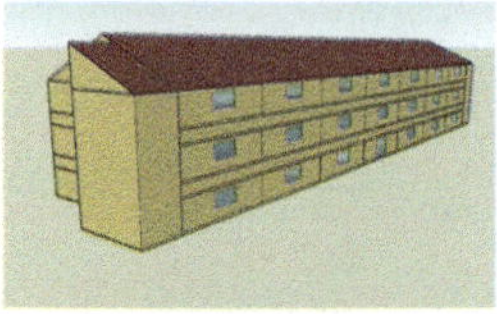	1-8	[A1]
	Office (Battalion HQ)	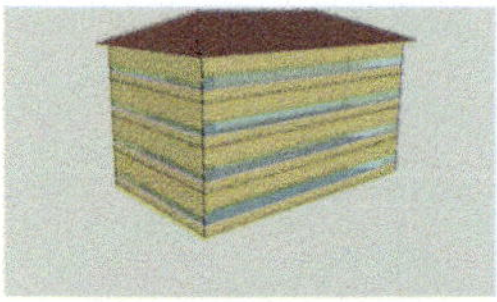	1-8	[A1]
TTU, Estonia	Public housing		6A	[A2]
M.E. Group, USA	Dormitory		5B	[A3]
Reading University, UK	Administrative building	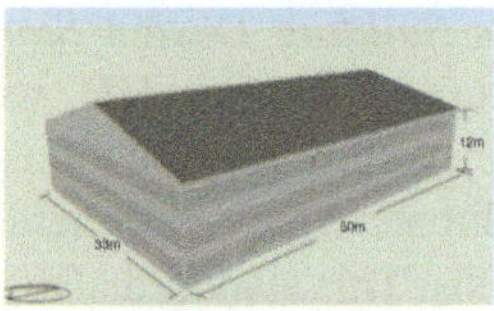	4A, 5A	[A4, A5]
Chongqing University, China	Administrative building	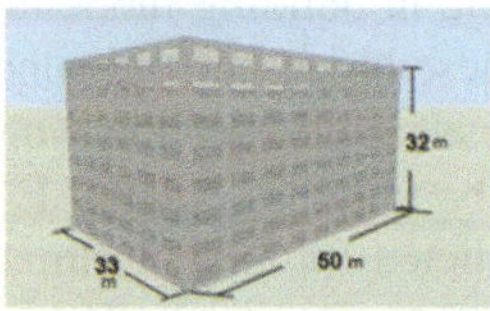	2a, 3a, 3c, 4a, 7	[A4, A5]
KEA and PHI, Germany	Office building		5A	[A6]

(continued)

Table A.1 (continued)

Team	Buildings modeled		Climate zone (DOE)	Results
AEE, Austria	Dormitory		4A and 7	[A6]
SBI, Denmark	School		5A	[A2]
RTU, Latvia	Dormitory	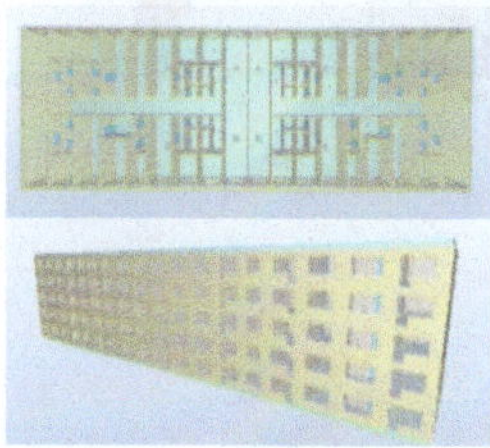	6A	[A7]
NREL, USA	Educational	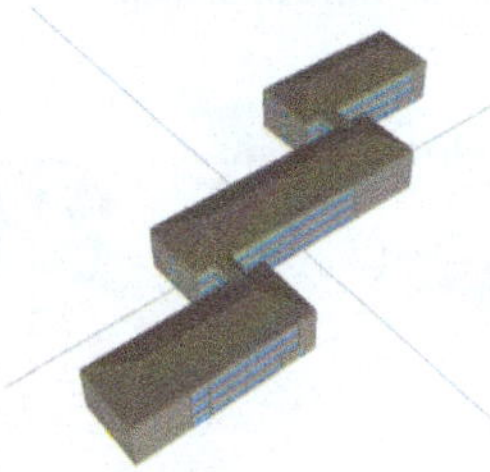	3C	[A5]

meet the minimum current standards (usually related to energy efficiency of fans, motors, chillers, furnaces, lighting fixtures, etc.) listed in Table A.3. Building use schedules and plug loads remain the same as in Scenario 1.

- *Scenario 3*. In this scenario, the characteristics of the core technology bundle are optimized to achieve about 50% of energy use reduction against the baseline or current national minimum building energy use requirement for existing buildings (whichever is more stringent).
- *Scenario 4*. This scenario optimizes the characteristics of the core technology bundle listed in Table 5.1 and uses additional EEMs (e.g., reduction in plug loads and domestic hot water use, etc.) to achieve the current "national dream" EUI levels in renovated buildings (e.g., the Passive House Standard), required in national regulations "if LCC effective."

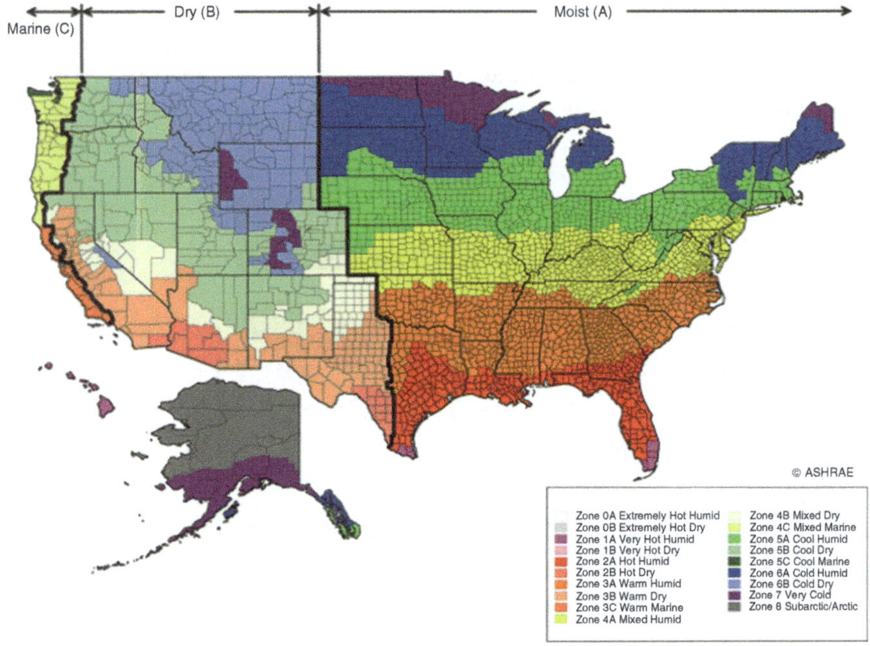

Fig. A.1 Climate zones for the USA in Standard 169-2013

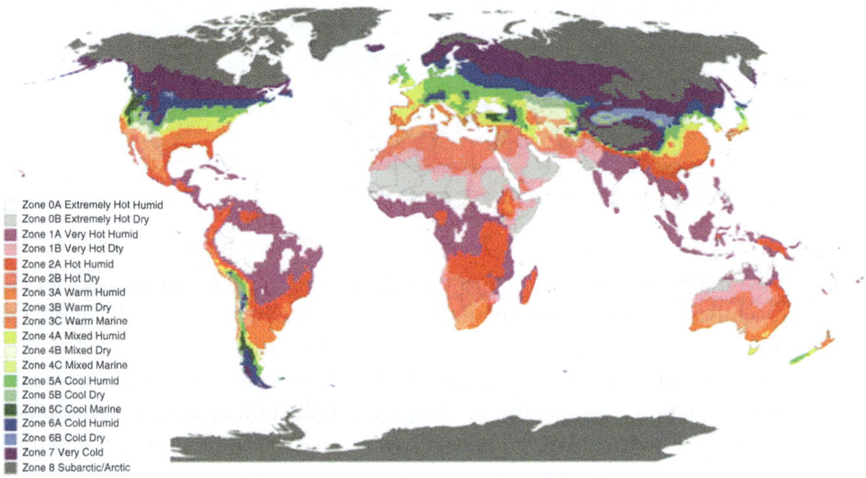

Fig. A.2. Climate zones worldwide in Standard 169-2013

The parameters for individual technologies for Scenario 1 were selected based on the national standards listed in Table A.3.

The parameters for individual technologies for Scenario 2 were selected based on Current National Standards for Renovation Projects, listed in Table A.4.

Table A.2 International locations' climate zones (based on DOE classification) used in the Annex 61 DER building envelope analysis

Country	Climate zone (s)	Representative city
Austria	5A and 6A	Graz, Obertauern
China	2A, 3A, 3C, 4A, 7	Beijing, Guangzhou, Harbin, Kunming, Shanghai
Denmark	5A	Copenhagen
Estonia	6A	Tartu
Germany	5A	Wuerzburg
Latvia	6A	Riga
UK	4A	London
China	4a	Beijing
USA	1a-8b	Miami, Houston, Phoenix, Memphis, El Paso, San Francisco, Baltimore, Albuquerque, Seattle, Chicago, Colorado Springs, Burlington, Helena, Duluth, Fairbanks

Table A.5 lists the current "national dream" requirements to renovated buildings used for Scenario 4.

This scenario optimizes the characteristics of the core technology bundle and uses additional EEMs to gain greater energy savings than can be achieved by using a "core technologies bundle" alone. The use of some of these measures may depend on the end user, rather than on the contractor (e.g., purchasing and installation more energy-efficient appliances and other plug loads, separate power lines, and timers to turn off some of electrical appliances). Other measures might include those that are specific to a particular building type (e.g., water saving shower heads and clothes washers, which can significantly reduce domestic hot water usage) or measures specific to the project (e.g., use of such low exergy heating and cooling systems as indirect evaporative cooling, or radiant heating and cooling systems, or by reusing heating and cooling return water energy and other waste streams).

Based on the results of these studies, the levels of the building envelope insulation and window types required to achieve DER in different climate conditions have been identified and are summarized in Tables 6.1, 6.2, and 7.4. These values were selected based on the performance of technology bundles (not on the economics of individual measures) for different climate conditions and individual country energy prices and on minimum national requirements for these technologies. These values are therefore equal to or more stringent than required by Standards listed in A.3. For example, insulation values of building envelope elements (Table A.6, also presented in Tables 6.1 and 6.2 [Chap. 6]) for the USA are more stringent than those listed in ASHRAE Standard 90.1 (2013), ASHRAE Standard 189.1 (2013), or the ASHRAE Advanced Energy Design Guides; they are not, however, as aggressive as those based on the Passive House Institute Standard.

The summary of the modeling results conducted under Annex 61 shows that, by using only "core technology bundles" in major renovation projects with low internal

Table A.3 National standards used to simulate pre-renovation building performance

Country	Building energy	Building envelope	HVAC	Lighting
Austria	No legal requirements before 1980	No legal requirements before 1980	No legal requirements before 1980	No legal requirements before 1980
	Pre-1980 building standard is based on generic data	Pre-1980 building standard is based on generic data	Pre-1980 building standard is based on generic data	Pre-1980 building standard is based on generic data
China	No legal requirements before 1980	JGJ 24-86	GBJ 19-1987	GBJ 133-90
		GB 50189-2005 (the 2005 standard contained 1980s building envelope information)	HVAC design guide (2008, second edition, edited by Lu Yaoqing).	HVAC design guide (2008, 2nd ed., edited by Lu Yaoqing)
Denmark	None	Danish Building Regulation 1961	None	None
Estonia	Ordinance No. 63. RT I, 18.10.2012, 1, 2012	EVS 908-1	EVS-EN 13779	Ordinance No. 70. RT I, 09.11.2012, 12
	Ordinance No. 68. RT I, 05.09.2012, 4, 2012	EVS-EN ISO 6946	EN 12237	
		EVS-EN ISO 13370	Ordinance No. 70. RT I, 09.11.2012, 12	
		EVS-EN ISO 10211		
		EVS-EN ISO 10077		
		EVS-EN 1026		
		EVS-EN 12207		
		EVS-EN 12208		
Germany	Real situation before refurbishment	Real situation before refurbishment	Real situation before refurbishment (most commonly ventilation using windows)	Real situation before refurbishment
Latvia	None.	SNiP, Soviet construction codes and regulations on building envelope and building physics	SNiP, Soviet construction codes and regulations on HVAC	None
UK	No legal requirements before 1980	Building regulation 1976	No legal requirements before 1980	No legal requirements before 1980
USA	ASHRAE Std 90.1 1975	ASHRAE Std 90.1 1975	ASHRAE Std 90.1 1975	ASHRAE Std 90.1 1975

Table A.4 Current national standards for renovation projects

Country	Building energy	Building envelope	HVAC	Lighting
Austria	OIB Directive No. 6	OIB RL 6, 2011	EN 1507, EN 12237	EN 12464-1 and 12464-2
			ÖNORM H 5057, OIB RL 6, 2011	EN 15193
China	GB 50189-2015	GB 50189-2015	GB 50736-2012	GB 50034-2013
		GB/T 7016-2008	GB 50189-2015	GB 50189-2015
Denmark	Danish Building Regulation 2010	Danish Building Regulation 2010	Standard 447	DS/EN ISO 12464-1
	DS Standard 418		Standard 452	
Estonia	Ordinance No. 63. RT I, 18.10.2012, 1, 2012	EVS 908-1	EVS-EN 13779, EN 12237	Ordinance No. 70. RT I, 09.11.2012, 12
	Ordinance No. 68. RT I, 05.09.2012, 4, 2012	EVS-EN ISO 6946	Ordinance No. 70. RT I, 09.11.2012, 12	
		EVS-EN ISO 13370		
		EVS-EN ISO 10211		
		EVS-EN ISO 10077		
		EVS-EN 1026		
		EVS-EN 12207		
		EVS-EN 12208		
Germany	DIN 18599-1; EnEV 2014	EnEV 2014, DIN 18361	EnEV 2014, DIN V 1859	DIN 18599-4
		DIN 18355, DIN V 18599/2	DIN 1946-6	DIN 5035 T 1-14
		DIN 4102	DIN EN 13779	
		DIN 4108	DIN 4108-6	
		DIN EN 13163	DIN 4701-10	
		DIN EN 13165		
Latvia	Law on the Energy Performance of Buildings	Latvian Construction Standard LBN 002-15	Latvian Construction Standard LBN 003-15	The regulation of the Cabinet of Ministers No.359
	The regulation of the Cabinet of Ministers No.348, No.383, and No.382	LVS-EN ISO 6946	The regulation of the Cabinet of Ministers No. 310	LVS-EN 15193/AC
		LVS-EN ISO 13370	LVS CR 1752	
		LVS-EN ISO 10211	LVS-EN ISO 7730	
		LVS-EN ISO 10077		

(continued)

Table A.4 (continued)

Country	Building energy	Building envelope	HVAC	Lighting
		LVS-EN 1026		
		LVS-EN 12207		
		LVS-EN 12208		
Sweden	BBR BFS 2011:6	BBR BFS 2011:6	BBR BFS 2011:6	BBR BFS 2011:6
	SFS 2006:985		EVP BFS 2011:11	SS-EN 12464-1
			SS-EN 12237	AFS 2009:2
			AFS 2009:2	
UK	BS EN 15603:2008	Building Regulations 2010-Conservation of Fuel and Power: Part L. (incorporating 2010, 2011, 2013, and 2016 amendments)	BS EN 15603:2008.	Building Regulations 2010-Conservation of Fuel and Power: Part L. (incorporating 2010, 2011, 2013, and 2016 amendments)
USA	ASHRAE Std 90.1 201.	ASHRAE Std 90.1 2010.	ASHRAE Std 90.1 2010.	ASHRAE Std 90.1 +IESNA recommended practices, 10th ed. 2010
	ASHRAE Std 100 2015			

loads, it is possible to reduce building site energy by about 50% compared to pre-renovation baseline.

Energy reduction (~40%) in hot and warm climates (c.z. 1–3) will be less dramatic due to the need for humidity control and significant cooling via plug loads. In cold and moderate climates, achieving 50% or better site energy use reduction does not present a problem. DER using only core technology bundles also results in significant source energy use reduction (35% and better). Modeling results have demonstrated that further site energy use reduction (up to 80% in moderate climates, i.e., achievement of the "national dream") is technically possible with the use of some additional energy efficiency technologies and plug load control. Source energy is significantly reduced (60–70%) as well. Use of building-dedicated renewable energy sources (e.g., photovoltaic [PV] and solar water heating) or heat pumps will further reduce both building site and source energy.

In building simulations conducted for locations in China, the pre-renovation baseline (based on pre-1980s' design) was develetersoped for a naturally ventilated office building with inferior insulation levels and poorer building airtightness, as compared to European countries and the USA. This resulted in lower insulation levels of the building envelope and window characteristics required for the DER scenario (50% energy use reduction compared to the baseline) presented in Tables A.7 and A.8. If param similar to those adopted by Western countries in similar climate conditions were used for DER in China, this will result in a greater energy use reduction in all climates.

Table A.5 National requirements to high-performance buildings after major renovation projects ("national dream" buildings)

Country	Building energy	Building envelope	HVAC	Lighting
Austria	OIB Directive No. 6	OIB Directive No. 6	OIB Directive No. 6	OIB Directive No. 6
China	GB 50189-2015	GB 50189-2015	GB 50189-2015	GB 50034-2013
Denmark	Danish Building Regulation 2015	Danish Building Regulation 2015	Standard 447	DS/EN ISO 12464-1
	DS Standard 418		Standard 452.	
Estonia	Ordinance No. 63, RT I, 18.10.2012, 1, 2012	EVS 908-1	EVS-EN 13779	Ordinance No. 70, RT I, 09.11.2012, 12
	Ordinance No. 68, RT I, 05.09.2012, 4, 2012	EVS-EN ISO 6946	EN 12237	
		EVS-EN ISO 13370	Ordinance No. 70, RT I, 09.11.2012, 12	
		EVS-EN ISO 10211		
		EVS-EN ISO 10077		
		EVS-EN 1026		
		EVS-EN 12207		
		EVS-EN 12208		
Germany	Passive house standard (no national requirement)	Passive house standard (no national requirement)	Minimum mechanical ventilation to assure hygienic needs (20 m^3/h/person) — additionally mech. ventilation with heat recovery to assure comfortable supply air temperatures (no national requirement).	Super-efficient LED lighting with max 3 W/m^2 installed electrical power for illumination of 100 lux (limit value); 2.5 W/m^2 for 100 lux (target value) (no national requirement)
Latvia	The regulation of the Cabinet of Ministers No.348	LVS-EN ISO 6946	The regulation of the Cabinet of Ministers No. 310	LVS-EN 15193/AC
	LVS-EN ISO 13790	LVS-EN ISO 13370	LVS CR 1752	
	Passive house planning package	LVS-EN ISO 10211	LVS-EN ISO 7730	
		LVS-EN ISO 10077		
		LVS-EN 1026		
		LVS-EN 12207		

(continued)

Table A.5 (continued)

Country	Building energy	Building envelope	HVAC	Lighting
		LVS-EN 12208		
UK	BS EN 15603:2008	Building Regulations 2010, *Conservation of Fuel and Power*, Part L. (incorporating 2010, 2011, 2013 and 2016 amendments)	BS EN 15603:2008	Building Regulations 2010, *Conservation of Fuel and Power*, Part L. (incorporating 2010, 2011, 2013 and 2016 amendments)
USA	ASHRAE Std 90.1 2010 – 30%			

Analysis conducted by ERDC researchers (Case et al. 2017) shows that while 50% or better energy use reduction is possible in buildings with low internal loads and ventilation requirements using only bundles of "core technologies," this approach does not work as well in buildings with high internal loads and ventilation requirements, e.g., in dining facilities, data centers, and hospitals. For example, in dining facilities, the application of the bundle of "core technologies" results in total site energy usage reduction only by 15–39% (depending on the climate zone) and heating-only energy use reduction by 29–64% in the modeled building, which is much smaller than is expected from a DER project. Based on the modeling results, significant energy reduction in a building with high internal loads and high ventilation needs requires aggressive measures to reduce those loads. For example, in dining facilities, use of higher-efficiency kitchen equipment results in about a 20% reduction in electrical and natural gas appliance energy usage. Energy reduction of approximately 50% in dining facilities is possible by switching to all electric appliances (although at high cost for appliance replacement and with implications for increased source energy usage depending on the source of electricity); ventilation demand reduction by approximately 50% in the kitchen and servery area by the use of improved kitchen hood designs and system controls and, finally, approximately 30% savings in domestic hot water requirements can be achieved by energy recovery in the sanitation area. With the use of these (and other) technologies, it is feasible to reduce site energy by approximately 50% or more in all climate zones except 1A (Table A.8).

Cost for deep retrofits can be very site and building specific. The savings shown in Table A.8, expressed as energy use intensity, can be used to calculate a budget for retrofits based on an individual country or region's energy and equipment costs. In Chicago (climate zone 5A), for example, assuming commercial electric rates are between \$0.06/kWh and \$0.15/kWh and natural gas rates between \$0.027/kWh and \$0.041/kWh, the change in electrical and natural gas EUI (not shown) yields an annual savings between \$34,169 and \$60,554. Drawing on the authors' experience

Table A.6 Wall and roof insulation values

Climate Zone	1A	2A	2B	3A	3B	3C	4A	4B	4C	5A	5B	6A	6B	7A	8A
Walls (in order from most to least stringent)															
Wall insulation (optimized R-value)	R-19 +R7.5ci	R-19–R1.5ci	R-19 +R1.5ci	R-19–R20ci	R-19–R20ci	R-19 +R10ci	R-19 +R25ci	R-19 +R25ci	R-19 +R20ci	R-19 +R30ci	R-19 +R30ci	R-19 +R40ci	R-19 +R40ci	R-19 +R50ci	R-19 +R50ci
90.1-2010 addenda bb — steel-framed walls	R-13 +R7.5ci	R-13–R7.5ci	R-13 +R7.5ci	R-13 +R7.5ci	R-13–R7.5ci	R-13 +R7.5ci	R-13 +R10.0ci	R-13 +R10.0ci	R-13 +R10.0ci	R-13 +R10.0ci	R-13–R10.0ci	R-13 +R10.0ci	R-13 +R10.0ci	R-13 +R10.0ci	R-13 +R10.0ci
189.1-2009 — steel-framed walls	R-13 +R7.5.0ci	R-13–R5.0ci	R-13 +R5.0ci	R-13 +R5.0ci	R-13–R5.0ci	R-13 +R5.0ci	R-13 +R10.0ci	R-13 +R10.0ci	R-13 +R10.0ci	R-13 +R10.0ci	R-13–R10.0ci	R-13 +R10.0ci	R-13 +R10.0ci	R-13 +R10.0ci	R-13 +R10.0ci
ASHRAE AFDG — steel-framed walls	R-13.0	R-13.0	R-13.0	R-13 +R3.8ci	R-13–R3.8ci	R-13 +R3.8ci	R-13 +R7.5ci	R-13 +R7.5ci	R-13 +R7.5ci	R-13 +R7.5ci	R-13–R7.5ci	R-13 +R7.5ci	R-13 +R7.5ci	R-13 +R7.5ci	R-13 +R21.6ci
90.1-2007 — steel-framed walls	R-13.0	R-13.0	R-13.0	R-13 +R3.8ci	R-13–R3.8ci	R-13 +R3.8ci	R-13 +R7.5ci	R-13 +R7.5ci	R-13 +R7.5ci	R-13 +R7.5ci	R-13 +R7.5ci	R-13–R7.5ci	R-13 +R7.5ci	R-13 +R7.5ci	R-13 +R7.5ci

(continued)

Table A.6 (continued)

Climate Zone	1A	2A	2B	3A	3B	3C	4A	4B	4C	5A	5B	6A	6B	7A	8A
Roofs (in order from most to least stringent)															
Roof insulation (optimized R-value)	R-25	R-30	R-30	R-35	R-35	R-25	R-45	R-45	R-35	R-55	R-55	R-70	R-70	R-80	R-90
90.1-2010 addenda bb — Roof insulation above deck	R-20	R-25	R-25	R-25	R-25	R-25	R-30	R-30	R-30	R-30	R-30	R-30	R-30	R-35	R-35
189.1-2009 — Roof insulation above deck	R-20	R-25	R-25	R-25	R-25	R-25	R-25	R-25	R-25	R-25	R-25	R-30	R-30	R-35	R-35
ASHRAE AFDG — Roof insulation above deck	R-15	R-20	R-20	R-20	R-20	R-20	R-20	R-20	R-20	R-20	R-20	R-20	R-20	R-20	R-20
90.1-2007 — Roof insulation above deck	R-15	R-15	R-15	R-20	R-20	R-20	R-20	R-20	R-20	R-20	R-20	R-20	R-20	R-20	R-20

Table A.7 Potential for site and source energy use reduction (compared to the baseline) for DER projects using core bundles of technologies and beyond

	Baseline			Base case		DER			HPB	
Climate zone	Total site EUI (100%) kWh/m^{2}year (kBtu/ft^2 year)	Site EUI for heating (100%) kWh/m^2 year (kBtu/ft^2 year)	Source EUIt, (100%) kWh/m^2 year (kBtu/ft^2 year)	Site energy use reduction, %	Source energy reduction, %	Site energy use reduction, %	Site heating energy use reduction, %	Source energy use reduction, %	Site energy use reduction, %	Source energy reduction, %
Public housing, Austria										
5A	218 (69)	152 (48)	210 (67)	38	31	50	73	64	55	68
7	253 (80)	184 (58)	235 (75)	47	36	50	68	62	55	68
Office building, China										
2A	105 (33)	3 (1)	105 (33)	37	37	47	56	47	54	54
3A	119 (38)	25 (8)	119 (38)	38	38	51	62	51	65	65
3C	77 (24)	8 (3)	77 (24)	36	36	47	64	47	69	69
4A	201 (64)	117 (37)	201 (64)	42	42	53	71	41	62	55
7	306 (97)	239 (76)	306 (97)	32	33	50	62	38	67	59
School building, Denmark										
6A	EZ	210 (67)	314 (99)	19	16	56	67	45	82	63
Dormitory, Estonia										
6A	153 (49)	213 (68)	225 (71)	29	22	47	69	37	70	58
Office building, Germany										
5A	256 (81)	220 (70)	307 (97)	40	27	55	58	53	81	76
Office building, UK										
4A	155 (49)	89 (28)	291 (92)	20	16	51	84	32	58	42
5A	201 (64)	135 (43)	341 (108)	23	20	60	83	42	67	52
Barracks, USA										
1A	1 (0)	398 (126)	1154 (366)	17	19	39	59	42	59	59
2A	33 (10)	380 (121)	1025 (325)	17	18	41	84	42	60	59

(continued)

Table A.7 (continued)

Climate zone	Baseline: Total site EUI (100%) kWh/m^{2}year (kBtu/ft^2 year)	Baseline: Site EUI for heating (100%) kWh/m^2 year (kBtu/ft^2 year)	Baseline: Source EUIt, (100%) kWh/m^2 year (kBtu/ft^2 year)	Base case: Site energy use reduction, %	Base case: Source energy reduction, %	DER: Site energy use reduction, %	DER: Site heating energy use reduction, %	DER: Source energy use reduction, %	HPB: Site energy use reduction, %	HPB: Source energy reduction, %
2B	17 (5)	365 (116)	1008 (320)	17	18	40	80	42	61	61
3A	65 (21)	394 (125)	965 (306)	19	18	45	84	42	63	59
3B	37 (12)	326 (103)	812 (258)	15	14	39	82	37	60	57
3C	35 (11)	273 (87)	634 (201)	12	9	33	70	31	46	37
4A	103 (33)	397 (126)	869 (276)	20	16	48	85	25	65	59
4B	86 (27)	333 (106)	745 (236)	16	12	42	88	35	62	56
4C	111 (35)	330 (105)	678 (215)	18	12	44	86	35	62	55
5A	160 (51)	422 (134)	872 (277)	21	17	51	87	42	67	60
5B	133 (42)	362 (115)	733 (233)	18	13	52	88	37	65	57
6A	212 (67)	448 (142)	839 (266)	22	16	55	88	44	70	61
6B	192 (61)	414 (131)	773 (245)	21	14	53	89	41	69	60
7	283 (90)	508 (161)	878 (279)	24	18	59	88	47	73	63
8	417 (132)	630 (200)	978 (310)	24	18	64	92	52	77	67
Office building, USA										
1A	24 (7)	261 (83)	815 (259)	30	27	48	91	45	66	64
2A	60 (19)	285 (90)	814 (258)	32	28	46	63	43	70	65
2B	81 (26)	314 (100)	862 (273)	36	29	49	87	41	73	91
3A	82 (26)	288 (91)	771 (245)	34	28	47	63	43	71	64
3B	68 (22)	251 (80)	680 (216)	30	23	51	92	41	66	58
3C	45 (14)	183 (58)	507 (161)	26	16	41	96	30	59	51
4A	96 (30)	271 (86)	685 (217)	35	26	50	89	38	69	60

4B	71 (22)	227 (72)	593 (188)	31	21	50	95	37	63	54
4C	76 (24)	206 (65)	513 (163)	31	18	48	96	33	63	52
5A	107 (34)	270 (86)	656 (208)	35	25	50	87	37	69	58
5B	83 (26)	223 (71)	552 (175)	31	20	50	95	35	64	53
6A	121 (39)	265 (84)	606 (192)	36	23	52	88	36	69	55
6B	118 (38)	254 (81)	575 (182)	34	22	51	88	34	68	55
7	145 (46)	278 (88)	594 (189)	39	24	54	87	36	71	55
8	218 (69)	340 (108)	634 (201)	42	27	59	83	39	76	58

Table A.8 Energy use reduction in dining facilities (EUIh, heating energy use intensity; EUIt, total energy use intensity)

Climate zone	Baseline			Base case		DER			HPB	
	Site EUIh kWh/m^2 year (kBtu/sq ft year)	Site EUIt kWh/m^2 year (kBtu/sq ft year)	Source EUIt kWh/m^2 year (kBtu/sq ft year)	Site energy %	Source energy %	Site energy %	Site heating energy %	Source energy %	Site energy %	Source energy %
1A	29 (9,198)	604 (191)	1616 (512)	2%	3%	15%	29%	16%	40%	40%
2A	147 (46,626)	706 (224)	1687 (535)	11%	9%	22%	45%	20%	48%	36%
2B	111 (35,208)	744 (236)	1897 (601)	10%	9%	22%	43%	22%	50%	40%
3A	307 (97,377)	840 (266)	1766 (560)	16%	12%	17%	43%	23%	57%	45%
3B	201 (63,755)	749 (237)	1704 (540)	16%	12%	26%	52%	23%	51%	42%
3C	196 (62,169)	645 (205)	1371 (434)	8%	7%	26%	29%	14%	46%	32%
4A	459 (145,590)	964 (306)	1832 (581)	20%	15%	30%	47%	25%	63%	43%
4B	333 (105,624)	854 (271)	1753 (556)	22%	16%	30%	53%	25%	58%	45%
4C	434 (137,660)	897 (284)	1665 (528)	19%	14%	27%	43%	22%	61%	44%
5A	572 (181,432)	1071 (340)	1932 (612)	19%	17%	31%	45%	42%	67%	50%
5B	470 (149,079)	972 (308)	1833 (581)	24%	18%	33%	52%	23%	64%	48%
6A	733 (232,500)	1215 (385)	2041 (647)	21%	17%	33%	45%	28%	71%	54%
6B	681 (216,006)	1177 (373)	2035 (645)	24%	19%	35%	50%	29%	69%	53%
7	938 (297,524)	1420 (450)	2257 (715)	22%	19%	36%	47%	31%	75%	58%
8	1376 (436,453)	1863 (590)	2731 (866)	18%	17%	39%	64%	34%	82%	66%

in DER of dining facilities, retrofitting a dining facility of the size modeled for this paper would cost on the order of $30,000 (all costs are in US dollars) for a DCKV and $100,000+ for a heat recovery system. It would not be unreasonable for a comprehensive upgrade to cost over $200,000. On the cook line, replacement of the appliances could range from $100,000 to $200,000 (more if the dish room were included). Thus a deep retrofit could easily cost over $500,000 for ventilation and appliances. Depending on utility rates, therefore, simple payback would run between approximately 8 and 14 years.

The analysis for the high-performance building, an aggressive DER, included both ventilation and appliance replacement. Sensitivity analysis conducted without appliance replacement showed that appliance replacement contributed on the order of 3–5% improvement over ventilation alone, at a rather steep price tag. The reality of commercial kitchens, in the authors' experience, is that appliances last a very long time and replacing them is very expensive. Dishwashing equipment is often custom built for a kitchen and may have been installed while the building was constructed, making it very difficult (and expensive) to replace. Thus, for a particular DER, ventilation is more cost-effective to retrofit than appliances, and appliance replacement may not be economically feasible if has not been planned anyway.

Appendix B: Insulation Materials

Types of Thermal Insulation

Many insulation material are available on the market today. They differ by their origin, type, thermal conductivity, water vapor resistance, environmental impact, flammability and other factors (Table B.1). Information on the cost of insulation materials (excluding installation labor costs) developed for U-value = 0.2 W/(m^2 K) or = 0.035 BTU/hr/ft^2/°F is listed in Tables B.1, B.2, B.3, B.4 and B.5. U us = U si/5.678.

Costs for Thermal Insulation for Opaque Parts of Building Envelope

Costs for thermal insulation layer for several parts of building envelope (outside wall, roof, topmost ceiling under cold roof, flor slab) were calculated. For cost calculation the thickness of insulation was taken as if the U-value of all parts would be 0.2 W/m^2K so that different materials could be compared. For resulting thickness the costs were calculated out of basic costs (glue, nails, etc.) and costs of material (€/m^3), which gives costs depending on thickness of insulation layer (Wagner 2015). The cost of materials was taken from several databases of building

Table B.1 Characteristics of typical thermal insulation materials for building envelope

Picture	Name	Origin	Thermal Conductance W/(m K)	Water vapor diffusion resistance factor (μ)	Vapor Permeance (per-inch)	Fire behavior	Approx. cost, Eu/m² To be developed assuming U=0,2	Assembly type	Health hazard protection requirements	Embedded energy MJ/kg	Remark
	Eurema Stonewool (mineral wool)	Mineral	0.035–0.045	1	30+	Incombustible, melting point ≥ 1.000°C	10–20	Panel, roll	Dust protection for inside applications	15–25	rot proof
	Eurema Glasswool (mineral wool)	Mineral	0.032–0.040	1		Incombustible	10–20	Panel, roll	Dust protection for inside applications	15–50	Rot proof, up to 70% made by waste glass, compressible
	Ultimate (mineral wool)	Mineral	0.032–0.040	1		Incombustible, melting point ≥ 1.000°C	10–20	Panel, roll	Dust protection for inside applications		Rot proof, up to 70% made by waste glass, compressible
	EPS	Synthetic	0.035–0.040	20–70	2.7	Hardly inflammable	5–20	Panel		75–125	
	Graphite embedded EPS	Synthetic	0.032	30–70		Hardly inflammable	10–20	Panel			
	XPS	Synthetic	0.030–0.040	x		normally inflammable	18–30	Panel		75–125	
	PUR	Synthetic	0.022–0.040	x		Hardly inflammable	10–20	Panel		70–125	

	Polyisocyanurate (PIR)	Synthetic	0.023–0.028	82–10.000		Hardly inflammable	10–20	Panel			
	Wood fiber	Vegetable	0.040–0.055			Normally inflammable	40–50	Panel			
	Hemp fiber	Vegetable	0.040–0.045	1–2		Normally inflammable	20–30	Panel			
	CL Cellulose	Vegetable	0.038–0.069	1–2		Normally inflammable	15–25	Panel, roll, bulk	Dust protection for inside applications	1–25	12-20 mass-% boric oxide for fire protection
	Vacuum insulation panel	Synthetic	0.007	> 1.000.000		Normally inflammable	100–150	Panel			Prefabricated slabs, not able to be machined, damageable, to handle with care
	Expanded clay	Mineral	0.010–0.018	2–8		Noncombustible	20–45	Bulk			
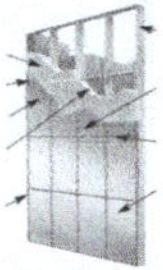	Kalwall	Synthetic	2.53–0.28 (with air gel)	NA	NA	B.12-D0		Panel	None		
	Fiber Reinforced Aerogel Blanket	Mineral/ Synthetic	0.015–0.018	5	14	Euroclass C, Euroclass A2	20–30	Roll, Panel	Dust protection for inside applications	56	European Technical Approval 11_0471

(continued)

Table B.1 (continued)

Picture	Name	Origin	Thermal Conductance W/(m K)	Water vapor diffusion resistance factor (μ)	Vapor Permeance (per-inch)	Fire behavior	Approx. cost, Eu/m² To be developed assuming U=0,2	Assembly type	Health hazard protection requirements	Embedded energy MJ/kg	Remark
	EPB expanded perlite	Mineral	0.04–0.06	3–8			<5	Panel, roll, bat		5–20	
	CG cellular glass	Mineral	0.035–0.055	infinite			<60	Panel		10–75	

Table B.2 Thermal insulation of outside walls

Material*	Thermal Conductivity (W/mK)	Thickness (cm)
inside plaster rendering	0.75	0.5
Brick wall	0.9	38.0
Outside rendering	0.75	1.0

*Starting point: original wall with U-value 1.2 W/m^{2}K (Wagner 2015)

Table B.3 Cost of thermal insulation for outside wall

Material	Thermal Conductivity (W/mK)	Thickness of Insulation (cm)	Min (€/m^2)	Average (€/m^2)	Max (€/m^2)
Mineral wool	0.035	14.5	26.93	37.15	42.44
XPS	0.035	14.5	27.96	35.40	41.76
EPS	0.040	16.5	37.28	49.06	61.46
PUR	0.028	11.5	46.62	60.20	65.75
wooden fiber board	0.090	37.0	65.77	90.47	89.34

Table B.4 Cost of thermal insulation of the topmost ceiling under cold roof

Material*	Thermal Conductivity (W/mK)	Thickness of Insulation (cm)	Min (€/m^2)	Average (€/m^2)	Max (€/m^2)
Mineral wool	0.040	15.0	9.70	11.67	13.70
Cellulose fiber (flakes)	0.045	17.0	12.15	15.50	19.59
XPS	0.040	15.0	13.02	15.12	17.16
EPS	0.040	15.0	13.56	16.98	19.93
Cellulose fiber as boards	0.040	15.0	19.39	23.57	29.34
XPS high-density board	0.035	13.0	21.92	28.,00	34.29
Cork boards	0.045	17.0	35.,78	39.10	41.81
Mineral wool plus wooden board to walk on	0.035	13.0	36.15	41.77	51.34
Wooden fiber board	0.040	15.0	37.7	41.21	43.80
coconut fiber	0.040	15.0	39.6	43.34	45.97
EPS + wooden board to walk on	0.035	13.0	39.72	47.62	56.99
Mineral cement boards	0.045	17.0	76.31	91.23	109.72

*Starting point: Original topmost ceiling with U-value 0.8 W/m^{2}K (Wagner 2015), similar to the roof

costs doing a linear interpolation. According to (Wagner 2015) not only average costs, but also minimal and maximal costs are given (Tables B.6, B.7, B.8, B.9 and B.10).

Table B.5 Thermal insulation of tilted roof with insulation between rafters

Material*	Thermal Conductivity (W/mK)	Thickness of Insulation (cm)
Wooden fiber board	0.65	5.0
Mineral wool (flakes) between rafters	0.06/0.374 (rafter)	5.5
Gypsum board below rafter	0.7	5.0

*Starting point: Original roof with U-value of 0.8 W/m^{2}K (Wagner 2015)

Table B.6 Cost of thermal insulation of tilted roof

Material	Thermal Conductivity (W/mK)	Thickness of Insulation (cm)	Min (€/m^2)	Average (€/m^2)	Max (€/m^2)
Cellulose fiber (flakes)	0.045	16.5	11.08	14.01	16.91
Mineral wool (flakes)	0.035	13.0	11.41	14.40	16.73
wooden fiber (flakes)	0.045	16.5	14.36	17.79	20.70
Mineral wool	0.055	21.0	16.83	21.62	26.35
Hemp fiber	0.040	15.0	25.98	28.37	32.34
Cotton	0.040	15.0	27.14	31.45	37.32
EPS	0.035	14.,0	28.53	31.41	34.64
Flax	0.040	15.0	31.65	38.05	45.38
Sheep's wool	0.040	15.0	37.82	45.21	54.14

Table B.7 Cost of thermal insulation of tilted roof with insulation below rafters

Material*	Thermal Conductivity (W/mK)	Thickness of Insulation (cm)	Min (€/m^2)	Average (€/m^2)	Max (€/m^2)
EPS	0.040	15.0	12.04	13.78	15.02
Mineral wool	0.040	16.0	12.25	15.63	16.87
PUR	0.030	12.0	17.74	21.46	26.39
Wooden fiber board	0.040	15.0	29.35	31.39	37.06
Cork boards	0.045	16.5	30.98	35.56	37.67
Sheep's wool	0.040	15.0	33.55	39.36	48.72

*Starting point as above: Original roof with U-value of 0.8 W/m^{2}K (Wagner 2015)

Insulation Materials Assembly

Thermal insulation materials are basically available in five different types, regardless of their wide range of basic materials and application variations:

- Boards and Batts
- Insulated metal panels

Table B.8 Cost of thermal insulation of tilted roof with insulation on top of rafters

Material*	Thermal Conductivity (W/mK)	Thickness of Insulation (cm)	Min (€/m^2)	Average (€/m^2)	Max (€/m^2)
PUR-foam	0.030	11.0	24.36	28.15	31.38
EPS	0.031	15.5	24.4	29.53	36.07
Wooden fiber board (rigid)	0.040	15.0	29.65	33.58	36.62
Cork boards	0.045	16.5	31.83	35.09	38.86
PUR boards	0.024	9	34.5	40.26	47.33
Phenolic foam boards	0.022	8	44.99	53.24	64.07
Wooden fiber board (soft)	0.045	16.5	45.01	53.33	63.22

*Starting point: Original roof with U-value of 0.8 W/m^{2}K (Wagner 2015)

Table B.9 Cost of thermal insulation of flat roof

Material*	Thermal Conductivity (W/mK)	Thickness of Insulation (cm)	Min (€/m^2)	Average (€/m^2)	Max (€/m^2)
Mineral wool	0.040	13.0	11.76	16.06	18.91
EPS	0.035	15.0	18.47	21.81	27.02

*Starting point: Original roof with U-value of 0.8 W/m^{2}K (Wagner 2015)

Table B.10 Cost of thermal insulation of cellar ceiling

Material*	Thermal Conductivity (W/mK)	Thickness of Insulation (cm)	Min (€/m^2)	Average (€/m^2)	Max (€/m^2)
EPS	0.035	13.5	24.43	29.12	35.77
Mineral wool	0.035	13.5	26.41	32.07	38.91
Wooden fiber board (soft)	0.040	15.5	28.35	31.16	34.99
Phenolic foam boards	0.022	8.5	31.52	35.22	38.17
PUR multilayer compound boards	0.024	9.0	60.54	70.79	82.13

*Starting point: Original roof with U-value of 0.8 W/m^{2}K (Wagner 2015)

- Translucent structural composite sandwich panel
- Foams
- Loose-fill
    - Blow-in
    - Stuff-in

Shape and dimensions of individual thermal insulations depends on their material properties as well as the intended use. However the necessary installation procedures vary significantly with the actual type of material supplied. Most common are boards and matts. The following paragraphs give a short introduction to the available types mentioned above.

Boards and Batts/Blanket from Mineral Wool

"Classic" mineral fiber insulation is composed of mineral wool, i.e., glass wool and rock wool, which are noncombustible and fire resistant.

Dimensional stability is used as the criterion to distinguish between "boards" and "matts": boards mostly are rigid and keep their shape. Usually they feature a rebate to reduce thermal bridging between neighboring boards. Matts are soft and also available as rolls; they always have to be mechanically fixed, with the exception of so-called clamping felts. These elastic mineral wool matts have to be cut in boards and fixed between the rafters of a pitched roof.

This material is used for thermal insulation of façade/building envelope with lightweight or massive building constructions (except where the insulation would be exposed to soil); roof insulation both sloped and flat roofs; thermal insulation under floating floors. Technical installations like air ducts and heating pipes may also be insulated by boards and matts from mineral fibers. They also may serve as acoustical absorbers for room acoustics purposes.

Blanket insulation (Batts and Rolls) are available in widths suited to standard spacing of wall studs, attic trusses or rafters, and floor joists. Continuous rolls can be hand-cut and trimmed to fit. They are available with or without facings. Manufacturers often attach a facing (such as Kraft paper, foil-kraft paper, or vinyl) to act as a vapor barrier and/or air barrier. Batts with a special flame-resistant facing are available in various widths for basement walls and other places where the insulation will be left exposed. A facing also helps facilitate fastening during installation. However, unfaced batts are a better choice when adding insulation over existing insulation.

Glass Wool/Fiberglass Matt and Light Weight Board

Standard fiberglass matts and light weight boards are made from lime, sand and recycling glass and come in a variety of sizes, thicknesses, and densities. They have a thermal resistance or R-value between R-2.9 and R-3.8 per inch of thickness. High-performance (medium-density and high-density) fiberglass blankets and batts have R-values between R-3.7 and R-4.3 per inch of thickness. See Table B.11 for an overview of these characteristics. The data in Table B.11 pertains to fiberglass batts only. Determine actual thickness, R-value, and cost from manufacturer and/or local building supplier.

Table B.11 Typical fiberglass batt insulation characteristics

Thickness mm (in.)	R-Value	Cost Eu/m² (cents/sq ft)
3½	11	12–16
3⅝	13	15–20
3½ (high density)	15	34–40
6–6¼	19	27–34
5¼ (high density)	21	33–39
8–8½	25	37–45
8 (high density)	30	45–49
9½ (standard)	30	39–43
12	38	55–60

Table B.12 Typical stone wool/rock wool batt insulation characteristics

Thickness mm (in.)	R-Value	Cost Eu/m² (cents/sq ft/in.)
3½	14.7	15–19
3⅝	15.225	15–19
3½ (high density)	14–14.7	30–70
6–6¼	25.2–26.25	15–19
5¼ (high density)	21–22.05	30–70
8–8½	33.6–35.7	15–19
8 (high density)	32–34.4	30–70
9½ (standard)	39.9	15–19
12	50.4	15–19

Stone Wool/Rock wool Batt and Semi-Rigid Board

Stone wool insulation comes in a variety of sizes, thicknesses, and densities. Low-, medium-, and high-density stone wool has an R-value of 3.8-4.3 (Table B.12). It is available in both batt and semi-rigid boards with various facing options. Stone wool insulation is noncombustible, dimensionally stable, hydrophobic, vapor permeable, and inorganic. It is used for a variety of applications including fireproofing, acoustical barriers, interior and exterior insulation on roofs, walls, below grade, and subslab.

Fiber Insulation Made from Natural Raw Materials

Another group of fiber insulation is mainly made from renewable sources, e.g., wood, flax, hemp, coco, sheep wool, cellulose. Cork insulation is available as rigid boards. To make these insulation materials usable for building constructions in terms of mold and fire protection, they must contain minimal particles of chemical substances.

Silica Aerogel Blanket Insulation

Aerogel blanket insulation's superior thermal performance arises from the nano-porous entrapment of air within the silica aerogel matrix, it does not rely on maintaining a vacuum or heavy gas molecule in place and therefore can be cut and fixed with conventional insulation fixings.

Applications include, but are not restricted to, the following:

1. Timber and Lightweight steel framing
2. Internal Wall Insulation (Solid Wall)
3. External Wall Insulation (Solid Wall)
4. Terrace, Balcony and gutter insulation
5. Floor insulation (domestic loading)
6. Roof insulation – pitched (internal/external) and flat
7. Thermal bridging treatments in domestic, multi-family and commercial construction.

Valuable benefits for the user arise when:

- Aerogel blanket insulations are used to meet code or sufficient thermal resistance in space challenged applications where traditional insulation materials will not fit or encroach excessively on valuable living space. This application provides twin benefits for the occupant and the investor as gross internal area is maximized for both parties.
- Aerogel blanket insulation is used to address residual thermal bridges, which would otherwise compromise the health of the occupant or the building fabric itself.
- Thermal upgrade options are limited due to the space constraints or building response to thermal upgrade. Aerogel blanket insulation is hydrophobic yet vapor permeable, meaning that the building fabric can respond to its natural environment post upgrade. This is particularly important for heritage buildings.

In retrofit scenarios the economic value propositions are related to: (1) the value of space and (2) the costs arising from having to revisit the project to address problems due to condensation, mold, fabric damage, and occupant displacement. The incremental cost of applying an aerogel blanket based solution is outweighed by the costs of remedying problems, hotel charges, loss of income etc. In high-density occupation, the ROI can be measured in months.

In new construction, the use of super insulating aerogel blanket insulation (Annex 65 terminology for advanced porous materials) can facilitate additional floors in multi-story construction products where height would otherwise be consumed by insulating materials (terraces and balconies).

Rigid Foam Boards

Foam boards — rigid panels of insulation — can be used to insulate almost any part of the building, from the roof down to the foundation. They provide good thermal resistance, and reduce heat conduction through structural elements, like wood and steel studs. The most common rigid foam boards are boards made from polystyrene (PS). Due to a somewhat different production process we distinguished between EPS and XPS. Polyurethane and polyisocyanurate foams (PUR, PIR) may be manufactured both from plastics and from renewable raw materials. Product R-value is determined at 75 °F in a laboratory. Different blowing agents for the various foam insulations perform very differently at high and low "real-world" temperatures.

Like fiber insulation, rigid foams cover a wide range of possible uses. An advantage is that they may be used in damp and wet conditions, e.g., as ground insulation in direct contact with soil.

This material is used for thermal insulation of the entire building envelope including façade and ground; insulation of sloped and flat roofs (over-rafter insulation); insulation under floating floors.

Boards Made from Mineral Materials

Boards made from foamed glass are mainly manufactured from recycling glass. Mineral foam boards are made from lime, sand, cement and water. Calcium silicate boards are made from calcium silicates mixed with cellulose. From their production process all three incorporate a large number of small pores. They are able to absorb moisture and have good dimensional stability. This material is used for thermal insulation of building envelope in massive buildings; interior insulation of walls, floors and ceilings; insulation of sloped and flat roofs (over-rafter insulation).

Insulated Metal Panels

Insulated metal panels (Fig. B.1) typically mount to cross steel struts (by others) that attach to structural steel (by others). These products are suitable for large scale industrial and commercial applications where both vertical and horizontal panels are required. Advantages include:

- Single component wall insulated metal wall panels provide exterior weather barrier, insulating core and interior vapor barrier all-in-one.
- Polyisocyanurate foam core retains original insulating value over time.
- Side joint clip system eliminates the need for exposed fasteners.
- Panels are lightweight and easy to install under most weather conditions.
- Wide variety of profiles and textures.

Fig. B.1 Insulated metal panel: (**a**) cross-section, (**b**) front view

- Panels are available in lengths of up to 52 ft to minimize the number of stack joints required.
- Accessory items including metal flashings and aluminum extrusions are also available.

Table B.13 lists panel thermal conductivity.

Wall panels are mechanically attached on one side only, with the other side free to slide along the tongue and groove joint configuration. In addition to this built-in slip joint, IMPs comply with air and water infiltration testing in accordance with ASTM E-283/331.

Table B.14 lists approximate unit cost ($\$/ft^2$) for installing metal insulation panels (rough estimates for comparison purposes only; includes cladding, insulation, air/water/vapor barrier, furring supports).

Table B.13 Insulated metal thermal properties

Panel Thickness	R-7.5 per inch	U-value	R-8 per in.	U-value
2 in.	R-14.4	0.094	R-16	0.085
2.5 in.	R-28	0.072	R-20	0.065
3 in.	R-21.6	0.050	R-24	0.045
4 in.	R-28.8	0.034	R-32	0.031
5 in.	R-36.0	0.028	R-40	0.025
6 in.	R43.2	0.023	R-48	0.021

Table B.14 Cost comparison for single skin and insulating metal panels

Type of Panel	Material Cost, $\$/ft^2$	Labor Cost, $\$/ft^2$	Total, $\$/ft^2$
Commercial wall panels (vertical panels)	6–8	3–6	9–14
Architectural wall panel (horizontal panels)	10–15	6–9	16–24

Insulating Diffuse Light Transmitting Building Panels

The primary component of the insulating, diffuse light transmitting building panel system is a translucent structural composite sandwich panel formed by permanently bonding specially formulated fiberglass sheets constructed of interlocking thermally broken extruded aluminum "I-beams." The panels are factory prefabricated to the exact size and configuration for each project. Panels can be flat or curved, and opening or fixed glazed window units can be incorporated. The panel is generally 70 mm thick (100 mm is also available) and by providing various densities of insulation the thermal resistance and therefore "U" values can be varied. One of the panel manufacturers, Kalwall®, has panels with different densities of IG-fiber and with aerogel insulation. Table B.15 lists insulation values assessed according to American Standards C-236, E 1423 and C1199 by an independent laboratory certified by the American NFRC.

Table B.15 Insulation values

Face Sheet Combinations		Panel 'U' Value – Walls – W/m^2K Panel Thickness 70mm										Panel 'U' Value – Walls – W/m^2K Panel Thickness 100mm						Internal Reaction To Fire
	Thermally Broken	2.57		1.25		0.78		0.56		0.28*		2.85		0.83		0.45		
	Standard	2.74		1.56		1.20		0.99		0.72*		N/A		N/A		N/A		
Exterior	Interior	Light Trans	Solar Factor	Light Trans	Solar Factor	Light Trans	Solar Factor	Light Trans	Solar Factor	Light Trans	Solar Factor	Light Trans	Solar Factor	Light Trans	Solar Factor	Light Trans	Solar Factor	
White SW-E	White 8-3C	20	0.31	14	0.16	8	0.10	5	0.07	12	0.13	20	0.31	12	0.13	5	0.06	B-s2,d0
Crystal SW-E	White 8-3C	30	0.38	20	0.19	9	0.12	6	0.07	15	0.18	30	0.38	13	0.15	5	0.07	B-s2,d0

Face Sheet Combinations		Panel 'U' Value – Roofs – W/m^2K Panel Thickness 70mm										Panel 'U' Value – Roofs – W/m^2K Panel Thickness 100mm						Internal Reaction To Fire
	Thermally Broken	2.86		1.31		0.80		0.57		0.28*		3.22		0.86		0.46		
	Standard	3.08		1.67		1.26		1.03		0.74*		N/A		N/A		N/A		
Exterior	Interior	Light Trans	Solar Factor	Light Trans	Solar Factor	Light Trans	Solar Factor	Light Trans	Solar Factor	Light Trans	Solar Factor	Light Trans	Solar Factor	Light Trans	Solar Factor	Light Trans	Solar Factor	
White SW-E	White 8-3C	20	0.28	14	0.15	8	0.09	5	0.06	12	0.13	20	0.28	12	0.12	5	0.05	B-s2,d0
Crystal SW-E	White 8-3C	30	0.35	20	0.18	9	0.11	6	0.07	15	0.18	30	0.35	13	0.14	5	0.06	B-s2,d0

*Kalwall +Aerogel™

Notes:

1. Solar Factor is total Solar Heat Transmittance.
2. U-values are calculated according to CIBSE Guide Volume A Chapter 3. (The difference between wall and roof values is because surface heat transfer is different for horizontal and the more severe vertical upward heat transfer. This is fully explained in the reference.)
3. Light transmissions are expressed in percentage terms.
4. Light transmissions over 30% are not recommended for normal use.
5. All performance figures based on 600 mm by 300 mm internal panel grid.
6. Fire performance in accordance with BS EN 13501-1.
7. Other face sheet combinations available on request.

Kalwall systems are resistant to most acidic or alkaline atmospheres. Environments usually considered to be corrosive such as paper or metal processing plants, seaside locations, and swimming pools do not affect Kalwall sandwich panels or the Kalwall Corrosion Resistant Finish on the Clamp-tite System. Generally, any building atmosphere, where people are exposed, without the need for any protective clothing, can be considered to have no adverse effect on Kalwall Systems. Kalwall is a product made from two specially formulated resin bonded glass-fiber reinforced face sheets thermally bonded under pressure to an interlocking, aluminum "I" section beam grid. Because of its structure, the product exhibits fire resistant properties that exceed any other translucent panels. The requirements of most building regulations for fire safety are well within its design capabilities. Kalwall is a thermoset material, which unlike thermoplastics such as polycarbonate or acrylic material does not melt or drip in a fire.

In-Situ (Spray-on) Plastic Foams

In-situ foams are produced on site from liquid components by means of transportable equipment. Common in-situ foams are polyurethane (PUR) and urea formaldehyde foams.

In-situ PUR foams may be used as spray-on insulation layer in walls (internal) or roofs (internal and external). In-situ PUR foams are mostly used to fill gaps and cavities; and for on-site thermal insulation of technical or industrial installations (cold and hot pipes etc.).

Loose-Fill Insulation from Fibrous and Mineral Materials

Blow-in Insulation

Loose-fill insulation consists of small particles of fibers, foam, or other materials such as mineral wool, grass fiber, wood fiber, hemp, and cellulose fiber jam and thus prevent through-flow. They need to be compressed to a certain density or blown into

Table B.16 Recommended specifications by loose-fill insulation material

	Cellulose	Fiberglass
R-value/in.	3.2–3.8	2.2–2.7
Inches (cm) needed for R-38	10–12 (25–30)	14–17 (35–43)
Density in lb/ft^3 (kg/m^3)	1.5–2.0 (24–36)	0.5–1.0 (10–14)
Weight at R-38 in lb/ft^2 (kg/m^2)	1.25–2.0 (6–10)	0.5–1.2 (3–6)
OK for ½-in. drywall, 24 in. on center?	No	Yes
OK for ½-in. drywall, 16 in. on center?	Yes	Yes
OK for ⅝-in. drywall, 24 in. on center?	Yes	Yes

cavities by special blow-in equipment to prevent gaps. These small particles form an insulation material that can conform to any space without disturbing structures or finishes. This ability to conform makes loose-fill insulation well suited for retrofits and locations where it would be difficult to install other types of insulation. Some materials are free flowing like e.g., perlite; granules made from EPS, XPS, lightweight silicate foam, and aerogels. However, if the cavity is not fully closed, the materials might flow through.

The most common types of materials used for loose-fill insulation are produced using recycled waste materials. Cellulose is primarily made from recycled newsprint. Most fiberglass contains 20–30% recycled glass. Mineral wool is usually produced from 75% post-industrial recycled content. For comparison, Table B.16 lists these three materials.

Blow-in thermal insulation is suitable for sloped roofs, domes, timber frame constructions, technical installation cavities, as core insulation in retrofit, and as insulation of the topmost floor slab. Cellulose, fiberglass, and rock wool are typically blown in by experienced installers skilled at achieving the correct density and R-values. Polystyrene beads, vermiculite, and perlite are typically poured.

Stuff-in Insulation

Stuff-in insulation is manufactured from plant and animal fibers, like e.g., flax, hemp, coco, sheep wool, also from mineral wool. Loose fiber materials are used only additionally to other thermal insulation materials, e.g., to prevent air-untight joints and thermal bridges between door and window frames and the surrounding building elements, mostly in the context of building retrofit.

Other Thermal Insulation Materials

Some more thermal insulation materials are available on the market, like compound systems (e.g., VIP vacuum insulation panels), or insulation plaster (plaster with lightweight aggregates like EPS), which have not been included in this overview.

Appendix C: Lighting Retrofit

Walls Insulation

Adhesive and Mechanical Bonding

Depending on application, insulation materials can be attached using any of several different methods, for example, through the use of adhesives, mechanical fasteners, face stapling of installation between studs, and machines to blow loose-fill insulation into uninsulated walls.

Adhesive bonding. The insulating materials can be bonded on both absorbent and nonabsorbent substrates. The adhesive can be used, e.g., on concrete, wooden materials, brickwork, and trapezoidal sheet metal profiles. New, non-talcumed bitumen sheeting with a full-surface, firmly adhering mineral sprinkling is also a suitable substrate. The following insulation materials can be installed using adhesives [Henkel AG & Co. KGaA]: rigid polystyrene (PS) foam (boards, roll-on membranes, folding membranes); rigid polyurethane (PUR) foam; rigid phenolic resin (PF) foam; and mineral fiber insulation materials. When using other insulating products than those mentioned above and when bonding mineral fiber and laminated insulation materials with each other, it is necessary to carry out preliminary adhesion tests and observe the insulation manufacturer's instructions. Figure C.1 shows examples of insulation installation using adhesives.

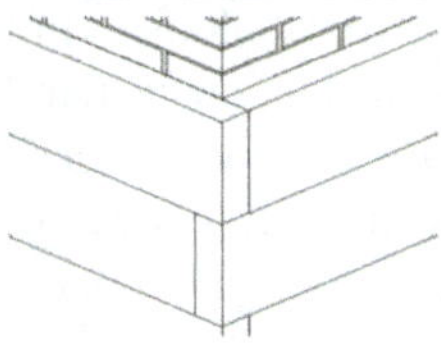	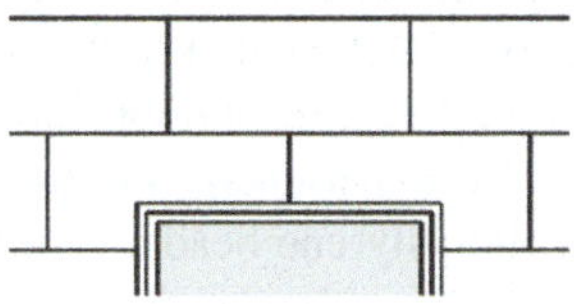
Interlock corner boards for more stable compound	Notch insulation boards at the corner of window openings

Ribbon-and-dab method; effective adhesive bonding surface > 40%	Partial-area bonding (effective adhesive bonding surface > 60%; application possible with EPS here). Manual adhesive application to entire surface

Fig. C.1 Examples of insulation installation using adhesives

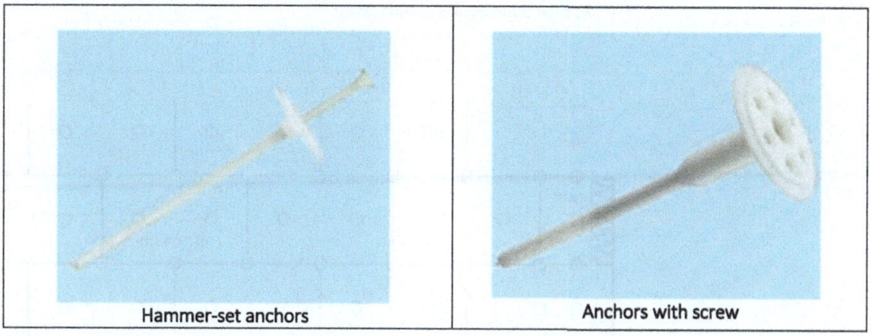

Fig. C.2 Screw-type or hammer-set anchors used to secure insulation materials to brick and concrete walls

Adhesive Bonding

Anchors Anchors are usually screw-type or hammer-set devices used to secure insulation materials to brick and concrete walls (see Fig. C.2). These are basically stainless steel or coated screws that cut their own thread when driven into the masonry. They are highly resistant to rust and provide excellent holding. When choosing the length of anchor, make sure that the screw embeds itself at least 1–1.5 in. into the wall after going through all the components that need to be adhered to the wall (Fig. C.3).

Anchor quantity for structural mechanical fastening. Table C.1 and Figs. C.3, C.4, C.5 and C.6 list and show Rockwool recommendations for statically relevant mechanical fixing and Fig. C.7 shows alternative installation methods including clip and rail, screw fastened with long screws, and brick ties. However, national regulation must be adhered to at all time. For higher buildings or higher wind loads, please seek professional advice.

Table C.1 Rockwool recommendations for statically relevant mechanical fixing

Height of building above ground level [m]		**up to 10 m**	**up to 18 m**	**up to 25 m**
constructional anchors		**4 – 6**	**4 – 6**	**4 – 6**
load bearing anchors				
Wind zone and location		Wind suction and anchor quantity per m²		
very low	wind load [kN/m²]	0.738	0.959	1.106
wind load	anchors	**6**	**8**	**8**
low	wind load [kN/m²]	0.959	1.18	1.328
wind load	anchors	**8**	**8**	**10**
medium	wind load [kN/m²]	1.18	1.328	1.623
wind load	anchors	**8**	**10**	**12**

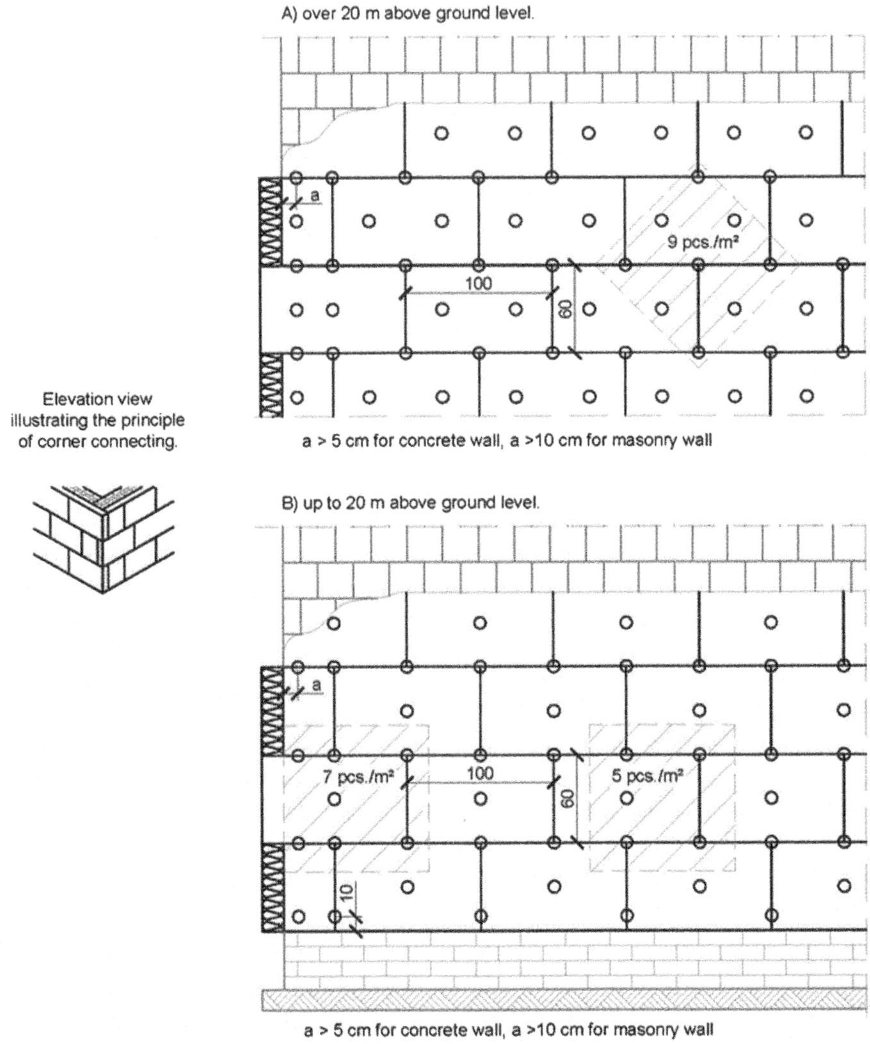

Fig. C.3 Example of mechanical fastening patter for large mineral wool boards

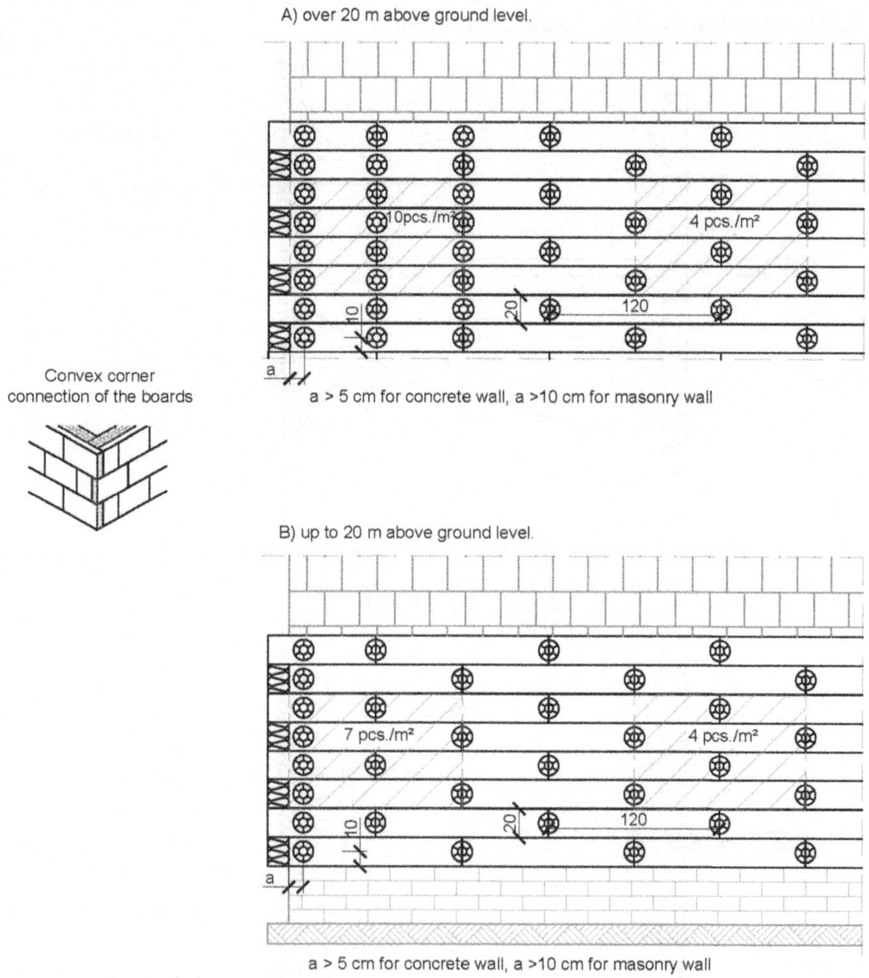

Fig. C.4 Example of mechanical fastening for small mineral wool boards

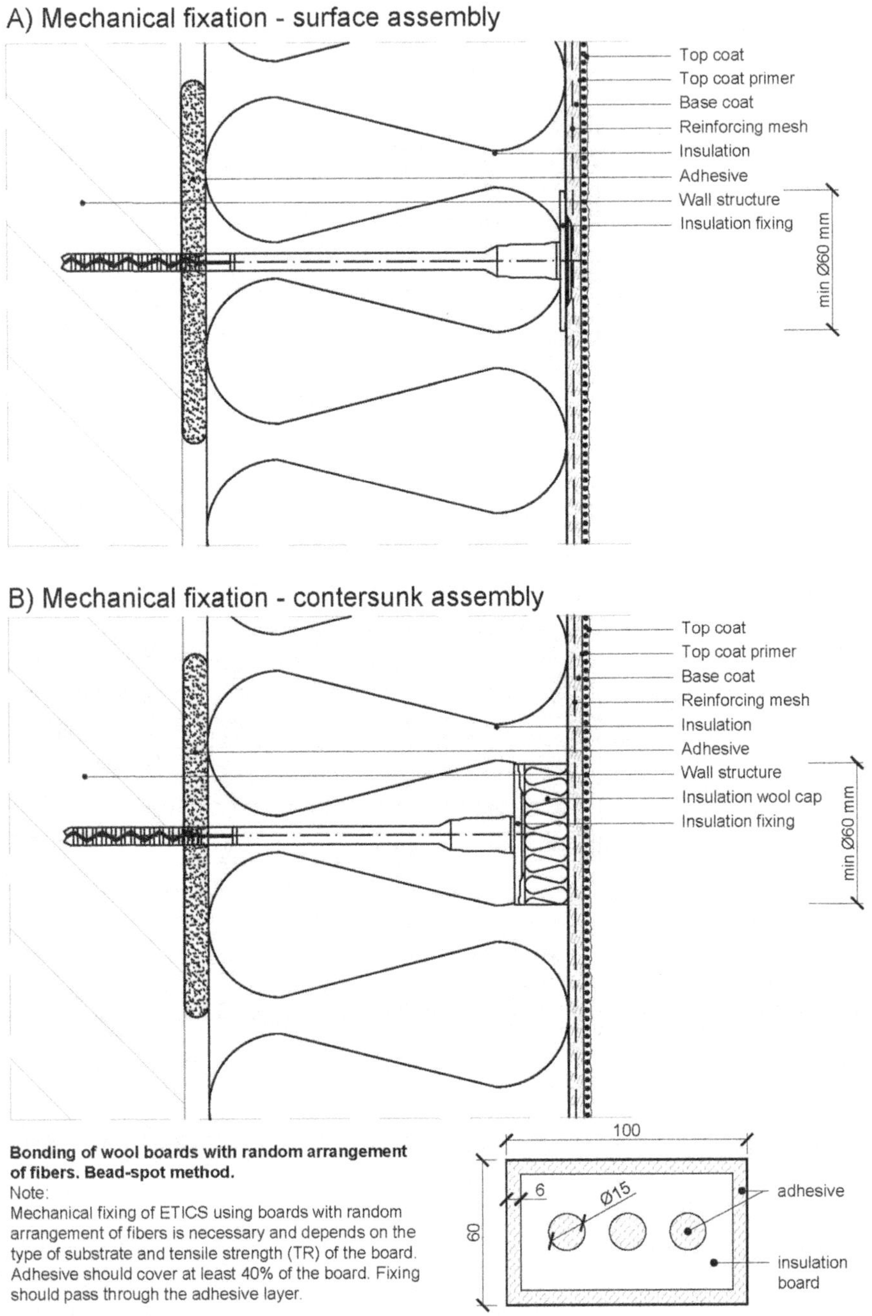

Fig. C.5 Example combined adhesive and mechanical fastening of mineral wool boards with random arrangement of fibers

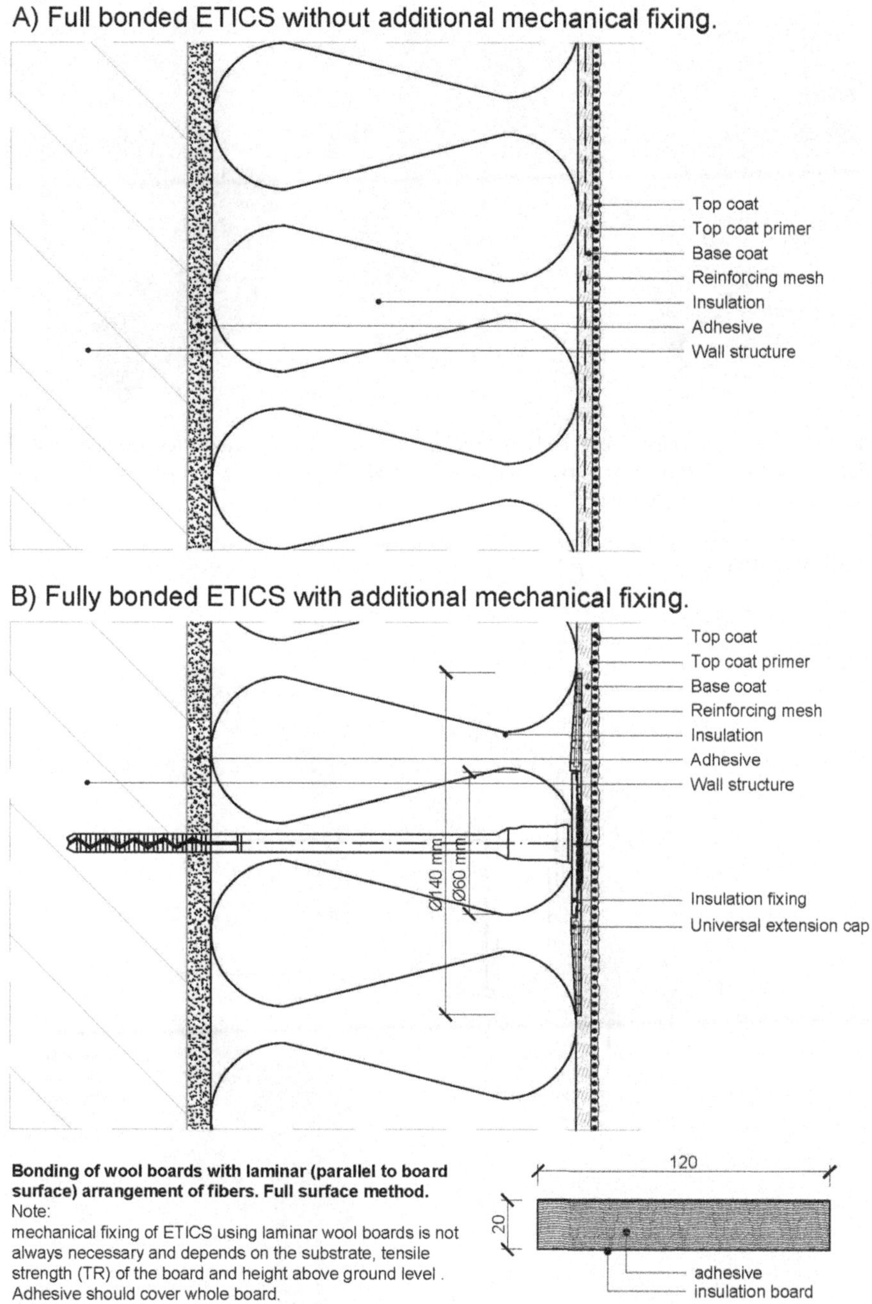

Fig. C.6 Example adhesive and combined adhesive and mechanical fastening of mineral wool

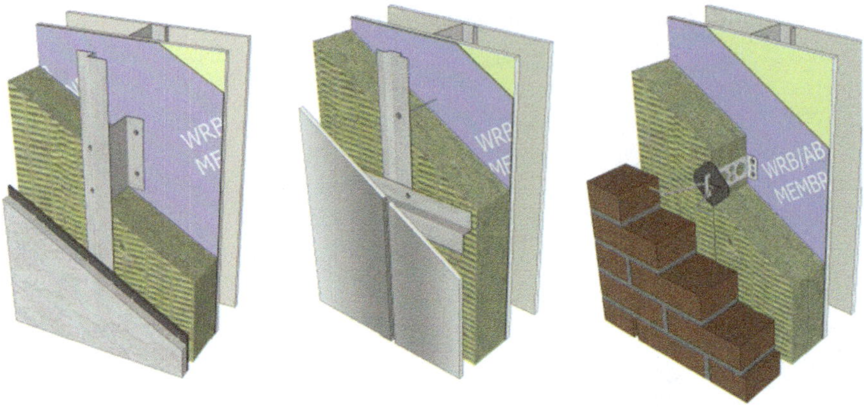

Fig. C.7 Non-proprietary bracket and clip fastening systems – exterior cavity insulation over steel-framed structure (EIFS). Alternative installation methods including clip and rail, screw fastened with long screws, and brick ties

Wall Systems

<table>
<tr><th>System</th><th>EIFS</th><th>ETICS</th></tr>
<tr><td rowspan="2">Drawings</td><td rowspan="2">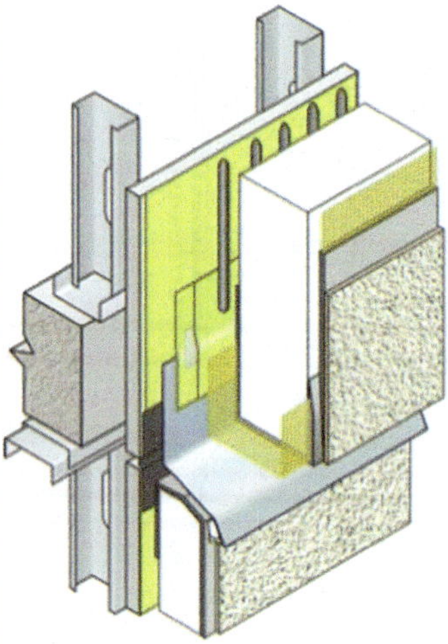</td><td>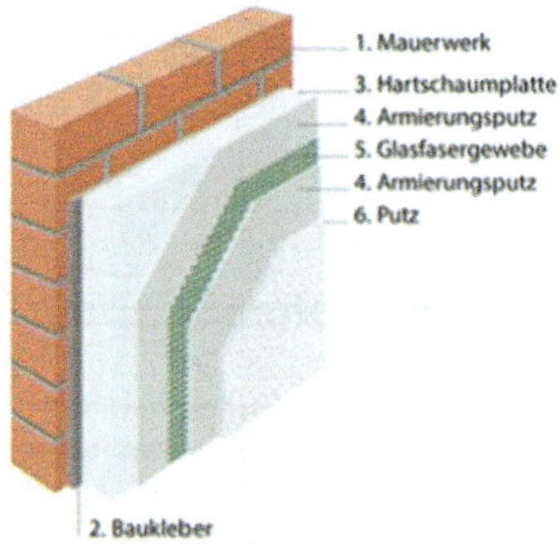
</td></tr>
<tr><td>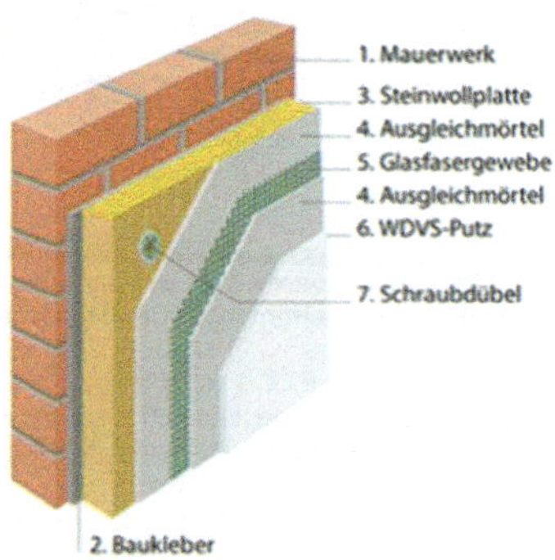
</td></tr>
</table>

(continued)

System	EIFS	ETICS
Layers	Frame	1 = loadbearing wall (masonry, concrete, OSB boards etc.)
	Board	2 = glue mortar; point-strip application with all boards except lamella mineral wool (full area application)
	Liquid-applied water barrier (yellow)	3 = thermal insulation material (MF, EPS, PUR, PF, WF)
	Adhesive (installed as vertical ribbons for drainage)	4 = first plaster layer
	Thermal insulation material	5 = mesh
	Mesh	6 = final layer of plaster
	External finish	7 = mechanical fastener (anchor) (if any)
Frame/contact to wall	EIFS applied to Frame construction (steel, wood)	ETICS applied directly to wall (glue mortar), no frame
	Frame construction is the norm in North America	Point-strip application of glue mortar
		 1. Kleber aufbringen Bild 1 von 5
		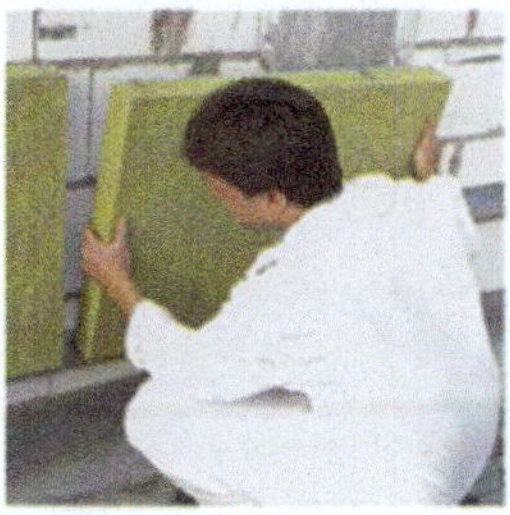 2. Dämmstoff Bild 2 von 5
Water-resistant barrier	The use of an integral liquid-applied water-resistant barrier (yellow in the drawing above) behind the EIFS is a code requirement	No additional water-resistant barrier. Water resistance is provided by plaster layers 4 and 6. Water can/should never get between the boards and the wall. The thermal insulation material is either water repellent by nature (like EPS, XPS, PUR); if not (e.g., MW), only boards that have been treated hydrophobic in the factory shall be used

(continued)

<table>
<tr><th>System</th><th>EIFS</th><th>ETICS</th></tr>
<tr><td>Connecting membrane at floor line</td><td>Use of a connecting membrane at the floor line to maintain the airtightness of the design and the continuity of the water resistive barrier</td><td>Not used/not necessary</td></tr>
<tr><td>Flashing</td><td>Use of flashing through the EIFS (not necessarily metal) to allow drainage of incidental water from behind the EIFS. The locations of required details are covered in the installation standard and the design application standard</td><td>Not used/not necessary</td></tr>
<tr><td rowspan="5">Wrapping</td><td rowspan="5">Pre-wrapping of the mesh around the bottom edge of the insulation. The mesh and the base coat are installed on a strip of insulation in advance of installation on the wall and allowed to cure. This is different from standard back-wrapping where the mesh is installed on the substrate first, ahead of the insulation. With back-wrapping, the insulation is installed and the mesh is pulled around the bottom of the insulation and troweled into the wet base coat. If this were done, it would interfere with the free flow of water from behind the system; therefore the prep wrapping is done. This is covered in the installation standard</td><td>Usually horizontal metal rails or metal frames or metal/plastic frames are used at the bottom of the ETICS system, as a means of mechanical protection and as a sealing against insects</td></tr>
<tr><td>Metal frame (thermal bridge effect due to metal part penetrating from outside surface to loadbearing wall)</td></tr>
<tr><td>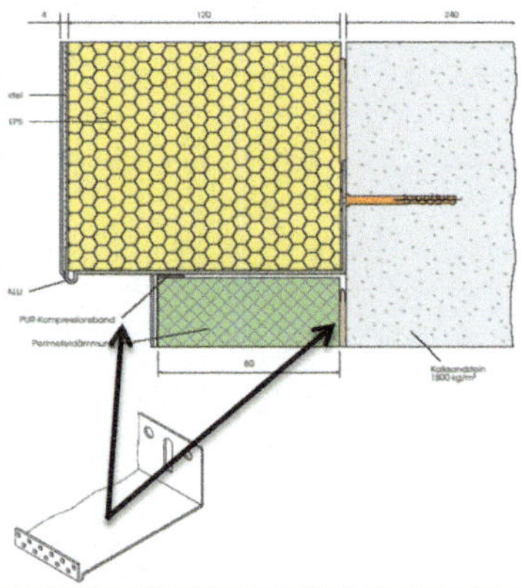
</td></tr>
<tr><td>Combination of plastic rail and small metal frames (avoiding the thermal bridge effect at the bottom of the ETICS)</td></tr>
<tr><td>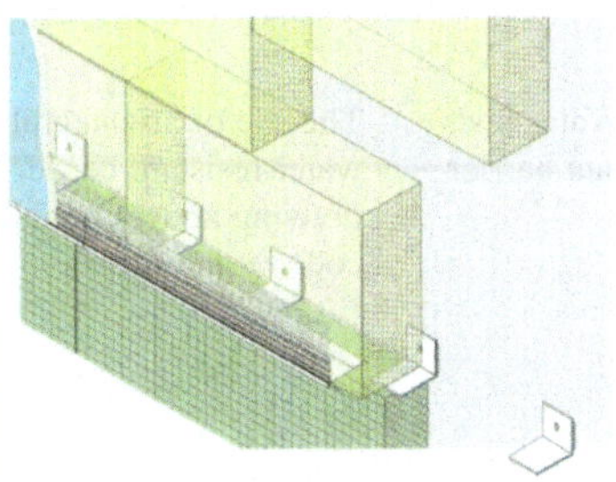</td></tr>
</table>

(continued)

System	EIFS	ETICS
2: Adhesive/ glue mortar	The adhesive is installed in vertical ribbons to encourage drainage. There are alternatives for use of geometrically designed cavities (grooves) in the insulation, but always a clear means of drainage must be maintained. The requirements for drainage in North America are severe. This has developed as a response to the water intrusion issues in Canada and the United States. Similar methods have been adopted in Sweden for the same reason. The drainage requirements are developed for the protection of wood as a worst-case scenario. Only 40 gm/ m^2 of water may be retained after an hour of running water through the sample. Details of the test are in the standard. This is considered to be the equivalent of no retention so there would be no threat to the wood substrate	Glue mortar installed in point-strip-wise or full surface. No air gaps within the systems
		The air volumes between the boards and the wall are considered sealed; to prevent air flow between boards and wall, the mortar has to be applied point-strip-wise or over the full rear face of the board. Water inflow at the junction between boards and wall is prevented by means of roof overhang, sealing of window sills, use of special sealing strips around the window frame to seal against the facing plaster, etc.
7: Mechanical fasteners	Mechanical attachment of rigid insulation is not part of the standard, but work is underway to have that included	With most fibrous insulation materials, anchors attach the insulation to the wall, and the glue mortar acts merely as a convenience during installation. However, with special lamella mineral wool boards, the bond is achieved by the glue mortar, and there is only one anchor per board as additional safety. With foamed plastics insulation boards, there are systems with and without need for mechanical fasteners
		The use of anchors (if at all) is regulated in the installation standard of the individual system. If necessary, an average 5–8 fasteners per 1 m^2 are required. With building height >20 m, mechanical fasteners are compulsory due to higher wind suction loads with higher buildings
Rendering	Not applicable	Mineral rendering is applied in 2 layers (lower and upper render) directly to the insulation boards. A glass-fiber meshing is incorporated in the lower layer and around all corners of the system (especially around the window openings) to prevent cracks to build in the render. The lower render is to be water-resistant, i.e., to prevent rain

(continued)

System	EIFS	ETICS
		water and driving rain from penetrating the system
Approval		ETICS are approved only as complete systems (glue mortar, thermal insulation material, mesh, anchors (if any), lower and upper plaster layers). They may only be used in the combination given in the technical approval. However, in real building life, some builders mix between the components from different systems because of price and availability; this may lead to cracks and faults and therefore malfunction of the system
		Before installation always check the manufacturer's installation manual and do stick to it!!
Not to be confused with	Not to be confused with rear-ventilated façades (German: VHF = vorgehängte hinterlüftete Fassade). Ventilation layer between thermal insulation and cladding	
	Rear-ventilated façades are NOT embraced by the term "ETICS"	
	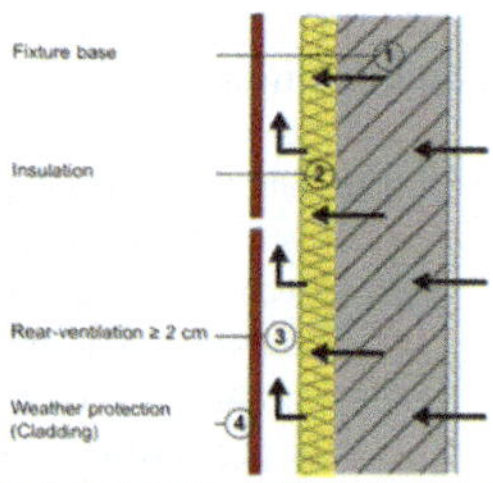	
	Drawing: FVHF (www.fvhf.de)	
Sources	Pictures from Sto Corp.	Pictures from www.heizkosten-einsparen.de (website of the German ETICS trade association).
	Text: John Edgar, Sto Corp.	Text: Martin H. Spitzner, München

Source document: ISO/TC163 SC3 N0714 Comparison EIFS-ETICS

Mineral Wool Cavity Wall Installation using Clip/Girt Systems (Example for EIFS)

Full-length z-girt systems are not recommended due to high thermal performance loss caused by highly conductive steel material. Rather, clip and girt systems or thermally broken clip systems are the preferred method.

For cavity wall insulation, proprietary clips/brackets or non-proprietary standoffs designed to support vertically or horizontally installed z-girts or L-channels can be used. Insulation is not required to provide any structural support with this type of

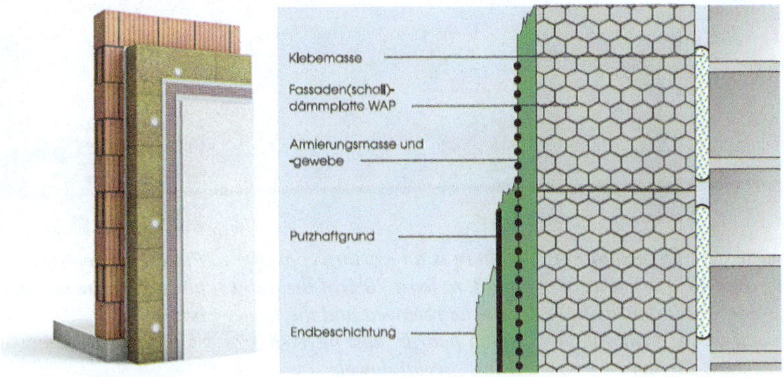

Picture 1a and 1b Principle of ETICS construction

system. Z-girts and fastening system should be designed to support wind loads and total applied dead loads of cladding independently of the insulation.

The insulation can be pinned to the wall if additional retention to the substrate is required. Adjust the number of pins or screws in insulation retention system appropriately based on the level of support provided by clip and girt system.

Masonry Wall with Thermal Insulation Composite System (ETICS)

Made of assembled bricks with a layer of pressure-resistant insulation and different layers of a composite system on the outside. This wall system can be applied in new and in existing buildings. The insulation layer can be fixed mechanically or by mortar (glued), a combination of both or by rails. The glued fixation is the low-cost solution. Since the surface of walls in renovation mostly is not stable, it is recommended to fix the insulation boards mechanically. The number and performance of the dowels per square meter depend on the wall construction and the materials. Only accredited ETIC systems consisting of insulation, fixations, mesh, mortar, and facing may be used. The suppliers of ETICS are responsible for the accreditation of the system. The following figure shows the sequence of ETICS installation:

		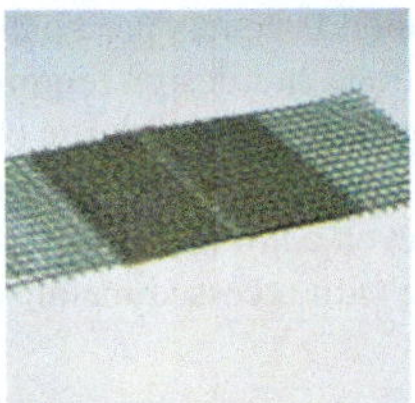
Before starting renovation with an ETICS the stability of the façade has to be proved	*The façade has to be cleaned intensively*	*If there are fractures, a special mesh has to be installed as next*

(continued)

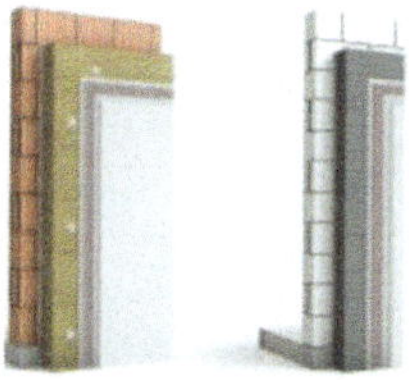		
In the next step, the insulation layer of rigid EPS or mineral wool (Stone wool) boards has to be glued on the wall, minimum thickness 60 mm	*If there is an existing coat of paint, at least 70% of the paint must be removed and the insulation boards must be fixed using special dowels*	*Put on the reinforcement plaster on the insulation boards*

Put the reinforcement mesh into the plaster and flatten it	*Optionally you can roughen the surface*	*Prime the basic plaster*

Put on the final plaster coating	*Optionally structuring of the surface*	*View of a renovated façade*

ETICS with Backup Wall (Fig. C.8)

Mineral wool semi-rigid boards typically require a minimum of five mechanically attached insulation fasteners per board. Designers can increase number of fasteners to meet specific conditions and design requirements (including wind-driven rain-loads). Types of fasteners that can be used include plastic cap nails, impaling pins, and insulation fasteners. Screws and washers are not recommended because of their big point-wise thermal bridging effect, which may lead to an uneven color appearance on the render. Best is to use special recessed ETICS fasteners (plastic casing with recessed metal screw) covered by an insulation disc

(continued)

The point-wise thermal bridging effect of mechanical fasteners is to be taken into account in the U-value of the façade. The thermal bridging effect is assessed by a 3D or circular numerical calculation according to ISO 10211 and denoted by X with units W/K (or Btu/Fh). The smaller X, the smaller is the thermal bridging effect. The U-value Ucorr with correction for the mechanical fasteners is calculated from the U-value without fasteners Uw/o and the number n of fasteners per m^2 and the point-wise thermal transmittance X of the fasteners: $Ucorr = Uw/o + n \cdot X$
Mechanical fasteners for ETICS with $X \leq 0.001$ W/K may be neglected in the thermal transmittance, i.e., no correction is necessary
ETICS are available from conventional insulation materials like:
PUR (including PIR) boards (only boards without aluminum foil to be used) with thermal conductivity 0.028 W(m·K)
Semi-rigid mineral wool boards and rigid lamella mineral wool boards, with $\lambda = 0.031 \ldots 0.035$ W/(m·K)
Polystyrene (EPS and XPS) with thermal conductivity of 0.030/0.035 W/(m·K)

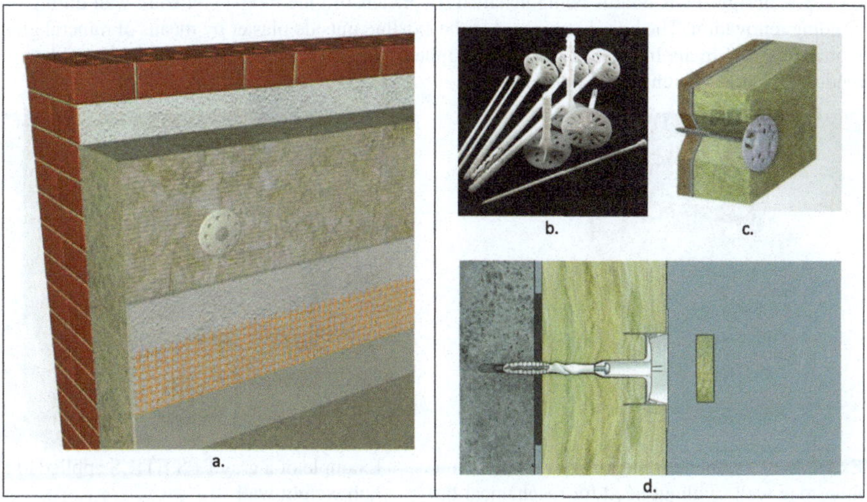

Fig. C.8 Mechanically attached insulation: (a) mechanically fastened mineral wool board, (b) fasteners, (c) cross-section, (d) recessed mechanical fastener with insulation disc

Examples: Solid or Lightweight Walls with ETICS

ETICS with rigid PUR boards being installed (by means of glue mortar) on solid wall during a building renovation. The boards are glued to the existing outside plaster by means of mineral glue mortar (point-strip application of the mortar to the rear face of the insulation boards), and later rendered. Source: puren (www.puren.de)

Example of a lamella mineral wool ETICS on a limestone wall (without need for mechanical fasteners, just 1 per m^2 as additional safeguard)

Example of a gray EPS ETICS applied to a lightweight wall

ETICS are available for traditional insulation materials like:

PUR (including PIR) boards (only boards without aluminum foil to be used) with thermal conductivity 0.028 W(m·K)

Semi-rigid mineral wool boards and rigid lamella mineral wool boards, with $\lambda = 0.031 \ldots 0.035$ W/(m·K);

Polystyrene (EPS and XPS) with thermal conductivity of 0.030/0.035 W/(m·K)

Application of Rigid Foam Boards (XPS, PUR) on Local Thermal Bridging Areas

Mechanically and or adhesively fixed to local thermal bridging areas. Maintain healthy frsi to avoid condensation and mold internally

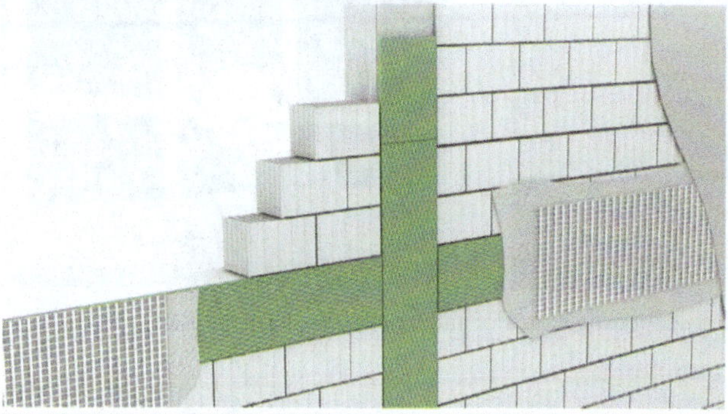

XPS boards reducing thermal bridges by concrete studs and ceilings in monolithic walls

Ventilated Façade insulation (VHF Façade)

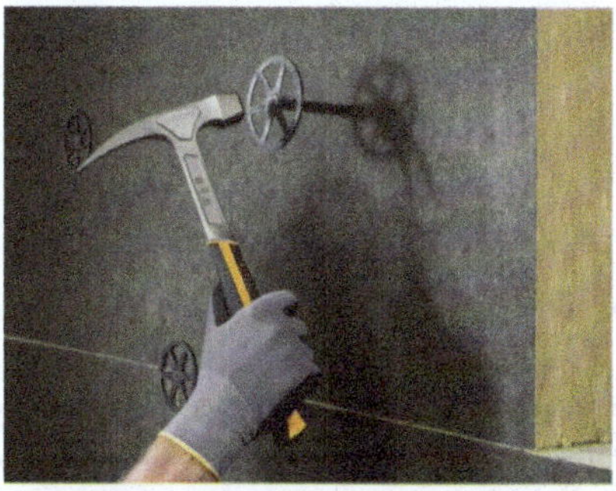	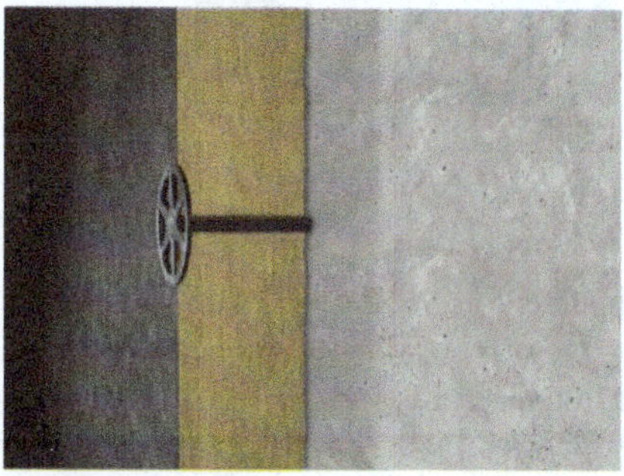
Fixation of insulation panels by special fastener	5 fasteners per m^2

Ventilated Façade with Wooden Supporting Structure and Wooden Cladding Profiles

Ventilated façade with wooden supporting structure and wooden cladding profiles

Façade insulation panels with facing glass tissue for higher weather resistance

Installation of Mineral Wool Insulation Using Strapping Attachment

Mineral wool (stonewool) insulation boards should be permanently mechanically supported by the strapping attachment if the attachment is designed to withstand live loads and total applied dead loads. During installation, minimal fasteners can be used to temporarily hold the product in place until the strapping is applied provided undesirable conditions are not expected in that time frame. For optimal performance, install strapping vertically to provide a drainage cavity; metal hat-channels or wood furring can be installed horizontally over the strapping to accommodate vertical cladding types.

The strapping attachment should be designed to withstand wind loads and total applied dead loads (insulation + fasteners + strapping + cladding). For mineral wool insulation (ROXUL® insulation), current research suggest limiting installed

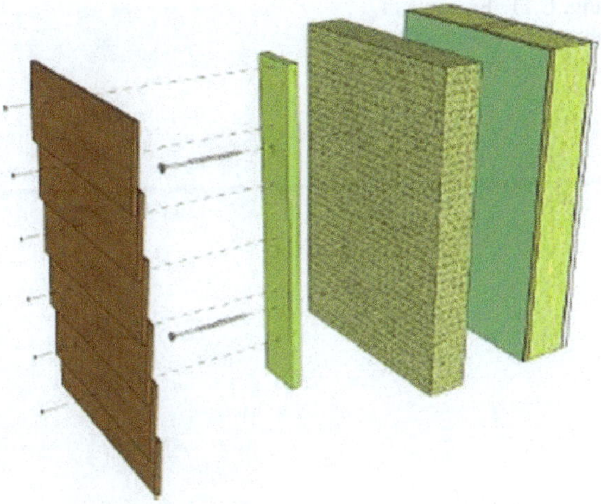

Fig. C.9 Installation layers of Wood frame wall with exterior mineral wool insulation and vertical wood strapping (ROXUL Inc.)

Fig. C.10 Exterior insulation (EIFS) fastened with long screws through wood strapping (wood frame structure)

cladding load to 10 lb./fastener (relative to fasteners attaching furring to the structure) over semi-rigid mineral wool of 8 lb/ft3 up to 4 inches thick. (Refer to ROXUL COMFORTBOARD Deflection Testing for more information) (Figs. C.9 and C.10)

There are a series of different thermally broken clip systems available in the market. The different systems will have different grades of performance, all however significantly higher than typical z-girt attachment methods. Manufacturer's instructions should be used for installation methods (Fig. C.11).

Fig. C.11 Example of thermally broken clip system (Cascadia Clip)

Cavity Insulation

Insulation of the cavity in brick walls has become popular, for example, in Denmark and in northern regions of Germany as it is normally fast and relatively cheap. Besides it does not cause changes neither to the inside nor to the outside of the building, and it is almost without nuisance for the occupier. In existing buildings with cavity walls, the only option to put in a layer of insulation into the cavity is to fill in a loose material or foam. Therefore special machines to blow the insulation are needed (see figures below).

Framing Insulation: Batt Insulation

Batts and rolls are available in widths suited to standard spacing of wall studs, attic trusses or rafters, and floor joists. Note the batt insulation in between the roof rafters.

Basement Insulation

External basement wall insulation allows for mitigation of thermal bridges and energy reduction for temperature control in the basement and is a practical and effective alternative to underfloor insulation. For existing buildings, this is typically accomplished by excavation along the building basement perimeter (depending on the structural possibilities), creating a clean wall surface (plaster base coat) before applying the vertical seal, bonding perimeter insulation panels to the basement wall, and inserting an expanding foam seal on the upper side; attach drip edge before backfilling. In addition to thermal performance improvement, this will help with waterproofing. For more details, see Appendix "Remediation of thermal bridges" (Fig. C.12).

Basement walls are insulated over the water barrier using water-resistant XPS insulation boards. These boards are bonded jointless close to each other with a special waterproof adhesive. XPS boards are highly pressure resistant

Fig. C.12 Installation of aerogel blanket insulation

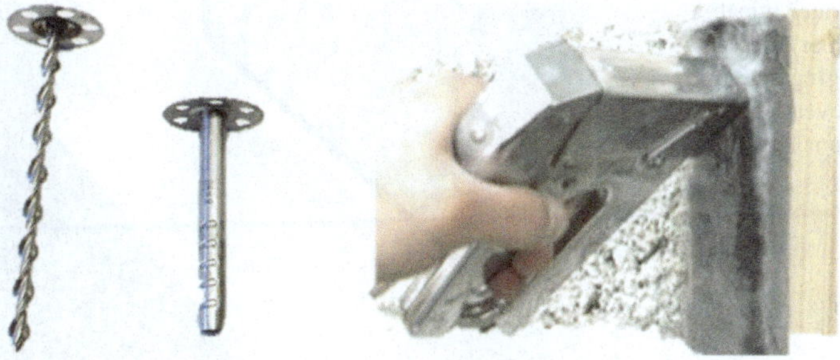

Fig. C.13 Stainless steel mechanical fixings, staple fixing

Aerogel blanket insulation's superior thermal performance arises from the nanoporous entrapment of air within the silica aerogel matrix; it does not rely on maintaining a vacuum or heavy gas molecule in place and therefore can be cut and fixed with conventional insulation fixings.

Aerogel blanket insulation may be adhesively and or mechanically attached to the working surface. In most cases, a steel mechanical fixing will be mandated by local fire regulations. Use of a low thermal conductivity fixing (stainless steel) is recommended to limit point thermal bridging. Aerogel blanket strips can be stapled to studwork before final fixing of plasterboard etc (Fig. C.13).

Mechanical fixings should be selected according to the type and condition of the underlying substrate. To avoid pull through, the fixing should accommodate a large diameter (3/4 in. 0 mm) head or retaining washer. This applies equally to IWI and EWI attachments. Typically systems of this nature have local building control certification – refer to the system provider's instructions for appropriate fixing information.

Aerogel blanket insulation may be adhesively attached using water- or solvent-based adhesives including lime and cementitious coatings. Base coats, renders, or plasters should be supported using a mesh or scrim that is mechanically affixed to the substrate. The system provider's guidelines should be consulted (Figs. C.14 and C.15).

Fig. C.14 Mechanical fixing with mesh to support EWI render

Fig. C.15 Mechanical fixing (Polymer) of IWI thermal bridging assembly at window reveal (Aerogel Blanket, Vapor Barrier + MGO board)

Examples of Aerogel Blanket Insulation Installation

Timber framing (Fig. C.16)

Fig. C.16 Mechanically fixed to battens in strip or blanket format to increase R-value and reduce repeat thermal bridging through studs

Internal Wall Insulation (IWI on Solid Wall) (Fig. C.17)

Fig. C.17 Mechanical and adhesive fix to internal wall. Finish with wet plaster/lime or rigid facing board according to customer needs. Perform robust risk assessment to determine hygro-thermal impacts of IWI upgrade

External Wall Insulation (EWI on solid wall) (Fig. C.18)

Fig. C.18 Mechanically fixed with low thermal conductivity fixings (stainless steel and polymer). Finished with siding, brick slips, render, and mesh or cladding. Preserve building character due to thinner insulation profile

Terrace, Balcony, and Gutter Insulation (Fig. C.19)

Fig. C.19 Loose lay, adhesive, or mechanical fix to terrace or balcony. Increase R-value without increasing thickness above safety height for fall/water ingress. System is compatible with traditional membrane and ballast systems

Floor Insulation (Domestic Loading) (Fig. C.20)

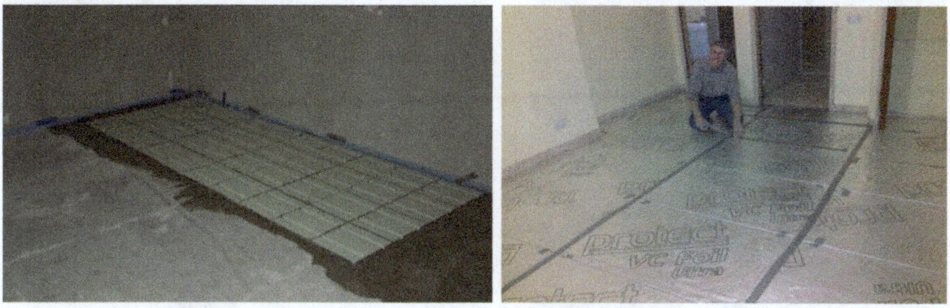

Fig. C.20 Loose lay or panel application, suitable for domestic loadings. Address heat loss through concrete slab without requiring door/threshold adjustments

Roof Insulation: Pitched (Internal/External) and Flat (Fig. C.21)

Fig. C.21 Mechanically fixed to/through battens. Hygro-thermal risk assessment is recommended, particularly in heritage applications

Thermal Bridging Treatments in Domestic, Multi-family, and Commercial Construction

Exterior Wall Insulation from Inside

Spray Foam

Liquid foam insulation, combined with foaming agent[1], can be applied for walls as a pressure-sprayed (foamed-in-place) product. The product will expand and harden as the mixture cures, conforming to the shape of the cavity, filling and sealing it thoroughly. Due to its properties, many applications also consider the spray foam to act as the vapor and air control layer[2].

Slow-curing liquid foams are also available. These foams are designed to flow over obstructions before expanding and curing, and they are often used for empty wall cavities in existing buildings. There are also liquid foam materials that can be poured from a container (Fig. C.22).

Installation of most types of liquid foam insulation requires special equipment and certification and should be done by experienced installers. Following installation, an approved thermal barrier equal in fire resistance to half-inch gypsum board must cover all foam materials. Also, some building codes do not recognize sprayed-

---

[1]Spray foams can contain toxic ingredients (Refer to https://www.epa.gov/saferchoice/spray-polyurethane-foam-spf-insulation-and-how-use-it-more-safely). They are also highly combustible and can cause increased fire spreads (Refer to http://www.federated.ca/files/pdfs/57-en-Polyurethanesolidgasoline.pdf).

[2]Improper installation and shrinkage of the spray foam can result in compromised air barrier system.

Fig. C.22 Spray foam being applied in a wood frame wall assembly

foam insulation as a vapor barrier, so installation might require an additional vapor retarder.

Mass Wall: Existing Brick Masonry Walls

The optimal solution for retrofitting existing masonry buildings would be to add exterior insulation. That being said, retrofitting to the interior is the most common practice as there are typically space and landmarks prohibiting from adding anything to the exterior side. When retrofitting to the interior side of a brick masonry building, special attention must be taken where the brick can be susceptible to freeze-thaw cycles (especially in cold climates).

Interior Retrofit

Spray Foam and Foamed-in-Place Insulation

Due to its vapor impermeability, and air barrier characteristics, the more common approach for interior retrofits of masonry buildings is to use spray foam insulation. However, as stated above, there are risks in accounting the spray foam as an air barrier system. Two common approaches are practiced, spray foam only with an interior stud frame or spray foam in conjunction with batt insulation in between the stud framing. Potential concerns about SPF include the health risks associated with its toxic ingredients (see http://www.epa.gov/saferchoice/spray-polyurethane-foam-spf-insulation-and-how-use-it-more-safely, Highly Combustible: "liquid gasoline":http://www.federated.ca/files/pdfs/57-en-Polyurethanesolidgasoline.pdf). Also, improper installation and shrinkage can result in a compromised air/vapor barrier (Figs. C.23 and C.24).

Note that, if a permeable insulation is used to the interior side in conjunction with a spray foam, caution must be made to ensure airtightness and avoid a double vapor barrier.

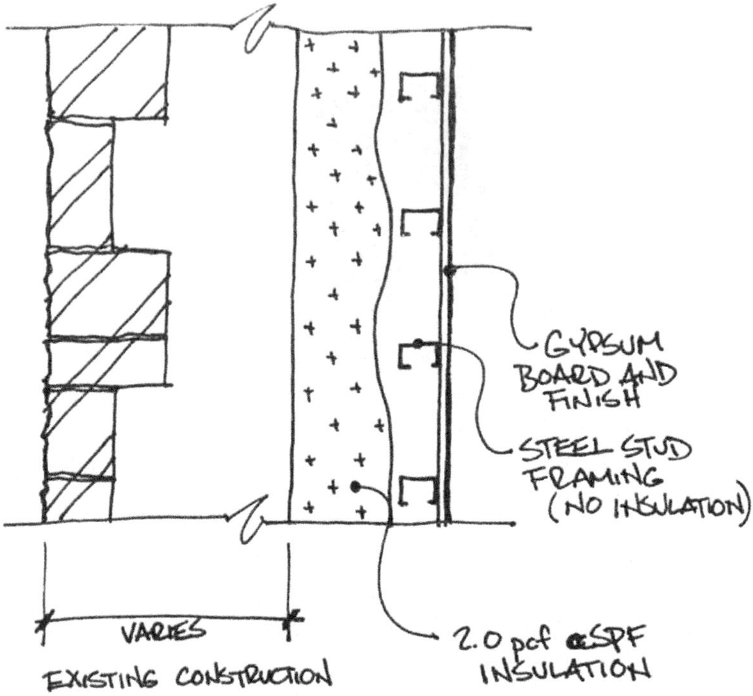

Fig. C.23 Existing masonry with spray foam interior insulation

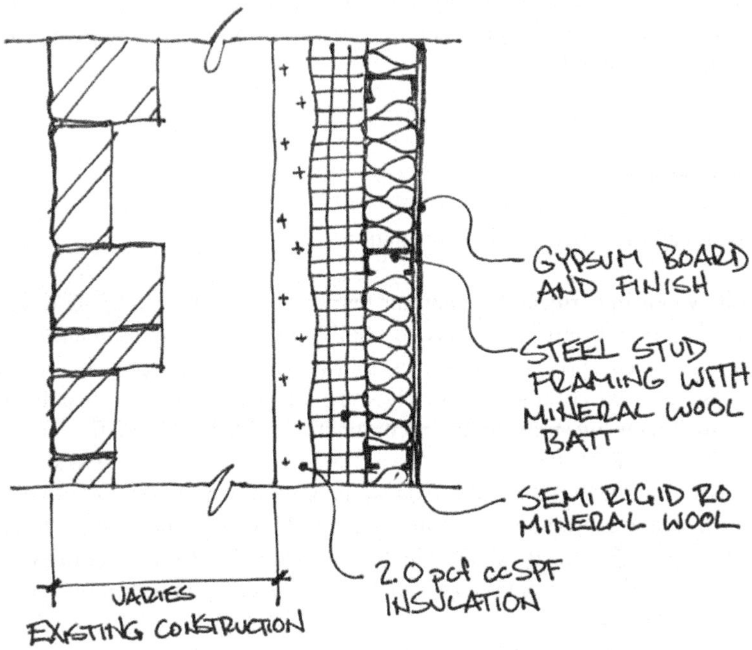

Fig. C.24 Existing masonry with hybrid interior solution (spray foam and mineral wool insulation)

Liquid foam insulation materials can be sprayed, foamed-in-place, injected, or poured. Some installations can have twice the R-value per inch of traditional batt insulation, and can fill even the smallest cavities, creating an effective **air barrier**.

Liquid foam insulation – combined with a foaming agent – can be applied using small spray containers or in larger quantities as a pressure-sprayed (foamed-in-place) product. Both types expand and harden as the mixture cures. They also conform to the shape of the cavity, filling and sealing it thoroughly.

Slow-curing liquid foams are also available. These foams are designed to flow over obstructions before expanding and curing, and they are often used for empty wall cavities in existing buildings. There are also liquid foam materials that can be poured from a container.

Installation of most types of liquid foam insulation requires special equipment and certification and should be done by experienced installers. Following installation, an approved thermal barrier equal in fire resistance to half-inch gypsum board must cover all foam materials. Also, some building codes do not recognize spray foam insulation as a vapor barrier, so installation might require an additional **vapor retarder**.

Liquid foam insulation products and installation usually cost more than traditional batt insulation. However, liquid foam insulation has higher **R-values** and forms an air barrier, which can eliminate some of the other costs and tasks associated with weatherizing a home, such as caulking, applying housewrap and vapor barrier, and taping joints. When building a new home, this type of insulation can also help reduce construction time and the number of specialized contractors, which saves money.

Board and Batt Insulation from Inside

Applying exterior wall insulation from the inside of the building is a much more ambitious task than simply adding insulation to the building's exterior due to building physics; temperature and humidity inside the construction changes because the insulation layerprevents heated air from penetrating the wall to enter the building's interior. The insulation must be secured to ensure that humidity that enters in the winter season can evaporate in the summer period. The outside of the wall must be protected against driving rain. This application must be done by expert craftsmen who can do the recommended physical calculations to anticipate damage by humidity.

There are in general two ways to realize an exterior wall insulation from inside:

8. Pressure-resistant insulation boards

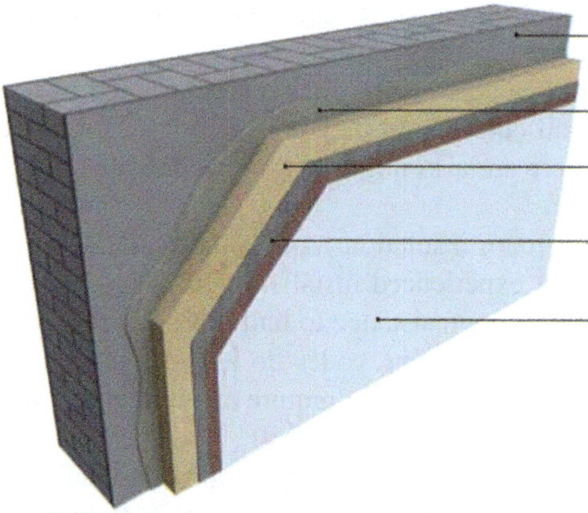

(1) Existing wall; (2) glue and reinforcement mortar; (3) mineral wool insulation board; (4) reinforcement mortar and mesh; (5) final rendering

The advantage of this option is that it requires no substructure. This construction effectly excludes thermal bridges without any additional provisions.

Step-by-step pictures

1. Put the glue/mortar on the backside of the insulation board

2. Glue the pressure-resistant boards on the existing wall

3. Put the reinforcement mortar on the insulation layer and press mesh in it

4. After drying of the reinforcement mortar (1 week), raise the final rendering

9. Flexible insulation boards or batts

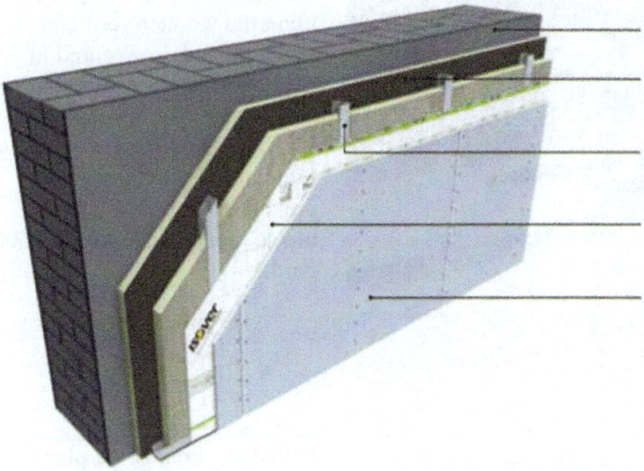

(1) Existing wall; (2) first mineral wool insulation layer (boards) to prevent thermal bridges; (3) metal or wooden studs as substructure, flexible mineral wool batts in between; (4) vapor barrier with flexible density (smart vapor retarder); (5) gypsum boards, open to diffuse

More sophisticated options will require expert work especially if airtightness is needed. Loss of interior space is unavoidable.

Step-by-step pictures

1. Build the substructure with distance to the existing wall (first put every second stud) and put the first insulation layer between wall and substructure. Then put the missing studs in front of the insulation layer

2. Put flexible mineral wool batts between the studs

3. Fix the vapor retarder on the studs by tape or hook and loop

4. Install the gypsum boards by screwing it into the studs (Source: All pictures from www.isover.de)

Roofs
Example: Flat lightweight roof with PUR over-rafter insulation

PUR rigid boards are placed upon the wooden roof construction and are secured in place (during the installation) by stripes of PUR spray foam. The boards are then covered by bitumen felt (and later with gravel, concrete pavement blocks or substrate for the roof garden)

PUR rigid boards are placed upon the wooden roof construction in two layers and are secured in place (during the installation) by stripes of PUR spray foam

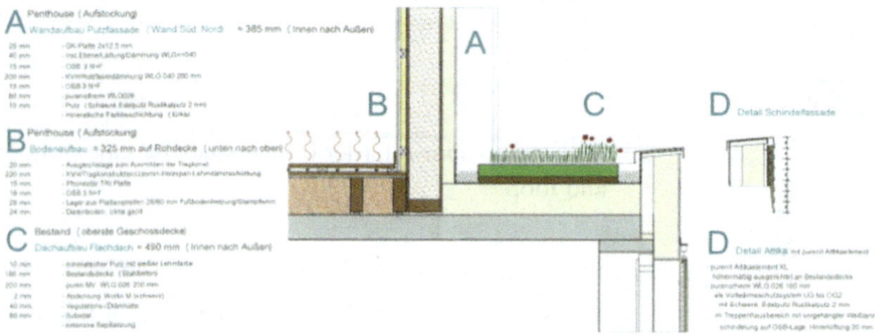

Cross-section through the roof layout (C)

PUR insulation on a flat lightweight roof (actually a building retrofit with installation of a new upper level in lightweight construction, with penthouse and partially green roof) Pictures used courtesy of Puren (www.puren.de)

Example: Flat concrete roof with external PUR insulation

PUR rigid boards applied to a flat concrete roof in low-energy building and then covered by bitumen felt. Pictures: Puren (www.puren.de)

Example: Flat roof insulation (stonewool) on a metal profile

Flat roof insulation (stonewool) on a metal profile with waterproofing

Flat roof insulation (stonewool) with direct glued bitumen sheeting

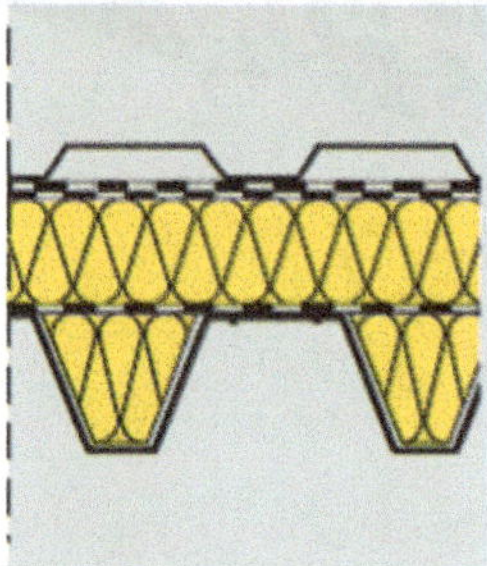

Flat roof insulation with glasswool between two profiled metal sheets

Example: Inverted flat roof (Umkehrdach): Flat roof insulation (XPS) on top of waterproofing

Flat roof with insulation layer (XPS) on top of waterproofing (Umkehrdach)

Pitched Roof Insulation

In wooden pitched roof constructions, insulation can be installed above or/and between the rafters. It depends on the:

1. Flexible insulation (batts) between the rafters

In wooden pitched roof constructions, insulation can be installed above and/or between the rafters, e.g., shown in the figure below.

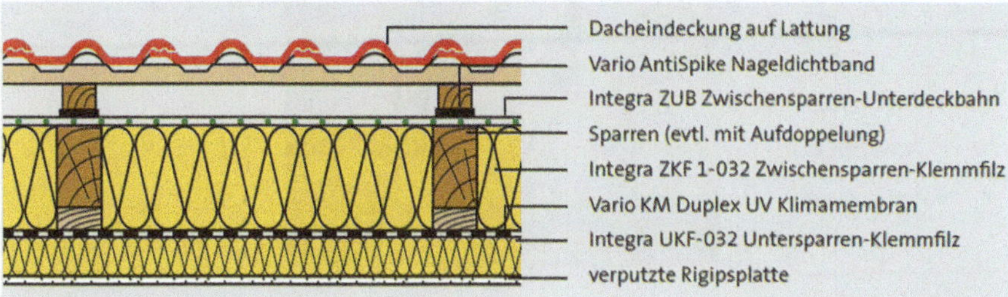

You can fill the existing empty space without losing space inside the building or extend the high of the roof construction to the outside. To minimize thermal bridges from the rafters, a thin additional layer of mineral wool is usually installed between the substructures of the internal finish. Air tightness is a very important parameter for durable function of the roof construction. It is recommended to use air tightness systems with complementary components.

Example: Insulation of Wood Frame Walls with a Lightweight Mineral Wool Insulation

No additional attachments are needed, just clamping between studs.

Pitched Roof Insulation Step-by-Step

	1. Cut from roll, width = distance between rafters +1 cm
	2. Install insulation panel between rafters; it clamps without additional fixation
	3. Install vapor barrier beyond rafters

(continued)

4. Adapt vapor barrier to window

5. Adapt vapor barrier to pipes by using special tapes and glues. Follow the manufacturer's recommendations.

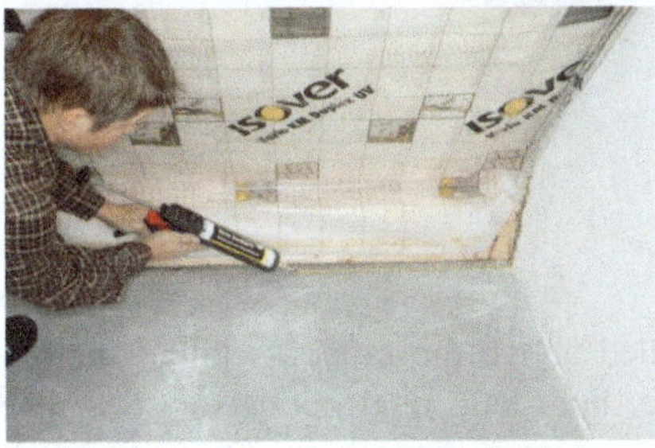

6. Seal vapor barrier to floor and walls

7. After installing and tightening the vapor barrier, attach it to the wooden studs to form a tight interior lining

8. Install additional thin insulation layer (max. 20 % of R-value) beyond rafters and vapor barrier

(continued)

	9. Use the space from the inside for electronic installation without damaging vapor retarder/ airtightness

Pressure-Resistant Insulation (Boards) Above the Rafters

In this option the insulation layer is installed above the rafters, without thermal bridges. The rafters are visible from inside. For static rand optical reasons, it is recommended to have a wooden construction as a first layer. On a first rain protection, the insulation boards have to be fixed by screws in the rafters beyond. Then the substructure and the roof covering follow.

Example: Sloped Roof with PUR Over-Rafter Insulation

	PUR rigid boards being installed on a lightweight roof and secured into place (during roof installation) by PUR spray foam stripes, beginning with the lowest board
	The boards come with an installed secondary water barrier. At the joints between the boards, the barriers are taped to each other (top over bottom barrier)
	A plastic rain barrier is installed on top of the insulation boards

(continued)

	Finished roof with PV panels installed. The PV panels also act as mechanical fastening for the PUR boards and the rain barrier
PUR insulation on a sloped roof	
Pictures: Puren (www.puren.de) ((more pictures?))	

Combination of Flexible Insulation (Batts) between the Rafters and Pressure-Resistant Insulation (Boards) Above the Rafters

For retrofit a combination of the both options of insulation between and above the rafters very often is used. In those situations you remove the roof covering and the substructure from the outside. In many cases you find an existing pitched roof construction – maybe with or without a thin layer of old mineral wool insulation between the rafters. In those cases you have to check the old insulation. If it is still dry and nearly clean, you should leave it and add an additional mineral wool layer to fill the space between the rafters completely. If the existing insulation is wet or

Table C.2 Combination of insulation layers between and above the rafters, vapor retarder flat above the rafters

Ebene Verlegung

U-Werte [W/(m²K)]

Bauteil [schematische Darstellung] Hinweise	ISOVER Dämmstoffdicke [mm]
Ergänzende Aufsparrendämmung mit Zwischensparrendämmung und ebener Verlegung der Vario KM Supraplex	

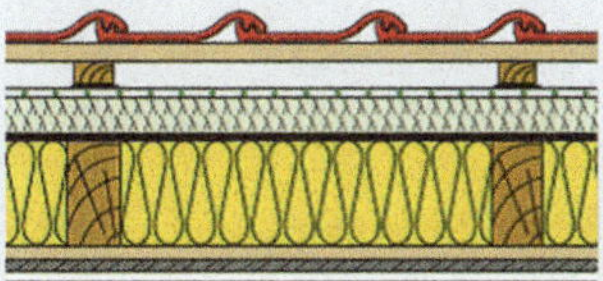

Zwischensparrendämmung als Sparrenvolldämmung bei 14,3% Holzanteil (Sparrenabstand 60 cm i. L.); rauminnenseitige Bekleidung mit 30 mm TRaglattung und 12,5 mm Gipsplatte

Integra AP PIR		Integra ZSF-032 100	120	140	160	180	200
100	vlies	0,15	0,14	0,13	×	×	×
	alu	0,14	0,13	0,12	×	×	×
120	vlies	0,14	0,13	0,12	×	×	×
	alu	0,13	0,12	0,11	×	×	×
140	vlies	0,12	0,12	0,11	0,10	0,10	×
	alu	0,11	0,11	0,10	0,10	0,09	×
160	alu	0,10	0,10	0,09	0,09	0,09	0,08
180	alu	0,09	0,09	0,09	0,08	0,08	0,08
200	alu	0,09	0,08	0,08	0,08	0,07	0,07

Integra AP PIR		Integra ZKF-035 100	120	140	160	180	200
100	vlies	0,16	0,15	0,14	×	×	×
	alu	0,14	0,14	0,13	×	×	×
120	vlies	0,14	0,13	0,12	×	×	×
	alu	0,13	0,12	0,11	×	×	×
140	vlies	0,13	0,12	0,11	0,11	0,10	×
	alu	0,12	0,11	0,10	0,10	0,09	×
160	alu	0,10	0,10	0,10	0,09	0,09	0,08
180	alu	0,10	0,09	0,09	0,08	0,08	0,08
200	alu	0,09	0,08	0,08	0,08	0,08	0,07

× = Diese Konstruktionen sind feuchtephysikalisch nicht möglich

damaged, you have to remove it and renew the complete insulation. To fulfill the requirements for airtightness, you have to install a vapor retarder from outside. This is an expert work. See the step-by-step description that follows. After installing the temporary rain shield layer, the additional pressure-resistant insulation layer can be installed above the rafters. This insulation layer can be mineral wool (rigid stone wool boards), PIR foam boards, or wood fiber boards. An expert construction engineer should calculate the appropriate combination of the different insulation layers and the position of the vapor retarder to best match the physics aspects of the building. Therefore you should contact the producers of those insulation systems during your planning period. Tables C.2 and C.3 summarize the capabilities of these insulation systems.

Table C.3 Combination of insulation layers between and above the rafters, vapor retarder around rafters

Schlaufenförmige Verlegung U-Werte [W/(m²K)]							
Bauteil [schematische Darstellung] Hinweise	ISOVER Dämmstoffdicke [mm]						
Ergänzende Aufsparrendämmung mit Zwischensparrendämmung und schlaufenförmiger Verlegung der Klimamembran Vario KM Duplex UV	Integra AP PIR vlies	Integra ZSF-032					
		100	120	140	160	180	200
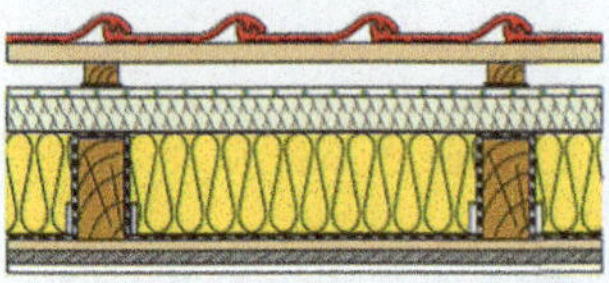	60	0,21	0,19	0,17	0,16	0,15	0,14
	80	0,18	0,16	0,15	0,14	0,13	0,12
	100	0,15	0,14	0,13	0,13	0,12	0,11
	120	0,14	0,13	0,12	0,11	0,11	0,10
	140	0,12	0,12	0,11	0,10	0,10	0,09
Zwischensparrendämmung als Sparrenvolldämmung bei 14,3 % Holzanteil (Sparrenabstand 60 cm i. L.); rauminnenseitige Bekleidung mit 30 mm HWL und 15 mm Kalkgipsputz		Integra ZKF-035					
		100	120	140	160	180	200
	60	0,21	0,19	0,18	0,16	0,15	0,14
	80	0,18	0,17	0,15	0,14	0,14	0,13
	100	0,16	0,15	0,14	0,13	0,12	0,12
	120	0,14	0,13	0,12	0,12	0,11	0,11
	140	0,13	0,12	0,11	0,11	0,10	0,10

Example for Flexible Insulation (Batts) Between the Rafters and XPS Boards Above the Rafters

Step-by-Step Description

	1. Remove existing substructure and damaged materials
	2a. Put flexible insulation (mineral wool) between the rafters; be sure to completely fill the empty space

(continued)

	3a. Install a vapor barrier and temporary rain shield layer above the rafters
	4a. Fix the vapor retarder on the rafters
	5a. Tape the parts of the vapor retarder
	6a. Install the insulation layer of PIR, wood fiber, or mineral wool boards above the rafters
	2b. Put a thin insulation layer on the existing inside lining. This protects the following vapor retarder from damaging by nails or screws

(continued)

	3b. Install the vapor retarder by passing it around the rafters and staple/tacker it without air gaps to the rafter
	4b. Fix the vapor retarder close into the corner of the rafters by suing small battens
	5b. Bring in the insulation between the rafters
	6b. Install the insulation layer of PIR, wood fiber, or mineral wool boards above the rafters
	7. Screw the substructure for the new roof covering
	8. Now the new roof covering can be installed

Source: All pictures from www.isover.de

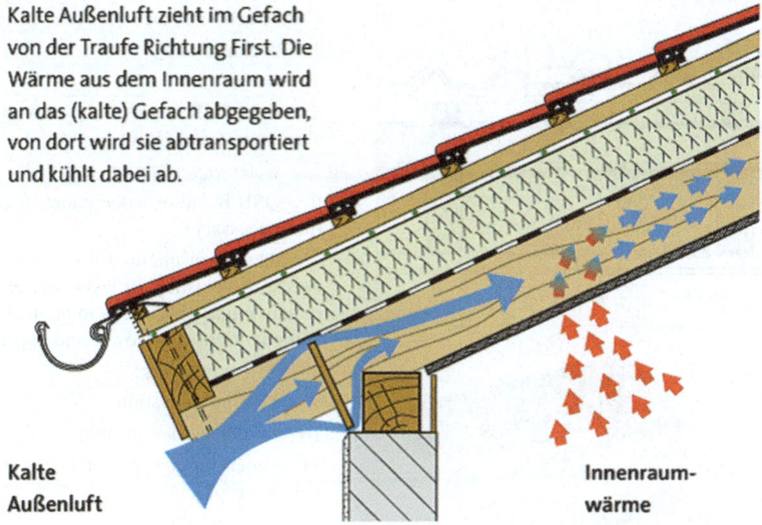

Note: in retrofit, to prevent cold air to get between the new insulation layer above the rafters and the inside lining, it is recommended to not leave parts or all of the former insulation layer empty (i.e., any space between the inside lining, the rafters and the over-rafter insulation, where the blue dots are in the above drawing), but rather leave in the old insulation and fill in mineral wool batts on top until the between-rafter space is filled. This insulation layer in addition effects good sound protection.

Example: Sloped Roof with (Old and/or New) Flexible Insulation Between the Rafters and PUR Over-Rafter Insulation

Numerical analysis shows that in Continental-Central Europe climate, with 80 mm or more of PUR (without or with aluminum lining) rigid over-rafter insulation, no moisture buildup occurs within the roof, regardless of the thickness and quality of the (old + new) between-rafter insulation and regardless whether the existing inner water vapor barrier is functioning or not. Thinner PUR over-rafter boards or more severe locations (e.g., high above sea level) should be assessed individually before installation. As an example, the following sketches and table give moisture damage-free combinations of mineral wool insulation between the rafters and PUR boards above the rafters for Germany, as well as the achievable roof U-values.

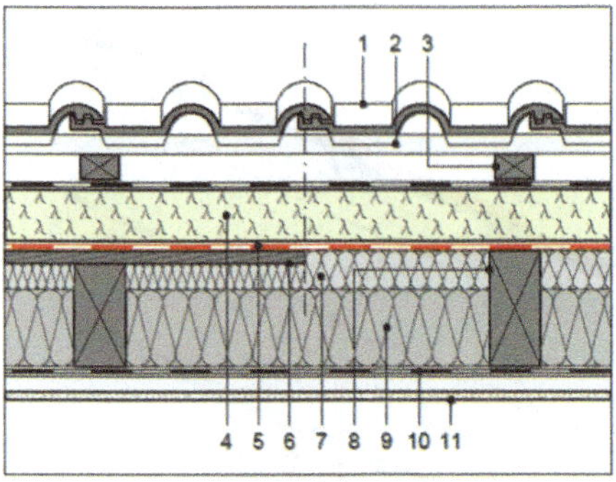

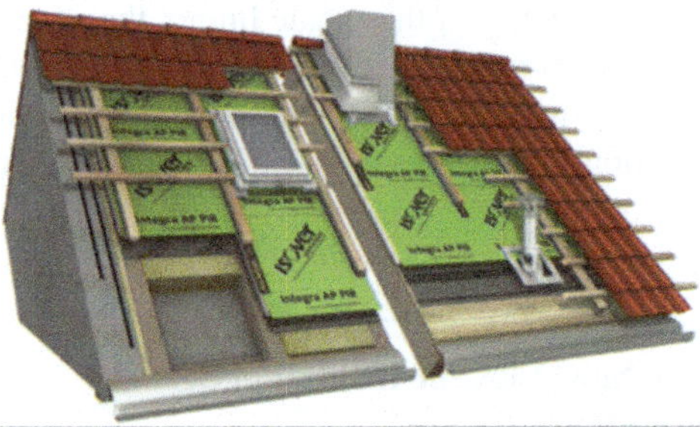

Cross-section of the roof after retrofit
1 = decking (new)
2, 3 =
4 = rigid PUR boards (with or without aluminum foils) (new)
5 = water vapor barrier (new)
6 = OSB board or wood panels (existing (not necessary)
7 = thermal insulation fill-up (new) (not necessary; existing air layer may also remain empty, but needs to be made air- and windtight at its lower and upper end)
8 = rafters (existing)
9 = existing insulation
10 = vapor barrier, airtight
11 = inner lining (gypsum board or wooden profiles)

Steildachsanierung mit PU-Aufsparrendämmung diffusionsfähig, mit Vollsparrendämmung, ohne oder mit raumseitiger Dampfbremse bis s_d = 10 m					
Dicke des Sparrens	Zwischen sparren-Dämmung neu (035) vorh (040) Dicke in mm	U-Wert in W/(m²·K)			
		≤ 0,10	≤ 0,14	≤ 0,16	≤ 0,18
		PU-Aufsparrendämmung, WLS 026-029, Dicke in mm (in Klammern: U-Wert in W/(m²K))			
20 cm	60 140	140 (0,10)	100 (0,12)	80 (0,14)	80 (0,14)
18 cm	40 140	160 (0,10)	100 (0,13)	80 (0,15)	80 (0,15)
16 cm	40 120	180 (0,10)	120 (013)	100 (0,14)	80 (0,16)
14 cm	20 120	180 (0,10)	120 (0,13)	100 (0,15)	80 (0,17)
14 cm	60 80	180 (0,10)	120 (0,13)	100 (0,15)	80 (0,17)
12 cm	— 120	180* (0,11)	120 (0,14)	100 (0,16)	80 (0,18)
12 cm	40 80	180 (0,10)	120 (0,14)	100 (0,16)	80 (0,18)

Tabelle 16: Zusammenfassung der Ergebnisse für Steildachsanierung mit PU-Aufsparrendämmung diffusionsfähig, mit Vollsparrendämmung, ohne oder mit raumseitiger Dampfbremse bis s_d = 10. Tendenziell ist die hygrothermische Situation um so besser, je dicker die PU-Aufsparrendämmung ist; dickere PU-Aufsparrendämmungen als angegeben dürfen deshalb immer verwendet werden.

* Der U-Wert der Spalte wird bei dieser PU-Dicke knapp nicht eingehalten, kann aber durch die Berücksichtigung zusätzlicher Schichten im U-Wert (z.B. eine 24 mm Luftschicht zwischen der Lattung unter der raumseitigen Bekleidung, und/oder eine dickere raumseitige Bekleidung) leicht eingehalten werden.

This section shows examples of drawings showing installation of wall insulation, with an emphasis on reducing or preventing thermal bridging at the joints. The drawings may serve as a showcase on how in principle the insulation layer is installed gap-free at the joints to prevent thermal bridges. It is based on materials provided by Rockwool®. Examples include insulation installation details of:

- Wall foundation
- Wall-to-roof connection
- Wall-to-window connection
- Wall to balcony and terrace connection
- Convex and concave wall corners
- Dilatation of the flat wall with vertical expansion joint profile
- Dilatation of the corner wall
- Arcades
- Roof parapets

Insulation of the Foundation

With a Base Profile

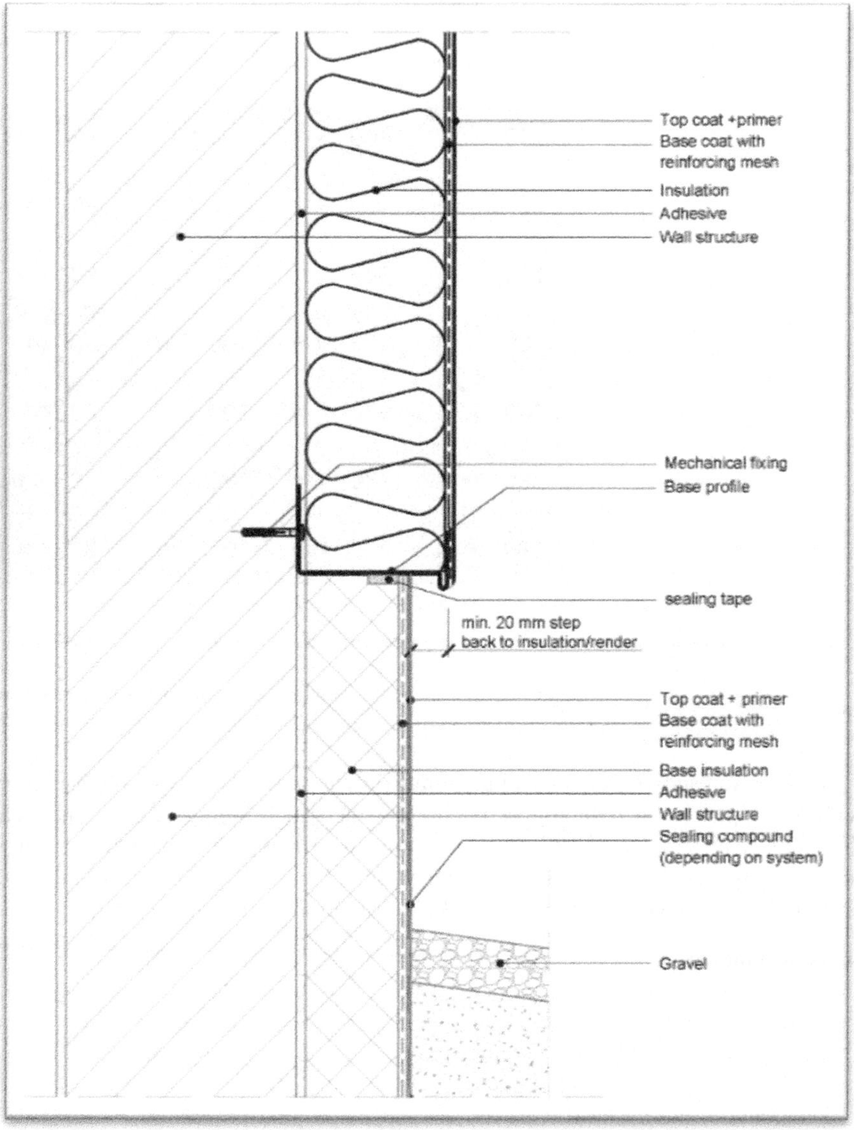

Without a Base Profile

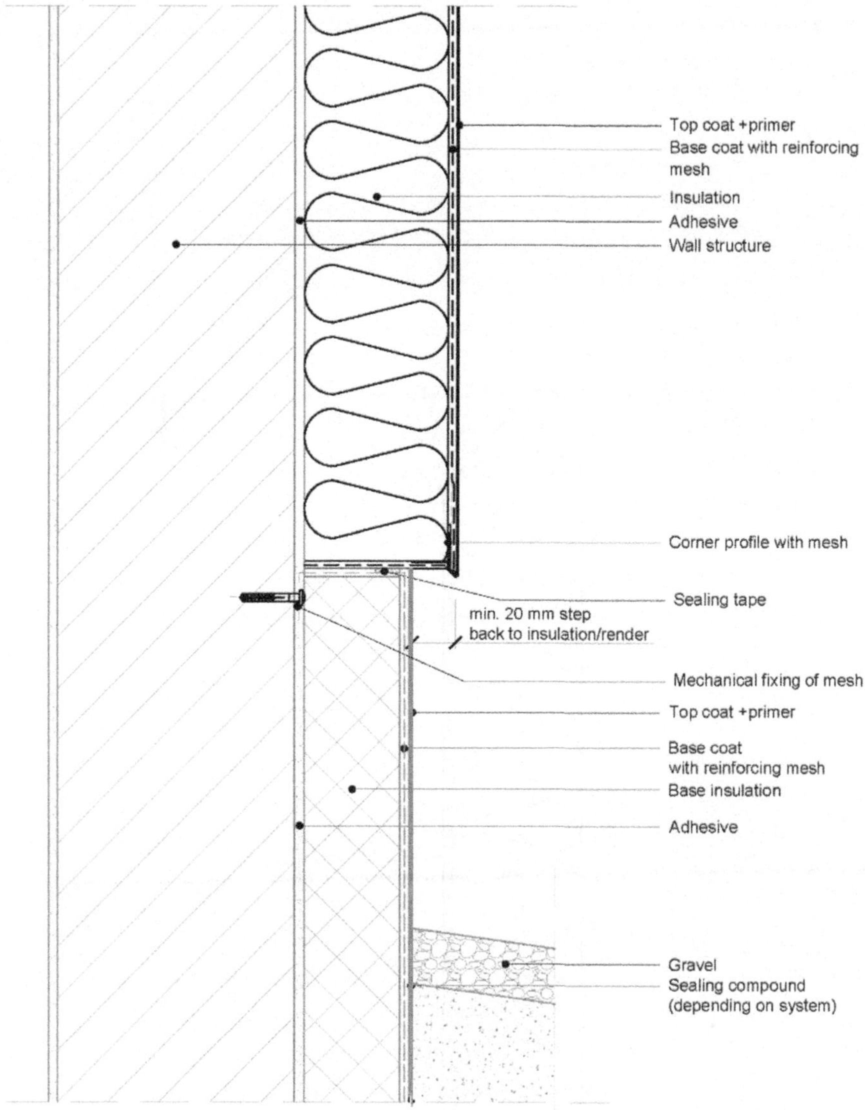

Wall Connection to Warm Roof

Without Soffits

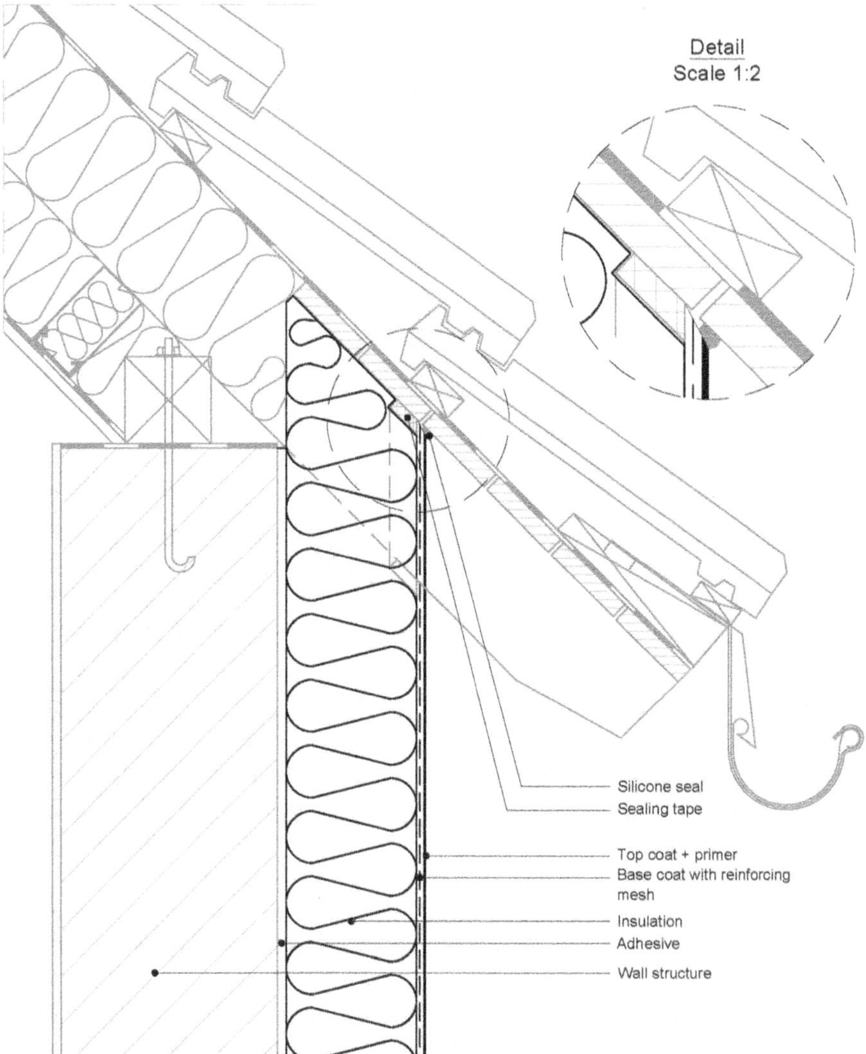

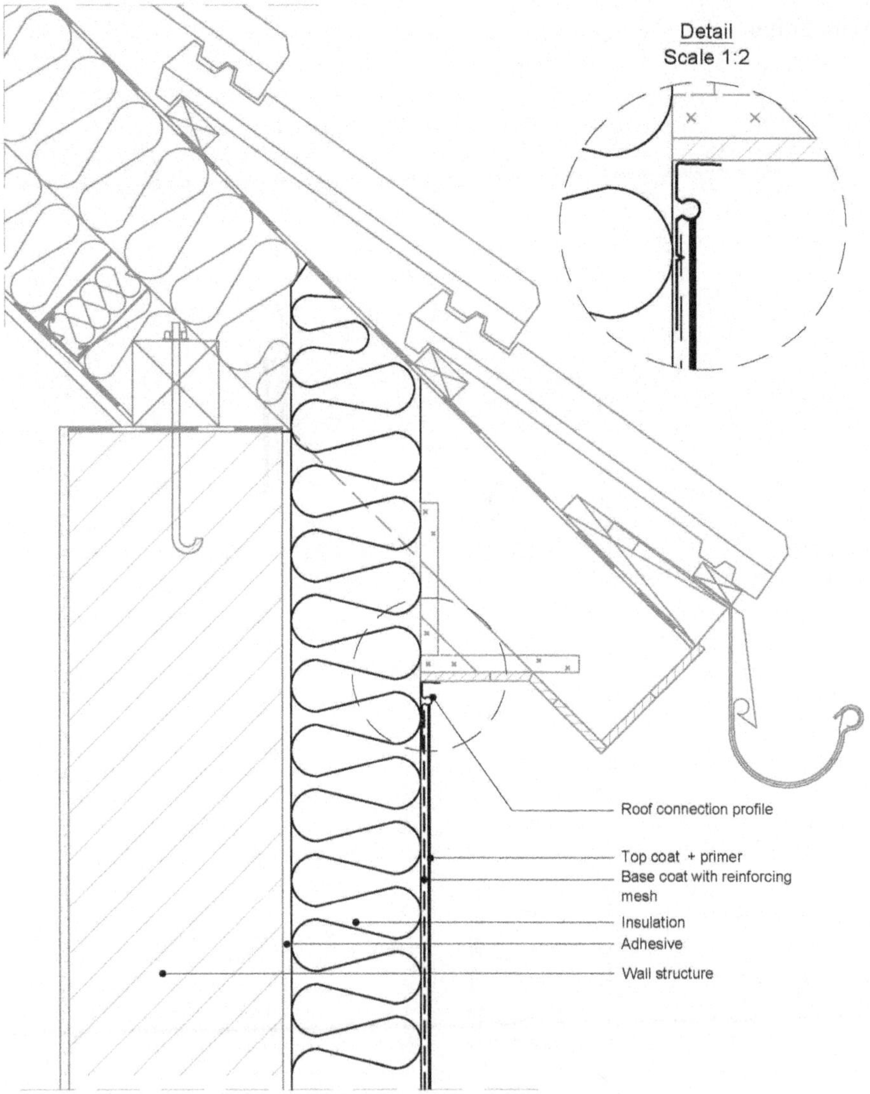
Detail
Scale 1:2
Roof connection profile
Top coat + primer
Base coat with reinforcing mesh
Insulation
Adhesive
Wall structure

With Soffits

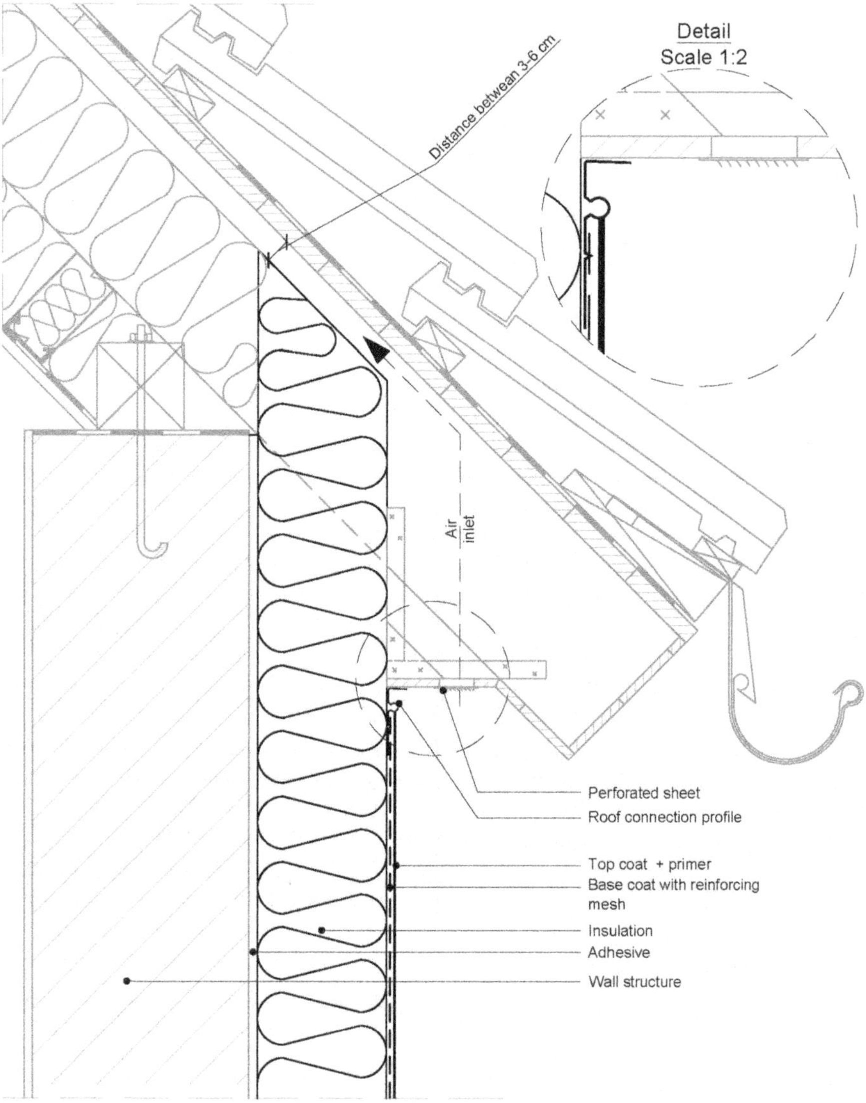

Back-Ventilated Roof Without Soffits

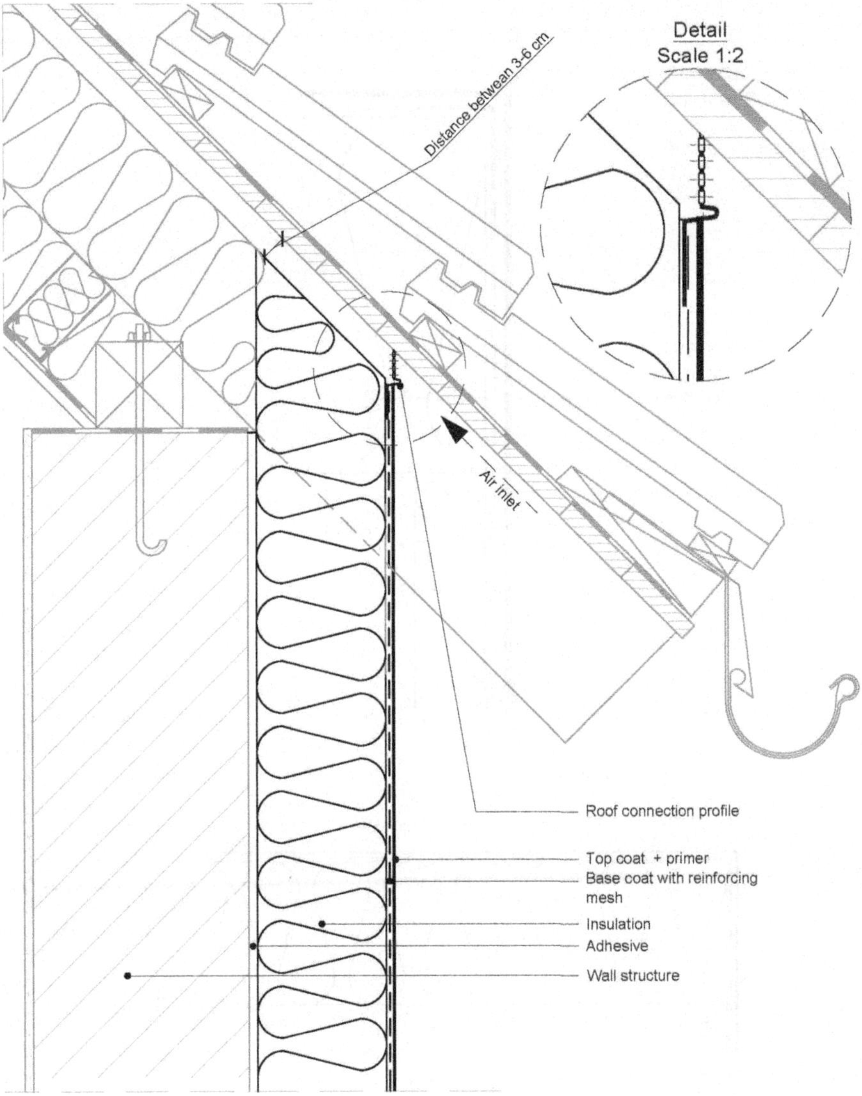

Insulation of the Window

Mounted in the Middle of the Wall, Section

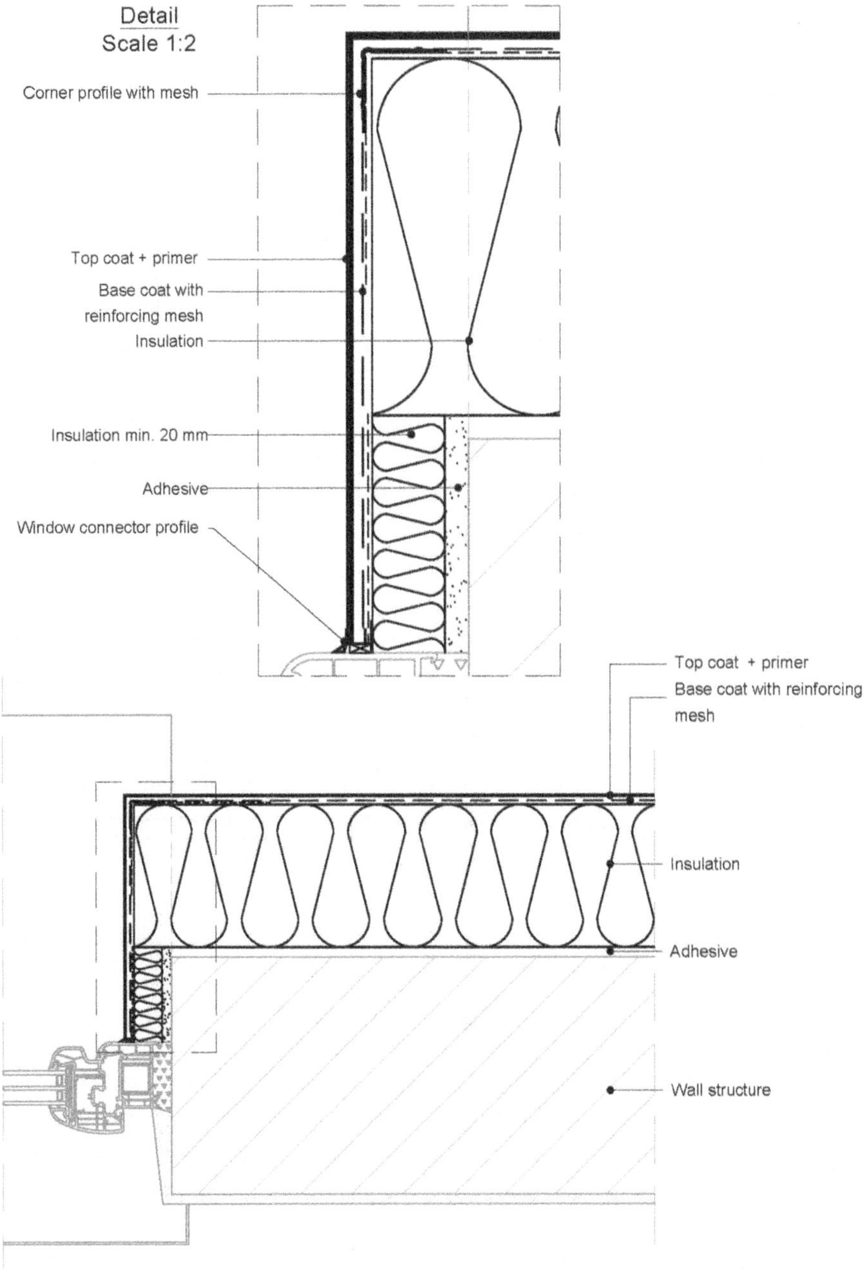

Mounted Equal to the Wall, Section

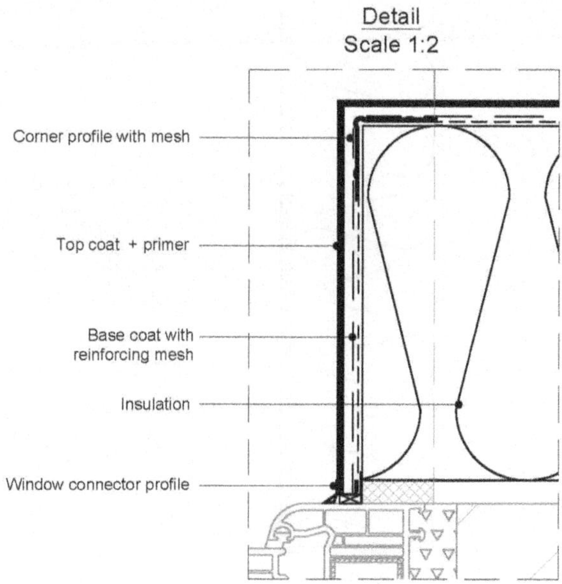

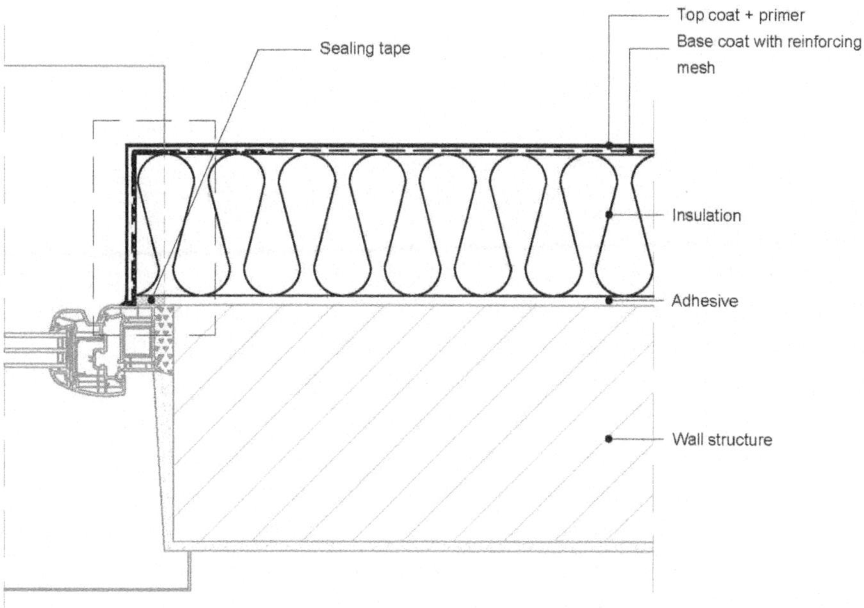

Mounted Equal to the Wall, Cross-Section

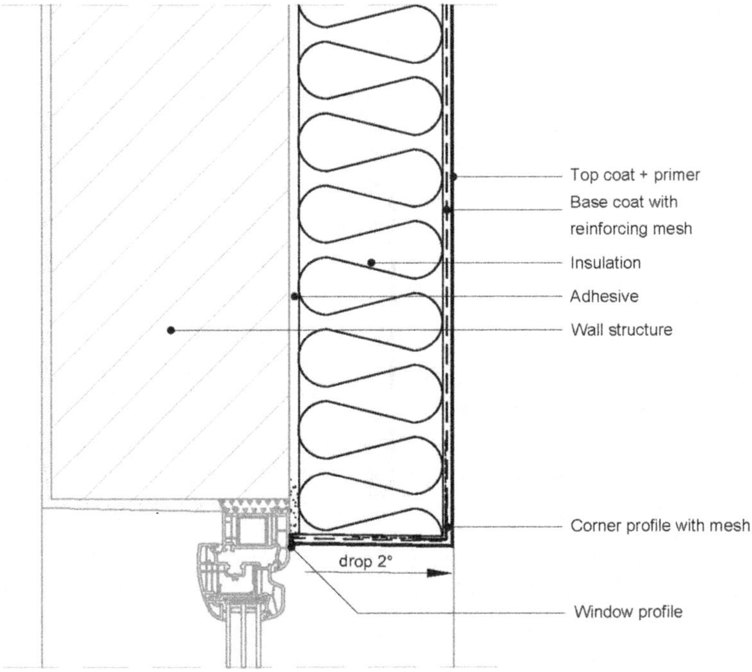

Mounted in the Insulation Layer, Section

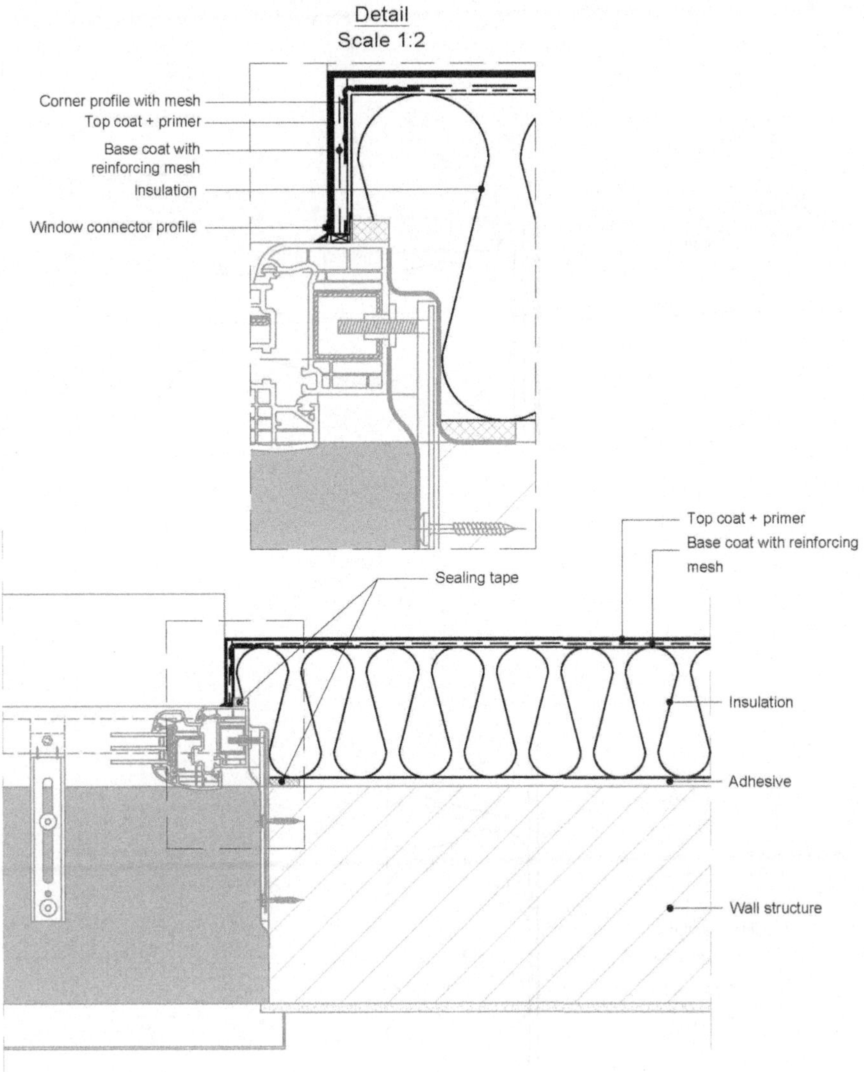

Mounted in the Insulation Layer, Insulation of the Windowsill, Cross-Section

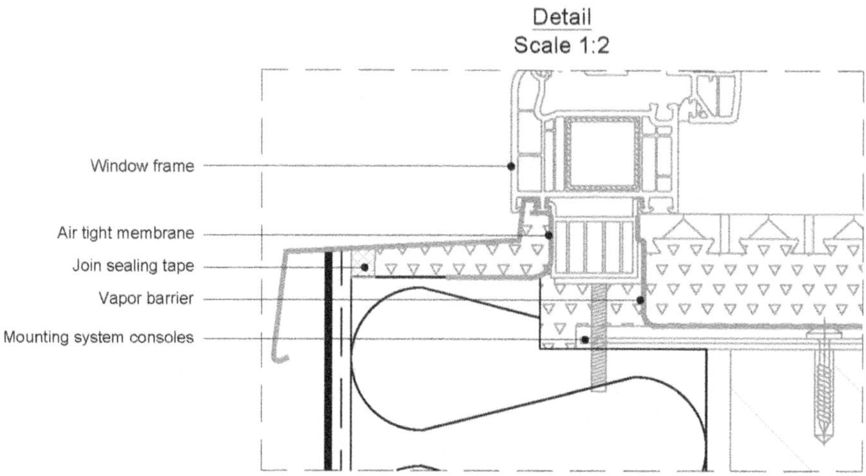

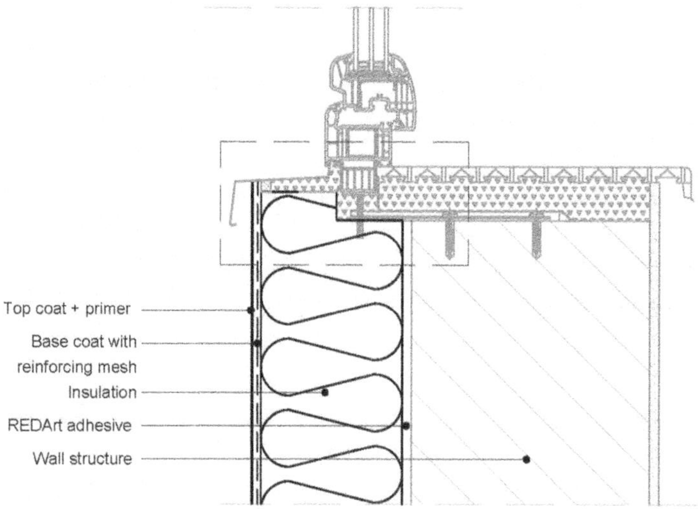

Mounted in the Middle of the Wall, Section

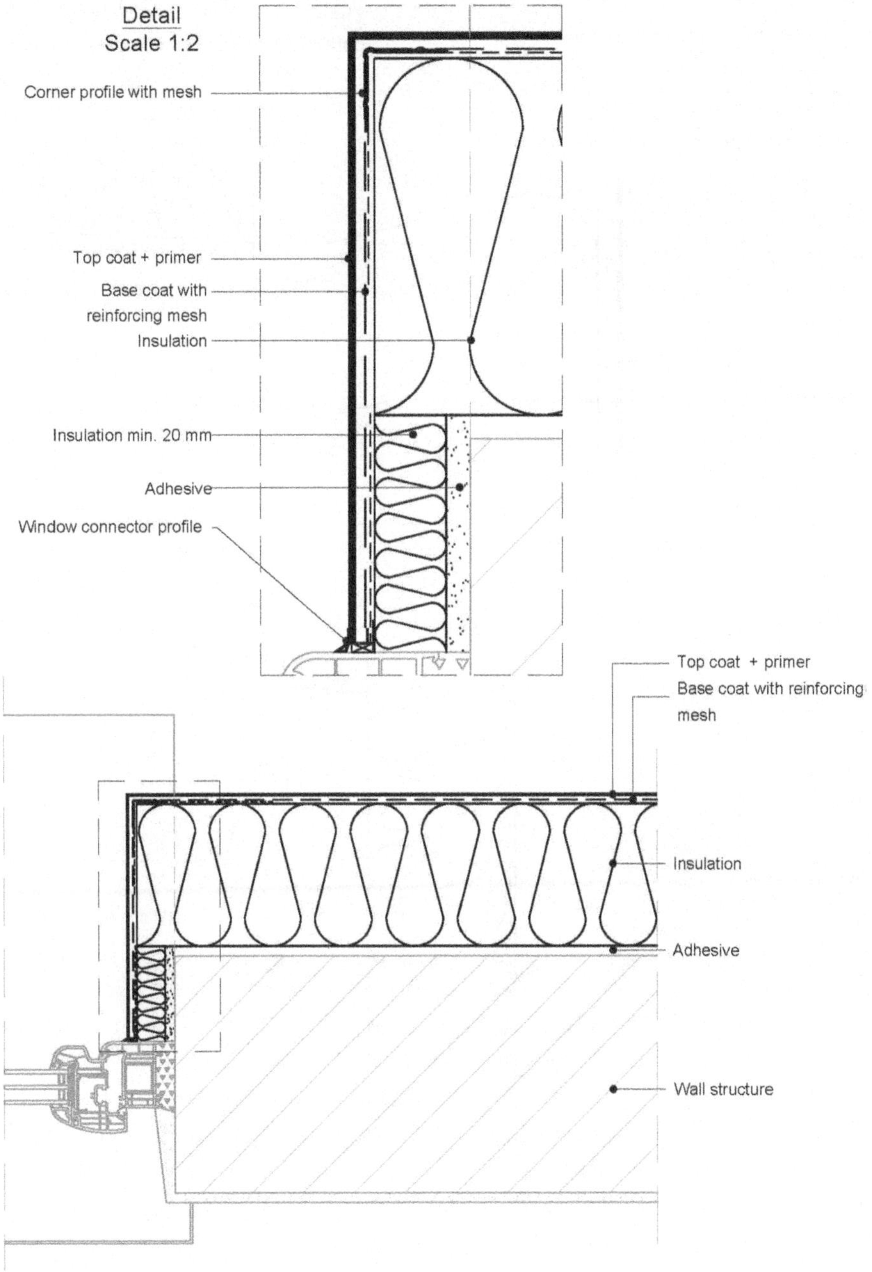

Mounted in the Middle of the Wall, Cross-Section

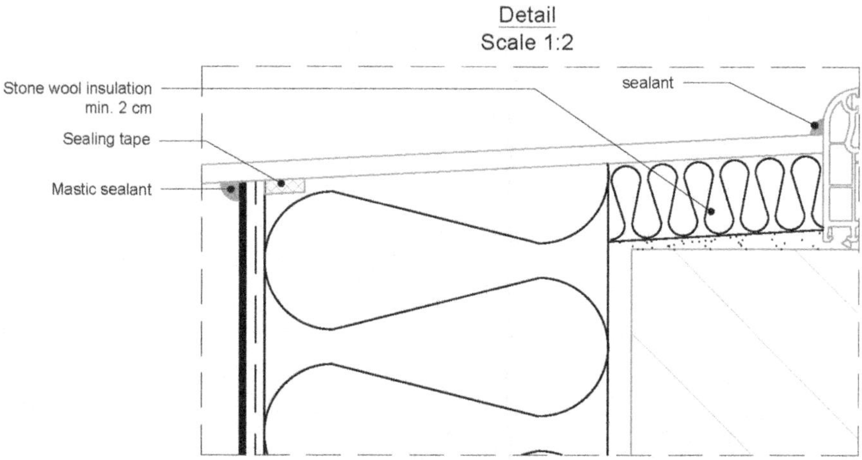

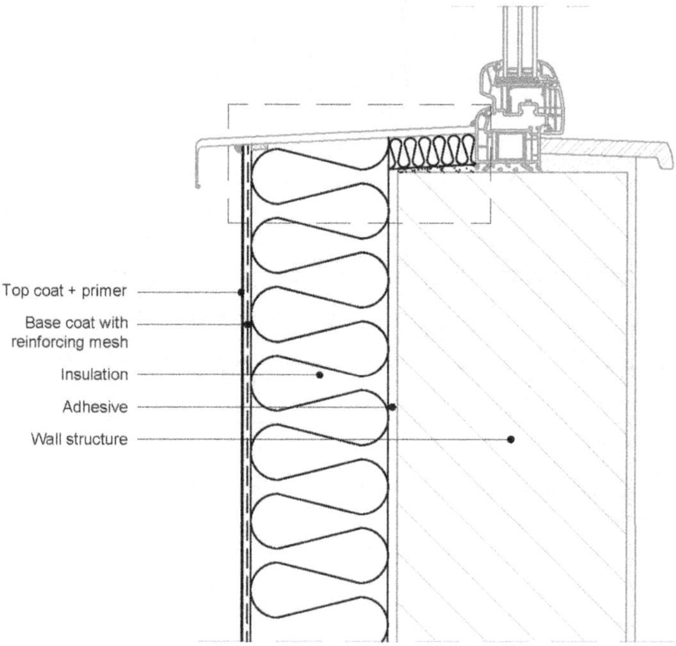

Insulation of the Shutters

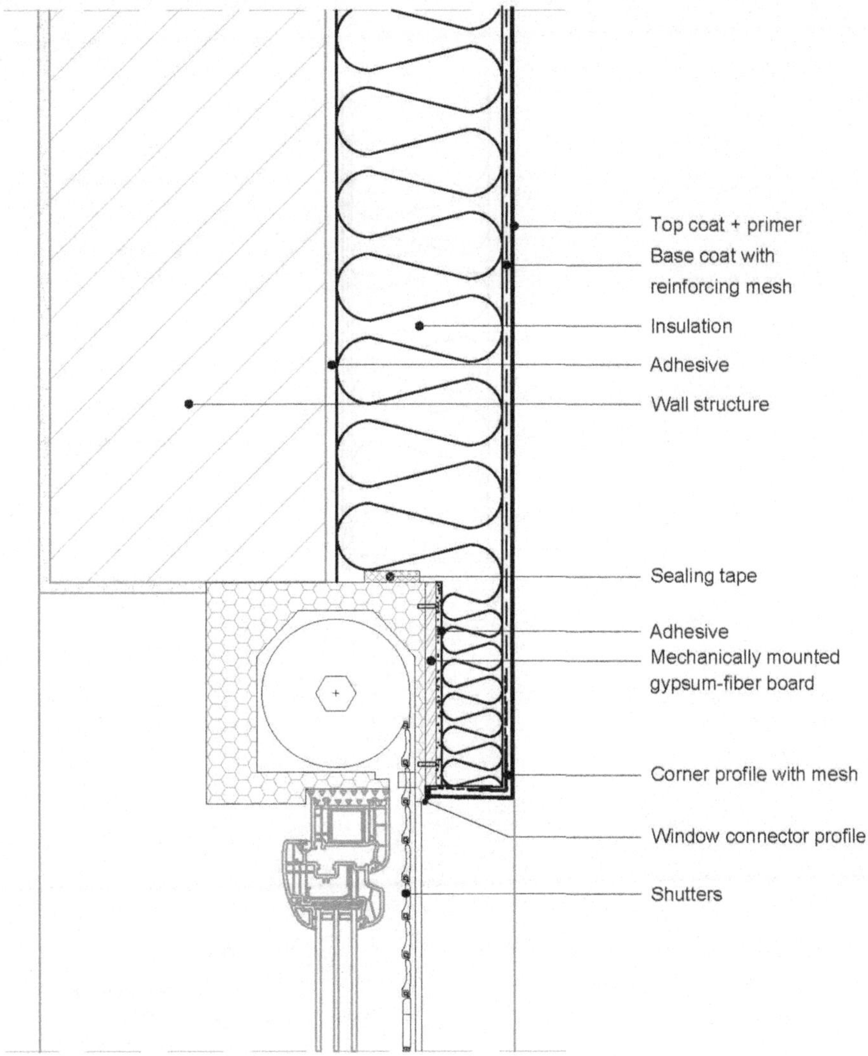

Insulation of the Balcony and Terrace

Detail without Base Profile

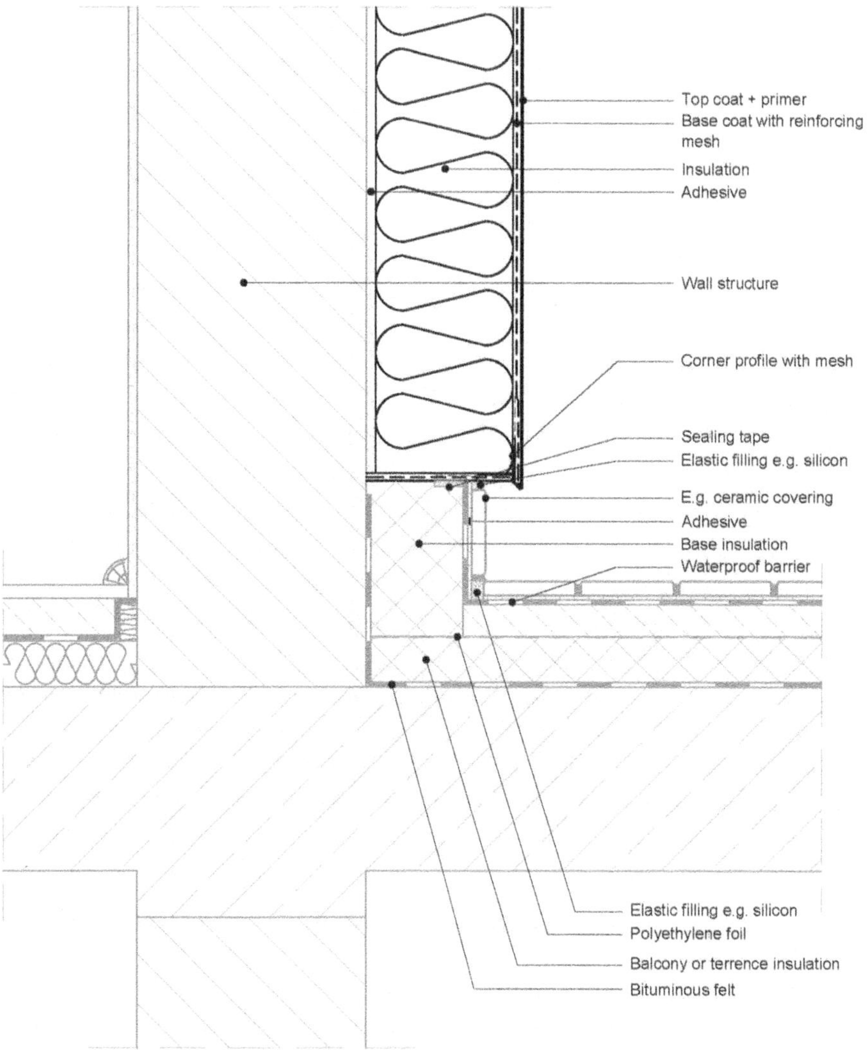

Detail with Base Profile

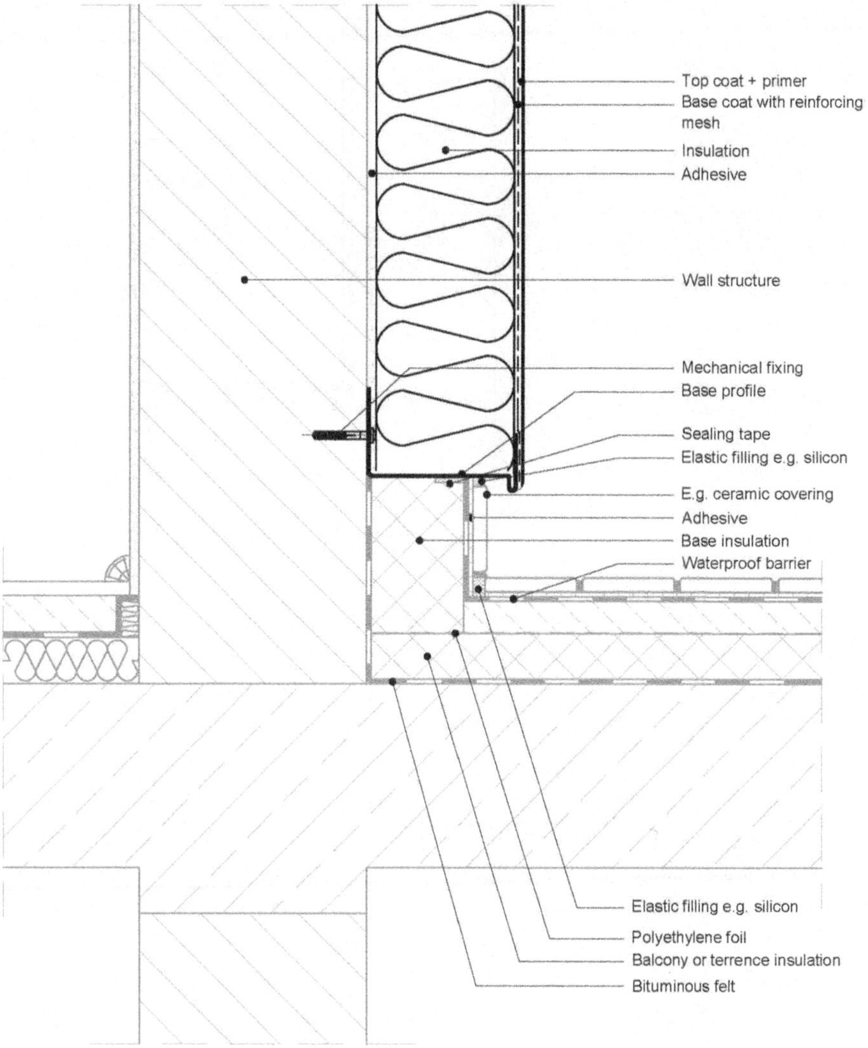

Insulation of the Corner

Convex

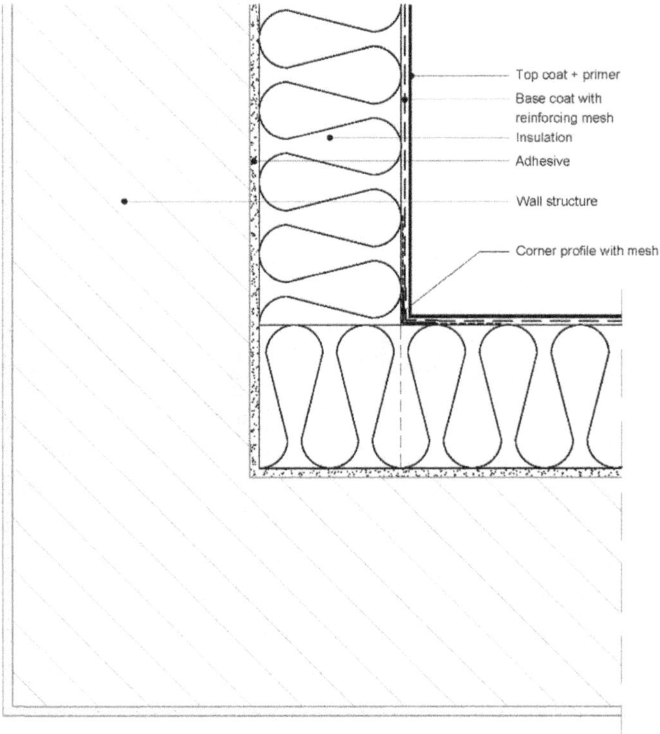

Concave

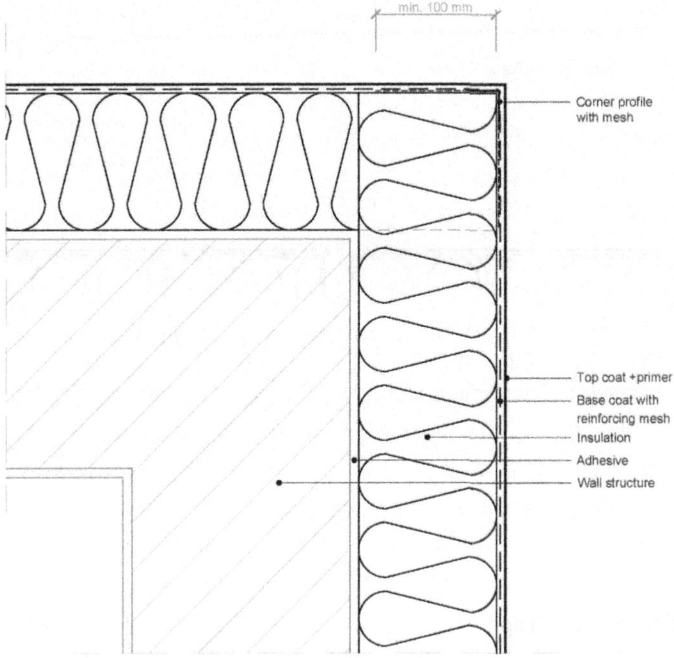
min. 100 mm
Corner profile
with mesh
Top coat +primer
Base coat with
reinforcing mesh
Insulation
Adhesive
Wall structure

Dilatation of the Flat Wall

With Vertical Expansion Joint Profile

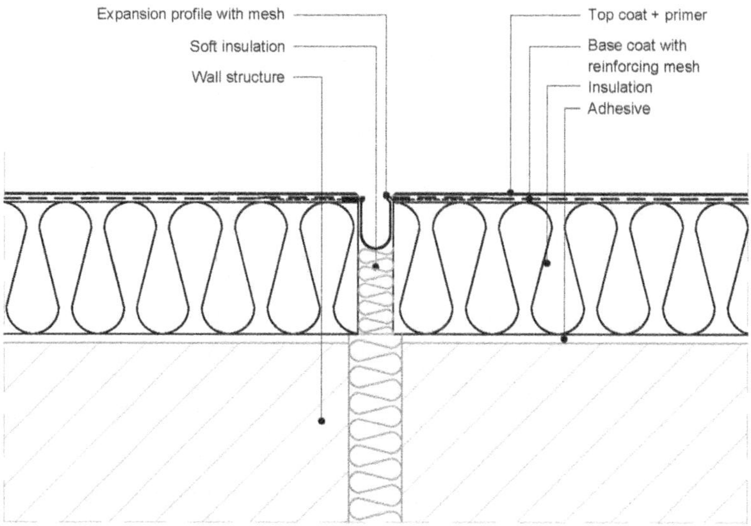

With Vertical Sealing Tape

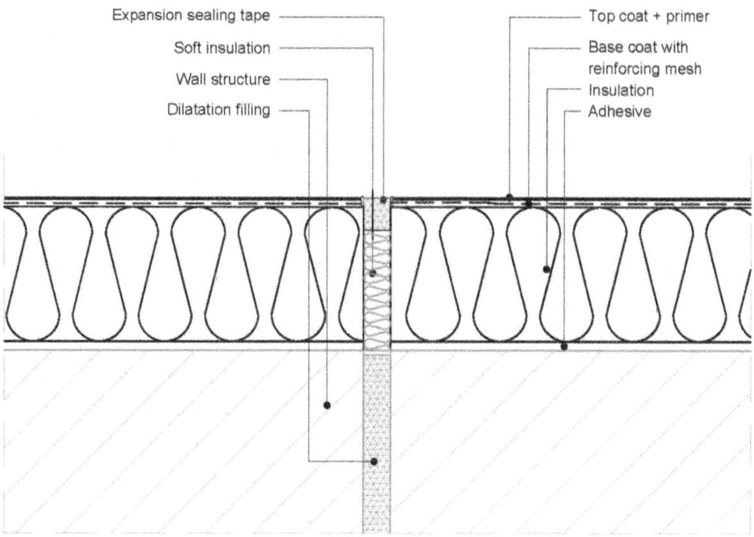

Dilatation of the Corner Wall

With Vertical Expansion Joint Profile

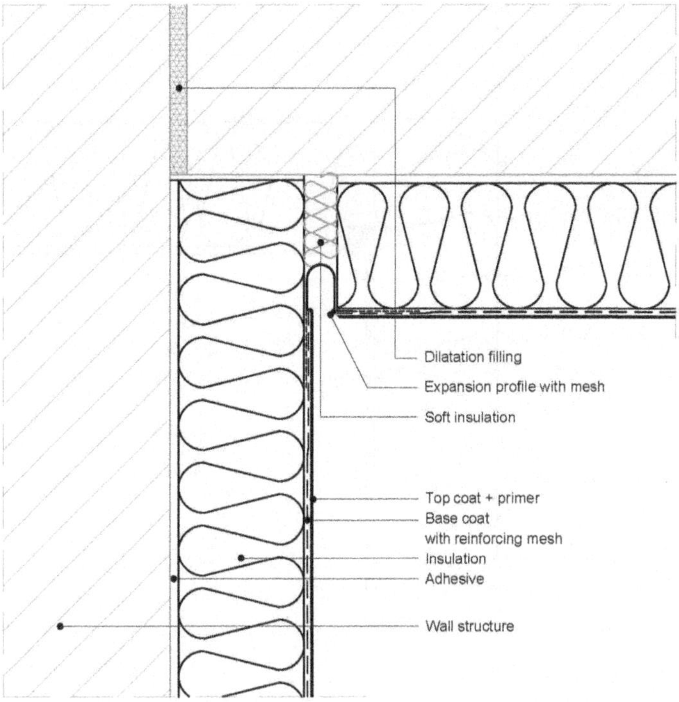

With Vertical Sealing Tape

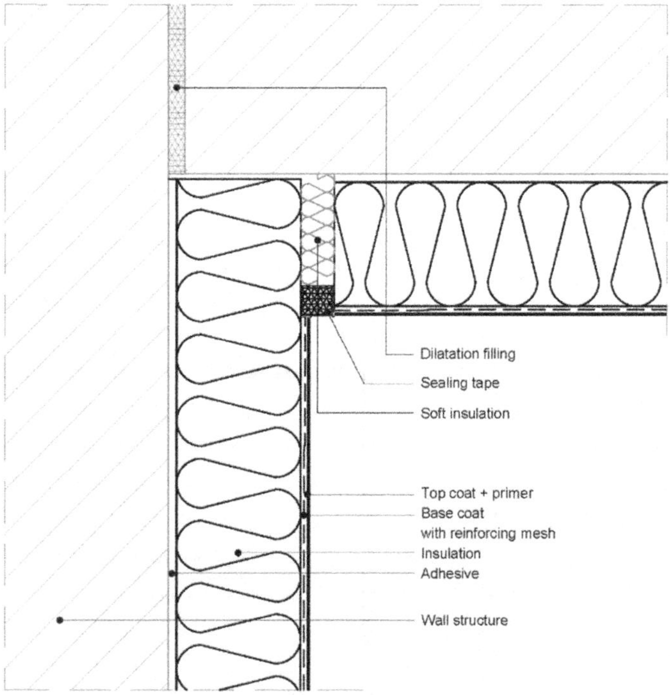

Arcade

With Corner Profile

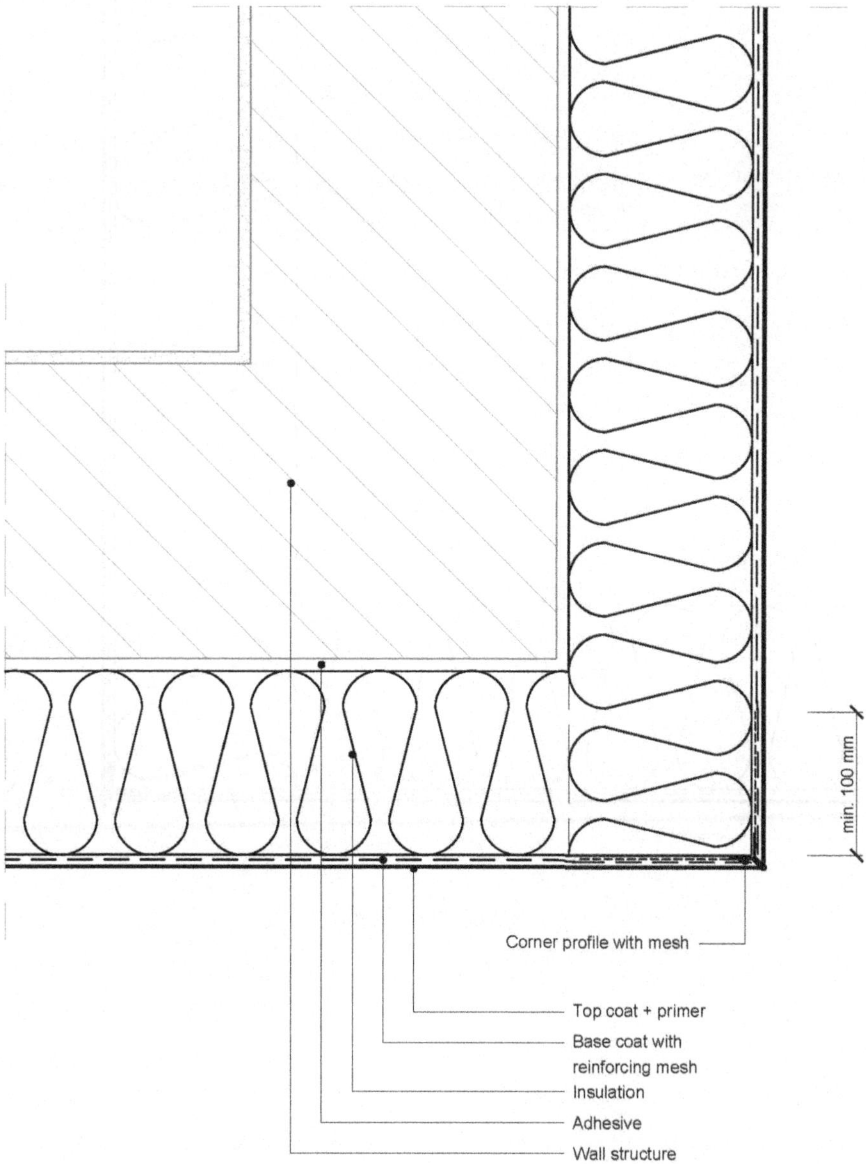

With Corner Profile (Drip Nose)

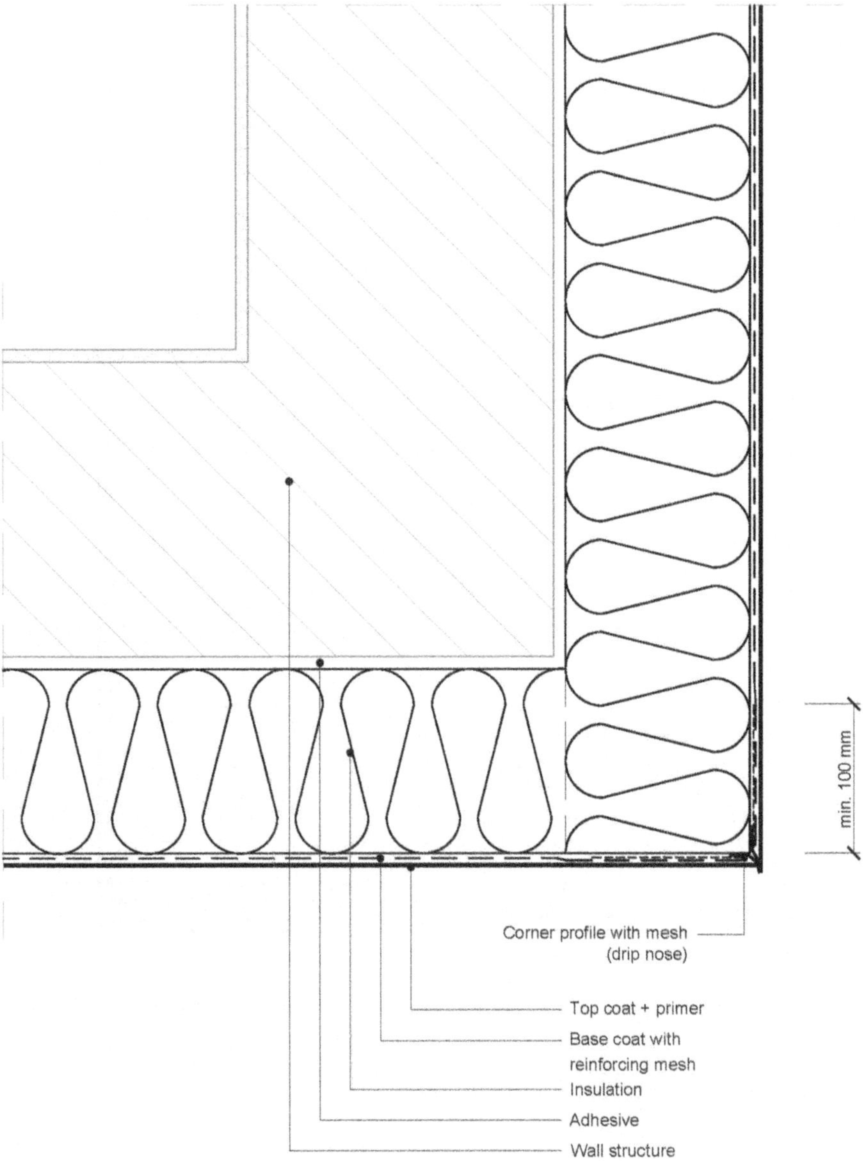

Formation of Drip Edge

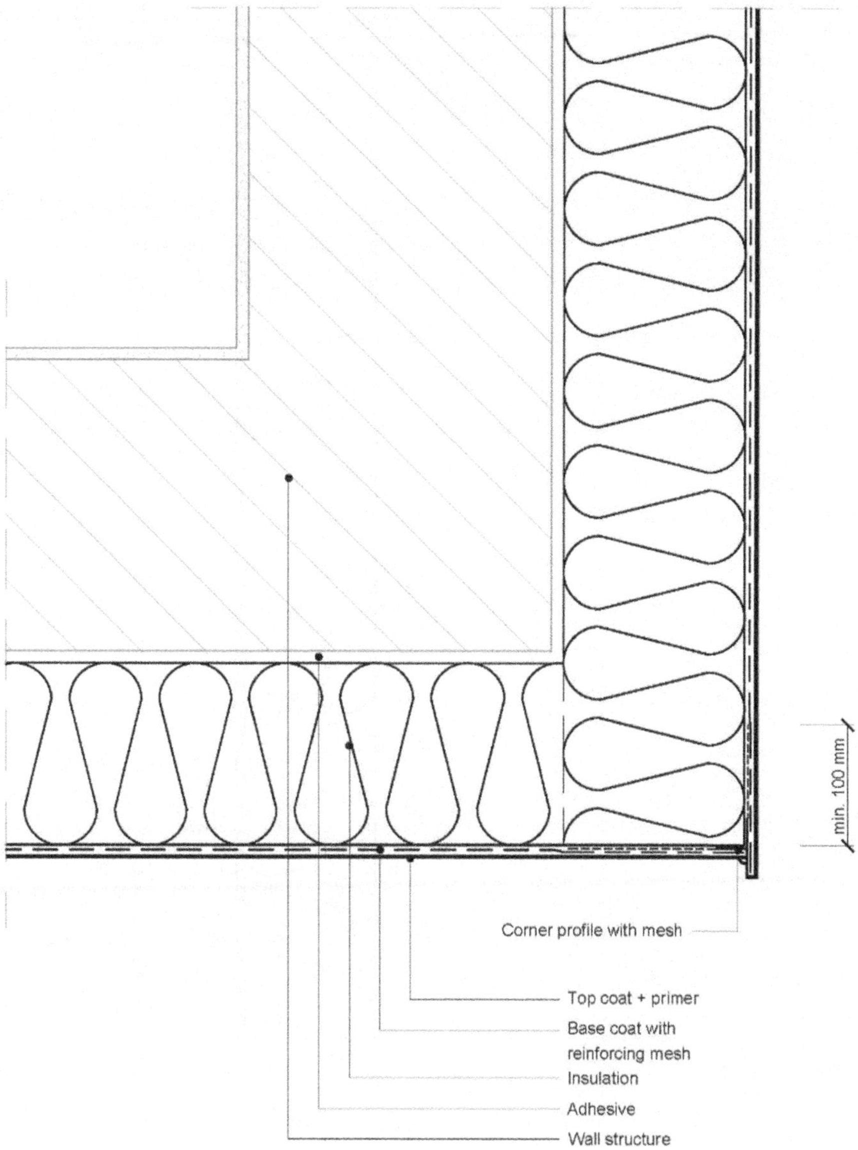

Insulation of the Roof Parapet

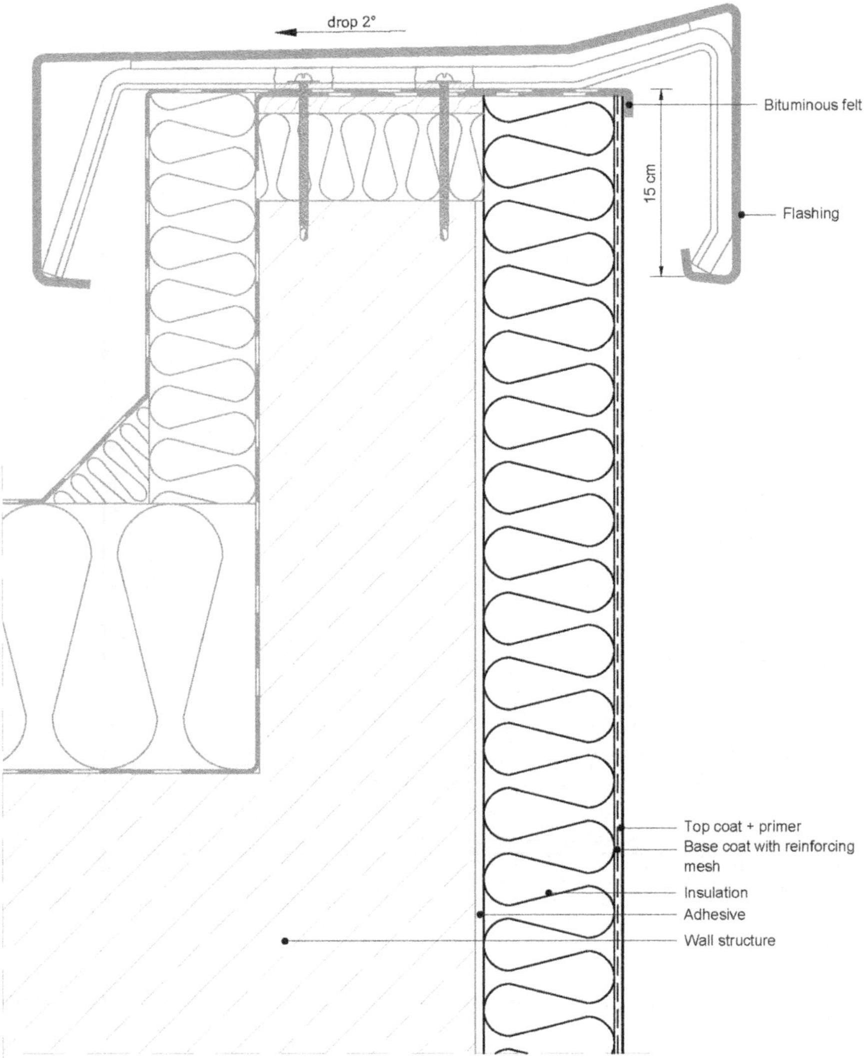

Appendix D: Remediation of Thermal Bridges

This appendix contains examples of thermal bridges in existing buildings and possible solutions for their treatment during major renovation projects. It includes sketches of commonplace thermal bridges for masonry and steel stud buildings with interior and exterior insulation scenarios. Examples show a step-by-step remediation process, and list psi values for pre- and post-renovation situations. The purpose of this Appendix is to provide information that informs designers on the effects of thermal bridges, that enables them to reference their project details to details presented in this document, and that clarifies how to execute those details. The content of the Appendix C is based on studies conducted by the US Army ERDC team and its contractors (Sect. "Sequencing of thermal bridge mitigation") (Pagan-Vazquez et al. 2015a, b; Lawton Mark et al. 2014). It also contains examples of building envelope renovation with some typical details reproduced with permission from "Details for Passive Houses: Renovation: A Catalogue of Ecologically Rated Constructions for Renovation" (Sect. "Examples of thermal bridges mitigation with the building envelope renovation") (IBO 2016). Section "Construction details" contains typical architectural details showing thermal bridges mitigation. Note that the content of Appendix E, "Window Installation Guidance," is relevant to this Appendix content, which addresses the treatment of thermal bridges with window replacement.

Sequencing of Thermal Bridge Mitigation

Section "Sequencing of thermal bridge mitigation" includes sequencing of thermal bridge mitigation for the following details:

D.1.1 Parapet with Concrete Roof Joint with a Wall Insulated on Exterior
D.1.2 Parapet with Concrete Roof Junction with a Wall Insulated on Interior
D.1.3 Parapet with OWSJ and a Deck (Exterior Wall Insulation)
D.1.4 Parapet with OWSJ and a Deck (Interior Wall Insulation)
D.1.5 Parapet with OWSJ and a Deck (Steel Stud Buildings)
D.1.6 At Grade Transition: Exterior Insulated Buildings
D.1.7 At Grade Transition: Interior Insulated Buildings
D.1.8 Through Slab Projections with Interior Insulated Walls: Internal Walls and Internal Floors
D.1.9 Flanking
D.1.10 Blast Resistant Door Jamb

Parapet with Concrete Roof Joint with a Wall Insulated on Exterior

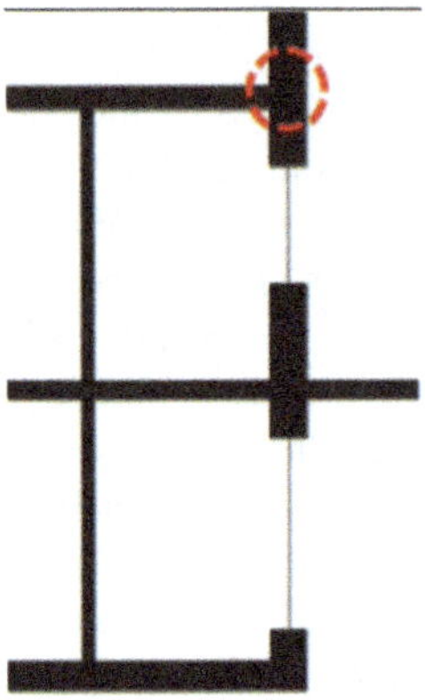	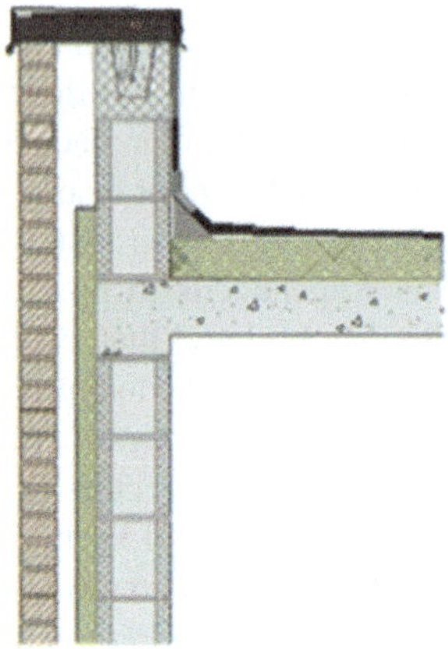
Remediation options	

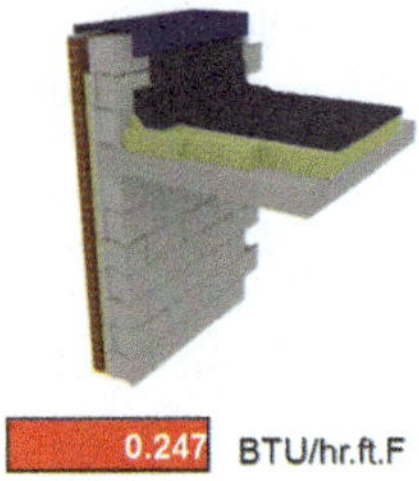	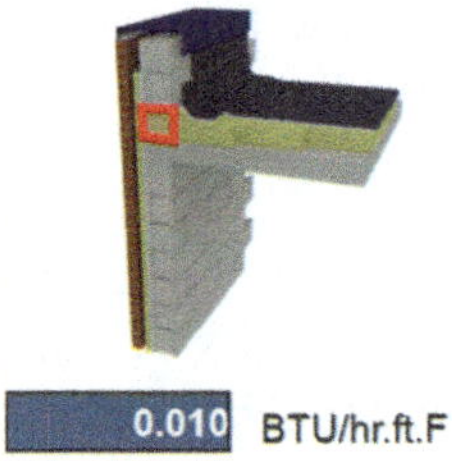	
Before renovation (example for CMU wall	Option 1: insert thermal break	Option 2: Wrap the parapet

Solution for Option 1: Create Continuity of Wall and Roof Insulation

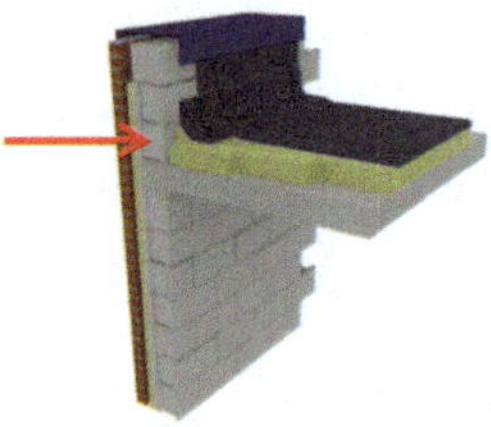

(continued)

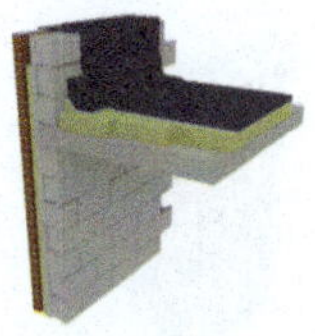		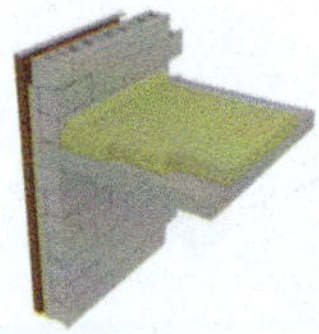	
Remove the parapet capping	Remove flashing and roofing asphalt to expose the CMU parapet	Remove roofing asphalt sufficiently to expose the insulation	Cut back and remove some of the roof insulation. The CMU parapet is now fully exposed

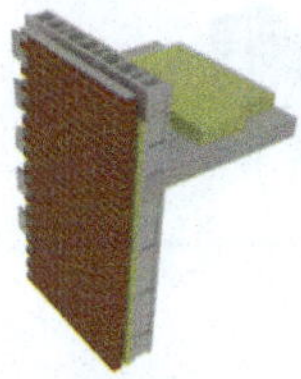	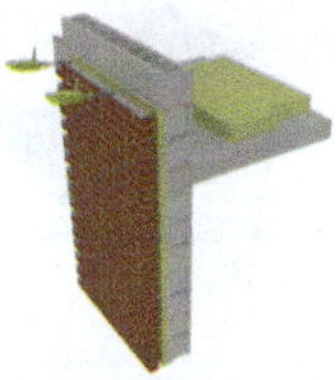	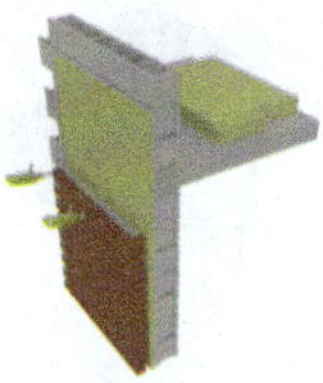	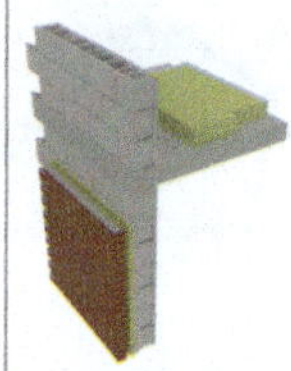
The exterior brick skin is the next element to be dealt with	Remove brick veneer until insulation is exposed	Remove top insulation board to expose the wall-to-roof junction	

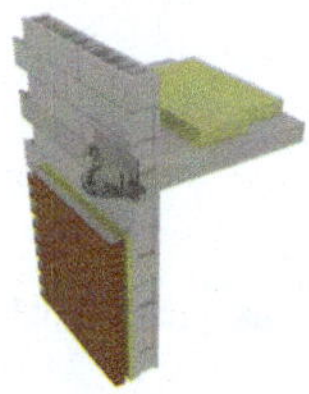	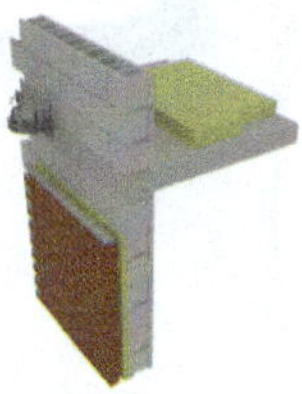	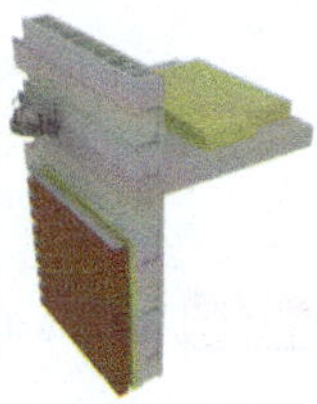	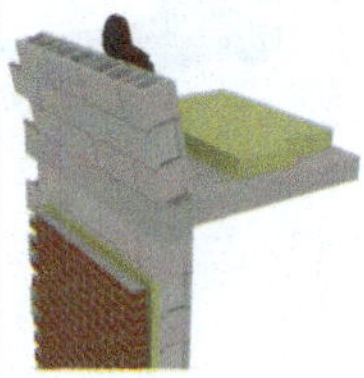
Remove the CMU parapet			

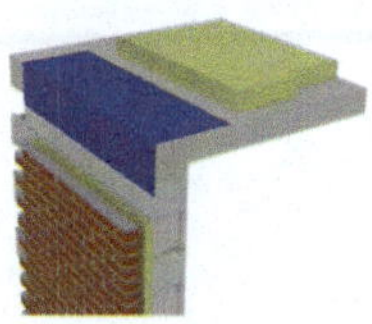	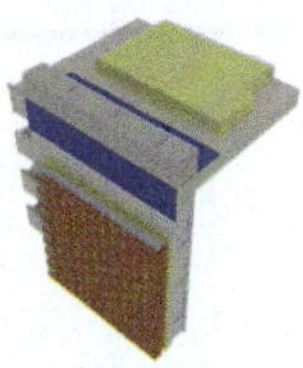	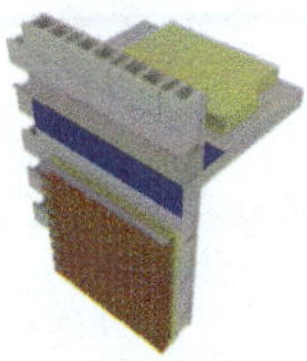	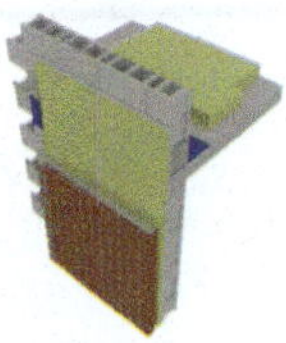
Add an airtightness barrier while the parapet is removed	Add a course of autoclaved aerated concrete (AAC) blocks. Anchor bolts or structural dowels should be added if needed to reinforce the junction	Two courses of CMU can then be added as normal in line with original building profile	Extend external insulation to provide complete overlap with newly inserted thermal break

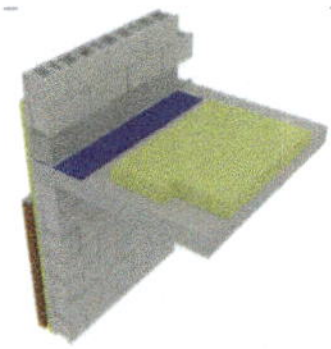	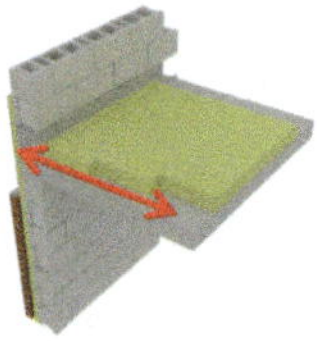	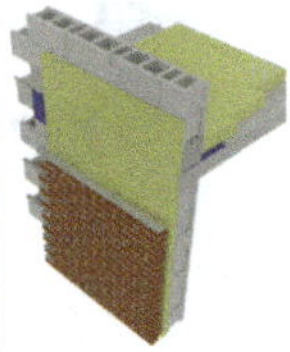	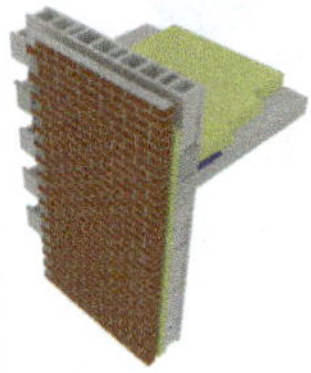
Reinstate insulation on the roof to abut the inserted thermal break	The wall and roof insulation are now "thermally connected"	The exterior brick veneer can now be reinstated	

	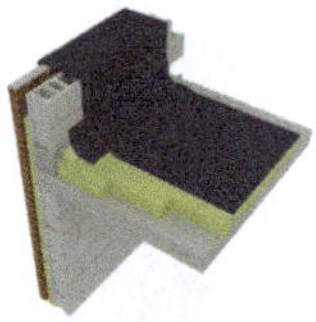		
Reinstate flashing	Add roof covering waterproof layers	Add finish covering waterproof layers	Finally, reinstate parapet capping

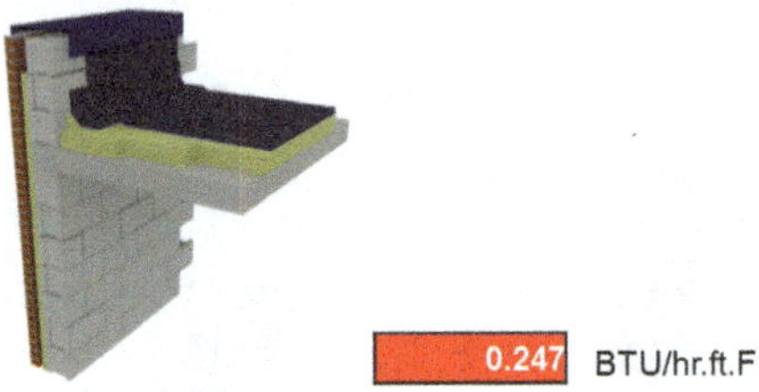	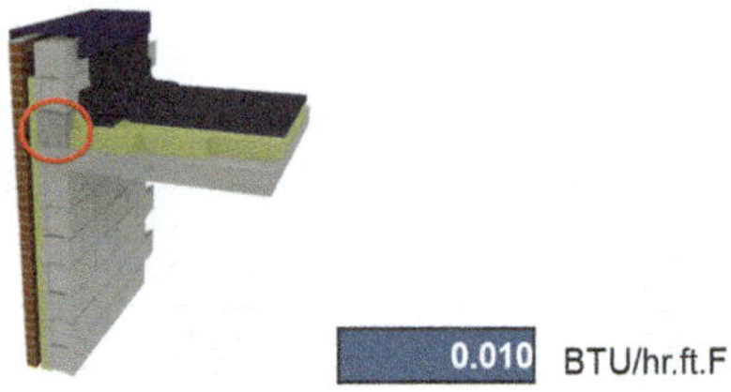
Old junction	New junction

Solution for Option 2: Wrap the Parapet

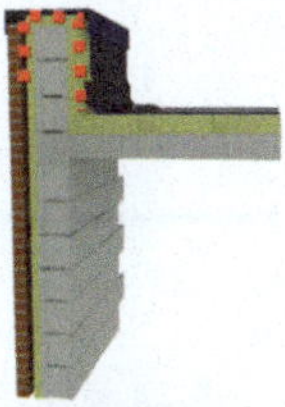

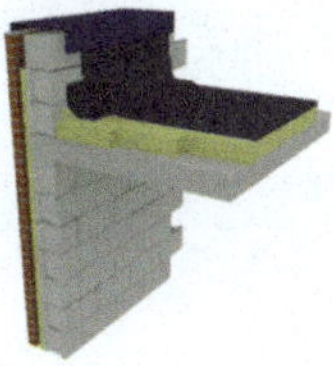

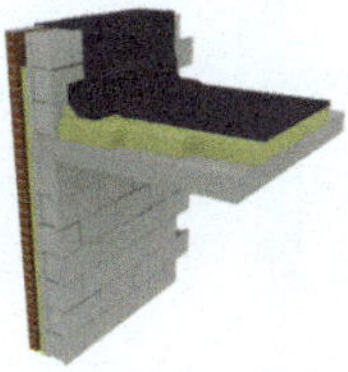

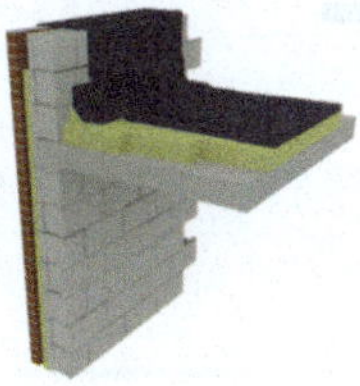

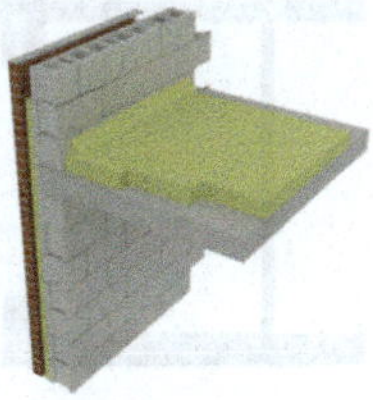

Remove capping, flashings, and roof coverings to expose CMU wall and roof insulation

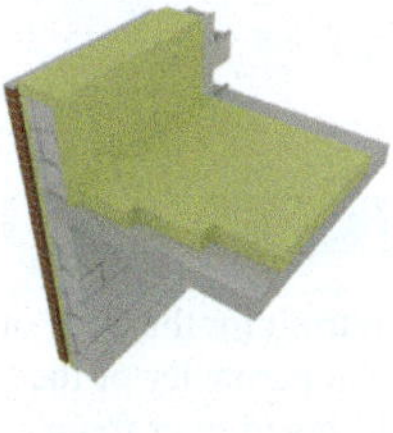

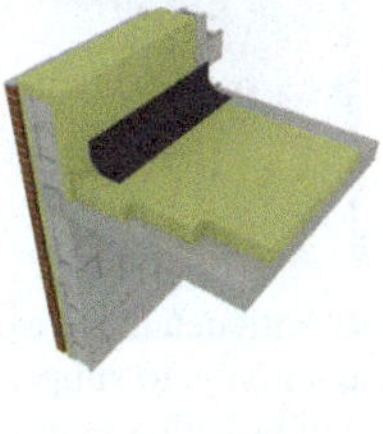

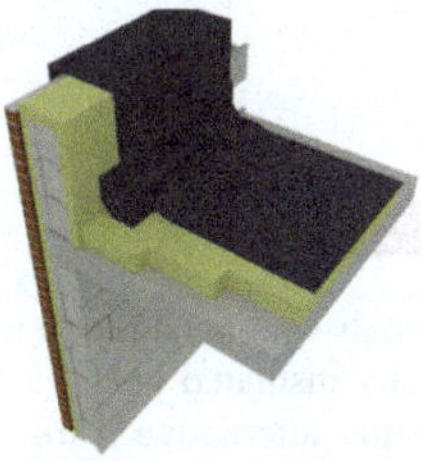

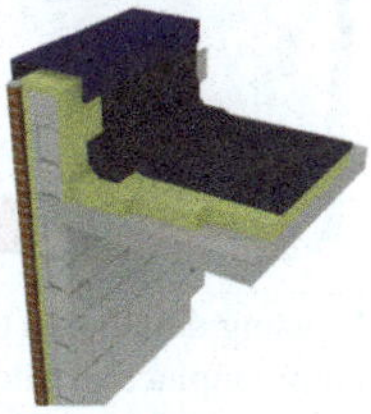

Add rigid insulation to the rear and top of the parapet as well as the cavity if possible

All waterproofing, flashings, and coping needs to be reinstated

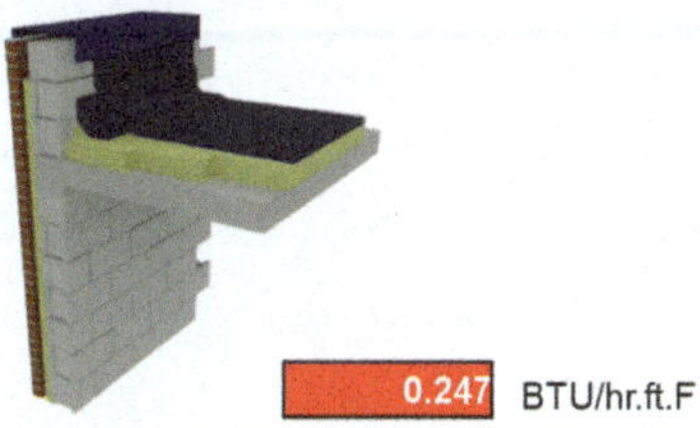

Old junction

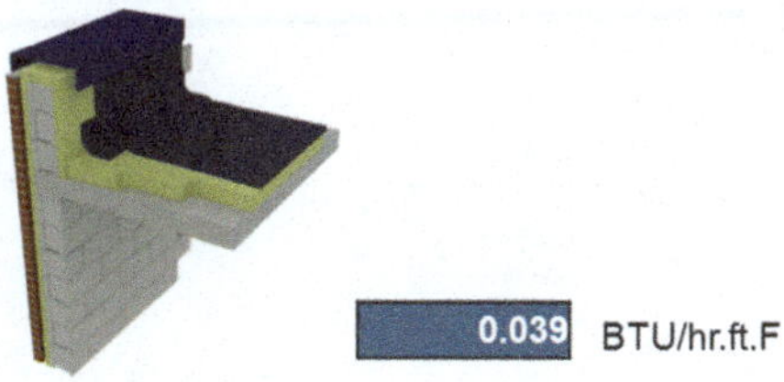

New junction

Parapet with Concrete Roof Junction with a Wall Insulated on Interior

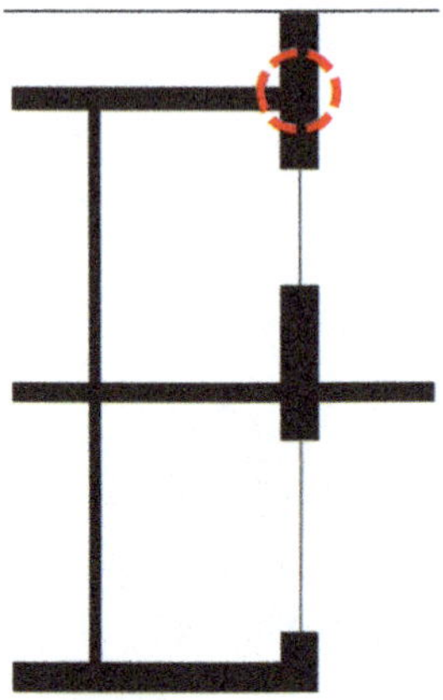

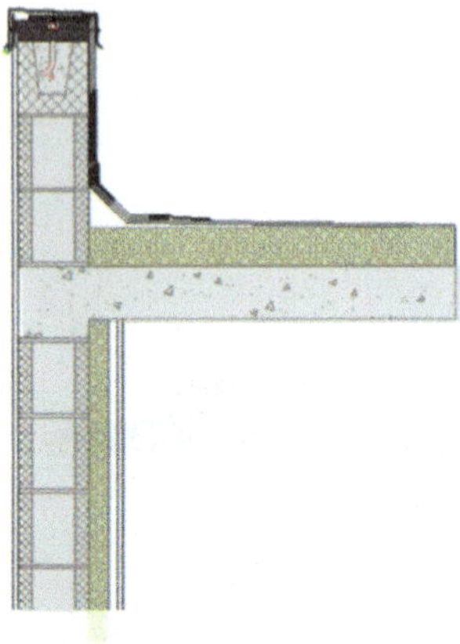

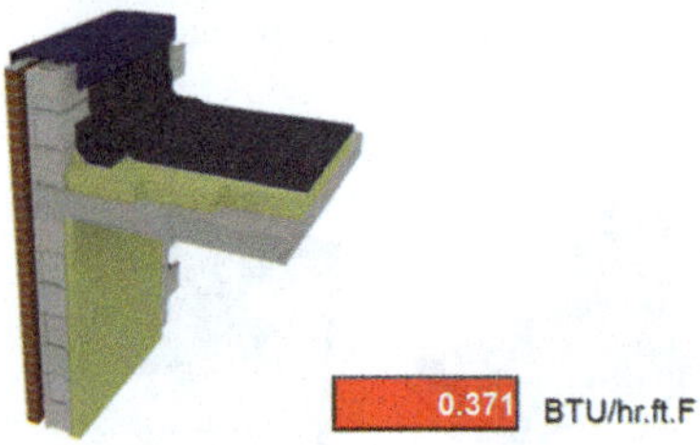

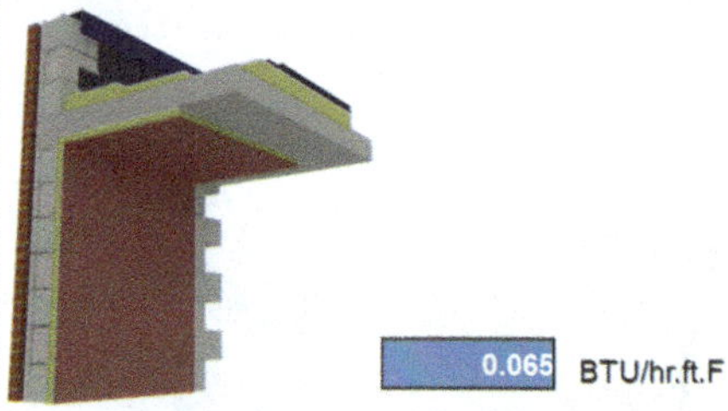

Existing detail: this through-wall penetration is commonplace in internally insulated walls, roofs, and floors. The better alternative is to insulate externally using the details in the previous section (although it may require replacement of the brick skin)	Retrofit detail: the easiest retrofit for this type of junction is to simply line the perimeter of the ceiling with a gypsum wall board over foam insulation. Just remember all internal walls and floors will also cause thermal bridges and space in the room is lost very quickly

Parapet with OWSJ and a Deck (Exterior Wall Insulation)

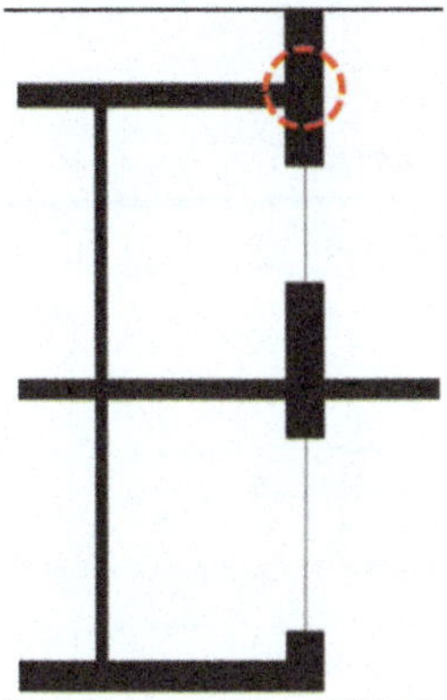

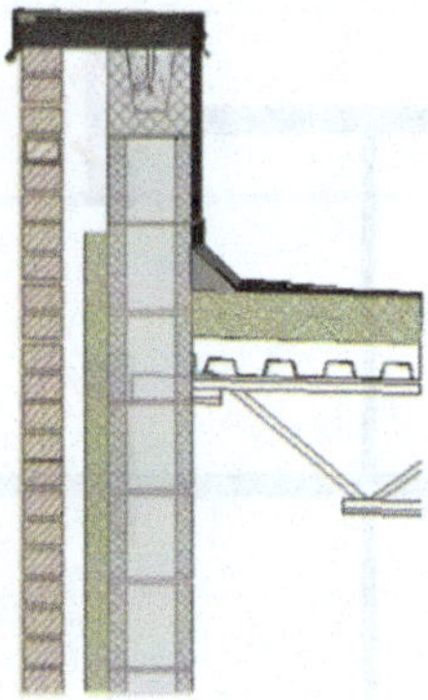

Same renovation principles apply: wrap or add thermal break

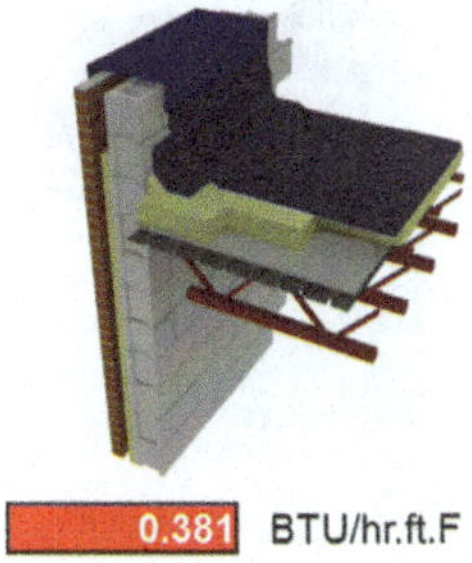

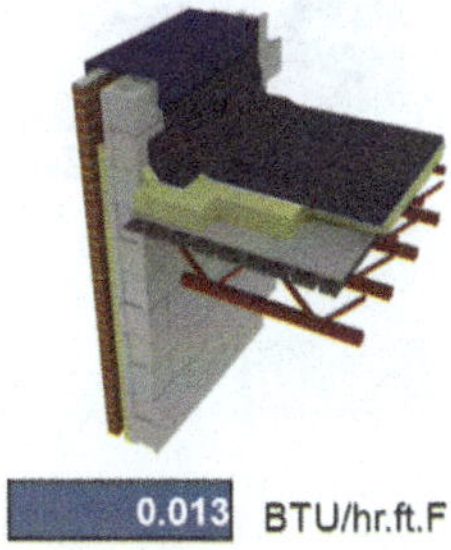

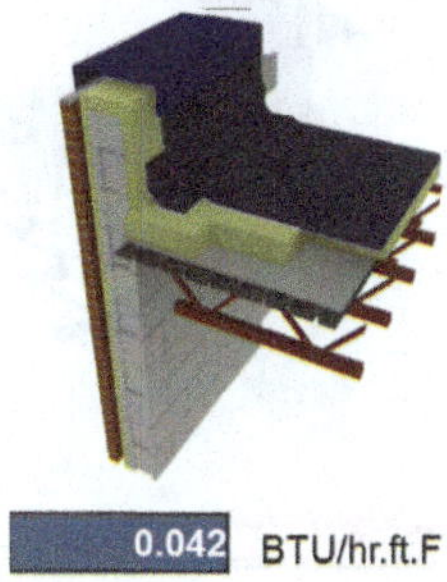

Alternative retrofit option: add spray foam internally

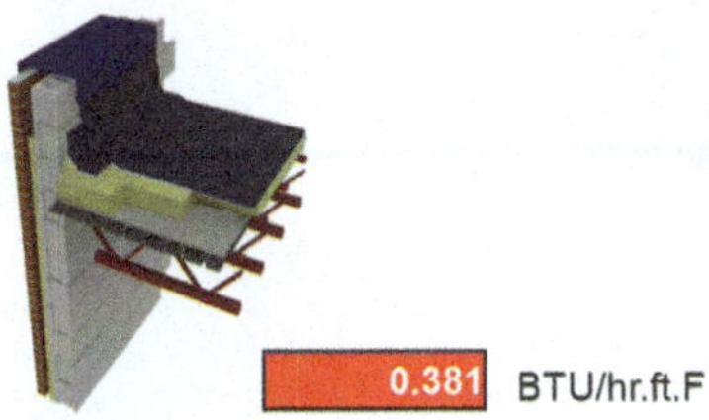

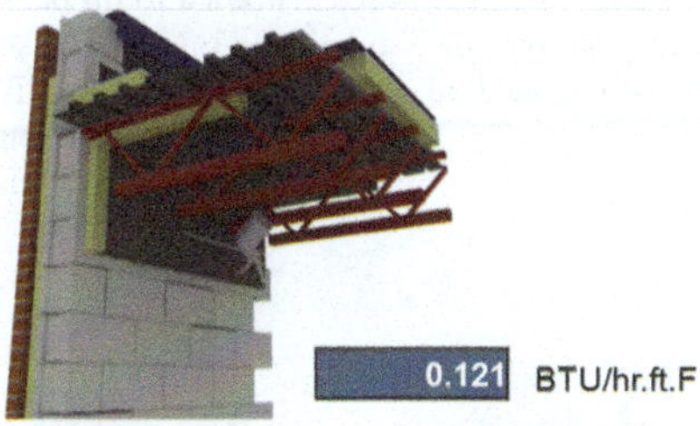

Existing situation

Add first layer of spray foam insulation

Add finish layer of spray-on fire protection

Parapet with OWSJ and a Deck (Interior Wall Insulation)

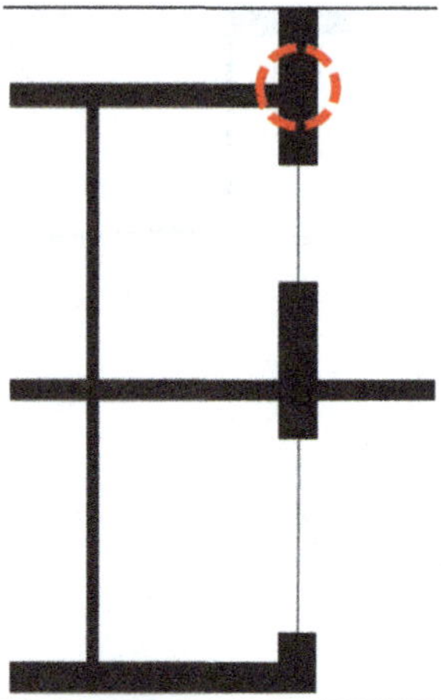	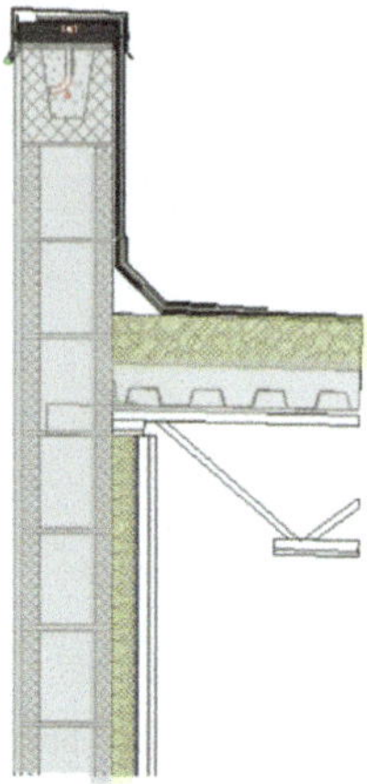	
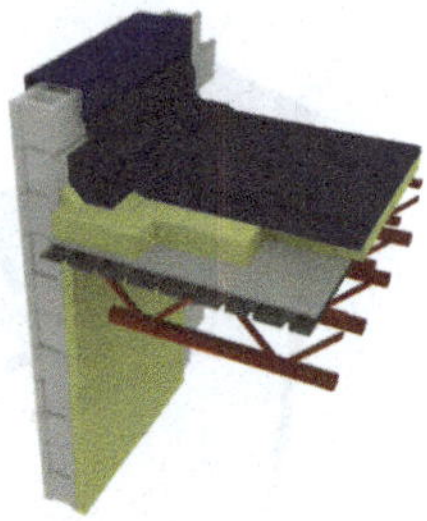 0.318 BTU/hr.ft.F		 0.121 BTU/hr.ft.F
Exiting detail: the high conductivity of the steel OWSJ will yield even worse psi values. Especially in the case of exposed CMU, an external retrofit is a much more robust method	Retrofit detail: a spray foam lining on the inside is a quick and easy retrofit. Spray foam is needed to fill between the studs and braces. The studs and braces still cause a relatively significant thermal bridge, but the psi is reduced nevertheless	In most cases, a layer of fireproofing covering the spray foam is needed

Parapet with OWSJ and a Deck (Steel Stud Buildings)

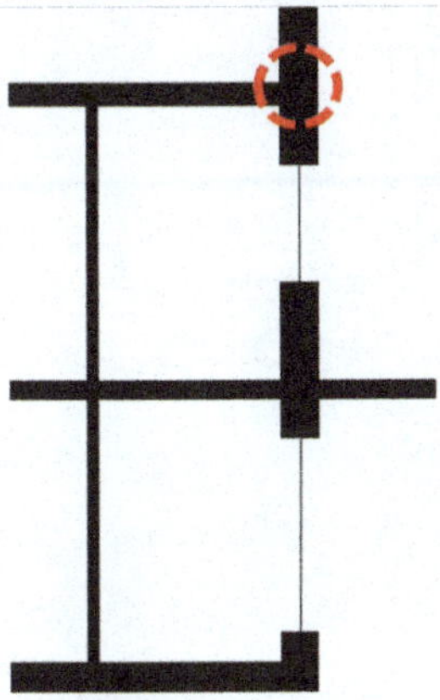

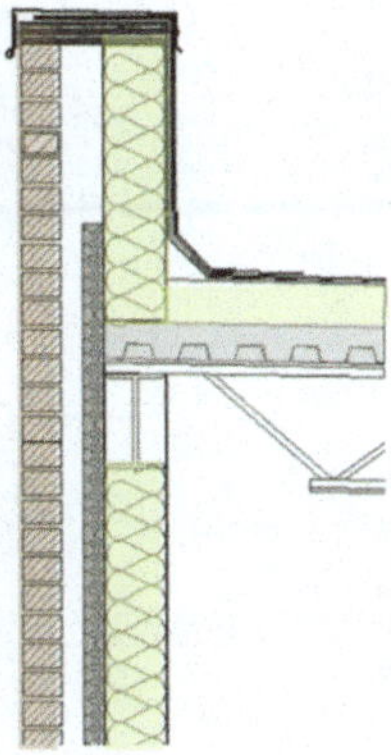

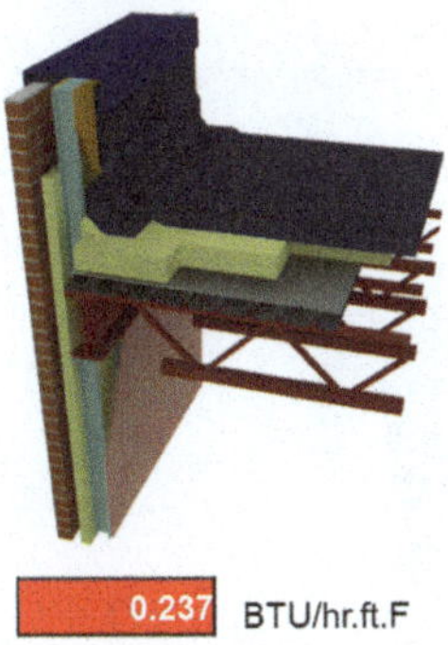

Existing detail: this common detail occurs widely in steel stud buildings

Retrofit option 1: Foam internally.

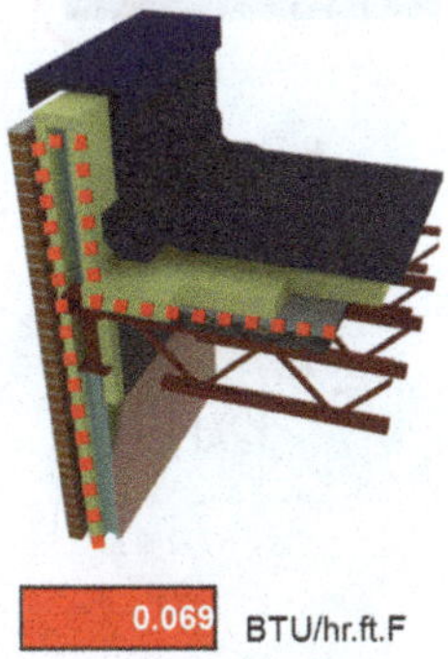

Retrofit option 2: Wrap the parapet.

At Grade Transition: Exterior Insulated Buildings

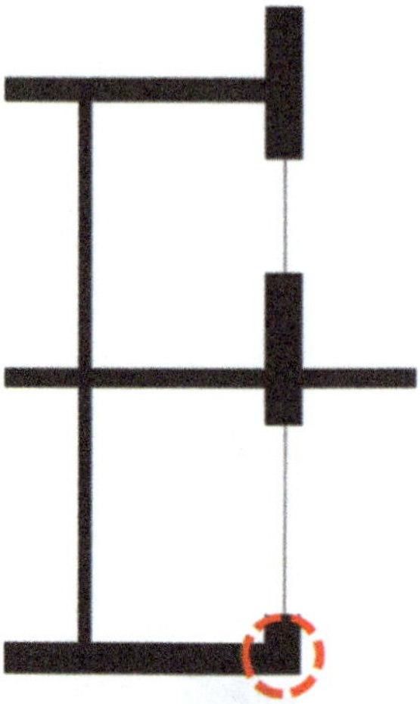	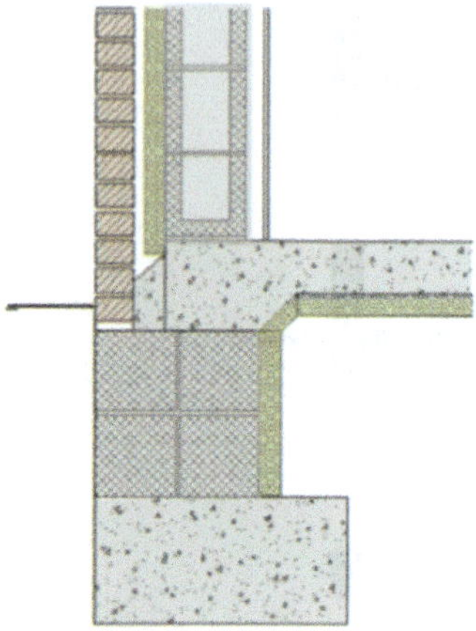
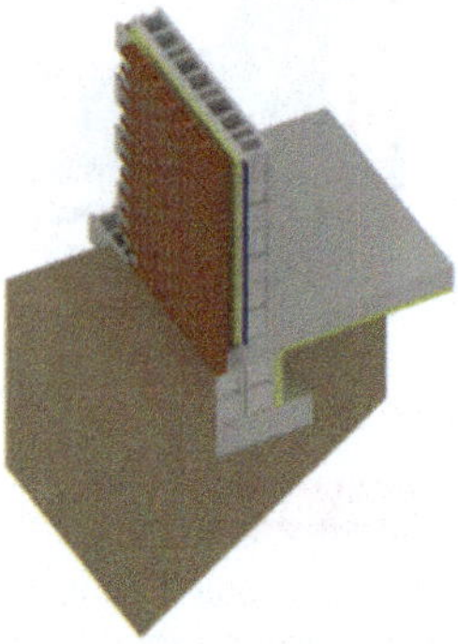	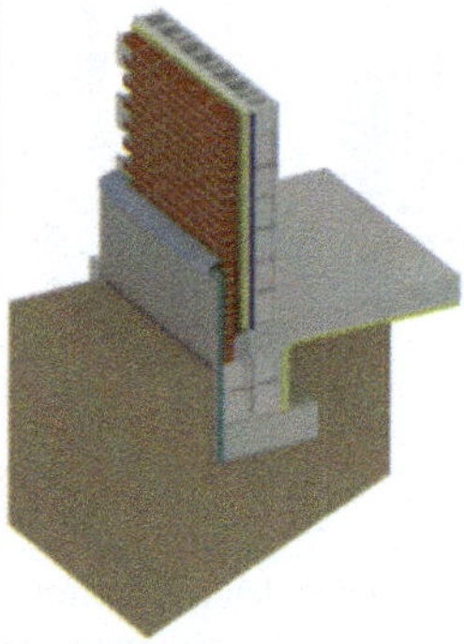
Existing detail: continuity of the insulation layer at the perimeter is important to reduce heat flow and rule out surface condensation on the internal surfaces. More so than mid-slab in large buildings. Unless insulation is added to the slab, the calculated psi value will not make much sense as the slab losses are so great. It may be more important to measure internal surface temperatures if there is no insulation in the floor. Retrofit construction: if the slab edge cannot be retrofit, then insulating the external perimeter can help to contain heat within the ground and raise internal surface temperatures	Retrofit construction sequence: the ground is excavated at the perimeter of the slab to expose the area for upgrade

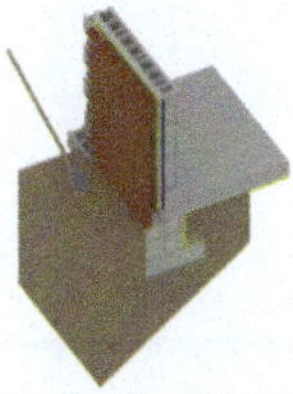		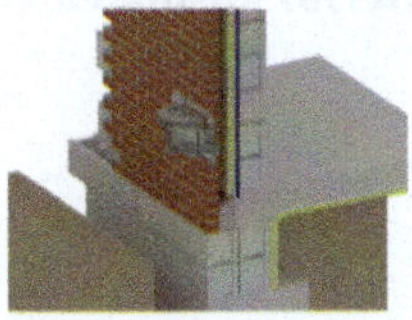	
The ground is excavated at the perimeter of the slab to expose the area for upgrade		A chase is cut in the mortar joint allowing for the flashing to be added later	

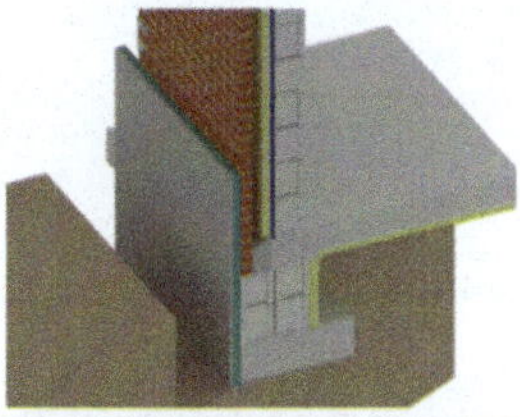		
A cementitious board is prefixed to a grooved, water-resistant XPS board (e.g., PERIMATE™) and affixed to the perimeter of the slab	A bead of sealant is added to the chase	A metal or UV-resistant flashing is bedded into the bed of sealant

	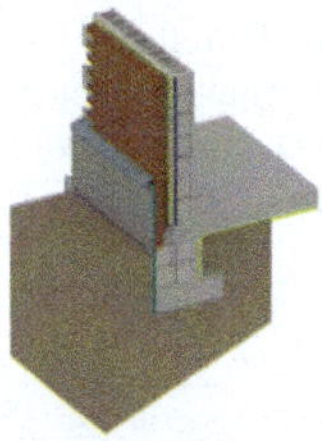	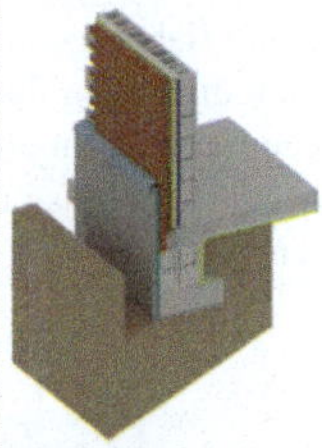
Add an additional bead of sealant to provide watershed	The remaining soil can then be added. This may also be an opportunity to add a gravel drain depending on site conditions	

At Grade Transition: Interior Insulated Buildings

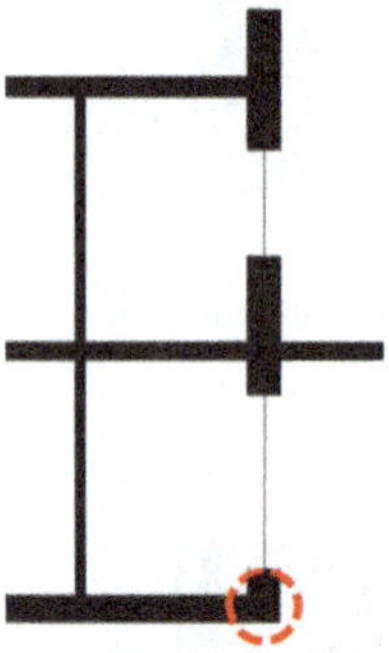

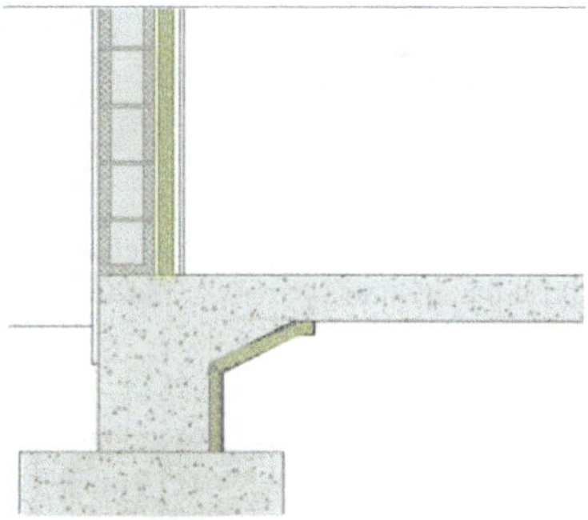

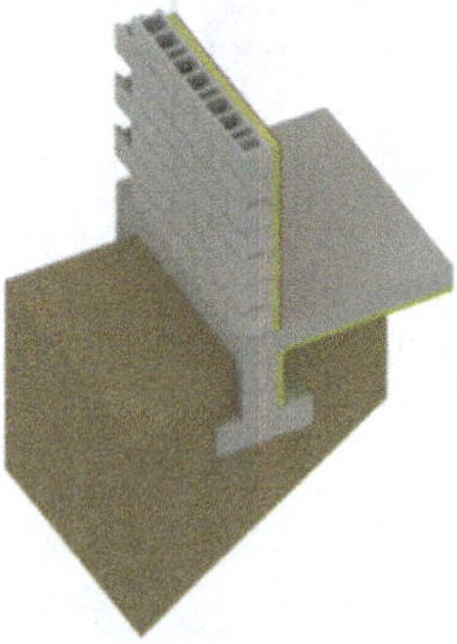

If the slab edge cannot be retrofit, then insulating the external perimeter can help to contain heat within the ground and raise internal surface temperatures

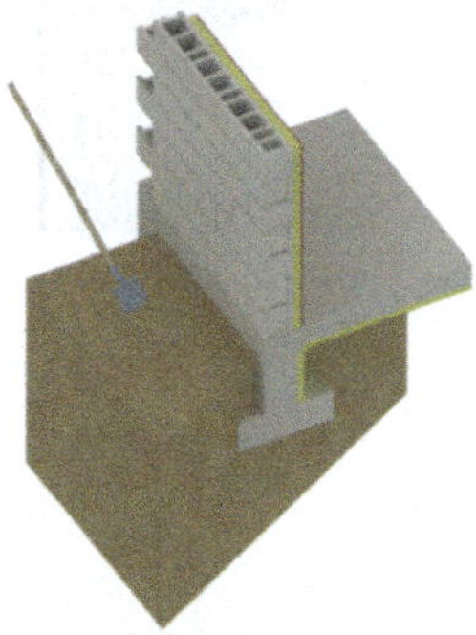

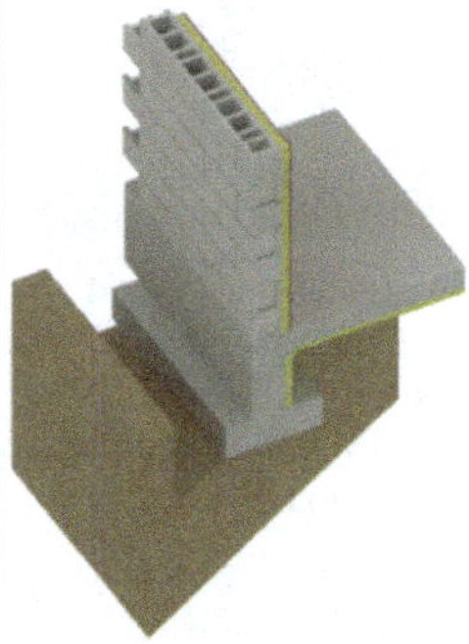

The ground is excavated at the perimeter of the slab to expose the area for upgrade

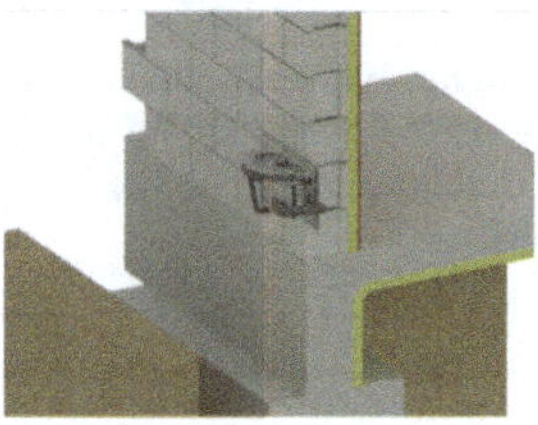

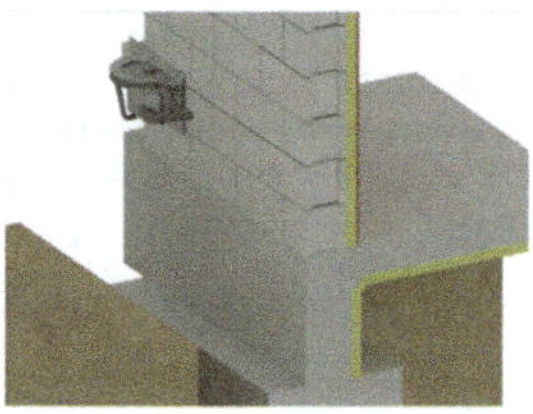

A chase is cut in the mortar joint allowing for the flashing to be added later

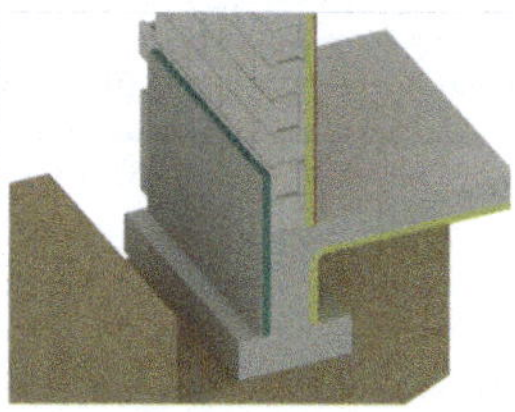

A cementitious board is prefixed to a grooved, water-resistant XPS board (e.g., PERIMATE™) and affixed to the perimeter of the slab

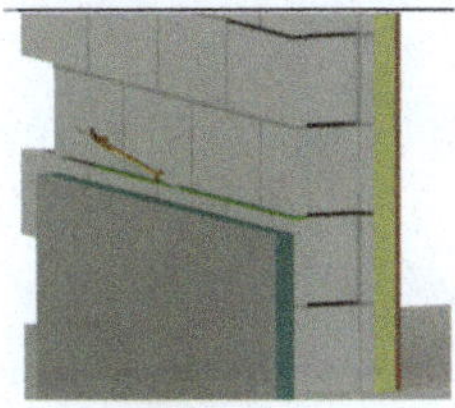	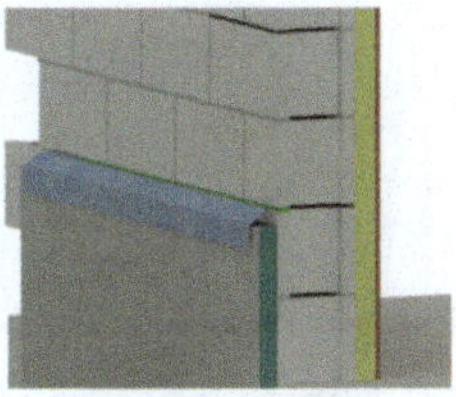	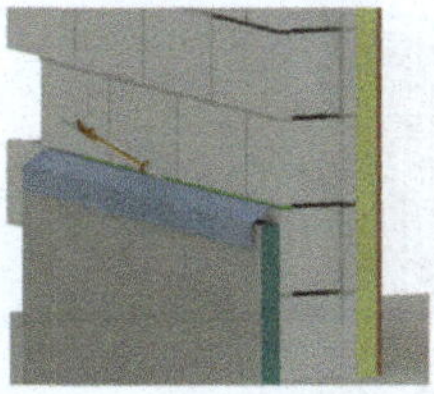
A bead of sealant is added to the chase	A flashing is bedded into the bed of sealant	Add an additional bead of sealant to provide watershed

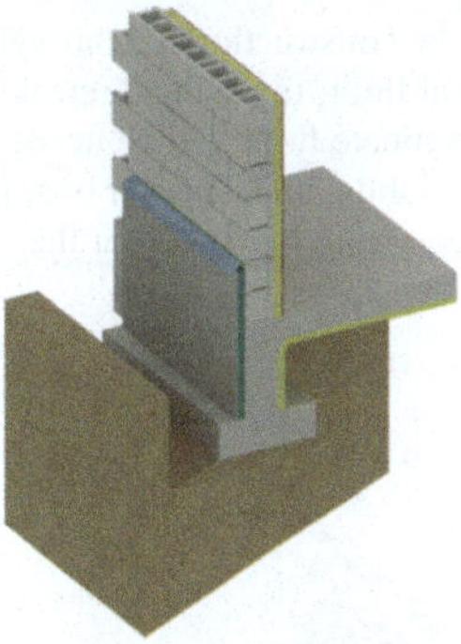	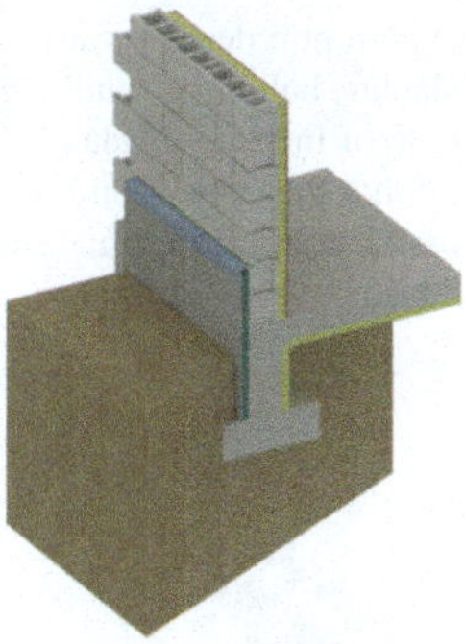
The remaining soil can then be added. This may also be an opportunity to add a gravel drain depending on site conditions	

Through Slab Projections with Interior Insulated Walls: Internal Walls and Internal Floors

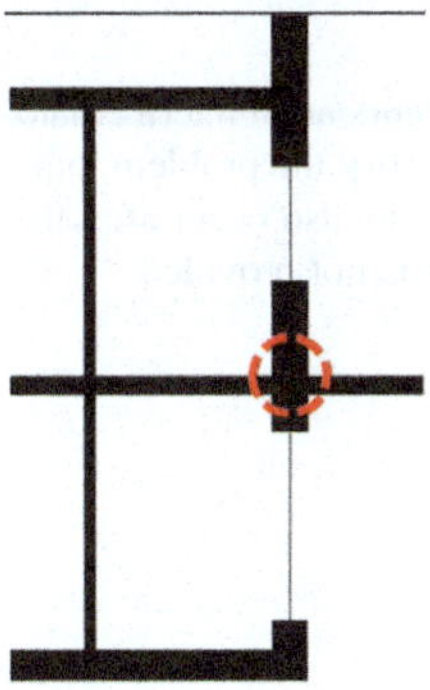	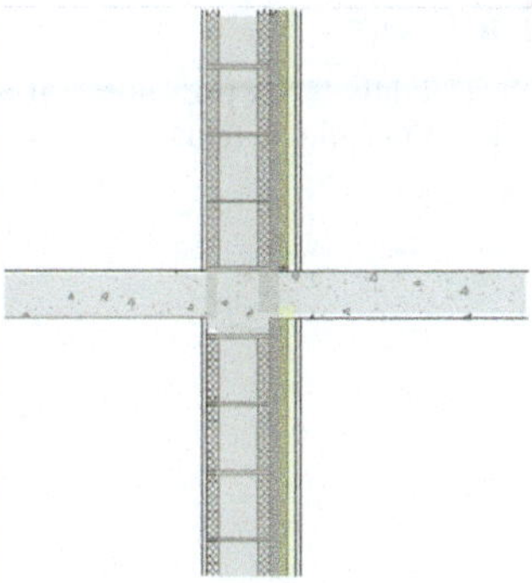

(continued)

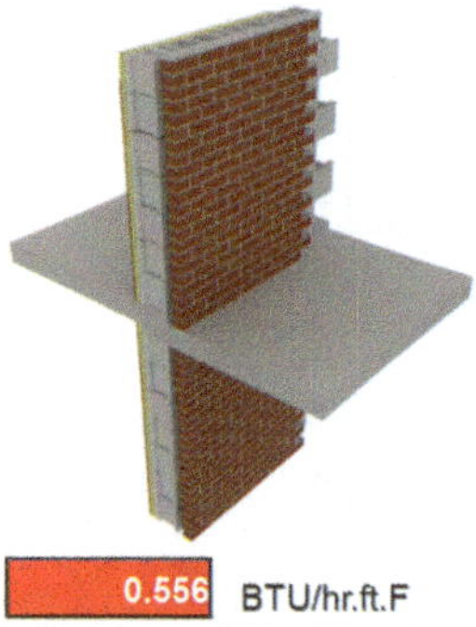

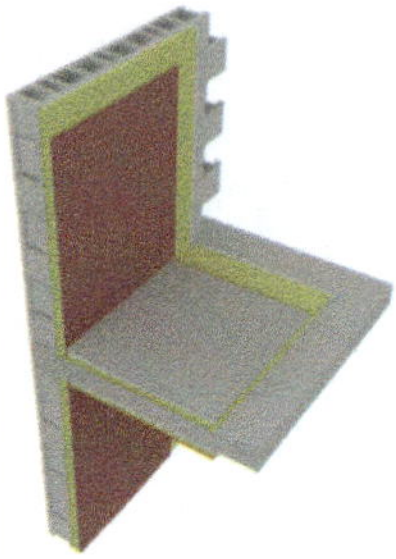

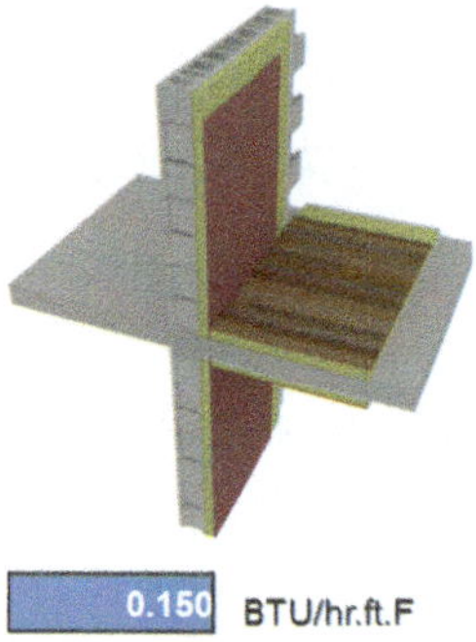

Existing detail: this common detail occurs widely at canopy shades, balconies, and vestibules. In the case of internal insulation, it is not critical that the projection be removed, but the psi value resulting in lengthening the heat flow path is still quite poor

Retrofit detail: whether the construction is a through slab projection or internal floor, the psi and retrofit are much the same. A laminate floor will be needed to cover the flanking insulation, and a plaster board-backed insulation will generally be needed at the ceiling level

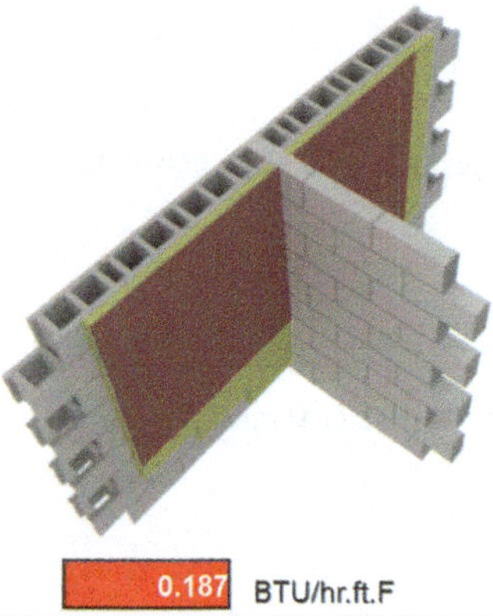

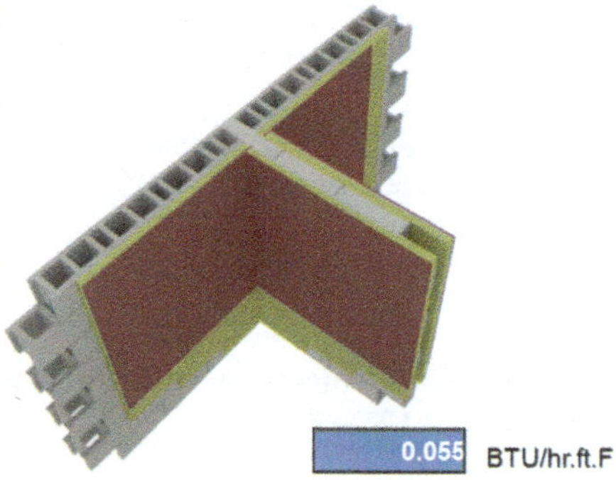

Internal walls have a slightly lesser psi value, but nevertheless a lot of magnitude (almost as much as windows)

If a condensation risk occurs at an intermediate floor or wall, this may rectify the problem, but interstitial condensation can also occur after the retrofit if vapor tightness is not provided

Flanking

8″ 12″ 16″ 20″ 24″ 28″
8″ 12″ 16″ 20″ 24″ 28″
8″ 12″ 16″ 20″ 24″ 28″
8″ 12″ 16″ 20″ 24″ 28″
8″ 12″ 16″ 20″ 24″ 28″
8″ 12″ 16″ 20″ 24″ 28″

Blast-Resistant Door Jamb

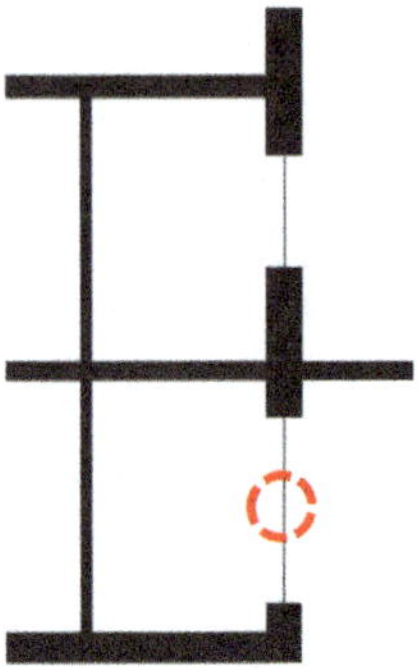	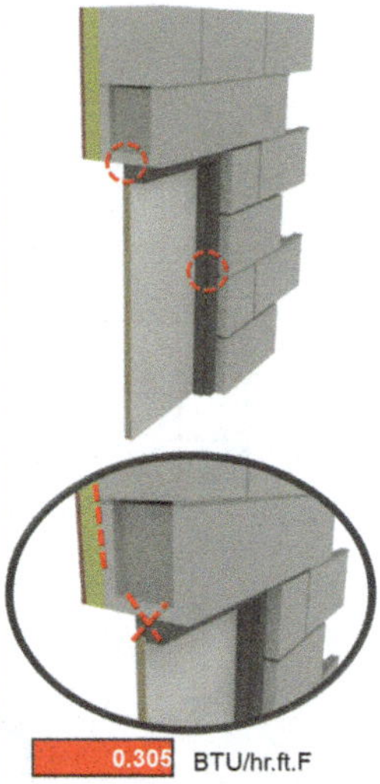	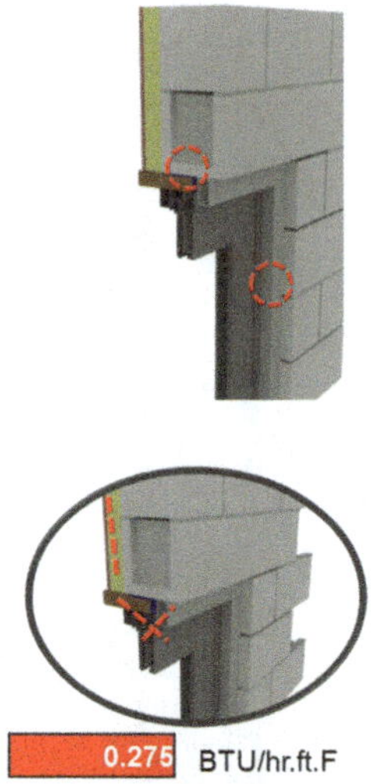
	Exiting detail: no alignment or connection of the thermal break to the insulation layer. No thermal break in the existing door will cause very high losses. As such, it is doubtful that the thermal bridge will be effective as there will be very large losses from the frame itself. Condensation may occur on the inside of this frame in cold weather	Retrofit: A 1.5-in.-thick timber break can be used to bridge the insulation layer to the thermal break in the door.

Examples of Thermal Bridges Mitigation with the Building Envelope Renovation

Section "Examples of thermal bridges mitigation with the building envelope renovation" includes the following examples of thermal bridge mitigation, reproduced with permission from "Details of Passive Houses Renovation" (2016):

D.2.1 Building Renovation: Brick Masonry Wall Insulated with ETICS and a Concrete Attic Slab Insulated on the Upper Side

D.2.2 Building Renovation: Cavity Brick Wall Insulated Using Cavity Insulation in Addition to External Insulation; Top Floor Reinforces Concrete Ceiling Slab Insulated on the Upper Side under the Tiled Roof with Eaves

D.2.3 Building Renovation: Concrete Composite Wall Insulated by Prefabricated Timber Panels Filled in with Cellulose Insulation and a Flat Roof Insulation

D.2.4 Building Renovation: Clay Block External Wall Insulated Using Prefabricated Timber Construction Filled in On-Site with Cellulose Insulation; Roof Insulated Using Prefabricated Insulated Panels

D.2.5 Building Renovation: Concrete Composite Block Wall Insulated with ETICS and a Flat Roof Insulated under Vapor Pressure-Equalized Metal Roof

D.2.6 Building Renovation: Concrete Composite Block Wall Insulated with ETICS and a Flat Roof Insulated under Vapor Pressure-Equalized Metal Roof

D.2.7 Building Renovation: Basement Ceiling Reinforced Concrete Slab Insulated from the Bottom Side

D.2.8 Building Renovation: Concrete Wall Insulated with Prefabricated Timber Panels and Insulation of the Basement Reinforced Concrete Slab Ceiling

D.2.9 Building Renovation: Brick Masonry Wall Insulated with ETICS and a Concrete Hollow Block Basement Ceiling Insulated on the Lower Side

Building Renovation: Brick Masonry Wall Insulated with ETICS and a Concrete Attic Slab Insulated on the Upper Side

This insulation concept is applicable to buildings with accessible attics, which can be used as a storage space.

Renovation concept: the external wall is thermally renovated using an external thermal insulation composite system	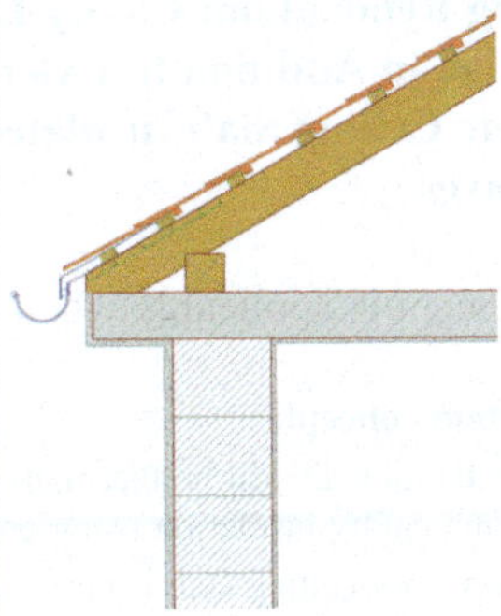
The external wall will be thermally renovated by means of an external thermal insulation composite system	
The top story ceiling is insulated with EPS insulation; a load distribution plate also makes the renovated attic accessible	
The air and vapor barriers is installed on the ceiling, which is connected with an airtight seal to the existing exterior plaster	
As a result of the attic floor insulation, the temperatures in the attic will be significantly lowered in the winter; this must be taken into account when considering future attic uses	
Note: the construction of the eaves gutter is carried out to retain the proportions in the exterior area. If this is not required for design reasons, the existing construction can be replicated with an edge gutter. As a result, the external wall connection "slides" approx. 30 cm upward	

(continued)

Construction process:	
Install vapor barrier, repair existing exterior plaster, and bond it with the vapor barrier in the roof area	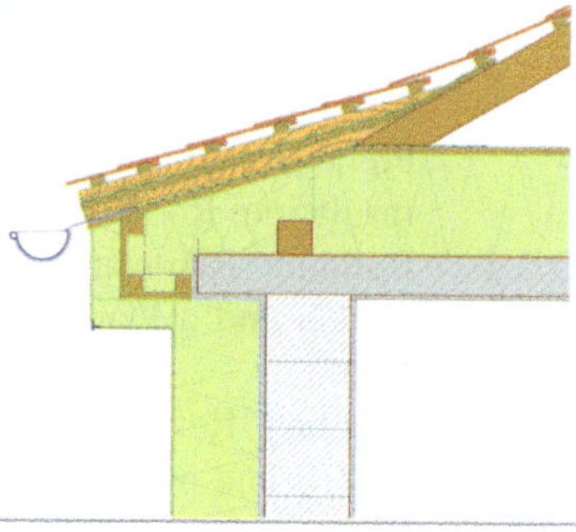
Install EPS insulation and an accessible load distribution plate	
Uncover the roof in the eaves area and double the rafters with fillets	
Install the cladding and vapor-permeable roofing sheet, and seal them in a windtight manner with the pore sealing panel and exterior plaster	

Wall structure		
Silicate plaster	0.6	cm
EPS – expanded polystyrene	30	cm
Lime cement plaster	2.5	cm
Crushed brick masonry	30	cm
Lime cement plaster	1.5	cm
U-value 0.12 [W/m²K]	Σ 64.6	cm

Roof structure		
Roof tile	2	cm
Batten 3/5	3	cm
Rafter, air space	13	cm
U-value - [W/m²K]	Σ 18.0	cm

Ceiling structure		
EPV	2.5	cm
Expanded polystyrene (EPS) W20 insulation panel	36	cm
Concrete slab	14	cm
Lime cement plaster	1.5	cm
U-value 0.10 [W/m²K]	Σ 54.0	cm

2-dimensional specific value		
Ψ-value interior/exterior	0.028	W/mK
3-dimensional specific values		
X-value interior/exterior	-0.141	W/K
f_{Rsi}	0.85	-

Building Renovation: Cavity Brick Wall Insulated Using Cavity Insulation in Addition to External Insulation; Top Floor Reinforces Concrete Ceiling Slab Insulated on the Upper Side under the Tiled Roof with Eaves

This concept is applicable when the roof is protected against driving rain

Renovation concept:	
The external wall will be thermally renovated by using interior and cavity insulation using perlite insulation	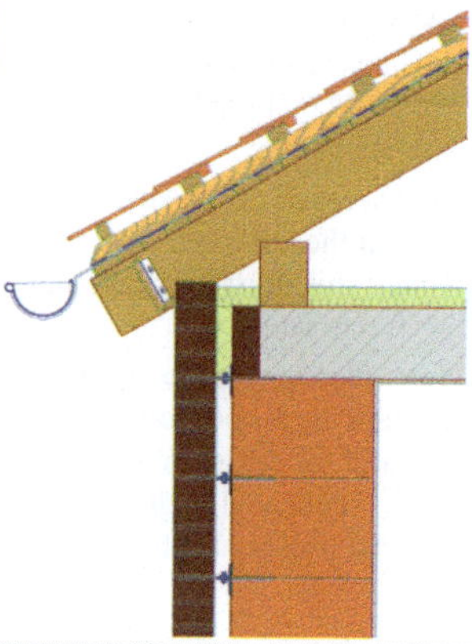
The top story ceiling slab is covered with pressure-resistant rock wool and an EPV panel. The roof space must remain well-ventilated	

(continued)

Construction process:	
Clean and hydrophobize the façade. Make sure that the existing masonry is dry enough	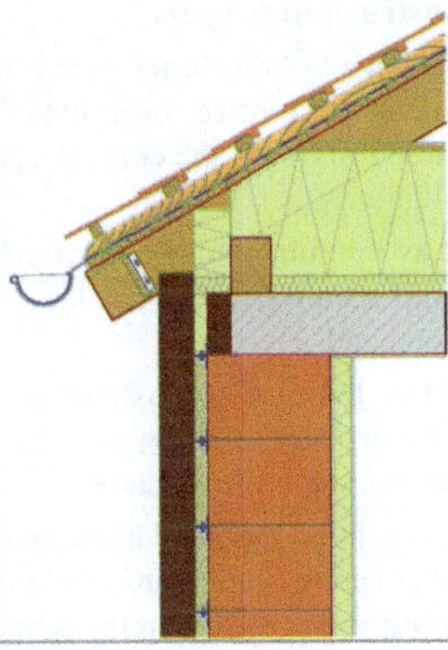
Fill the interstice of the cavity structure with perlite	
Clean the inner side of paint residues, etc.; fully bond, and install the interior insulation	
Tightly lay the rock wool; lay the EPV (elastomeric modified reinforced bitumen) panel on top of it	
Note: no air barrier is added. Make sure there is sufficient roof space ventilation	

Wall structure		
Clinker brick facing	10.5	cm
Expanded perlite, steel anchor	4	cm
Clay block	36	cm
Lime cement plaster	1	cm
Calcium silicate insulation board	5	cm
Lime cement plaster	1	cm
U-value 0.32 [W/m²K]	Σ 57.5	cm

Roof structure		
Roof tile	2	cm
Batten	3	cm
Counter batten	4	cm
Damp-proof membrane	0.5	cm
Timber cladding	2.4	cm
Rafter 14/6, air	14	cm
U-value 1.83 [W/m²K]	Σ 25.9	cm

Ceiling structure		
EPV panel	2.5	cm
Rock wool	36	cm
Glass wool	5	cm
Reinforced concrete	18	cm
Lime cement plaster	1	cm
U-value 0.09 [W/m²K]	Σ 62.5	cm

2-dimensional specific value		
Ψ-value interior/exterior	0.080	W/mK
3-dimensional specific values		
X-value interior/exterior	0.002	W/K
f_{Rsi}	0.79	-

Building Renovation: Concrete Composite Wall Insulated by Prefabricated Timber Panels Filled in with Cellulose Insulation and a Flat Roof Insulation

This insulation concept is applicable to buildings when the roof slab is structurally suitable.

Renovation concept:	
The parapet and timber construction of the roof is removed	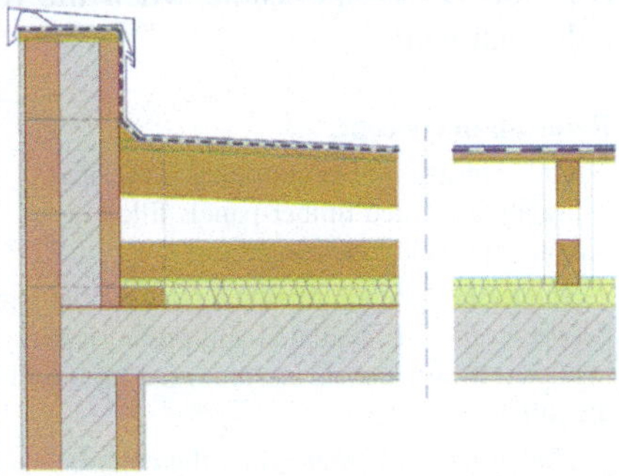
The external wall is thermally renovated using a prefabricated timber construction that is filled in with a cellulose insulation on-site	
The roof is renovated using a duo roof	

(continued)

Construction process:
Clean exterior plaster, repair missing parts. The exterior plaster constitutes the air barrier. Seal it airtight on the concrete grating with sealing tape
Lay bituminous vapor barrier/temporary seal; if required, apply the leveling layer beforehand
Place the timber construction, affix the temporary seal all the way to the outside and blow out the hollow spaces
Fully bond the insulation panels in the parapet/concrete grating area to safely rule out an upward vapor release in the separating layer of the existing plaster and the insulation panels
Note: as an alternative, the roof can also be renovated with a prefabricated timber construction with a vapor pressure-equalized seal

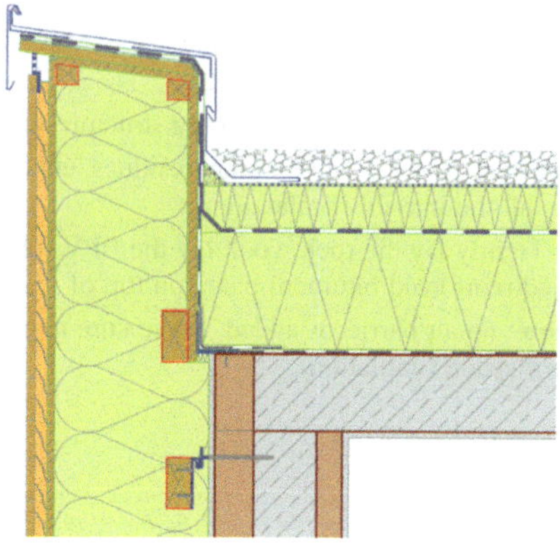

Wall structure		
Larch façade	2	cm
Batten, ventilated cavity	3	cm
MDF panel, tongued and grooved, waterproofed	1.6	cm
Cellulose fiber flakes between timber frame	36	cm
Lime cement exterior plaster	2.5	cm
Woodchip component	9	cm
Concrete	15	cm
Woodchip component	6	cm
Lime cement plaster	1.5	cm
U-value 0.10 [W/m²K]	Σ 76.6	cm

Roof structure		
Gravel	8	cm
Fleece (PP)	0.2	cm
Extruded polystyrene, CO_2-foamed (XPS)	10	cm
Polymer bitumen waterproofing membrane	0.02	cm
Vapor pressure compensation layer	0.02	cm
Expanded polystyrene (EPS) - W25 insulation panel	28	cm
Aluminum-polymer waterproofing membrane	0.3	cm
Vapor pressure compensation layer	0.2	cm
Reinforced concrete slab	18	cm
Gypsum plaster	1.5	cm
U-value 0.09 [W/m²K]	Σ 66.6	cm

2-dimensional specific values		
Ψ-value interior/exterior	-0.046	W/mK

Building Renovation: Clay Block External Wall Insulated Using Prefabricated Timber Construction Filled in On-Site with Cellulose Insulation; Roof Insulated Using Prefabricated Insulated Panels

This concept is applicable when the roof rafter is structurally adequate to carry additional weight.

Renovation concept:
The external wall is thermally renovated using prefabricated timber panels filled in on-site with cellulose insulation
The roof is opened from the outside; the gap between rafters is insulated and prefabricated insulated timber panels are mounted on top of the rafters
The air barrier is created by the exterior plaster, which is bonded in an airtight manner with the vapor barrier in the roof

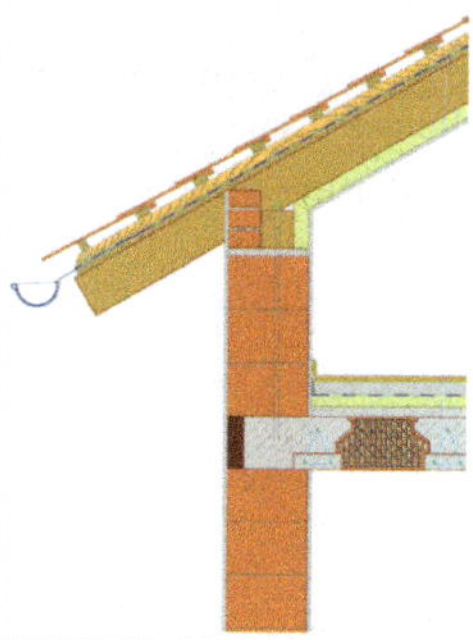

(continued)

Construction process:	
Uncover the roof, insulate the space between existing rafters, and install the timber prefabricated roof components	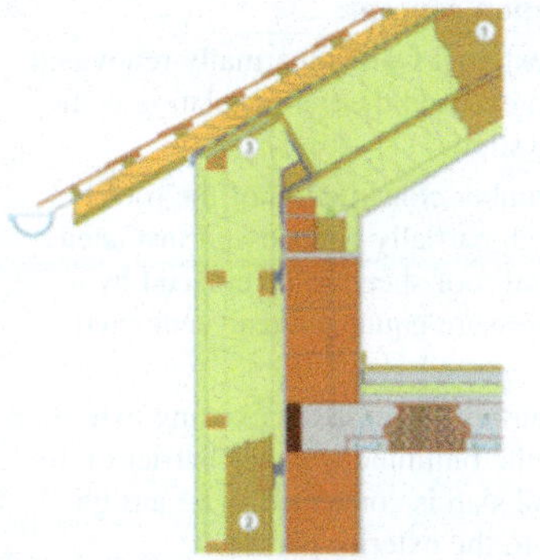
Build the canopy with suitably sized counter battens	
Bond the vapor barrier airtight with the existing exterior wall plaster	
Install the prefabricated wall components and fill them with cellulose flakes	(1) Prefabricated box beam roof element; (2) solid structural timber battens, (3) compressed insulation

Wall structure		
Exterior plaster	*0.6*	*cm*
Wood fiberboard	*4*	*cm*
Cellulose, posts, post mounting system and solid structural timber	32	cm
Lime cement plaster	*1.5*	*cm*
Clay block	*30*	*cm*
Lime cement plaster	*1.5*	*cm*
U-value 0.10 [W/m²K]	Σ 69.6	cm

Roof structure		
Roof tile	*2*	*cm*
Batten	*3*	*cm*
Counter batten	*8*	*cm*
PE roofing sheet, vapor-permeable	*0.2*	*cm*
Sub-roof panel – wood fiber	*2.2*	*cm*
Glass wool, box beam	*30*	*cm*
Vapor barrier	*0.2*	*cm*
OSB	*2.4*	*cm*
Glass wool, rafter 16/8	*16*	*cm*
Wood wool lightweight panel, magnesite-bonded (Heraklith)	*5*	*cm*
Plaster	*1.5*	*cm*
U-value 0.09 [W/m²K]	Σ 70.5	cm

Ceiling structure		
Flooring	*2*	*cm*
Sand-cement screed	*5*	*cm*
PE foil/separating layer	*0.2*	*cm*
EPS – expanded polystyrene	*4*	*cm*
Concrete topping	*4*	*cm*
Brick, reinforced concrete	*20*	*cm*
U-value 0.63 [W/m²K]	Σ 35.2	cm

2-dimensional specific values		
Ψ-value interior/exterior	*-0.035*	*W/mK*
f_{Rsi}	*0.95*	-

Building Renovation: Concrete Composite Block Wall Insulated with ETICS and a Flat Roof Insulated Under Vapor Pressure-Equalized Metal Roof

The roof insulation concept is applicable to buildings if the timber construction of the roof is in usable conditions, and if the reinforced concrete slab is sealed, the roof slab is structurally suitable.

Renovation concept:

The external wall is thermally renovated using an external thermal insulation composite system

The timber construction of the roof is preserved, partially opened and insulated. The metal roof sheeting is replaced by a vapor pressure-equalized/rear-ventilated seal

The airtight layer is the existing exterior plaster; the bituminous vapor barrier on the structural slab is connected in an airtight manner to the exterior plaster

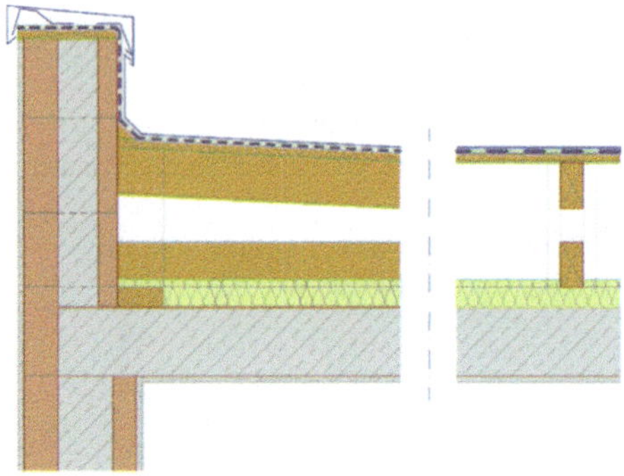

Construction process:

Remove metal roof sheeting, open the timber cladding in parts

Clean the exterior plaster, repair the missing parts. The exterior plaster constitutes vertical air barrier. Apply plaster base coat above the parapet to the top edge of the timber cladding

Keep existing thermal insulation, if possible. Insulate the interstice with fiber insulation material, close the openings again. Fasten the timber construction around the attic and insulate it with rock wool

Lay the vapor-permeable roofing sheet, weld the joints

Lay the counter battens, nail on the timber cladding, attach the roof seal and metal roof sheeting

Fully bond the insulation panels in the parapet area to safely rule out an upward vapor release in the separating layer of the existing plaster and the insulation panels

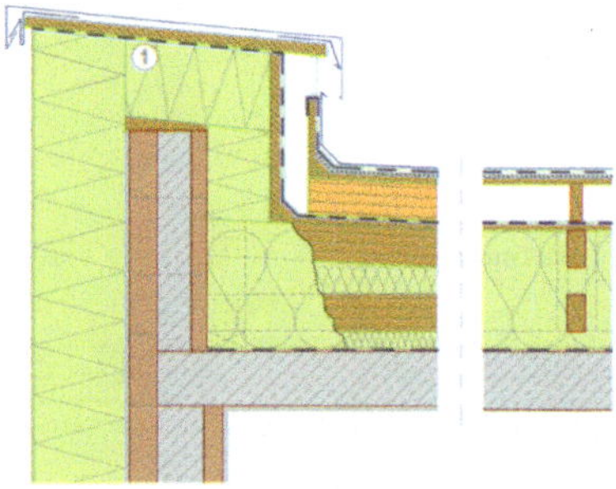

Note: as an alternative, remove the parapet and construct a lightweight parapet

1 – insulation material between timber battens

Wall structure		
Silicate plaster	0.6	cm
EPS – expanded polystyrene	30	cm
Lime cement exterior plaster	2.5	cm
Woodchip component	9	cm
Concrete	15	cm
Woodchip component	6	cm
Lime cement plaster	1.5	cm
U-value 0.11 [W/m²K]	Σ 64.6	cm

Roof structure		
Sheet metal roofing	0.2	cm
PE waterproofing membrane	1.8	cm
Fleece (PP)	0.2	cm
Ventilated cavity, battens 4/12	12	cm
Seal	0.2	cm
Timber cladding	2.4	cm
Glass wool, I-beam (timber flange + OSB beam)	40	cm
Vapor pressure compensation layer	0.2	cm
Reinforced concrete slab	18	cm
Gypsum plaster	1.5	cm
U-value 0.12 [W/m²K]	Σ 76.5	cm

2-dimensional specific value		
Ψ-value interior/exterior	0.020	W/mK
3-dimensional specific values		
X-value interior/exterior	0.038	W/K
f_{Rsi}	0.91	-

Building Renovation: Concrete Composite Block Wall Insulated with ETICS and a Flat Roof Insulated Under Vapor Pressure-Equalized Metal Roof

The roof insulation concept is applicable to buildings if the timber construction of the roof is in usable conditions, and if the reinforced concrete slab is sealed, the roof slab is structurally suitable.

Renovation concept:	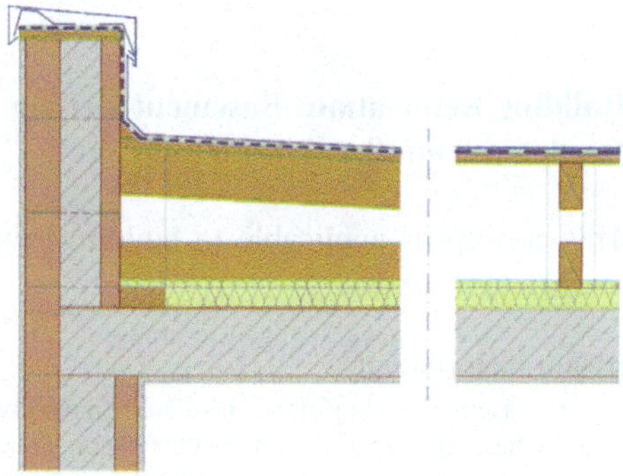
The external wall is thermally renovated using an external thermal insulation composite system.	
The timber construction of the roof is preserved, partially opened, and insulated. The metal roof sheeting is replaced by a vapor pressure-equalized/rear-ventilated seal	
The airtight layer is the existing exterior plaster; the bituminous vapor barrier on the structural slab is connected in an airtight manner to the exterior plaster	
Construction process:	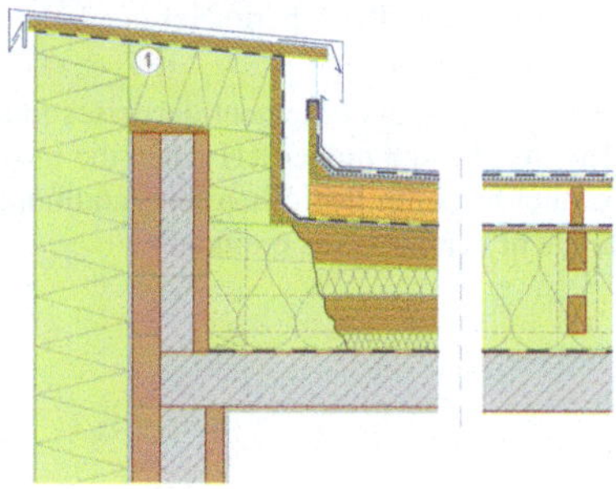
Remove metal roof sheeting, open the timber cladding in parts	
Clean the exterior plaster and repair the missing parts. The exterior plaster constitutes vertical air barrier. Apply plaster base coat above the parapet to the top edge of the timber cladding	
Keep existing thermal insulation, if possible. Insulate the interstice with fiber insulation material and close the openings again. Fasten the timber construction around the attic and insulate it with rock wool	
Lay the vapor-permeable roofing sheet and weld the joints	
Lay the counter battens, nail on the timber cladding, and attach the roof seal and metal roof sheeting	
Fully bond the insulation panels in the parapet area to safely rule out an upward vapor release in the separating layer of the existing plaster and the insulation panels	
Note: as an alternative, remove the parapet and construct a lightweight parapet	1 – insulation material between timber battens

Wall structure		
Silicate plaster	0.6	cm
EPS – expanded polystyrene	30	cm
Lime cement exterior plaster	2.5	cm
Woodchip component	9	cm
Concrete	15	cm
Woodchip component	6	cm
Lime cement plaster	1.5	cm
U-value 0.11 [W/m²K]	Σ 64.6	cm

Roof structure		
Sheet metal roofing	0.2	cm
PE waterproofing membrane	1.8	cm
Fleece (PP)	0.2	cm
Ventilated cavity, battens 4/12	12	cm
Seal	0.2	cm
Timber cladding	2.4	cm
Glass wool, I-beam (timber flange + OSB beam)	40	cm
Vapor pressure compensation layer	0.2	cm
Reinforced concrete slab	18	cm
Gypsum plaster	1.5	cm
U-value 0.12 [W/m²K]	Σ 76.5	cm

2-dimensional specific value		
Ψ-value interior/exterior	0.020	W/mK
3-dimensional specific values		
X-value interior/exterior	0.038	W/K
f_{Rsi}	0.91	-

Building Renovation: Basement Ceiling Reinforced Concrete Slab Insulated from the Bottom Side

This concept is applicable to buildings with no special requirements to basement temperature control.

Renovation concept:

The basement ceiling slab is insulated on the lower side with a thermal insulation composite system. To reduce thermal bridge losses and to increase the minimum temperature on the ground floor, perimeter insulation will be applied

The temperature in the basement decreases in the winter. This results in a higher RH when the moisture sources cannot be controlled. Therefore, additional intelligent basement ventilation is required

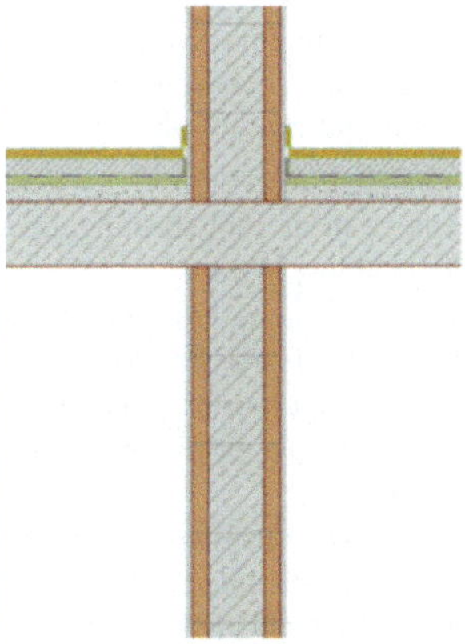

Construction process:

The reinforced concrete slab is a good air barrier by itself

Insulation panels are bonded to the basement ceiling and, depending on the subsurface, can be additionally fastened to them

Consider the use of an intelligent ventilation system in any basement intended for use as a high-quality storage space.

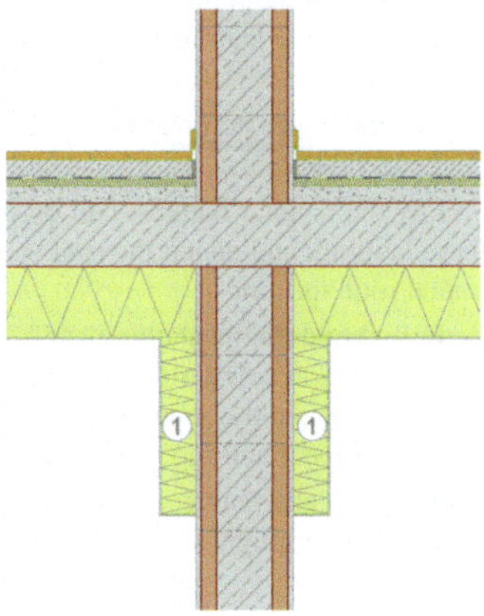

Note: perimeter wall insulation is recommended if the basement temperature can get very low

(1) Plastered EPS

Wall structure		
Lime cement plaster	1.5	cm
Woodchip component	4.5	cm
Concrete	16.0	cm
Woodchip component	4.5	cm
Lime cement plaster	1.5	cm
U-value 0.99 [W/m²K]	Σ 28.0	cm

Ceiling structure		
Flooring	2	cm
Screed	5	cm
PE foil/separating layer	0.2	cm
Impact sound insulation panel	2	cm
Fill	5	cm
Reinforced concrete slab	18	cm
EPS – expanded polystyrene	20.0	cm
Silicate plaster	0.5	cm
U-value 0.16 [W/m²K]	Σ 52.7	cm

2-dimensional specific values		
Ψ-value interior/exterior	0.009	W/mK
f_{RSi}	0.93	-

Building Renovation: Concrete Wall Insulated with Prefabricated Timber Panels and Insulation of the Basement Reinforced Concrete Slab Ceiling

This insulation concept is applicable to buildings with a basement that does not have high requirements to environmental control.

Pre-renovation issues: rising damp, efflorescence, the basement intended for a storage room cannot be used	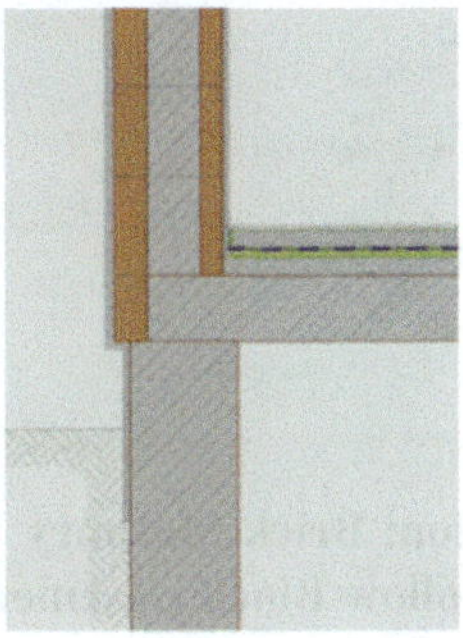
Renovation concept: The external wall has been thermally renovated by a prefabricated timber panels filled in with cellulose insulation on-site	
The basement ceiling slab is thermally renovated on the lower side with an external thermal insulation composite system	
XPS thermal insulation is attached to the base area and the area that is in contact with the ground	
The temperature of the basement space is expected to be below the temperature in the rest of the building. Due to the lack of moisture source control, the basement will have a higher RH	
Construction process:	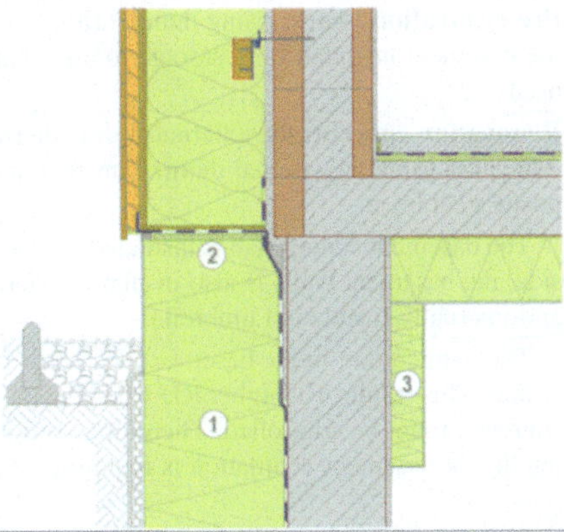
Seal the airtight layer (exterior plaster, fully filled if required) all the way to the lower edge of the ceiling. This also provides a clean surface for applying the seal.	
Bond the insulation panels on the basement ceiling, and, depending on the subsurface, additionally dowel them	
Tightly bond the seal on the complete surface up to at least 30 cm above ground level (splash zone)	
Bond and mechanically secure the drip edge to the end of the timber construction.	

(continued)

Bond the ECB membrane on the lower end of the timber construction (OSB panel)	
Press the perimeter insulation panels with expanding foam seals and affixed strips of fibrous insulation material upward; the top insulation panel should be completely bonded	
Consider the use of an intelligent ventilation system, particularly with high requirements on the storage quality of the basement	
Note: if a sealed surface (e.g., sidewalk) outside the building is connected directly (without a drainage layer) to the external wall, create a sloped surface inclined away from the building	

Wall structure		
Larch façade	2	cm
Timber battens	4	cm
Waterproofed MDF panel	2.2	cm
Cellulose fiber flakes between timber frame	36	cm
Lime cement exterior plaster	2.5	cm
Woodchip component	9	cm
Concrete	15	cm
Woodchip component	6	cm
Lime cement plaster	1.5	cm
U-value 0.10 [W/m²K]	Σ 78.2	cm

Ceiling structure		
Flooring	1	cm
Screed	5	cm
PE foil/separating layer	0.2	cm
Impact sound insulation panel	2	cm
Fill	5	cm
Reinforced concrete slab	18	cm
Expanded polystyrene	20	cm
Silicate plaster	0.5	cm
U-value 0.17 [W/m²K]	Σ 51.7	cm

2-dimensional specific values		
Ψ-value interior/exterior	-0.005	W/mK
Ψ-value interior/basement	0.032	W/mK
f_{Rsi}	0.91	-

Building Renovation: Brick Masonry Wall Insulated with ETICS and a Concrete Hollow Block Basement Ceiling Insulated on the Lower Side

This insulation concept is applicable to buildings with a low exposure to rising damp, low exposure to efflorescence, and when the ground must be dug up anyway to create a drainage.

Pre-renovation issues: rising damp, efflorescence, the basement intended for a storage room cannot be used	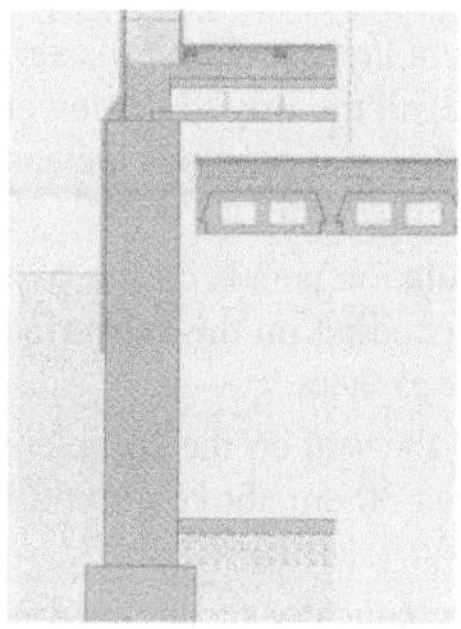
Renovation concept: the external wall is thermally renovated using an external thermal insulation composite system	
The basement ceiling slab is insulated on the lower side, the basement walls is also insulated externally and internally (insulation umbrella)	
The temperature in the basement decreases in the winter. This results in a higher RH when the moisture sources cannot be controlled. Therefore, additional intelligent basement ventilation is required	

(continued)

Construction process:	
Excavation depending on the structural possibilities	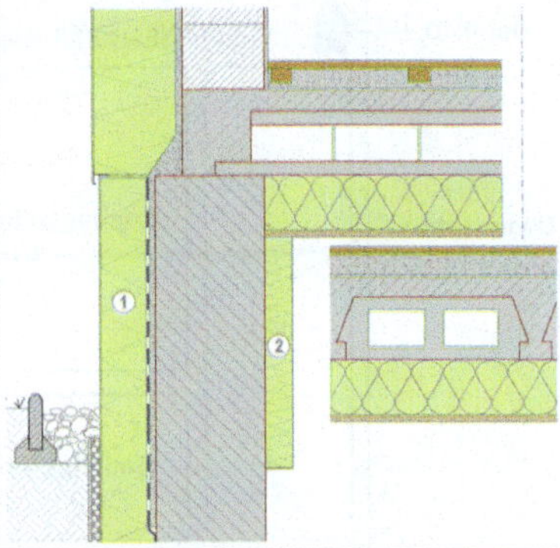
Create a clean surface (plaster base coat) before applying the vertical seal	
The exterior plaster of the basement wall is the airtight layer	
Fully bond perimeter insulation panels to the basement wall, insert an expanding foam seal on the upper side, attach drip edge, and plaster wall	
Consider the use of an intelligent ventilation system for a basement, particularly it is intended to be used as a high-quality storage space	1, XPS; 2, laminated mineral wool

Wall structure		
Silicate plaster	0.6	cm
EPS – expanded polystyrene	30	cm
Lime cement plaster	1.5	cm
Crushed brick masonry	30	cm
Lime cement plaster	2.5	cm
U-value 0.12 [W/m²K]	Σ 64.6	cm

Ceiling structure		
Parquet floor	2.5	cm
Fill, raft battens 5/8	5	cm
Fill	2.5	cm
Concrete topping	8	cm
Concrete hollow block ceiling	23	cm
Glass wool MW-W insulation felt	20	cm
Wood wool lightweight panel	2.5	cm
U-value 0.15 [W/m²K]	Σ 63.5	cm

2-dimensional specific values		
Ψ-value interior/exterior	0.036	W/mK
Ψ-value interior/basement	0.179	W/mK
f_{RSi}	0.86	-

Wall structure, in contact with the ground		
Reinforced concrete	40	cm
U-value 3.29 [W/m²K]	Σ 40.0	cm

Construction Details

This section lists contraction details for 7 types of buildings construction types and lists 30 different thermal bridge scenarios commonly found in buildings. Each scenario is supplemented with good practice commentaries (Fig. D.1), including practical applications, durability, and the retrofit construction process.

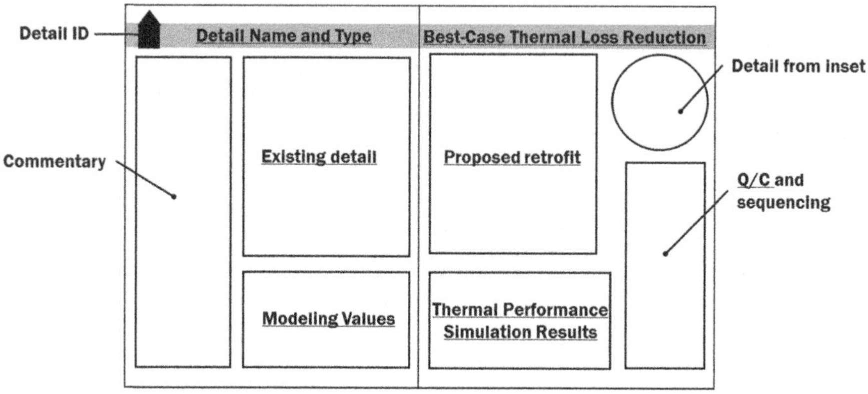

Fig. D.1 Thermal bridge mitigation catalogue page layout

Each scenario includes:

- **Commentary:** General background and details of the problems and mitigation method(s)
- **Detail identification (ID)**: Number, the detail name, and description of which correspond to the construction and building architectural type, respectively
- **Existing detail**: Illustration of the area on the structure experiencing the thermal bridging effects
- **Proposed retrofit**: Illustration of the proposed thermal bridge mitigation solution with noted changes
- **Modeling values**: Defining the material thermally relevant properties and thicknesses for the modeling of thermal bridging mitigation performance
- **Thermal Performance Simulation Results**: Noting quantitative improvements following the retrofit method
- **Q/C and sequencing**: QC or step-by-step description of how to transform existing detail into proposed solution
- **Detail from inset**: Enlarged view of region of interest providing detailed visual description of relevant components requiring special attention

The following scenarios are addressed in this section:

1. CMU or concrete wall with interior insulation:
    a. At grade (stem wall)
    b. At suspended slab (w/steel stud or exposed block)
    c. At parapet with concrete roof, concrete parapet

d. Steel roof joists at parapet
e. Window jamb
f. Window head
g. Window sill
h. Blast resistant window jamb
i. Door jambs to CMU
j. Thru slab projection, e.g., shade or balcony

2. CMU or concrete wall with exterior insulation:

   a. Roof parapet with concrete roof
   b. Roof parapet with OWSJ + deck
   c. At grade transition (stem wall)
   d. Window jamb
   e. Window head
   f. Window sill
   g. Blast-resistant window jamb
   h. Blast-resistant window head
   i. Suspended slab at shelf angle

3. Steel stud infill wall in steel or concrete frame:

   a. Roof parapet with steel frame
   b. Window jamb
   c. Window head
   d. Window sill
   e. Steel tube blast resistant curtain wall perimeter

4. Steel building with insulated metal panel:

   a. Eave detail

5. Precast sandwich panel:

   a. Roof of steel joists bearing on inner wythe of sandwich

6. Important clear wall details:

   a. 6 in. steel studs @ 16 in. with brick ties
   b. Concrete wall with interior steel stud assembly

7. Historical details with interior insulation:

   a. Window sill in solid brick masonry

1a

CMU or Concrete Wall with Interior Insulation

At-Grade Stem Wall

Commentary

Below-grade insulation may be expanded polystyrene or extruded polystyrene. The finish can be cement-based stucco with corrosion-resistant reinforcement, metal, or PVC sheets. An aesthetically appealing finish is also often desired.

In case of damp or irrigated landscaping, a waterproofing layer should be attached to the foundation before placing the insulation.

Insulation requires impact protection and, in the case of foam plastics, ultraviolet radiation protection. Appropriate protection can be selected based on stucco with corrosion-resistant mesh (polymer-modified reinforced with glass fiber), synthetic stucco and cement board materials.

Foam insulation can act as a protected pathway for termite access, so appropriate flashing, and termiticides should be used.

The exterior insulation top can be covered by inserting metal flashing into a reglet cut into the wall.

The thermal performance of this detail can be further improved by increasing the thickness of the insulation and covering more of the exterior, as well as by adding a skirt around the perimeter to reduce losses to the soil.

The reported Ψ-value does not include thermal effects related to the insulation protection or the top metal flashing.

Existing stem wall

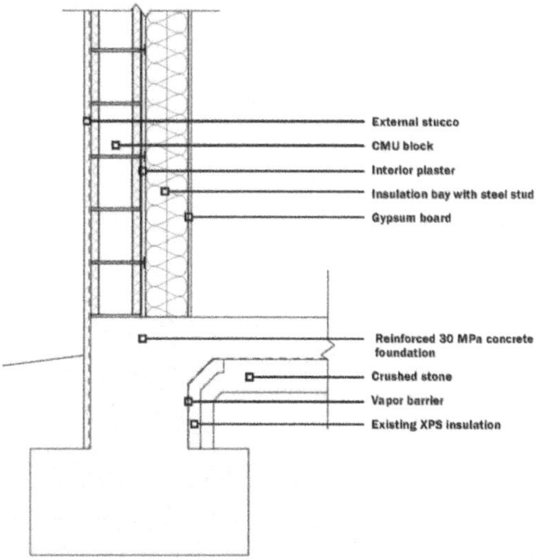

1a Stem Wall Modeling Values

Component	Thickness Inches (mm)	Conductivity Btu/h·ft·°F (W/m K)	Nominal Resistance hr-ft2·°F/Btu (m²K/W)	Density lb/ft³ (kg/m³)
Interior Film (Wall)	–	–	R-0.74 (0.13 RSI)	–
Interior Film (Floor)	–	–	R-0.97 (0.17 RSI)	–
External Stucco	1" (25)	0.8089 (1.4)	R-0.10 (0.018 RSI)	115(1850)
CMU	7 5/8" (194)	0.069 (1.2)	R-0.916 (0.161 RSI)	130 (2100)
Interior Plaster	3/4" (19)	0.8089 (1.4)	R-0.08 (0.014 RSI)	115(1850)
Insulation Bay with Steel Studs	6 1/8" (156)	0.0584 (0.10)	R-8.7 (1.539 RSI)	–
Gypsum Board	1/2" (13)	0.092 (0.16)	R-0.5 (0.08 RSI)	50 (800)
Exterior XPS Insulation	3 1/2" (89)	0.017 (0.029)	R-17.2 (3 RSI)	1.8 (28)
Reinforced 30 MPa concrete	–	1.04 (2.4)	–	150 (2400)
Existing XPS Insulation	2" (51)	0.017 (0.029)	–	1.8 (28)
Reinforced 30 MPa concrete	6.5" (165)	1.04 (2.4)	R-0.39 (0.069 RSI)	150 (2400)
Crushed Stone	5" (127)	0.9245 (1.6)	R-0.45 (0.079 RSI)	125(2000)
Exterior Film (Wall)	–	–	R-0.23 (0.04 RSI)	–

Best-Case Thermal-Loss Reduction: 0.176 Btu/h • ft • °F

Proposed retrofit

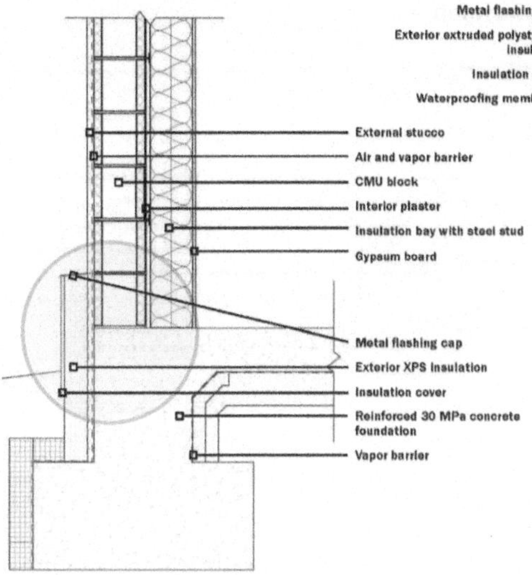

Detail from inset

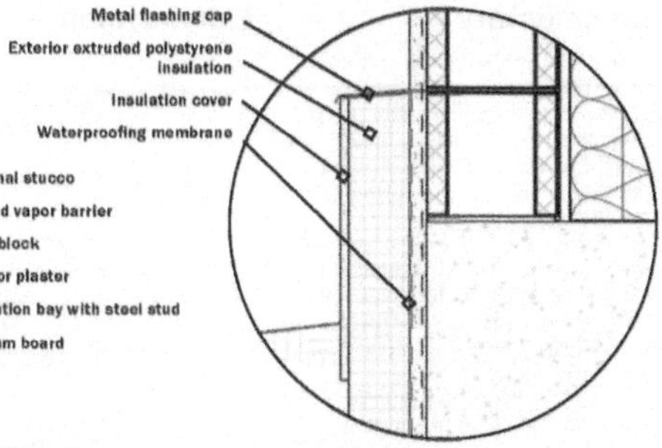

1a Thermal Performance Simulation Results		
Condition	**Clear Wall R-Value (W/m²K)**	**Linear Transmittance (Ψ) Btu/h • ft • °F (W/mK)**
Wall Clear Field	R-10.3 (0.552)	–
Floor Clear Field	R-0.84 (6.749)	
As-Built Slab (no exterior or interior insulation)	–	-0.359 (-0.622)
Retrofit with...		
12" exterior	–	-0.452 (-0.782)
12" exterior (x) and 12" exterior (y)	–	-0.471 (-0.815)
18" exterior (x) and 12" exterior (y)	–	-0.507 (-0.877)
24" exterior (x) and 12" exterior (y)	–	-0.535 (-0.926)
As-built transmittance – Lowest retrofit transmittance = Best-case thermal-loss reduction		
0.359-(-0.535) = 0.176 Btu/h·ft·°F		

Q/C and sequencing

1. Excavate around the entire building perimeter until reaching the foundation footing.
2. Dry-clean the exposed surfaces of the foundation wall and footing, and remove any other surface debris.
3. In the case of high moisture content soil or irrigated landscaping, apply a waterproofing material layer to prevent water infiltration.
4. Measure and cut the selected external foundation insulation to size, ensuring that it wraps the entire foundation footing and wall. This should extend from the foundation footing above grade.
5. Use adhesive to attach the selected exterior insulation to the foundation wall.
6. Protect or cover the foundation insulation.
7. Install flashing above the insulation by inserting through the reglet cut, making sure to entirely cover the selected insulation.

1b

CMU or Concrete Wall with Interior Insulation

Suspended Slab (Steel Stud or Exposed Block)

Commentary

It is relatively simple to add air and vapor barrier to the exterior of the existing building at the same time as insulation, thereby maintaining the existing structure and enclosure at more stable temperatures and much drier conditions.

The insulation choice is often limited to products that can resist moisture and some wind, meet fire codes, etc. Thus the choices tend to be fire-resistant foam plastics and stone wool semi-rigid boards.

In most of the cases the concrete wall itself, or at least its paint finish, will act as a vapor barrier at the exterior. It is usual to not add further waterproofing membrane at the concrete exterior.

The flashing at the insulation top edge should be steeply sloped and durable. Similarly, it is important for the drip edge at the bottom of the projecting insulation band to avoid collecting rain and melt water.

Although shown as concrete, the wall could also be precast, tilt-up, or CMU construction with almost no impact on thermal performance. It is assumed the wall is functional before the retrofit is applied.

Thermal performance can be further improved by increasing the thickness of the insulation and covering more (or all) of the exterior. The primary improvement is gained by increasing "h," and the entire wall should ideally be covered with exterior insulation to achieve high performance.

Existing slab

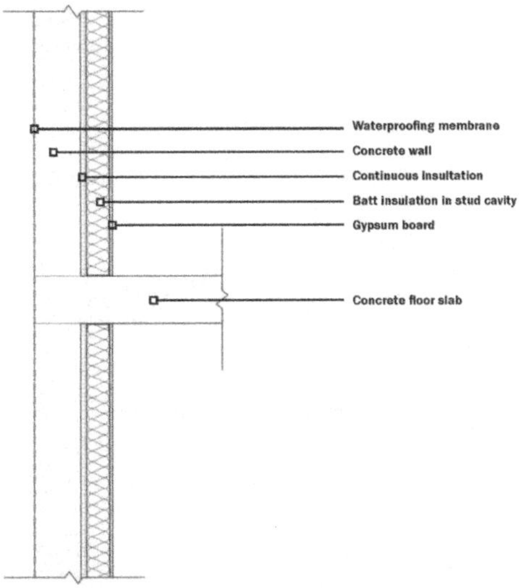

1b Suspended Slab Modeling Values

Component	Thickness Inches (mm)	Conductivity Btu/h • ft • °F (W/m K)	Nominal Resisitance hr·ft²·°F/Btu (m²K/W)	Density lb/ft³ (kg/m³)
Interior Film (right side)1	-	-	R-0.6 (0.11 RSI) to	-
R-0.9 (0.16 RSI)	-	0.578 (1)	R-0.523 (0.092 RSI)	110 (1800)
Gypsum Board	1/2" (13)	0.092 (0.16)	R-0.5 (0.08 RSI)	50 (800)
3 5/8" Steel Studs with Top and Bottom Tracks	18 Gauge	35.825 (62)	-	489 (7830)
Batt Insulation in Stud Cavity	3 5/8" (92)	0.024 (0.042)	R-12 (2.1 RSI)	0.9 (14)
Continuous Insulation	1" (25)	-	R-5 (0.88 RSI)	1.8 (28)
Concrete Wall/Floor Slab	8" (203)	1.04 (1.8)	-	140 (2250)
Exterior insulation	2" (51)	0.017 (0.029)	R-10 (1.75 RSI)	1.8 (28)
Exterior Film (left side)1	-	-	R-0.2 (0.03 RSI)	-
Brick Sill	3 5/8" (92)	0.578 (1)	-	110 (1800)
Exterior Film	-	-	R-0.23 (0.04 RSI)	-

Best-Case Thermal Loss Reduction: 0.334 Btu/h • ft • °F

Proposed retrofit

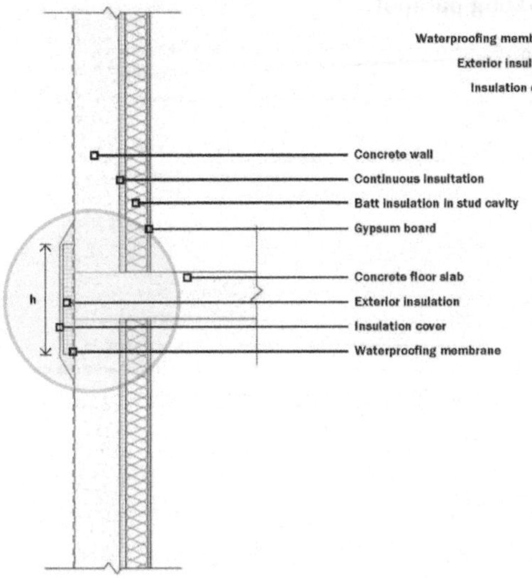

Detail from inset

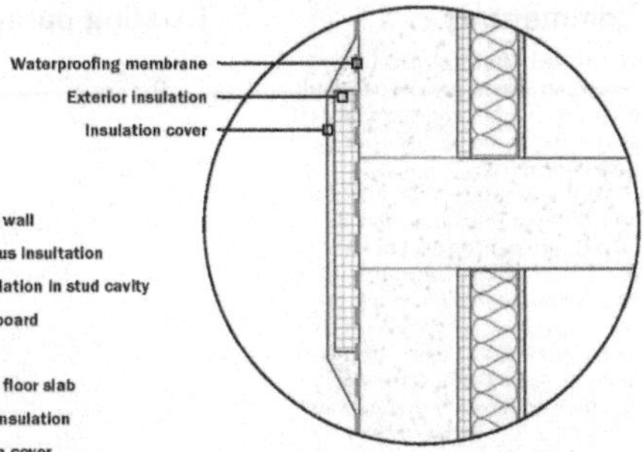

1b Thermal Performance Simulation Results			
	Condition	Clear Wall R-Value (W/m²K)	Linear Transmittance (Ψ) Btu/h • ft • °F (W/mK)
R5 Foam Insulation between Studs and Concrete	Wall Clear Field	R-15 (2.64 RSI)	-
	As-Built Slab (no exterior insulation)	-	0.494 (0.855)
	8" (h) external insulation	-	0.457 (0.79)
	18" (h) external insulation	-	0.388 (0.671)
	36" (h) external insulation	-	0.275 (0.476)
No Insulation between Studs and Concrete	Wall Clear Field	R-7.7 (1.36 RSI)	-
	As-Built Slab (no exterior insulation)	-	0.486 (0.841)
	8" (h) external insulation	-	0.423 (0.731)
	18" (h) external insulation	-	0.326 (0.564)
	36" (h) external insulation	-	0.152 (0.262)
As-built transmittance - Lowest retrofit transmittance = Best-case thermal-loss reduction 0.486-0.152 = 0.334 Btu/h·ft·°F			

Q/C and sequencing

1. Measure and cut the external selected insulation (i.e. stone wool semi-rigid boards) to an appropriate height, h, as indicated in the 1b thermal performance table (8–36 in.).
2. Cut a slit in the concrete wall, allowing the top insulation flashing cover to be inserted later. In the case of a CMU wall, cut a slit in the mortar joint.
3. Attach the exterior insulation to the wall covering the floor slab junction region by using appropriate insulation adhesive.
4. Attach the exterior insulation finish cover.
5. Add a bead of sealant into the slit cut in step 2.
6. Insert metal flashing into the slit and add additional bead of sealant to provide watershed.

1c

CMU or Concrete Wall with Interior Insulation

At Parapet with Concrete Roof

Commentary

Although shown as concrete, the wall could also be precast panel, tilt-up panel, or CMU construction. It is assumed that the wall is functional before the retrofit is applied.

The interior insulation will need air and vapor resistance, and should be pressure tight against the wall to prevent convective loops. It also must be a fire retardant or have a fire-resistive barrier. For effective convective loop suppression, there should be no noticeable gap between the insulation and the interfaces. Additional vapor barrier may be needed subject to climate or location requirements.

The thermal performance results incorporate the effects associated with the interior, exterior film coefficients, metal cap flashing and roof finish material.

Adding at least 24 in. of interior insulation along the interior ceiling will reduce the heat flow by one third. To accomplish this, interior building access is required.

Thermal performance can be further improved by increasing the insulation thickness and covering more or all the exterior. Thermal-breaking the heat flow by replacing the parapet base support CMU with a low-conductivity block will provide the best best parapet thermal performance.

Existing parapet

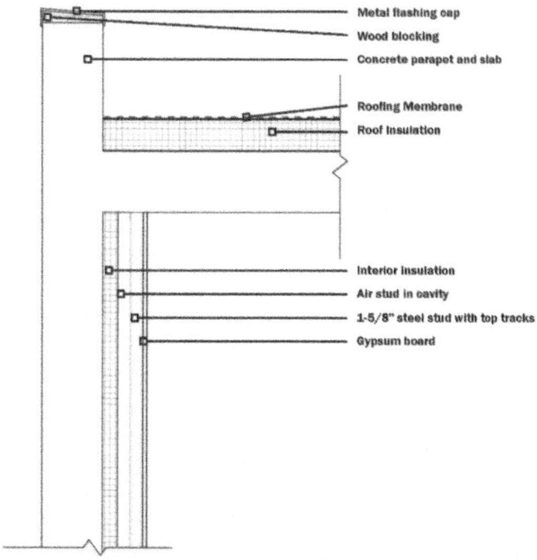

1c Parapet Modeling Values (Concrete Roof)

Component	Thickness Inches (mm)	Conductivity Btu/h•ft•°F (W/m K)	Nominal Resistance hr•ft²•°F/Btu (m²K/W)	Density lb/ft³ (kg/m³)
Interior Film (right side)	-	-	R-0.6 (0.11 RSI) to	-
R-0.7 (0.12 RSI)	-	0.578 (1)	R-0.523 (0.092 RSI)	110 (1800)
Gypsum Board	1/2" (13)	0.092 (0.16)	R-0.5 (0.08 RSI)	50 (800)
1 5/8" Steel Studs with Top Tracks	20 gauge	35.825 (62)	-	489 (7830)
Air in Stud Cavity	1 5/8" (42)	-	R-0.9 (0.16 RSI)	0.075 (1.2)
Interior Insulation	2" (50)	-	R-10 (1.76 RSI)	1.8 (28)
Roof Insulation	4" (102)	-	R-20 (3.5 RSI)	1.8 (28)
Wood Blocking	5/8" (16)	0.052 (0.09)	R-1 (0.18 RSI)	27.8 (445)
Concrete Slab & Parapet	8" (203)	1.04 (1.8)	-	140 (2250)
Metal cap flashing/ finish roof material is incorporated into exterior heat transfer coefficient	3 5/8" (92)	0.578 (1)	-	110 (1800)
Exterior Insulation	2" (51)	0.017 (0.029)	R-10 (1.75 RSI)	1.8 (28)
Interior Insulation	2" (51)	0.017 (0.029)	R-10 (1.75 RSI)	1.8 (28)
Exterior Film (left side)	-	-	R-0.2 (0.03 RSI)	-

Best-Case Thermal Loss Reduction: 0.304 Btu/h • ft • °F

Proposed retrofit

Detail from inset

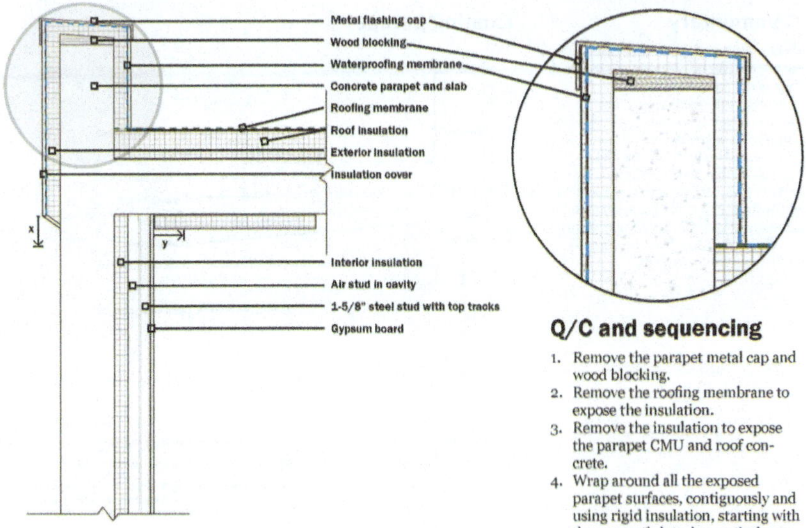

1c Thermal Performance Simulation Results

	Condition	Clear Wall R-Value (W/m²K)	Linear Transmittance (Ψ) Btu/h • ft • °F (W/mK)
R10 Wall Insulation	Wall Clear Field	R-13.5 (2.37 RSI)	-
	Roof Clear Field	R-21.4 (3.77 RSI)	-
	As-Built Slab (no exterior or interior insulation)	-	0.6 (1.038)
	12" exterior (x) and 12" interior (y)	-	0.174 (0.301)
No Wall Insulation	Wall Clear Field	2.79 (0.49)	-
	Roof Clear Field	21.41 (3.77)	-
	As-Built Slab (no exterior or interior insulation)	-	0.451 (0.78)
	12" exterior (x) and 12" interior (y)	-	-0.006 (-0.011)
	12" exterior (x) and 24" interior (y)	-	-0.085 (-0.049)
	12" exterior (x) and 0" interior (y)	-	0.081 (0.14)
	no exterior and 12" interior (y)	-	0.229 (0.397)
	no exterior and 24" interior (y)	-	0.147 (0.254)
As-built transmittance – Lowest retrofit transmittance = Best-case thermal-loss reduction			
0.451-0.147 = 0.304 Btu/h·ft·°F			

Q/C and sequencing

1. Remove the parapet metal cap and wood blocking.
2. Remove the roofing membrane to expose the insulation.
3. Remove the insulation to expose the parapet CMU and roof concrete.
4. Wrap around all the exposed parapet surfaces, contiguously and using rigid insulation, starting with the parapet's interior vertical surface, followed by the top and finally to the parapet's external vertical surface. The external vertical surface insulation should be extended 12 in. below the roof ceiling plane (see the proposed solution detail drawing).
5. Replace the insulation that was cut at the roof (step 3). The roof insulation should possess sufficient continuity with the insulation added to the parapet interior face in step 4.
6. Replace all the previously removed roofing membrane so that continuous protection to both the roof and the parapet is restored.
7. Reinstall the parapet wood blocking.
8. Reinsert the parapet metal cap.

1d

CMU or Concrete Wall with Interior Insulation

Parapet with Steel Roof Deck and Steel Joists

Commentary

For improved thermal performance, insulation can be added at the external face of the parapet.

For the interior flanking, use closed-cell spray foam to control vapor diffusion condensation in climate zones 5 and higher. This insulation type is capable of providing excellent airflow control, by conforming and adhering to rough surfaces, making them common choices for interior masonry retrofits. Like most plastics, they do not support mold growth.

Existing parapet

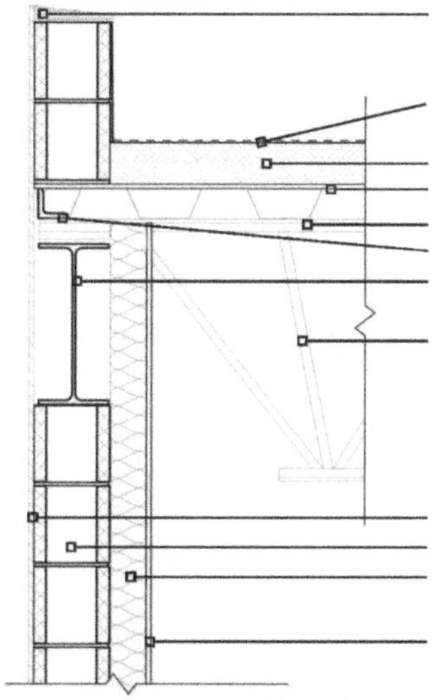

1d Parapet Modeling Values (Steel Roof Deck and Joists)				
Component	**Thickness Inches (mm)**	**Conductivity Btu/h • ft• °F (W/m K)**	**Nominal Resisitance hr·ft²·°F/Btu (m²K/W)**	**Density lb/ft³ (kg/m³)**
Gypsum Board	1/2" (13)	0.092 (0.16)	R-0.5 (0.08 RSI)	50 (800)
3 5/8" x 1 5/8" Steel Studs	18 Gauge	35.825 (62)	-	489 (7830)
Fiberglass Batt Insulation	3 5/8" (92)	0.024 (0.042)	R-12 (2.1 RSI)	0.9 (14)
CMU Block Wall	8" (203)	0.603 (1.044)	R-1.1 (0.19 RSI)	140 (2550)
Exterior Sheathing	1/2" (13)	0.092 (0.16)	R-0.5 (0.09 RSI)	50 (800)
Beam Air Cavities	-	-	R-0.9 (0.16 RSI)	-
Steel Beam (W410)	-	28.891 (50)	-	489 (7830)
Open Web Steel Joist	-	28.891 (50)	-	489 (7830)
Steel Deck	-	28.891 (50)	-	489 (7830)
Shelf Angle	-	35.825 (62)	-	489 (7830)
Roof Air Cavity	3" (76)	-	R-0.9 (0.16 RSI)	-
Gypsum Overlay Board	1/2" (13)	0.092 (0.16)	R-0.5 (0.08 RSI)	50 (800)
Exterior Insulation, Roof	4" (102)	-	R-20 (3.5 RSI)	1.8 (28)
Wood Blocking	5/8" (16)	0.052 (0.09)	R-1 (0.18 RSI)	27.8 (445)
Metal cap flashing/ finish roof material is incorporated into exterior heat transfer coefficient				
Interior Spray Foam	-	-	R-10 (1.75 RSI)	1.8 (28)
Exterior Insulation, Parapet	2" (51)	0.017 (0.029)	R-10 (1.75 RSI)	1.8 (28)

Best-Case Thermal Loss Reduction: 0.853 Btu/h • ft • °F

Proposed retrofit

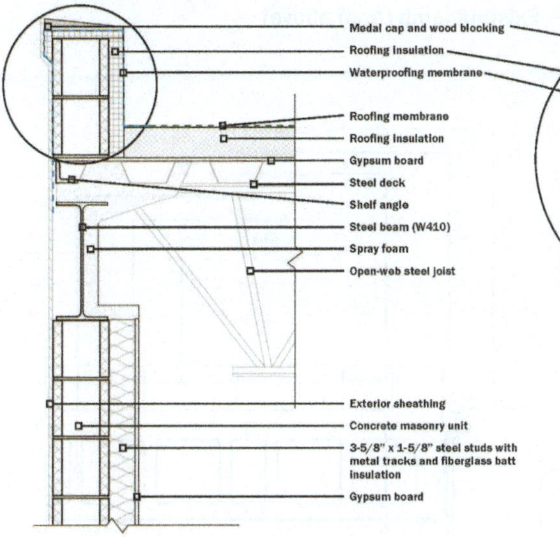

Detail from inset

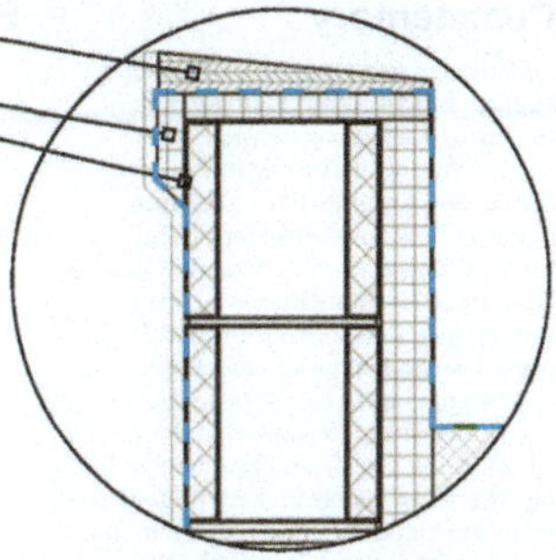

1d Thermal Performance Simulation Results

Condition	Clear Wall R-Value (W/m²K)	Linear Transmittance (Ψ) Btu/h • ft • °F (W/mK)
Wall Clear Field	R-8.9 (1.57 RSI)	-
Roof Clear Field	R-21.4 (3.77 RSI)	-
As-Built Slab, full stud height	-	0.574 (0.994)
As-Built Slab, low stud height		1.314 (2.274)
Retrofit with...		
R10 Parapet Insulation, full stud height	-	0.548 (0.948)
R10 Spray Foam, low stud height	-	0.461 (0.798)
As-built transmittance – Lowest retrofit transmittance = Best-case thermal-loss reduction **1.314-0.461 = 0.853 Btu/h·ft·°F**		

Q/C and sequencing

Exterior

1. Remove the parapet cap flashing and wood blocking.
2. Remove the roofing membrane to expose the CMU portion of roof parapet and part of roof insulation.
3. Cut back and remove roof insulation section to expose base of the CMU parapet and metal roof deck.
4. Add rigid insulation to vertical face of concrete parapet. The bottom must contact the roof and it should extend to the top of the parapet.
5. Add rigid insulation above the parapet top surface. Make sure it covers the top of the exterior sheathing and is contiguous with the parapet interior wall insulation.
6. Replace the previously removed roof insulation, making sure it is contiguous with parapet interior wall insulation at the parapet base.
7. Replace all previously removed waterproofing membrane (see Step 2) to restore continuous protection to roof and parapet.
8. Replace the wood blocking and flashing atop the parapet.

Interior

1. Apply layer of spray foam insulation along roof/wall intersection.
2. Apply a finish layer of spray-on fire protection over the insulation.

1e

CMU or Concrete Wall with Interior Insulation

Window Jamb

Commentary

Although shown as CMU, the wall's support function could also be precast, tilt-up, or concrete construction.

The hollow open space in window frames will promote natural convective heat flow between exterior and interior. This undesired heat flow can be reduced by filling these voids with factory-installed, custom-formed foam plastic or rigid stone wool sections.

Designers need to complete the exterior closure to ensure that the insulation and the air and vapor barrier layers are not visible and are protected from sun and direct rain impingement.

It is critical that the head flashing direct water outward at this location and it is also necessary to maintain air tightness here.

Heat flow can be reduced by a factor of four by wrapping a thin layer of insulation into the rough opening. The thermal performance can be further improved by increasing the thickness of the insulation, especially at the window rough opening return.

Note that these comments and the sequencing notes at far right also apply to items 1f and 1g.

Existing jamb (from above)

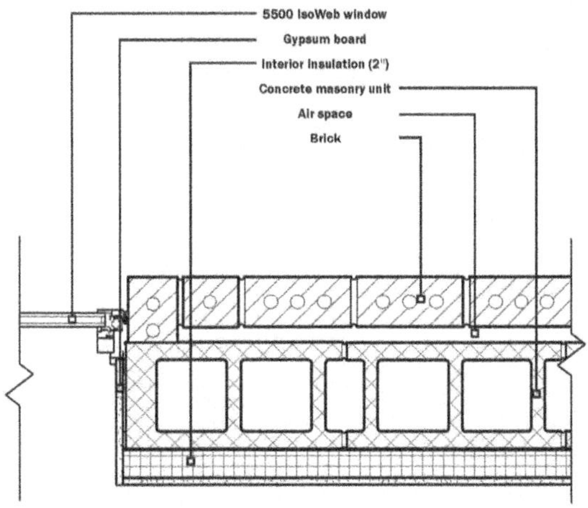

1e Window Jamb Modeling Values				
Component	Thickness Inches (mm)	Conductivity Btu/h • ft • °F (W/m K)	Nominal Resistance hr-ft²·°F/Btu (m²K/W)	Density lb/ft³ (kg/m³)
Interior Film	-	-	R-0.74 (0.13 RSI)	-
Brick	3 5/8" (92)	0.578 (1)	R-0.523 (0.092 RSI)	110 (1800)
Air Cavity	1" (25)	0.070 (0.122)	R-1.185 (0.209 RSI)	-
CMU	7 5/8" (194)	0.069 (1.2)	R-0.916 (0.161 RSI)	130 (2100)
Interior Insulation	2" (51)	0.0139 (0.024)	R-12 (2.11 RSI)	2 (32)
Gypsum Board	1/2" (13)	0.092 (0.16)	R-0.5 (0.08 RSI)	50 (800)
5500 IsoWeb Window	-	-	-	-
Plywood	3/4" (19)	0.058 (0.1)	-	40 (600)
Interior Insulation	1" (25)	0.0139 (0.024)	-	2 (32)
Air - Water Control Layer	-	-	-	-
Exterior Film	-	-	R-0.23 (0.04 RSI)	-

Best-Case Thermal Loss Reduction: 0.229 Btu/h • ft • °F

Proposed retrofit (from above)

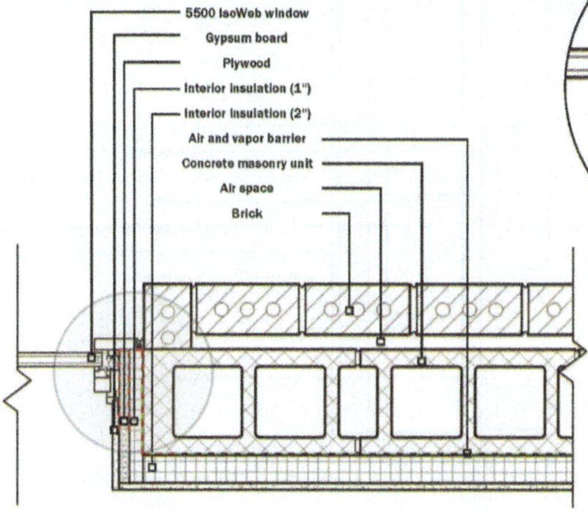

Detail from inset

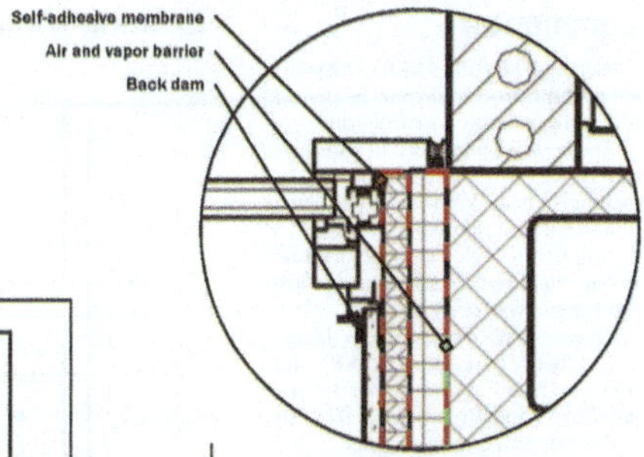

1e Thermal Performance Simulation Results		
Condition	Clear Wall R-Value (W/m²K)	Linear Transmittance (Ψ) Btu/h • ft • °F (W/mK)
Wall Clear Field	R-16 (0.351)	-
As-Built Fitting Situation	-	0.308 (0.533)
Proposed Fitting Situation	-	0.079 (0.136)
As-built transmittance – Lowest retrofit transmittance = Best-case thermal-loss reduction 0.308-0.079 = 0.229 Btu/h·ft·°F		

Q/C and sequencing

1. Remove old window and any caulk or seals (may require chiselling out the reveals to remove fixing straps).
2. Remove the sill brick.
3. Remove gypsum board at window opening (if needed the drywall at the reveal) and the mortar bed.
4. Add a plywood buck adhered to a high R/in. insulation and pre-wrapped in a self-adhesive membrane to all four sides of the reveal.
5. Apply the self-adhesive watershed membrane at the sill, and tape all joints to the reveals. The watershed membrane at the sill should run from the back dam (see detail 1g), cover the exposed CMU vertical face, and work as the base of the sill block.
6. Apply a bead of sealant to the back dam.
7. Reinsert the window into the window opening.
8. Replace the sill block (see detail 1g).
9. Apply snap-on trims. Fix a preformed housing around the reveals.
10. Add snap-on cover piece over the exterior-exposed window frame.

1f

CMU or Concrete Wall with Interior Insulation

Window Head

Commentary

Although shown as CMU, the wall's support function could also be precast, tilt-up, or concrete construction.

The hollow open space in window frames will promotet natural convective heat flow through it. This undesired heat flow can be reduced by filling these voids with factory-installed, custom-formed foam plastic or rigid stone wool sections.

Designers need to complete the exterior closure to ensure that the insulation and the air and vapor barrier are not visible, and are protected from sun and direct rain impingement.

It is critical that the head flashing direct water outward at this location, and it is also necessary to maintain air tightness here.

Heat flow can be reduced by a factor of four by wrapping a thin layer of insulation into the rough opening. The thermal performance can be further improved by increasing the thickness of the insulation, especially at the window rough opening return.

Note that these comments and the sequencing notes at far right also apply to items 1e and 1g.

Existing window head

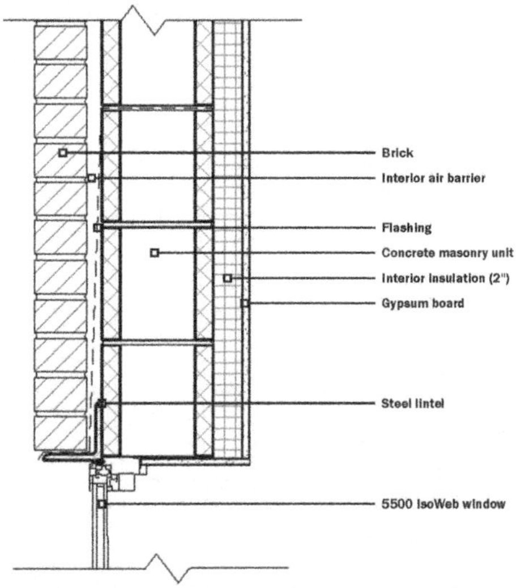

1f Window Head Modeling Values

Component	Thickness Inches (mm)	Conductivity Btu/h • ft • °F (W/m K)	Nominal Resisitance hr·ft²·°F/Btu (m²K/W)	Density lb/ft³ (kg/m³)
Interior Film	-	-	R-0.74 (0.13 RSI)	-
Brick	3 5/8" (92)	0.578 (1)	R-0.523 (0.092 RSI)	110 (1800)
Air Cavity	1" (25)	0.070 (0.122)	R-1.185 (0.209 RSI)	-
CMU	7 5/8" (194)	0.069 (1.2)	R-0.916 (0.161 RSI)	130 (2100)
Interior Insulation	2" (51)	0.0139 (0.024)	R-12 (2.11 RSI)	2 (32)
Gypsum Board	1/2" (13)	0.092 (0.16)	R-0.5 (0.08 RSI)	50 (800)
Air/Water Control Layer	-	-	-	-
Steel Lintel	-	27.7 (48)	-	450(7700)
5500 IsoWeb window	-	-	-	-
Plywood	3/4" (19)	0.058 (0.1)	-	38 (600)
Interior Insulation	1" (25)	0.0139 (0.024)	-	2 (32)
Exterior Film	-	-	R-0.23 (0.04 RSI)	-

Best-Case Thermal Loss Reduction: 0.229 Btu/h • ft • °F

Proposed retrofit

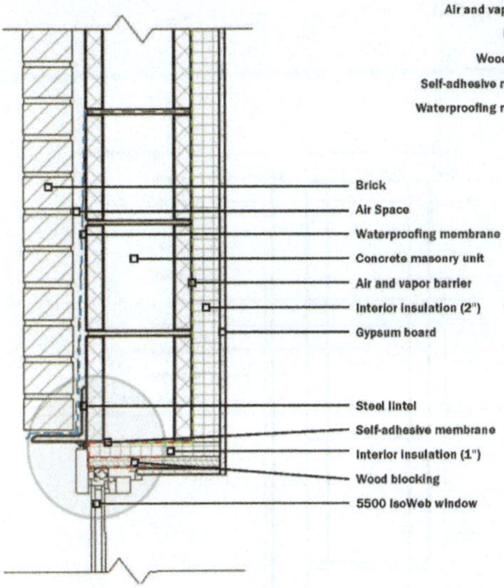

Detail from inset

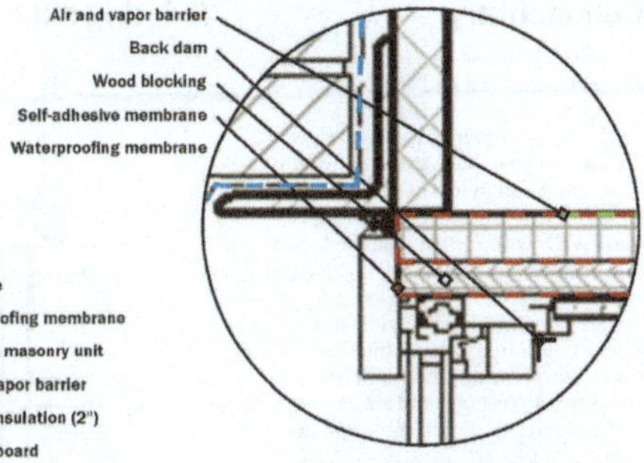

1f Thermal Performance Simulation Results		
Condition	Clear Wall R-Value (W/m²K)	Linear Transmittance (Ψ) Btu/h • ft • °F (W/mK)
Wall Clear Field	R-16 (0.351)	-
As-Built Fitting Situation	-	0.315 (0.546)
Proposed Fitting Situation	-	0.086 (0.149)
As-built transmittance – Lowest retrofit transmittance = Best-case thermal-loss reduction 0.315-0.086 = 0.229 Btu/h·ft·°F		

Q/C and sequencing

1. Remove old window and any caulk or seals (may require chiselling out the reveals to remove fixing straps).
2. Remove the sill brick.
3. Remove gypsum board at window opening (and, if needed, the drywall at the reveal) and the mortar bed.
4. Add a wood blocking adhered to a high R/in. insulation and prewrapped in a self-adhesive membrane to all four sides of the reveal.
5. Apply the self-adhesive watershed membrane at the sill and tape all joints to the reveals. The watershed membrane at the sill should run from the back dam, cover the exposed CMU vertical face, and work as the base of the sill block.
6. Apply a bead of sealant to the back dam.
7. Reinsert the window into the window opening.
8. Replace the sill block.
9. Apply snap-on trims. Fix a preformed housing around the reveals.
10. Add snap-on cover piece over the exterior-exposed window frame.

CMU or Concrete Wall with Interior Insulation

1g Window Sill

Commentary

Although shown as CMU, the wall's support function could also be precast, tilt-up, or concrete construction.

The hollow open space in window frames will promote natural convective heat flow through it. This undesired heat flow can be reduced by filling these voids with factory-installed, custom-formed foam plastic or rigid stone wool sections.

Designers need to complete the exterior closure to ensure that the insulation and the air and vapor control layers are not visible, and are protected from sun and direct rain impingement.

It is critical that the head flashing direct water outward at this location, and it is also necessary to maintain air tightness here.

Heat flow can be reduced by a factor of four by wrapping a thin layer of insulation into the rough opening. The thermal performance can be further increased by increasing the thickness of the insulation, especially at the window rough opening return.

Note that these comments and the sequencing notes at far right also apply to items 1e and 1f.

Existing sill

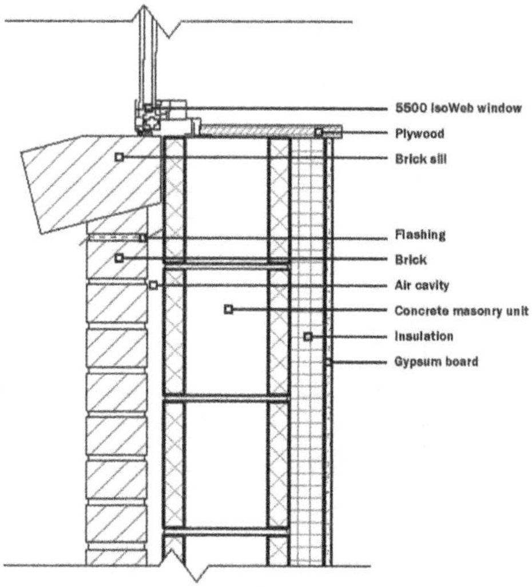

1g Window Sill Modeling Values

Component	Thickness Inches (mm)	Conductivity Btu/h•ft•°F (W/m K)	Nominal Resistance hr•ft^2•°F/Btu (m^{2}K/W)	Density lb/ft^3 (kg/m^3)
Interior Film	-	-	R-0.74 (0.13 RSI)	-
Brick	3 5/8" (92)	0.578 (1)	R-0.523 (0.092 RSI)	110 (1800)
Air Cavity	1" (25)	0.070 (0.122)	R-1.185 (0.209 RSI)	-
CMU	7 5/8" (194)	0.069 (1.2)	R-0.916 (0.161 RSI)	130 (2100)
Interior Insulation	2" (51)	0.0139 (0.024)	R-12 (2.11 RSI)	2 (32)
Gypsum Board	1/2" (13)	0.092 (0.16)	R-0.5 (0.08 RSI)	50 (800)
5500 IsoWeb Window	-	-	-	-
Aluminum Sill Flashing	12 Gauge	86 (160)	-	
Brick Sill	3 5/8" (92)	0.578 (1)	-	110 (1800)
Concrete Sill	-	1.4 (2.4)	-	110 (1800)
Air Water Control Layer	-	-	-	-
5500 IsoWeb Window	-	-	-	-
Plywood	3/4" (19)	0.058 (0.1)	-	40 (600)
Rigid Insulation	1" (25)	0.0139 (0.024)	-	2 (32)
Exterior Film	-	-	R-0.23 (0.04 RSI)	-

Best-Case Thermal Loss Reduction: 0.234 Btu/h • ft • °F

Proposed retrofit

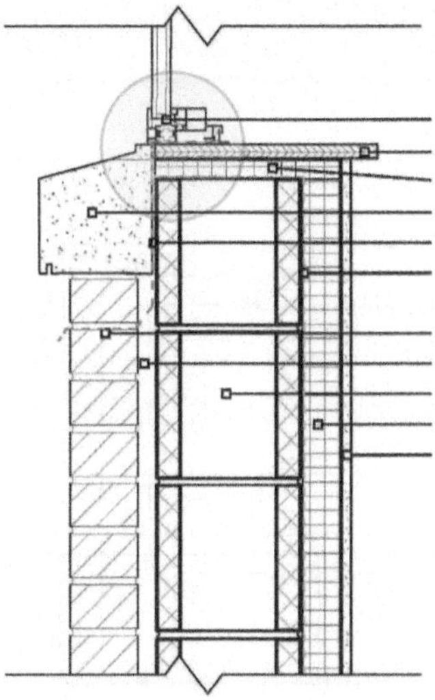

Detail from inset

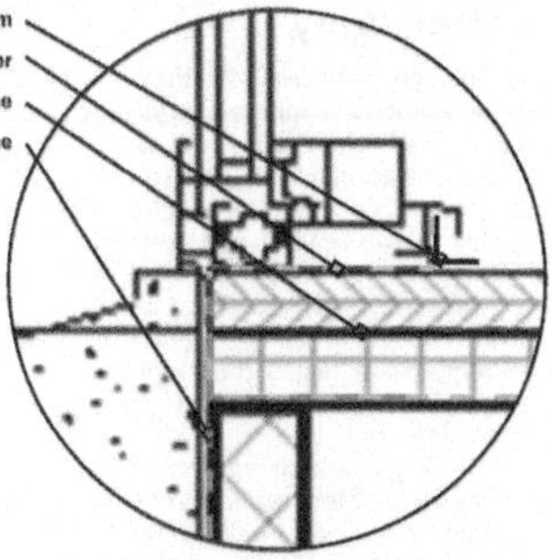

1g Thermal Performance Simulation Results		
Condition	Clear Wall R-Value (W/m²K)	Linear Transmittance (Ψ) Btu/h • ft • °F (W/mK)
Wall Clear Field	R-15.7 (0.369)	-
Existing Fitting Situation	-	0.322 (0.558)
Proposed Fitting Situation	-	0.088 (0.152)
As-built transmittance – Lowest retrofit transmittance = Best-case thermal-loss reduction 0.322-0.088 = 0.234 Btu/h·ft·°F		

Q/C and sequencing

1. Remove old window and any caulks or seals (may require chiselling out the reveals to remove fixing straps).
2. Remove the sill brick.
3. Remove gypsum board at window opening (and, if needed, the dry-wall at the reveal) and the mortar bed.
4. Add a plywood adhered to a high R/in. insulation and prewrapped in a self adhesive membrane to all four sides of the reveal.
5. Apply the self-adhesive watershed membrane at the sill and tape all joints to the reveals. The watershed membrane at the sill should run from the back dam, cover the exposed CMU vertical face, and work as the base of the sill block.
6. Apply a bead of sealant to the back dam.
7. Replace the sill block.
8. Reinsert the window into the window opening.
9. Apply snap-on trims. Fix a pre-formed housing around the reveals.
10. Add snap-on cover piece over the exterior-exposed window frame.

1h

CMU or Concrete Wall with Interior Insulation

Blast-Resistant Window Jamb

Commentary

Although shown as CMU, the support function could be provided by concrete or masonry construction.

To transfer high blast loads from the window frame to the structure, solid wood can be used. The insulation level provided by the wood is modest but can make a very significant reduction in thermal bridging. Alternative solutions could include the use of insulation at the jamb and widely spaced metal clips.

Providing continuity of the air barrier system and managing convection are important issues for this detail.

The primary risk in the existing condition is that uninsulated gypsum wallboard can often be damaged by condensation at these jamb locations in cold weather. The wood or insulation at the jamb in the proposed corrected detail must be protected from the exterior by an air and vapor barrier.

The performance of the detail could be increased by using thicker layers of insulation.

Existing jamb (from above)

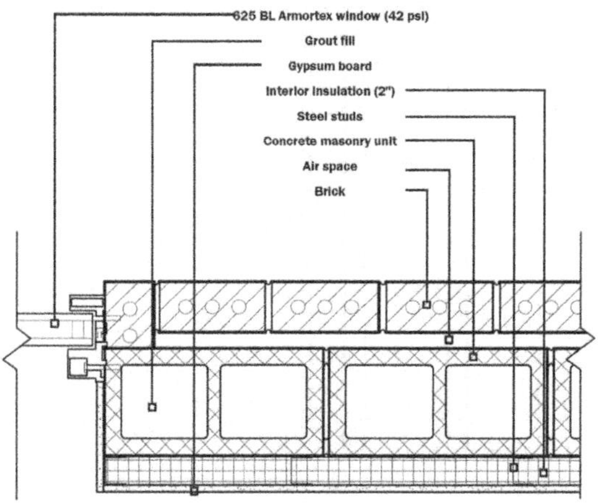

1h Blast-Resistant Window Jamb Modeling Values

Component	Thickness Inches (mm)	Conductivity Btu/h • ft • °F (W/m K)	Nominal Resistance hr-ft²·°F/Btu (m²K/W)	Density lb/ft³ (kg/m³)
Interior Film	-	-	R-0.74 (0.13 RSI)	-
Brick	3 5/8" (92)	0.578 (1)	R-0.523 (0.092 RSI)	110 (1800)
Air Cavity	1" (25)	0.070 (0.122)	R-1.185 (0.209 RSI)	-
CMU	7 5/8" (194)	0.069 (1.2)	R-0.916 (0.161 RSI)	130 (2100)
Grout Fill		0.092 (1.6)	-	125 (2000)
Insulation	-	0.021 (0.036)	-	0.8 (13)
Gypsum Board	1/2" (13)	0.092 (0.16)	R-0.5 (0.08 RSI)	50 (800)
Steel Studs	-	27.7 (48)	-	480(7700)
625 BL Armortex Window 42psi	-	-	-	-
Wood buck	-	0.006 (0.10)	-	32 (510)
Exterior Film	-	-	R-0.23 (0.04 RSI)	-

Best-Case Thermal Loss Reduction: 0.125 Btu/h • ft • °F

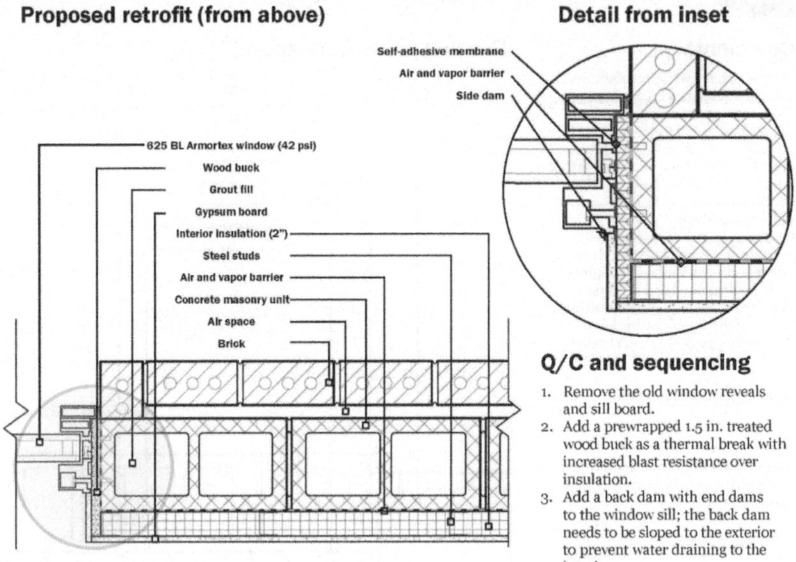

1h Thermal Performance Simulation Results		
Condition	Clear Wall R-Value (W/m2K)	Linear Transmittance (Ψ) Btu/h • ft • °F (W/mK)
Wall Clear Field	R-6 (0.947)	-
Existing Fitting Situation	-	0.235 (0.406)
Proposed Fitting Situation	-	0.110 (0.192)
As-built transmittance – Lowest retrofit transmittance = Best-case thermal-loss reduction 0.235-0.110 = 0.125 Btu/h-ft·°F		

Q/C and sequencing

1. Remove the old window reveals and sill board.
2. Add a prewrapped 1.5 in. treated wood buck as a thermal break with increased blast resistance over insulation.
3. Add a back dam with end dams to the window sill; the back dam needs to be sloped to the exterior to prevent water draining to the interior.
4. Add a bead of sealant to the back dam anchor.
5. Insert the window into position and bolted or strapped according to manufacturer's instructions.
6. Snap on trims can then be added to cover the insulation and membranes.
7. Attach the cover piece to the angled clips.

CMU or Concrete Wall with Interior Insulation

1i Blast Resistant Door Jamb to Concrete

Commentary

Although shown as poured concrete, the support structure could be CMU or brick masonry.

The losses from a standard steel door are very high and will lead to more heat losses and related problems than the thermal bridge at its edge. Due to the nature of thermal bridging, the ψ-value will always be very odd if there are one or two uninsulated components or thermal bridging elements. The relevance of thermal bridging is more important when analyzing an insulated element or structure. Thus the ψ-value for the existing jamb thermal bridge could be misinterpreted as being acceptable. The reality is that the heat flow at the intersection is actually very high, but the heat flow directly across the door surface is also very high.

To improve thermal performance, the thickness of the opening's perimeter insulation can be improved, and a better-insulated door can be installed (metal-clad wood, fiberglass, etc.). Providing continuity of the air barrier system and managing convection are important issues for this detail .

Existing jamb (from above)

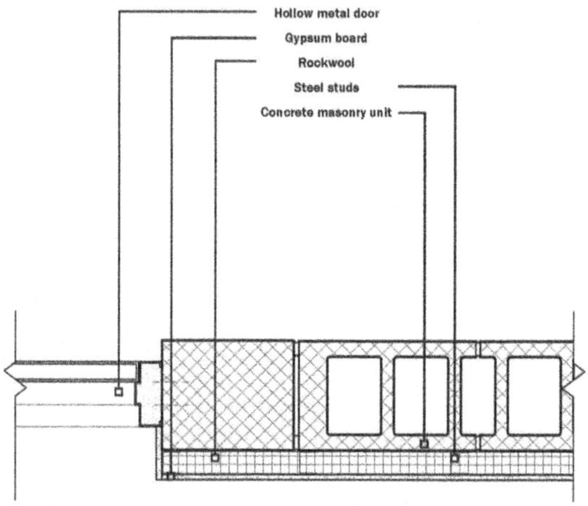

1i Door Jamb Modeling Values				
Component	Thickness Inches (mm)	Conductivity Btu/h • ft • °F (W/m K)	Nominal Resistance hr·ft²·°F/Btu (m²K/W)	Density lb/ft³ (kg/m³)
Interior Film	-	-	R-0.74 (0.13 RSI)	-
CMU	7 5/8" (194)	0.069 (1.2)	R-0.916 (0.161 RSI)	130 (2100)
Rockwool	2" (51)	0.021 (0.036)	R-6.26 (1.103 RSI)	2 (32)
Steel Studs	-	27.7 (48)	-	450(7700)
Gypsum Board	1/2" (13)	0.092 (0.16)	R-0.5 (0.08 RSI)	50 (800)
Hollow Metal Door	-	-	-	-
Kawneer AA®250/425 Blast R Door	-	-	-	-
Exterior Film	-	-	R-0.23 (0.04 RSI)	-

Best-Case Thermal Loss Reduction: 0.010 Btu/h • ft • °F

Proposed retrofit (from above)

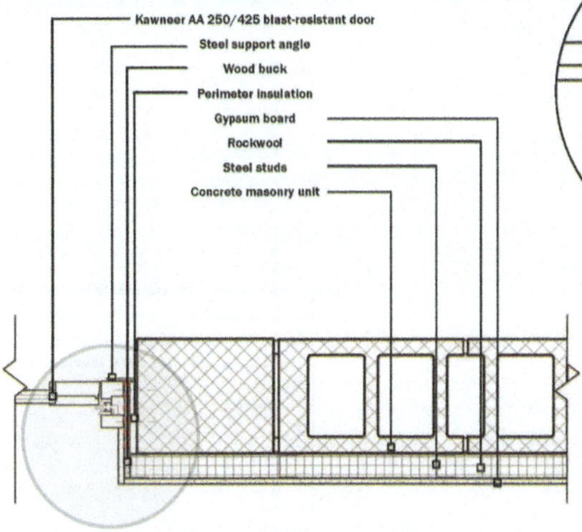

Detail from inset

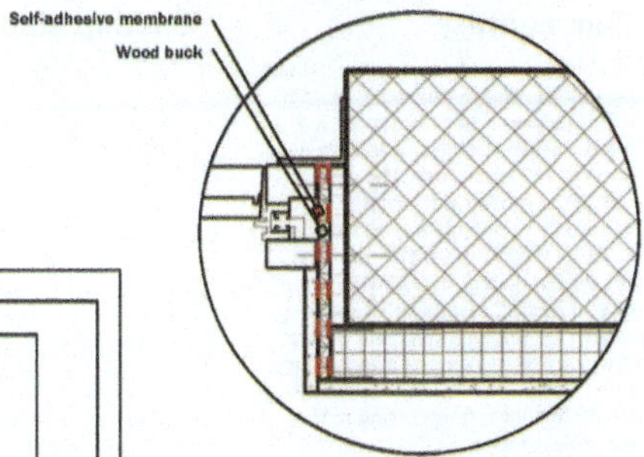

1i Thermal Performance Simulation Results		
Condition	**Clear Wall R-Value (W/m²K)**	**Linear Transmittance (Ψ) Btu/h • ft • °F (W/mK)**
Wall Clear Field	R-5.39 (0.950)	-
Existing Fitting Situation	-	0.113 (0.196)
Proposed Fitting Situation	-	0.103 (0.178)
As-built transmittance – Lowest retrofit transmittance = Best-case thermal-loss reduction		
0.113-0.103 = 0.010 Btu/h·ft·°F		

Q/C and sequencing

1. Remove the old blast-proof door and any caulks or seals.
2. Add a prewrapped 1.5 in. treated solid timber wood buck as a thermal break with increased blast resistance over insulation.
3. Verify that weather stripping for door is intact and undamaged; replace if needed.
4. Insert the blast-proof door into position and bolted or strapped according to manufacturer's instructions.
5. Snap on trims (steel support angles) can then be added to cover the insulation and membranes.
6. Attach the cover piece to the angled clips.

1j

CMU or Concrete Wall with Interior Insulation

Through-Slab Projection

Commentary

This type of detail is used in balcony projections and cast-in-place shading over windows. The floor is concrete, but the wall could also be load-bearing CMU or brick masonry. The primary risk in the retrofit is the potential for damage to the insulation on the top horizontal projection. This must be sloped and also protected with some type of roofing membrane. In most cases, the concrete wall itself or its exterior paint will act as a vapor/moisture barrier. It is usually not necessary to add further waterproofing to the concrete exterior.

The insulation choice is often limited to products that can resist moisture, some wind, meet fire codes, etc. The options tend to be fire-resistant foam plastics and stone wool semi-rigid boards. Increasing insulation thickness adds value to slab thermal performance. Through-slab projection thermal performance could be improved further by placing vertical and continuous insulation in the wall of 6–12 in. atop the slab projection.

Also not shown in the drawings, the thermal performance was computed by including an aluminum frame double glazed window into the Through-Slab Projection concrete wall model.

Existing slab

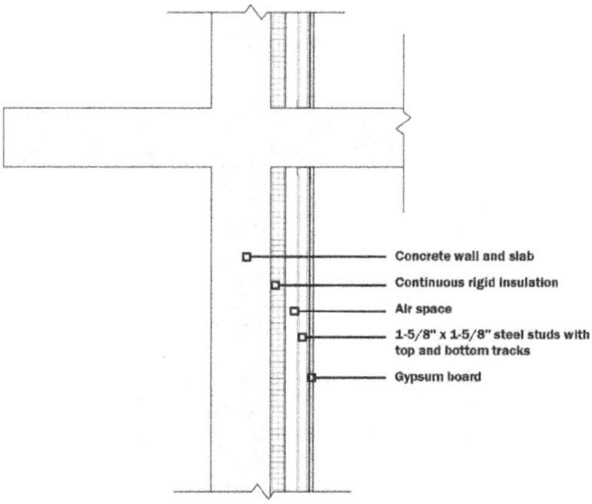

1j Through-Slab Modeling Values

Component	Thickness Inches (mm)	Conductivity Btu/h • ft • °F (W/m K)	Nominal Resistance hr-ft²·°F/Btu (m²K/W)	Density lb/ft³ (kg/m³)
Interior Film (right side)	-	-	R-0.6 (0.11 RSI) to	-
R-1.1 (0.20 RSI)	-	0.069 (1.2)	R-0.916 (0.161 RSI)	130 (2100)
Gypsum Board	1/2" (13)	0.092 (0.16)	R-0.5 (0.08 RSI)	50 (800)
1 5/8" x 1 5/8" Steel Studs with Top and Bottom Tracks	18 Gauge	35.825 (62)	-	489 (7830)
Air in Stud Cavity	1 5/8" (41)	-	R-0.9 (RSI-0.16)	0.075 (1.2)
Continuous Rigid Insulation	3" (76)	-	R-15 (2.6 RSI)	1.8 (28)
Concrete Wall/ Projected Floor Slab	8" (203)	1.04 (1.8)	-	140 (2250)
Metal sheet connected to studs	18 Gauge	35.825 (62)	-	489 (7830)
Wood Sill	1 ¼" (30)	0.052 (0.09)	-	31 (500)
Aluminum Flashing	16 Gauge	127.12 (220)	-	0.21 (900)
Exterior Insulation (h)	2" (51)	0.017 (0.029)	R-10 (1.75 RSI)	1.8 (28)
Exterior Film (left side)	-	-	R-0.2 (0.03 RSI)	-
5' (1.5m) x 6' (1.8m) Aluminum window: double glazed and thermally broken. U = 0.35 BTU/hrft²oF (2.0 W/m²K)				

Best-Case Thermal Loss Reduction: 0.173 Btu/h • ft • °F

Proposed retrofit

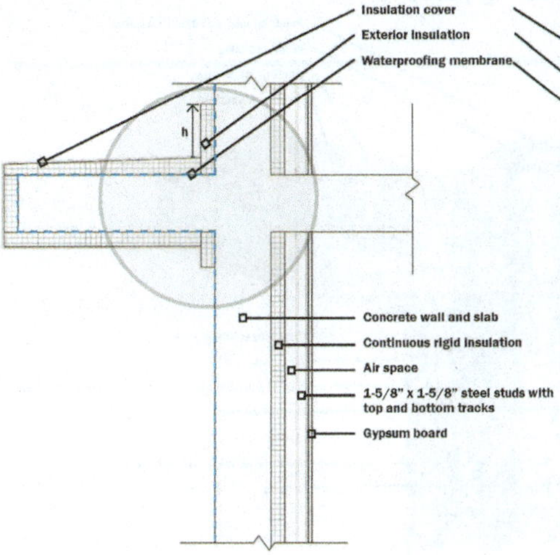

Detail from inset

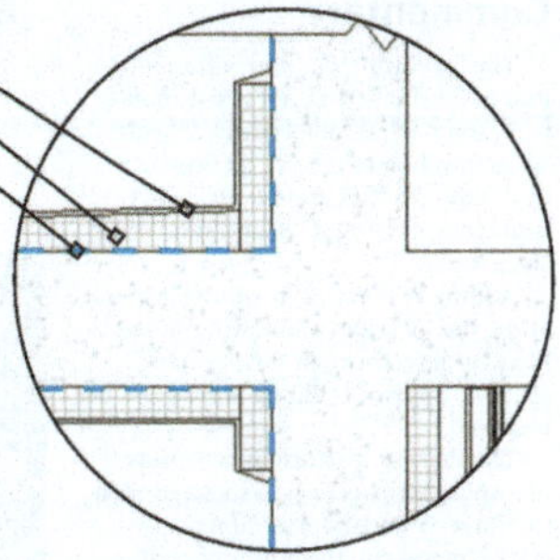

Q/C and sequencing

1. Clean the slab projection and make sure that it does not have any weather-inclemency residuals (i.e., water or snow)
2. Measure and cut the slab projection insulation (i.e., stone wool semi-rigid boards) as appropriate to wrap around (bottom, side, and top face) the slab. Make sure that the added top insulation is sloped, such that it will allow rain water, etc., to drain properly and not accumulate.
3. Attach the insulation by using appropriate insulation adhesive.
4. Protect the top of the slab projection by adding some type of roofing membrane.

1j Thermal Performance Simulation Results

Condition	Clear Wall R-Value (W/m²K)	Linear Transmittance (Ψ) Btu/h • ft • °F (W/mK)
Wall Clear Field	R-17.5 (0.308)	-
As-Built Fitting Situation	-	0.103 (0.178)
Wall Clear Field	R-17.5 (3.08 RSI)	-
As-Built Slab Projection (no exterior insulation)	-	0.539 (0.932)
Retrofit with...		
Insulation on slab only	-	0.454 (0.786)
Insulation on slab and 6" (h)	-	0.405 (0.701)
Insulation on slab and 12" (h)	-	0.366 (0.634)
As-built transmittance - Lowest retrofit transmittance = Best-case thermal-loss reduction		
0.539-0.366 = 0.173 Btu/h·ft·°F		

2a

CMU or Concrete Wall with Exterior Insulation

Roof Parapet with Concrete Roof

Commentary

The insulation choice is often limited to products that can resist moisture, some wind, meet fire codes, etc. Thus the choices tend to be fire-resistant foam plastics and board insulation made of EPS, XPS, and semi-rigid stone wool.

Adding at least 24 in. of insulation along the interior ceiling will reduce the heat flow through the parapet section. Interior building access will be required.

The thermal performance results incorporate the effects associated with the interior and exterior film coefficients, metal cap flashing and roof finish material.

Thermal performance can be further improved by increasing the insulation thickness. Thermally breaking the heat flow by replacing the parapet base support CMU with a low-conductivity block will provide the best thermal performance.

Existing parapet

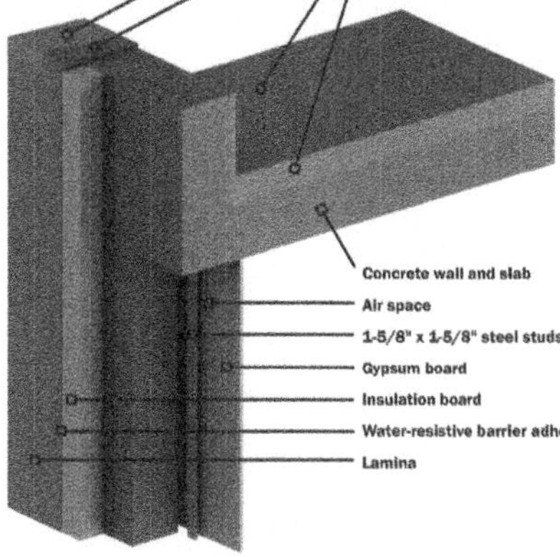

2a Parapet Modeling Values (Concrete Roof)

Component	Thickness Inches (mm)	Conductivity Btu/h • ft • °F (W/m K)	Nominal Resisitance hr-ft²•°F/Btu (m²K/W)	Density lb/ft³ (kg/m³)
R-0.7 (0.12 RSI)	-	0.578 (1)	R-0.523 (0.092 RSI)	110 (1800)
Gypsum Board	1/2" (13)	0.092 (0.16)	R-0.5 (0.08 RSI)	50 (800)
1 5/8" x 1 5/8" Steel Studs (16"o.c.) with Top Tracks	18 Gauge	35.825 (62)	-	489 (7830)
Air in Stud Cavity	1 5/8" (41)	-	R-0.9 (0.16 RSI)	0.075 (1.2)
Concrete Wall	8" (203)	1.04 (1.8)	-	140 (2250)
Insulation Board	4" (100)	0.023 (0.039)	R-15 (2.64 RSI)	1 (16)
Lamina	1/8" (4)	0.52 (0.9)	R-0.04 (0.01 RSI)	120 (1922)
Concrete Slab & Parapet	8" (203)	1.04 (1.8)	-	140 (2250)
Roof Insulation	4" (100)	-	R-20 (3.5 RSI)	1.8 (28)
Parapet Insulation	1" (25)	-	R-5 (0.88 RSI)	1.8 (28)
Parapet Insulation - Fully Insulated	3" (76)	-	R-15 (2.64 RSI)	1.8 (28)
Wood Blocking	5/8" (16)	0.052 (0.09)	R-1 (0.18 RSI)	27.8 (445)

Best-Case Thermal Loss Reduction: 0.106 Btu/h • ft • °F

Proposed retrofit

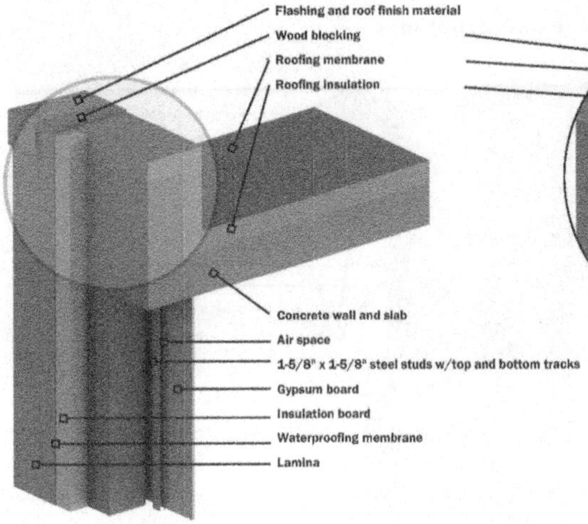

Detail from inset

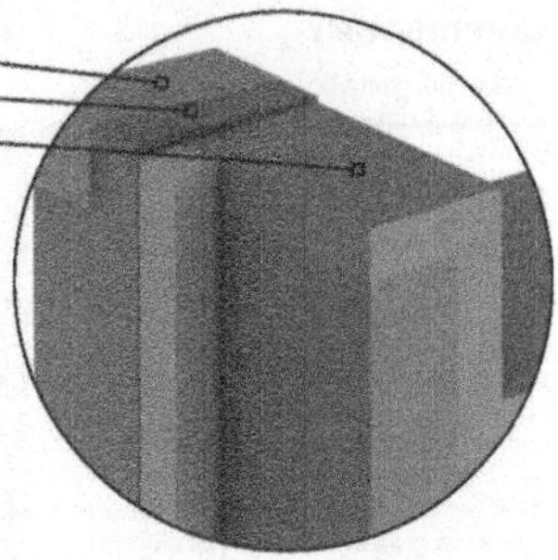

2a Thermal Performance Simulation Results		
Condition	**Clear Wall R-Value (W/m²K)**	**Linear Transmittance (Ψ) Btu/h • ft • °F (W/mK)**
Wall Clear Field	R-17.6 (3.10 RSI)	-
Roof Clear Field	R-21.9 (3.86 RSI)	-
As-Built Parapet	-	0.231 (0.400)
Retrofit with...		
Fully Insulated Parapet	-	0.125 (0.217)
As-built transmittance – Lowest retrofit transmittance = Best-case thermal-loss reduction		
0.231-0.125 = 0.106 Btu/h·ft·°F		

Q/C and sequencing

1. Remove the parapet metal cap (also known as parapet capping).
2. Remove parapet wood blocking.
3. Remove the roof covering (asphalt, roofing membrane, etc.) to expose the roof insulation.
4. Add rigid insulation to the rear vertical face of the parapet. This vertical piece of insulation must have contact at its bottom with the roof insulation to maintain continuity. The attached vertical insulation should be able to reach the top of the parapet.
5. Add rigid insulation on the top section of the parapet. Make sure that it is contiguous with the insulation placed in the rear vertical face of the parapet and the front exterior insulation board of the wall (already in place).
6. Replace the removed roof membrane system so that continuous protection to both the roof and the parapet is restored.
7. Reattach the wood blocking.
8. Reattach the parapet metal cap.

CMU or Concrete Wall with Exterior Insulation

2b Parapet with Steel Roof Deck and Joists

Commentary

The wall could be load-bearing concrete, brick masonry, or CMU as shown.

For the interior flanking, select low- to medium-density closed-cell foam to control vapor diffusion condensation in climate zones 5 and higher. It also should be able to withstand potential movement at the location of application. This insulation type is capable of providing excellent airflow control by conforming and adhering to rough surfaces, making them common choices for interior masonry retrofits. Like most plastics, they do not support mold growth.

The insulation should be sprayed to cover at least 1 ft below the lowest end and 1 ft away from the steel beam. Nevertheless, the foam insulation should be applied above the ceiling plane.

One possible improvement is to add insulation around the interior face and the top of the parapet using the parapet insulation wrapping strategy.

The thermal performance results incorporate the effects associated with the interior and exterior film coefficients, metal cap flashing, and roof finish material.

Existing parapet

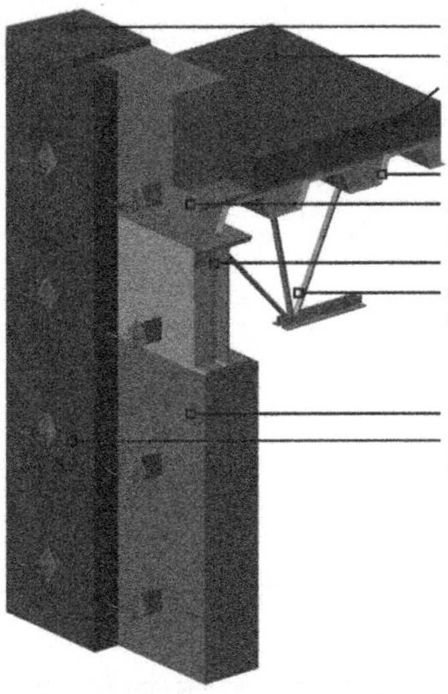

2b Modeling Values for Parapet (Steel Roof and Joists)				
Component	Thickness Inches (mm)	Conductivity Btu/h • ft • °F (W/m K)	Nominal Resisitance hr·ft²·°F/Btu (m²K/W)	Density lb/ft³ (kg/m³)
R-0.9 (0.16 RSI)	-	0.578 (1)	R-0.523 (0.092 RSI)	110 (1800)
CMU Block Wall	8" (203)	0.603 (1.044)	R-1.1 (0.19 RSI)	140 (2550)
Air Cavity	3 3/8" (85)	-	R-0.9 (0.16 RSI)	
Exterior Insulation, Wall	2" (51)	0.017 (0.029)	R-10 (1.75 RSI)	1.8 (28)
Masonry Ties @ 16" (406 mm) o.c.	14 gauge	28.891 (50)	-	489 (7830)
Steel Beam (W410)	-	28.891 (50)	-	489 (7830)
Open Web Steel Joist	-	28.891 (50)	-	489 (7830)
Steel Deck	-	28.891 (50)	-	489 (7830)
Shelf Angle	-	35.825 (62)	-	489 (7830)
Roof Air Cavity	3" (76)	-	R-0.9 (0.16 RSI)	
Gypsum Overlay Board	1/2" (13)	0.092 (0.16)	R-0.5 (0.08 RSI)	50 (800)
Roof Insulation	4" (102)	-	R-20 (3.5 RSI)	1.8 (28)
Wood Blocking	5/8" (16)	0.052 (0.09)	R-1 (0.18 RSI)	27.8 (445)
Interior Spray Foam	-	-	R-10 (1.75 RSI)	1.8 (28)

Best-Case Thermal Loss Reduction: 0.197 Btu/h • ft • °F

Proposed retrofit

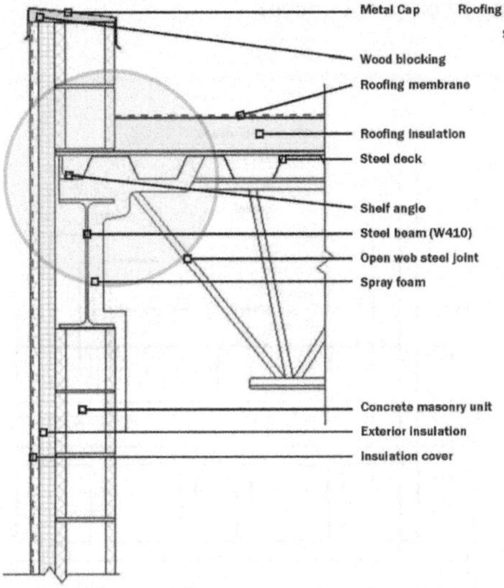

Detail from inset

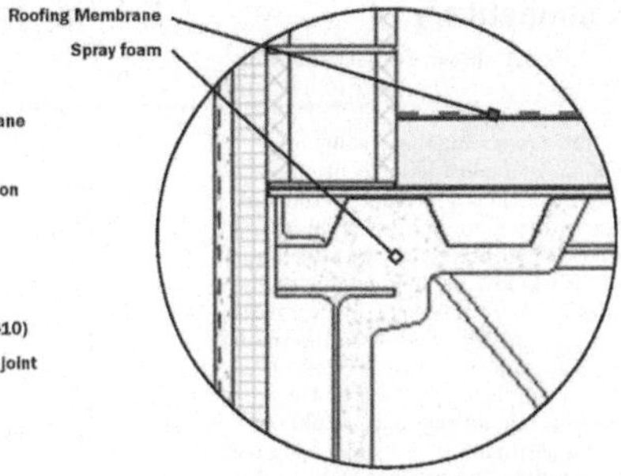

2b Thermal Performance Simulation Results		
Condition	**Clear Wall R-Value (W/m²K)**	**Linear Transmittance (Ψ) Btu/h • ft • °F (W/mK)**
Wall Clear Field	R-11.7 (2.00 RSI)	-
Roof Clear Field	R-21.4 (3.77 RSI)	-
As-Built Slab	-	0.318 (0.55)
Retrofit with...		
R10 Spray Foam	-	0.121 (0.209)
As-built transmittance – Lowest retrofit transmittance = Best-case thermal-loss reduction		
0.318-0.121 = 0.197 Btu/h·ft·°F		

Q/C and sequencing

Details

1. Apply a layer of spray foam insulation at the interior roof/wall intersection region. The foam insulation should be sprayed to cover at least 12 in. below and 12 in. away from the steel beam (that is, spray the area beneath the roof deck).
2. Apply a finish layer of spray-on fire protection over the insulation.

Note

For improved performance, refer to the Q/C and sequencing recommendations for exterior parapet wraps in item 1d (page 9).

CMU or Concrete Wall with Exterior Insulation

2d Window Jamb

Commentary

Although shown as CMU, this also applies to poured concrete or masonry construction.

After removing the existing brick sill, make the insulation continuous and aligned with the window thermal break. Key to the success of this detail is ensuring good structural attachment of the window and alignment with the window thermal break. This technique can improve the window air tightness and moisture-intrusion performance.

Sub-sill flashing is required to prevent rain intrusion. It should be formed with a raised vertical section at the back (i.e., a back dam) tall enough to allow the application of sealant between it and the window for crucial water and airflow control continuity.

The hollow space of open window frames promotes natural convective heat flow between building interior and exterior. This undesired heat flow can be reduced by filling these voids with factory-installed custom-formed foam plastic or rigid stone wool insulation sections.

The performance of the correct version can be improved only slightly from Ψ - 0.017 Btu/hr ft °F using thicker insulation and modifying details of the window sill attachment to the window and the alignment of the thermal break.

The Ψ value reported here does not include the metal angle back dam or the window opening wood buck thermal conductivity effects.

Note that these comments and the sequencing notes at far right also apply to items 2e and 2f.

Existing jamb (from above)

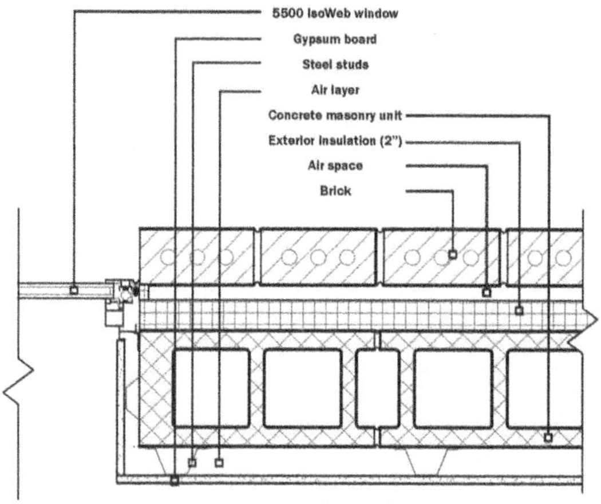

2d Modeling Values for Window Jamb

Component	Thickness Inches (mm)	Conductivity Btu/h • ft • °F (W/m K)	Nominal Resisitance hr-ft^2·°F/Btu (m^{2}K/W)	Density lb/ft^3 (kg/m^3)
Interior Film	-	-	R-0.74 (0.13 RSI)	-
Brick	3 5/8" (92)	0.578 (1)	R-0.523 (0.092 RSI)	110 (1800)
Air Cavity	1" (25)	0.070 (0.122)	R-1.185 (0.209 RSI)	-
Insulation	2" (51)	0.0139 (0.024)	R-11.99 (2.112 RSI)	2 (32)
CMU	7 5/8" (194)	0.069 (1.2)	R-0.916 (0.161 RSI)	130 (2100)
Air Layer with Steel Studs	1 3/4" (44)	0.2219 (0.384)	R-0.66 (0.116 RSI)	-
Gypsum Board	1/2" (13)	0.092 (0.16)	R-0.5 (0.08 RSI)	50 (800)
5500 IsoWeb Window	-	-	-	-
Steel Studs	-	27.7 (48)	-	-
Exterior Film	-	-	R-0.23 (0.04 RSI)	-

Best-Case Thermal Loss Reduction: 0.003 Btu/h • ft • °F

Proposed retrofit (from above)

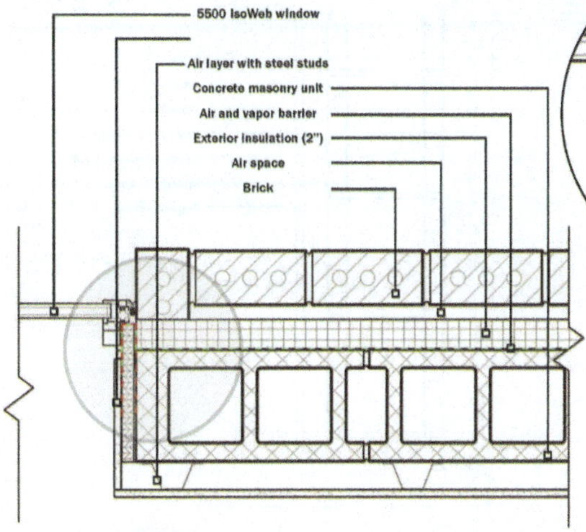

Detail from inset

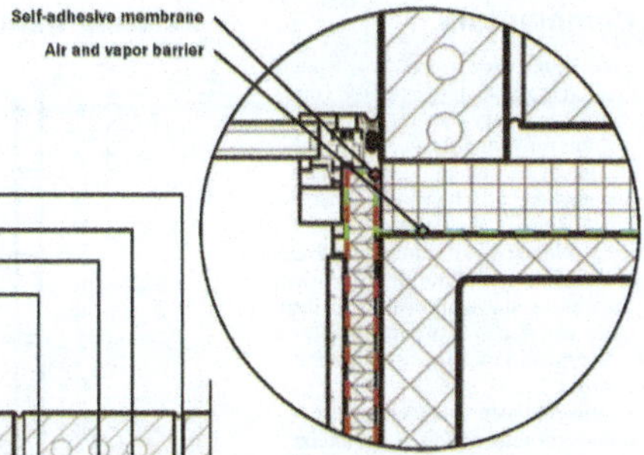

2d Thermal Performance Simulation Results		
Condition	**Clear Wall R-Value (W/m²K)**	**Linear Transmittance (Ψ) Btu/h • ft • °F (W/mK)**
Wall Clear Field	R-15.7 (0.369)	-
As-Built Fitting Situation	-	0.012 (0.021)
Proposed Fitting Situation	-	0.009 (0.015)
As-built transmittance – Lowest retrofit transmittance = Best-case thermal-loss reduction **0.012-0.009 = 0.003 Btu/h·ft·°F**		

Q/C and sequencing

1. Remove old window, including seals, window board, perimeter bricks, brick sill, and mortar bed.
2. Install plywood buck, sandwiched by high-R insulation prewrapped with membrane, around all sides of opening with angle brackets so it contacts window frame.
3. Cover any exposed wood surface and protruding part of wood buck with self adhesive membrane.
4. Add a back dam to the wood buck on window sill, then install concrete sill block.
5. Replace the 1 in. rigid insulation on top of wood buck, with a 45 degree cut for drainage.
6. Replace wall opening jamb bricks up to lintel level and attach lintel over window opening, adding side dam to sides of lintel.
7. On top of opening, connect wall barrier sheathing to lintel, attaching self-adhesive membrane from top of exposed wall to lintel.
8. Replace complementary rigid insulation (see Step 6) using 45 degree angle and contacting membrane.
9. Reinstall wall opening head bricks.
10. Press window into position against back dam and seal around frame.

2e

CMU or Concrete Wall with Exterior Insulation

Window Head

Commentary

Although shown as CMU, the modification is applicable to poured concrete or masonry structural systems.

After removing the existing brick sill, make the insulation continuous and aligned with the window thermal break. Key to the success of this detail is ensuring good structural attachment of the window and the alignment with the window thermal break. This technique can also improve the window air tightness and rain-intrusion performance.

Sub-sill flashing is required to prevent rain intrusion. It should be formed with a raised vertical section at the back (i.e., back dam) and be tall enough to allow the application of sealant between it and the window for continuity of air- and water-intrusion control.

The hollow space of open window frames will promote natural convective heat flow between the building interior and exterior. This undesired heat flow can be reduced by filling voids with factory-installed, custom-formed foam plastic or rigid stone wool insulation sections.

The performance of the corrected element can be improved only slightly from Ψ - 0.017 BTU/hr ft °F using thicker insulation and modifying details of the window sill attachment to the window and the alignment with the thermal break.

The reported Ψ-value does not include the metal angle back dam nor the window opening wood buck thermal conductivity effects.

Note that these comments and the sequencing notes at far right also apply to items 2d and 2f.

Existing head

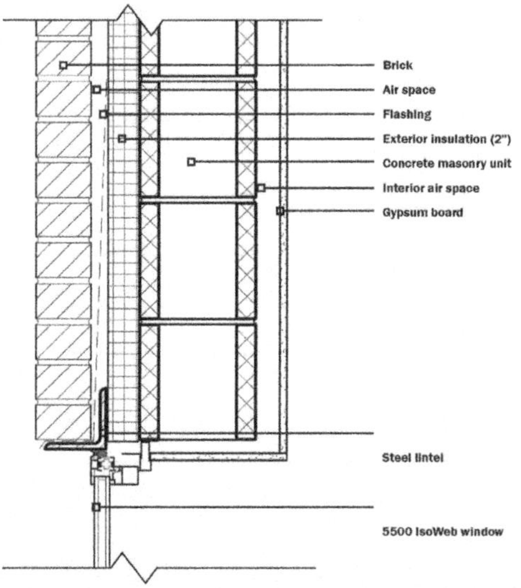

2e Modeling Values for Window Head

Component	Thickness Inches (mm)	Conductivity Btu/h•ft•°F (W/m K)	Nominal Resistance hr•ft²•°F/Btu (m²K/W)	Density lb/ft³ (kg/m³)
Interior Film	-	-	R-0.74 (0.13 RSI)	-
Brick	3 5/8" (92)	0.578 (1)	R-0.523 (0.092 RSI)	110 (1800)
Air Cavity	1" (25)	0.070 (0.122)	R-1.185 (0.209 RSI)	-
Insulation	2" (51)	0.0139 (0.024)	R-11.99 (2.112 RSI)	2 (32)
CMU	7 5/8" (194)	0.069 (1.2)	R-0.916 (0.161 RSI)	130 (2100)
Air Layer with Steel Studs	1 3/4" (44)	0.2219 (0.384)	R-0.66 (0.116 RSI)	-
Gypsum Board	1/2" (13)	0.092 (0.16)	R-0.5 (0.08 RSI)	50 (800)
5500 IsoWeb Window	-	-	-	-
Steel Lintel	1/4" (6.4)	27.7 (48)	-	-
Exterior Film	-	-	R-0.23 (0.04 RSI)	-

2f

CMU or Concrete Wall with Exterior Insulation

Window Sill

Commentary

Although shown as CMU, the structural system could also be poured concrete or masonry for this application.

After removing the existing brick sill, make the insulation continuous and align it with the window thermal break. Key to the success of this detail is ensuring good structural attachment of the window and the alignment with the window thermal break. This offers a chance to improve the window air tightness and resistance to rain intrusion.

Sub-sill flashing is required for preventing rain intrusion. It should be formed with a raised vertical back dam that is tall enough to allow the installation of sealant between it and the window for prevention of water and air intrusion.

The hollow space of open window frames will promote convective heat flow between the building interior and exterior. This undesired heat flow can be reduced by filling voids with factory-installed, custom-formed foam plastic or rigid stone wool insulation sections.

The performance of the corrected sill can be improved only slightly from Ψ - 0.017 BTU/hr ft °F using thicker insulation and modifying the details of sill attachment to the window and alignment with the thermal break.

The reported Ψ-value does not include the metal angle back dam or window-opening wood buck thermal conductivity effects.

Note that these comments and the sequencing notes at far right also apply to items 2d and 2e.

Existing sill

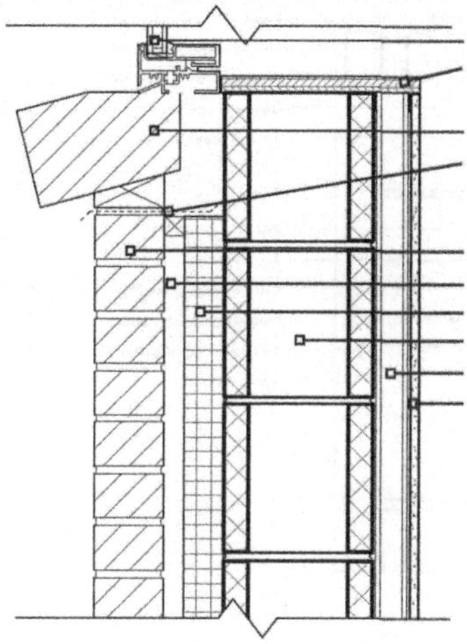

2f Modeling Values for Window Sill

Component	Thickness Inches (mm)	Conductivity Btu/h • ft • °F (W/m K)	Nominal Resisitance hr·ft²·°F/Btu (m²K/W)	Density lb/ft³ (kg/m³)
Interior Film	-	-	R-0.74 (0.13 RSI)	-
Brick	3 5/8" (92)	0.578 (1)	R-0.523 (0.092 RSI)	110 (1800)
Air Cavity	1" (25)	0.070 (0.122)	R-1.185 (0.209 RSI)	-
Insulation	2" (51)	0.0139 (0.024)	R-11.99 (2.112 RSI)	-
CMU Block	7 5/8" (194)	0.069 (1.2)	R-0.916 (0.161 RSI)	130 (2100)
Air Layer with Steel Studs	1 3/4" (44)	0.2219 (0.384)	R-0.66 (0.116 RSI)	-
Gypsum Board	1/2" (13)	0.092 (0.16)	R-0.5 (0.08 RSI)	50 (800)
5500 IsoWeb Window	-	-	-	-
Aluminum Sill Flashing	12 Gauge	160	-	-
Brick Sill	3 5/8" (92)	0.578 (1)	-	110 (1800)
Exterior Film	-	-	R-0.23 (0.04 RSI)	-

Best-Case Thermal Loss Reduction: 0.428 Btu/h • ft • °F

Proposed retrofit

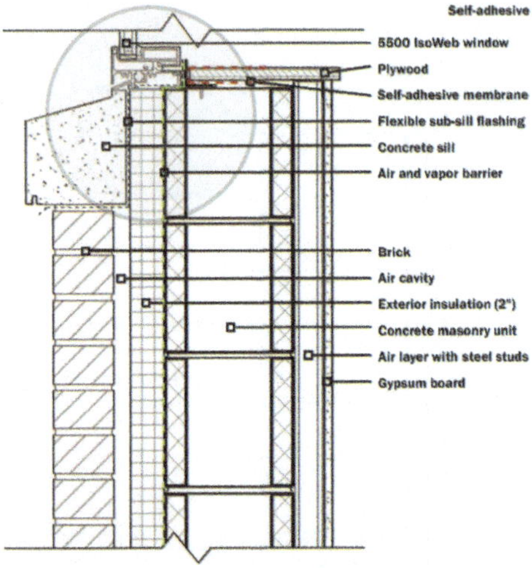

Detail from inset

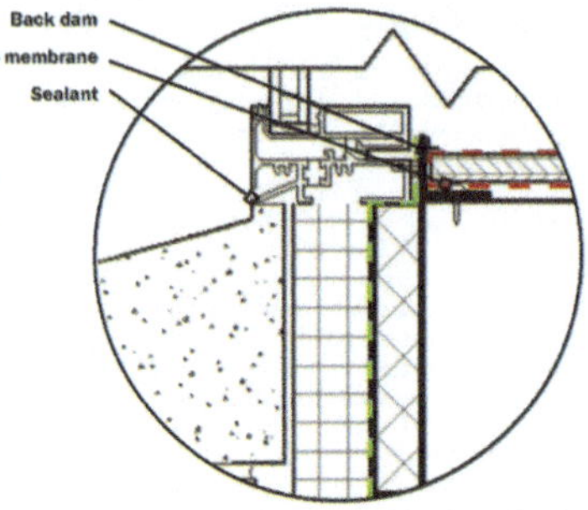

2f Thermal Performance Simulation Results		
Condition	**Clear Wall R-Value (W/m²K)**	**Linear Transmittance (Ψ) Btu/h • ft • °F (W/mK)**
Wall Clear Field	R-15.7 (0.369)	-
Existing Fitting Situation	-	0.445 (0.771)
Corrected Fitting Situation	-	0.017 (0.030)
As-built transmittance – Lowest retrofit transmittance = Best-case thermal-loss reduction **0.445-0.017 = 0.428 Btu/h·ft·°F**		

Q/C and sequencing

1. Remove old window, including seals, window board, perimeter bricks, brick sill, and mortar bed.
2. Install plywood buck, sandwiched by high-R insulation prewrapped with membrane, around all sides of opening with angle brackets so it contacts window frame.
3. Cover any exposed wood surface and protruding part of wood buck with peel-and-stick tape.
4. Add a back dam to the wood buck on the window sill, then install concrete sill block.
5. Replace wall opening jamb bricks up to lintel level and attach lintel over window opening.
6. On top of opening, connect wall barrier sheathing to lintel, attaching self-adhesive membrane from top of exposed wall to lintel (see detail 2e).
7. Replace complementary rigid insulation (see Step 6) using 45 degree angle and contacting the membrane (see detail 2e).
8. Reinstall wall opening head bricks (see detail 2e).
9. Press window into position against back dam and seal around frame.
10. Drill holes at sill to allow drainage.

2g

CMU or Concrete Wall with Exterior Insulation

Blast-Resistant Window Jamb

Commentary

The wall structure may be either concrete or masonry construction. During window replacement, care must be taken to leave weep holes at the sill to allow any trapped water to escape. The use of timber blocks (Step 3 in Q/C and sequencing notes) will allow the window to transfer compressive loads. Use stainless steel bolts to transfer the tension loads.

For this detail to perform in a durable manner, the timber blocking must be protected from excessive wetting using a properly lapped waterproofing membrane, preferably fully-adhered sheets or a fluid-applied barrier. To allow drying of the wood blocking in wet or cold climates, the membrane should also be vapor permeable.

To improve the performance further, use a higher-performance structural break such as high-strength (500 psi capable) polyurethane or 300 psi capable foam glass (also called "cellular glass").

Note that this item and item 2h are retrofits for a single window and must be applied at the same time.

Existing jamb (from above)

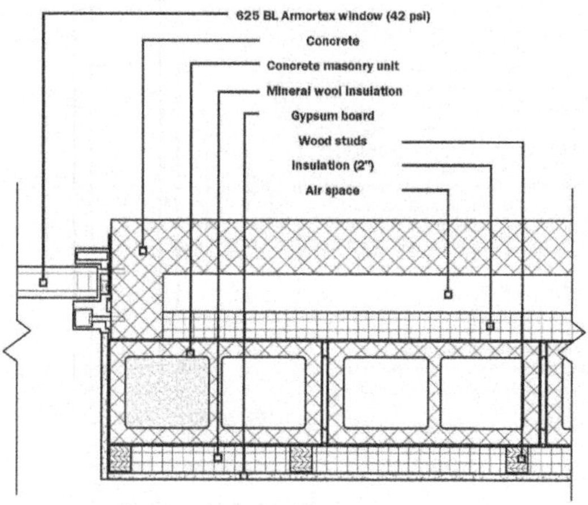

2g Modeling Values for Blast-Resistant Window Jamb

Component	Thickness Inches (mm)	Conductivity Btu/h • ft • °F (W/m K)	Nominal Resisitance hr·ft²·°F/Btu (m²K/W)	Density lb/ft³ (kg/m³)
Interior Film	-	-	R-0.74 (0.13 RSI)	-
CMU	7 5/8" (194)	0.069 (1.2)	R-0.916 (0.161 RSI)	130 (2100)
Air Cavity	2.5" (64)	0.1489 (0.258)	R-1.4 (0.246 RSI)	-
Insulation	2" (51)	0.0139 (0.024)	R-11.99 (2.112 RSI)	2 (32)
Grout	-	0.092 (0.16)	-	130(2100)
Mineral wool insulation	2" (51)	0.021 (0.036)	R-6.26 (1.103 RSI)	4(64)
Gypsum Board	1/2" (13)	0.092 (0.16)	R-0.5 (0.08 RSI)	50 (800)
625 BL Armortex Window 42psi	-	-	-	-
Exterior Film	-	-	R-0.23 (0.04 RSI)	-

2h

CMU or Concrete Wall with Exterior Insulation

Blast-Resistant Window Head

Commentary

The wall structure may be either concrete or masonry. During window replacement, care must be taken to preserve weep holes at the sill to allow any trapped water to escape.

The use of timber blocks (Step 4 of QC and sequencing) will allow the window to transfer compressive loads. Use stainless steel bolts to transfer the tension loads.

For this detail to perform in a durable manner, the wood blocking must be protected from excessive wetting using a properly lapped waterproofing membrane, preferably fully-adhered sheets or a fluid-applied barrier. To allow drying of the wood blocking in wet or cold climates, the membrane should be vapor permeable.

To improve performance further, use a higher-performance structural break such as high-strength (500 psi capable) polyurethane or 300 psi capable foam glass (also called "cellular glass").

Note that this item and item 2g are retrofits for a single window and must be applied at the same time.

Existing head

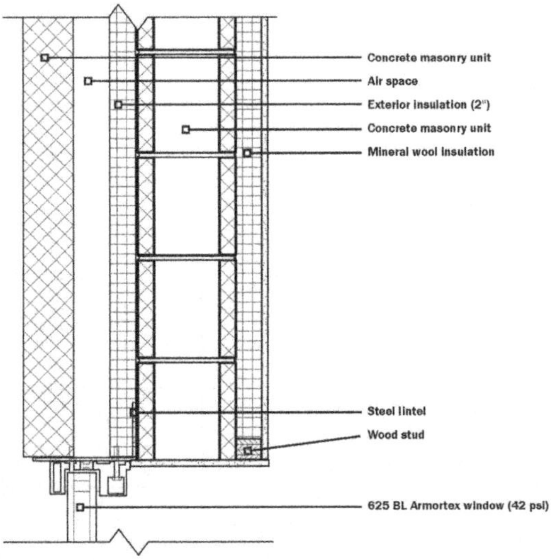

2h Modeling Values for Blast-Resistant Window Head

Component	Thickness Inches (mm)	Conductivity Btu/h • ft • °F (W/m K)	Nominal Resisitance hr·ft²·°F/Btu (m²K/W)	Density lb/ft³ (kg/m³)
Interior Film	-	-	R-0.74 (0.13 RSI)	-
CMU	7 5/8" (194)	0.069 (1.2)	R-0.916 (0.161 RSI)	130 (2100)
Air Cavity	2.5" (64)	0.1489 (0.258)	R-1.4 (0.246 RSI)	-
Insulation	2" (51)	0.0139 (0.024)	R-11.99 (2.112 RSI)	2 (32)
Grout	-	0.092 (0.16)	-	130 (2100)
Mineral wool insulation	2" (51)	0.0266 (0.046)	R-6.26 (1.103 RSI)	4 (64)
Gypsum Board	1/2" (13)	0.092 (0.16)	R-0.5 (0.08 RSI)	50 (800)
Steel Lintel	-	27.7 (48)	-	480 (7800)
625 BL Armortex Window 42psi	-	-	-	-
Timber Break	-	0.006 (0.10)	-	32 (510)
Exterior Film	-	-	R-0.23 (0.04 RSI)	-

CMU or Concrete Wall with Exterior Insulation

2i Suspended Slab at Shelf Angle

Commentary

This detail blunts the thermal bridge of the relieving or shelf angle often used in brick veneer construction by reducing the area of steel passing through the insulation. However, cross-sectional area reductions of 75–95% are often needed to reduce heat flow by even one half in the case of metals. Fortunately, this is often easy to accomplish. For example, in structural design, the use of knife-edge supports of steel can achieve this level of heat-flow reduction.

Significant further reductions could be made by spacing the standoffs farther apart (often 48 in. on center is found to be structurally sufficient) and replacing the standoff with stainless steel instead of carbon steel.

Avoiding the use of shelf angles by designing the brick veneer to be load bearing is often practical for buildings under 5 or 6 stories high. (Differential movement becomes more and more important as the veneer height increases.)

The 20 gauge steel flashing still acts as a thermal bridge, so replacing it with thermally nonconductive polymer or thinner stainless steel can improve the performance. For example, combining a reduction in area of metal with the use of stainless steel (which has one-third the conductivity of carbon steel and less than one-eighth the conductivity of aluminum) can often yield large reductions in heat flow.

Existing slab

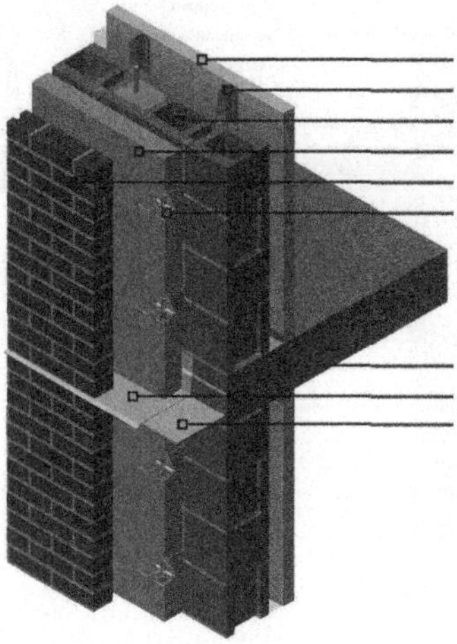

2i Modeling Values for Suspended Slab at Shelf Angle

Component	Thickness Inches (mm)	Conductivity Btu/h • ft • °F (W/m K)	Nominal Resistance hr-ft²·°F/Btu (m²K/W)	Density lb/ft³ (kg/m³)
Interior Film (right side)1	-	-	R-0.6 (0.11 RSI) to	-
R-0.9 (0.16 RSI)	-	0.578 (1)	R-0.523 (0.092 RSI)	110 (1800)
Gypsum Board	1/2" (13)	1.1 (0.16)	R-0.5 (0.08 RSI)	50 (800)
1 5/8" Steel Studs with Metal Tracks	20 gauge	430 (62)	-	489 (7830)
Air in Stud Cavity	1 5/8" (41)	-	R-0.9 (0.16 RSI)	0.075 (1.2)
Standard Concrete Block	7 5/8" (190)	3.5 (0.5)	-	119 (1900)
Cement Mortar	-	3.5 (0.5)	-	113 (1800)
Masonry Ties @ 16" (406) o.c.	14 gauge	347 (50)	-	489 (7830)
Insulation	Varies	-	R-10 (1.76 RSI)	1.8 (28)
Shelf Angle	3/8" (10)	347 (50)	-	489 (7830)
Flashing-steel	20 gauge	347 (50)	-	489 (7830)
Brick Veneer	3 5/8" (92)	5.4 (0.78)	-	120 (1920)
Concrete Slab	8" (203)	12.5 (1.8)	-	140 (2250)
Air Gap	1" (25)	-	R-0.9 (0.16 RSI)	0.075 (1.2)
Exterior Film (left side)1	-	-	R-0.2 (0.03 RSI)	-

Best-Case Thermal Loss Reduction: 0.072 Btu/h • ft • °F

Proposed retrofit

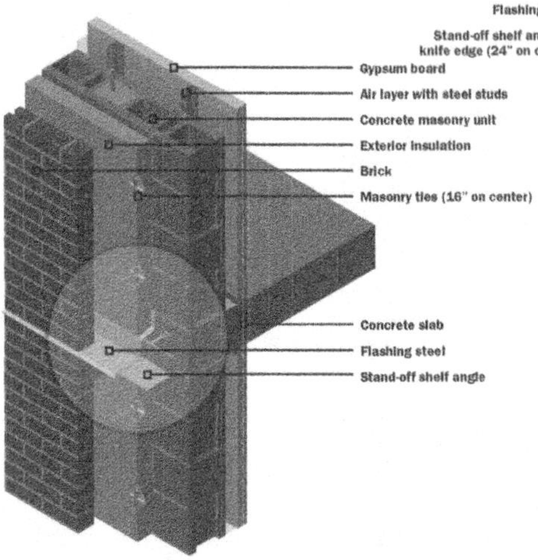

Detail from inset

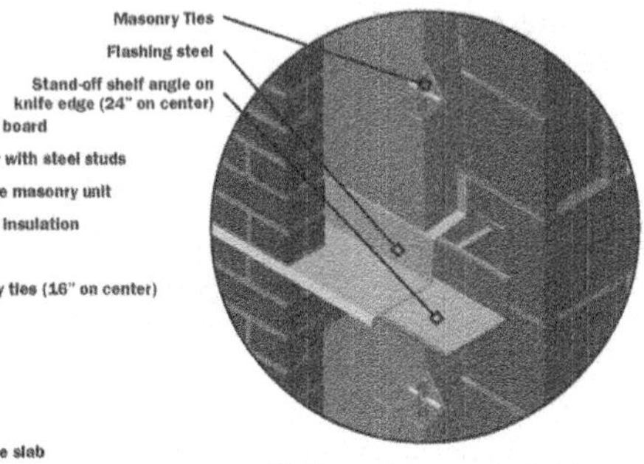

2i Thermal Performance Simulation Results		
Condition	**Clear Wall R-Value (W/m²K)**	**Linear Transmittance (Ψ) Btu/h • ft • °F (W/mK)**
Wall Clear Field	R-14.2 (2.50)	-
As-Built Slab, Continuous Shelf Angle	-	0.258 (0.446)
Retrofit with...		
Standoff Shelf Angle	-	0.186 (0.322)
As-built transmittance – Lowest retrofit transmittance = Best-case thermal-loss reduction		
0.258-0.186 = 0.072 Btu/h·ft·°F		

Q/C and sequencing

1. Remove the brick sections held by the shelf angle.
2. The flashing used to cover the shelf angle may have to be removed as part of this process, or it could also be retained and reused over the replacement shelf angle if practicable.
3. Remove the shelf angle.
4. Insert new type of shelf angle that greatly reduces thermal bridge effect. The one illustrated in the proposed solution image is often called a knife-edge shelf angle. It can support the same load as the conventional one, but the designer must confirm that the capacity of the replacement angle is equal to or greater than the original.
5. Apply insulation in and around the voids of the knife-edge shelf angle and make sure that full insulation continuity is achieved on the entire surface.
6. Reinstall existing flashing or replace with a material of lower thermal conductivity.
7. Reinstall the brick facade.

3b

Steel Stud Wall with Interior and Exterior Insulation

Window Jamb

Commentary

Key to the effectiveness of this detail is ensuring good structural attachment of the window and alignment of the window thermal bridge. Every window section is likely to have a slightly different solution, but all will have continuous air and vapor barrier clearly identified. Designers will also need to complete the exterior closure to ensure that the insulation and the air and vapor barriers are not visible and are protected from sun and direct rain infiltration. Also, it is critical that the head flashing seal against air infiltration and direct water to the outside.

Self-adhesive membranes are non-conductive and can be used to connect the water-control layer on the face of the wall to the metal flashing. This approach must ensure that the polymer flashing does not sag due to lack of support, which can trap water within the wall.

The hollow space in open window frames promote undesired natural convective heat flow between the interior and exterior. This can be reduced by filling voids with factory-installed, custom-formed foam plastic or rigid stone wool sections.

Often an overlooked principle, aligning the thermal control part of the window frame with the thermal control layer of the wall, is important to avoid cold-weather condensation and thermal bridging. In aluminum-framed windows, the thermal break provides a clear indication of the thermal-control layer. For fiberglass-, vinyl-, and wood-framed windows, the thermal resistance of the frame is more uniform, so thermal-control layer alignment is enhanced (as the frame is wider than the thermal break).

Note that retrofit items in 3b, 3c, and 3d are for one individual window and must be completed at the same time.

Existing jamb (from above)

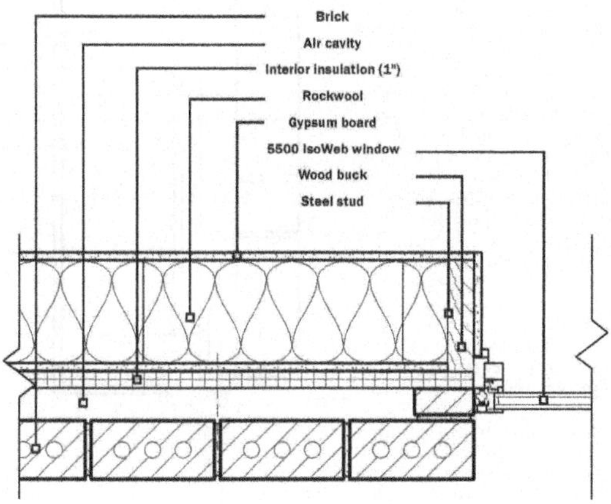

3b Modeling Values for Window Jamb

Component	Thickness inches (mm)	Conductivity Btu/h•ft•°F (W/m K)	Nominal Resisitance hr•ft²•°F/Btu (m²K/W)	Density lb/ft³ (kg/m³)
Interior Film	-	-	R-0.74 (0.13 RSI)	-
Brick	3 5/8" (92)	0.578 (1)	R-0.523 (0.092 RSI)	110 (1800)
Air Cavity	2" (51)	0.132 (0.23)	R-1.261 (0.222 RSI)	-
Insulation	1" (25)	0.0139 (0.024)	R-6 (1.055 RSI)	2 (32)
Rockwool	6 3/8" (162)	0.0208 (0.036)	-	4 (64)
Gypsum Board	1/2" (13)	0.092 (0.16)	R-0.5 (0.08 RSI)	50 (800)
Steel Studs	-	27.7 (48)	-	480(7800)
Steel Lintel	-	27.7 (48)	-	480(7800)
5500 IsoWeb Window	-	-	-	-
Wood buck	-	0.006 (0.10)	-	30(450)
Exterior Film	-	-	R-0.23 (0.04 RSI)	-

Steel Stud Wall with Interior and Exterior Insulation

3c Window Head

Commentary

Key to the effectiveness of this detail is ensuring good structural attachment of the window and alignment of the window thermal bridge. Every window section is likely to have a slightly different solution, but all will have continuous water- and air-control layers clearly identified. Designers will also need to complete the exterior closure to ensure that the insulation and the air/water control layers are not visible and are protected from sun and direct rain infiltration. Also, it is critical that the head flashing seal against air infiltration and direct water to the outside.

Self-adhered polymer membranes are nonconductive and can be used to connect the water-control layer on the face of the wall to the metal flashing. This approach must ensure that the polymer flashing does not sag due to lack of support, which can trap water within the wall.

The hollow space in open window frames promote undesired natural convective heat flow between the interior and exterior. This can be reduced by filling voids with factory-installed, custom-formed foam plastic or rigid stone wool sections.

Often an overlooked principle, aligning the thermal control part of the window frame with the thermal control layer of the wall, is important to avoid cold-weather condensation. In aluminum-framed windows, the thermal break provides a clear indication of the thermal-control layer. For fiberglass-, vinyl-, and wood-framed windows, the thermal resistance of the frame is more uniform, so thermal-control layer alignment is enhanced (as the frame is wider than the thermal break).

Note that retrofit items in 3b, 3c, and 3d are for one individual window and must be completed at the same time.

Existing head

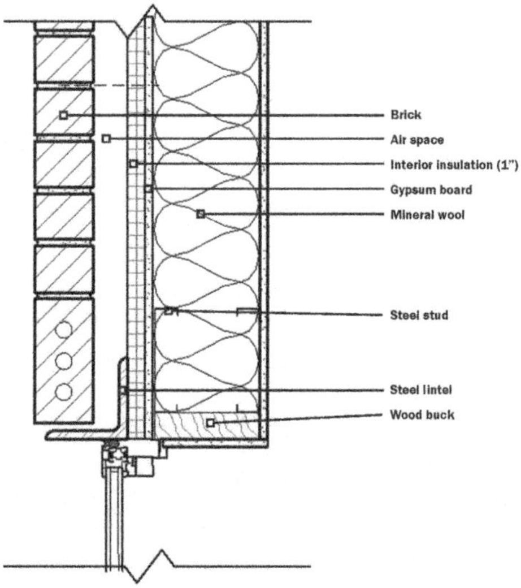

3c Modeling Values for Window Head

Component	Thickness Inches (mm)	Conductivity Btu/h • ft • °F (W/m K)	Nominal Resistance hr-ft²·°F/Btu (m²K/W)	Density lb/ft³ (kg/m³)
Interior Film	-	-	R-0.74 (0.13 RSI)	-
Brick	3 5/8" (92)	0.578 (1)	R-0.523 (0.092 RSI)	110 (1800)
Air Cavity	2" (51)	0.132 (0.23)	R-1.261 (0.222 RSI)	-
Insulation	1" (25)	0.0139 (0.024)	R-6 (1.055 RSI)	2(32)
Rockwool & Material Wool	6 3/8" (162)	0.0370 (0.064)	R-14.36 (2.53 RSI)	2(32)
Gypsum Board	1/2" (13)	0.092 (0.16)	R-0.5 (0.08 RSI)	50 (800)
Steel Studs	-	27.7 (48)	-	480(7800)
Steel Lintel	-	27.7 (48)	-	480(7800)
5500 IsoWeb Window	-	-	-	-
Wood buck	-	0.006 (0.10)	-	30(450)
Water/Air Control Barrier	-	-	-	-
Exterior Film	-	-	R-0.23 (0.04 RSI)	-

3d Steel Stud Wall with Interior and Exterior Insulation

Window Sill

Commentary

Key to the effectiveness of this detail is ensuring good structural attachment of the window and alignment of the window thermal bridge. Every window section is likely to have a slightly different solution, but all will have continuous air and vapor barrier clearly identified. Designers will also need to complete the exterior closure to ensure that the insulation and the air and vapor barrier are not visible and are protected from sun and direct rain infiltration. Also, it is critical that the head flashing seal against air infiltration and direct water to the outside.

Self-adhesive membranes are nonconductive and can be used to connect the water-control layer on the face of the wall to the metal flashing. This approach must ensure that the polymer flashing does not sag due to lack of support, which can trap water within the wall.

The hollow space in open window frames promote undesired natural convective heat flow between the interior and exterior. This can be reduced by filling voids with factory-installed, custom-formed foam plastic or rigid stone wool sections.

Often an overlooked principle, aligning the thermal control part of the window frame with the thermal control layer of the wall, is important to avoid cold-weather condensation. In aluminum-framed windows, the thermal break provides a clear indication of the thermal-control layer. For fiberglass-, vinyl-, and wood-framed windows, the thermal resistance of the frame is more uniform, so thermal-control layer alignment is enhanced (as the frame is wider than the thermal break).

Note that retrofit items in 3b, 3c, and 3d are for one individual window and must be completed at the same time.

Existing sill

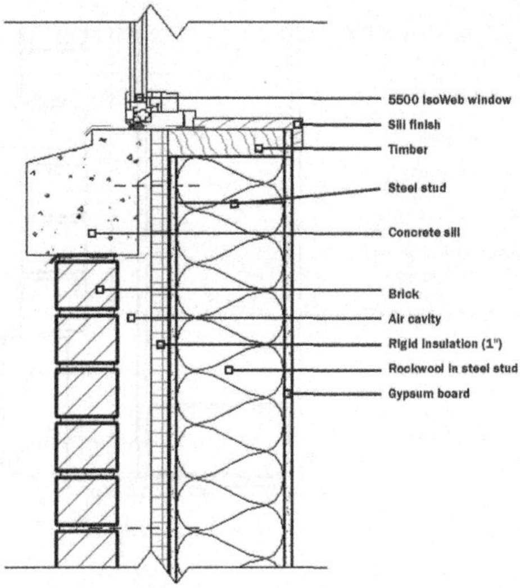

3d Modeling Values for Window Sill

Component	Thickness Inches (mm)	Conductivity Btu/h•ft•°F (W/m K)	Nominal Resistance hr-ft²•°F/Btu (m²K/W)	Density lb/ft³ (kg/m³)
Interior Film	-	-	R-0.74 (0.13 RSI)	-
Concrete Sill	-	1.4 (2.4)	-	150(2400)
Brick	3 5/8" (92)	0.578 (1)	R-0.523 (0.092 RSI)	110 (1800)
Air Cavity	2" (51)	0.132 (0.23)	R-1.261 (0.222 RSI)	-
Insulation	1" (25)	0.0139 (0.024)	R-6 (1.055 RSI)	2(32)
Mineral Wool with Steel Studs	6 3/8" (162)	0.0370 (0.064)	R-14.36 (2.53 RSI)	2(32)
Gypsum Board	1/2" (13)	0.092 (0.16)	R-0.5 (0.08 RSI)	50 (800)
Steel Studs	-	27.7 (48)	-	480(7800)
Timber	-	0.006 (0.10)	-	30(450)
5500 IsoWeb Window	-	-	-	-
Aluminum Sill Pan	-	92.45 (160)	-	175(2800)
Exterior Film	-	-	R-0.23 (0.04 RSI)	-

3e

Steel Stud Wall with Interior and Exterior Insulation

Steel-Tube Blast-Resistant Curtain Wall

Commentary

Although blast anchors are typically rather large pieces of thermally conductive steel or aluminum, they do not increase heat flow significantly if they are inside the thermal break. Metal components that bridge the insulation layer, such as lintels, can have a large impact and should be avoided where possible.

This detail shows the benefit of using loose lintels over the window instead of a penetrating metal plate. Loose lintels are also better able to accommodate differential movement between the brick and the support structure.

Additional benefits could be obtained by adding spray foam to the interior of the blast-resisting steel tube window support.

This sequencing assumes the use of a blast-resistant curtain wall section at the window opening perimeter

Existing curtain wall

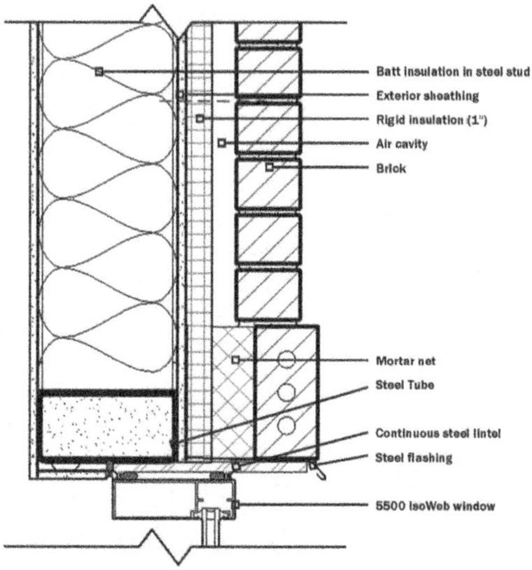

3e Modeling Values for Steel-Tube Blast-Resistant Curtain Wall

Component	Thickness Inches (mm)	Conductivity Btu/h•ft•°F (W/m K)	Nominal Resistance hr-ft²·°F/Btu (m²K/W)	Density lb/ft³ (kg/m³)
Interior Film (left side)	-	-	R-0.7 (0.12 RSI) to	-
R-0.9 (0.16 RSI)	-	0.578 (1)	R-0.523 (0.092 RSI)	110 (1800)
Gypsum Board	1/2" (13)	0.092 (0.16)	R-0.5 (0.08 RSI)	50 (800)
Fiberglass Batt Insulation	8" (90)	0.024 (0.042)	R-25 (4.4 RSI)	0.9 (14)
8" x 1 5/8" Steel Studs with Metal Tracks	18 gauge	35.825 (62)	-	489 (7830)
Exterior Sheathing	1/2" (13)	0.092 (0.16)	R-0.5 (0.08 RSI)	50 (800)
Exterior Insulation, Wall	1 1/2" (38)	0.017 (0.029)	R-7.5 (1.32 RSI)	1.8 (28)
Brick Ties	14 gauge	28.891 (50)	-	489 (7830)
Brick Veneer	3 5/8" (92)	0.451 (0.78)		
Steel Flashing	18 gauge	35.825 (62)	-	489 (7830)
Continuous Steel Lintel	3/8" (10)	28.891 (50)	-	489 (7830)
HSS Post	3/8" (10)	28.891 (50)	-	489 (7830)
Interior Spray Foam	8" (203)	-	R-40 (7.0 RSI)	1.8 (28)
Wood Blocking	1" (25)	0.052 (0.09)	-	27.8 (445)
Conventional Curtain Wall System with Double Glazed IGU. UCOG = 0.32 Btu/ft2.hr. °F (1.81 W/m2K)				
Exterior Film (right side)	-	-	R-0.2 (0.03 RSI)	-

Steel Stud Wall with Interior and Exterior Insulation

3f Steel Beam Penetration

Commentary

The specific steel section shown at right is a moderately large one. The heat flow varies little for reasonable ranges in steel size and thickness.

Heat transfer could be further reduced by avoiding the use of a large penetrating structural steel element. However, as this is a rather large beam, it is unlikely that there would be many penetrations in a large building. Some beams that penetrate the enclosure could be designed as propped cantilevers (e.g., at canopies or balconies), which would allow for significant reductions in heat flow by reducing the steel area penetrating the insulation through the use of bolts in tension and shear rather than hot-rolled sections.

The primary reason to apply the illustrated solution could be to solve a condensation problem by reducing the local heat loss on the interior. Buildings in cold climates with high interior humidity are greatly affected by otherwise-modest thermal bridges. To control condensation in cold climates, structural connections made of metal will often benefit from being insulated on the interior with an air- and vapor-impermeable insulating layer such as closed-cell spray foam.

In many penetrations, water and air intrusion also need to be addressed. These control layers are usually not provided by proprietary thermal breaks, and will require additional design effort. Vapor diffusion may also allow cold-weather condensation, so vapor-impermeable insulation should be used.

It is critical that these measures be assessed to confirm they have sufficient fire resistance, structural strength, and stiffness.

Existing beam penetration

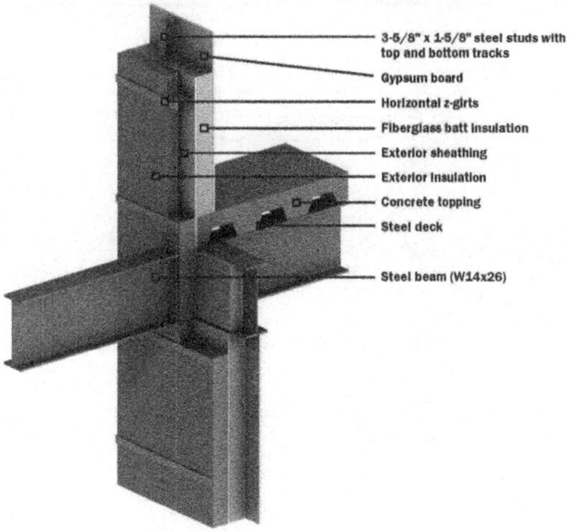

3f Modeling Values for Steel Beam Penetration

Component	Thickness Inches (mm)	Conductivity Btu/h•ft•°F (W/m K)	Nominal Resistance hr·ft²·°F/Btu (m²K/W)	Density lb/ft³ (kg/m³)
Interior Film (right side)	-		R-0.6 (0.11 RSI) to	-
R-0.9 (0.16 RSI)		0.578 (1)	R-0.523 (0.092 RSI)	110 (1800)
Gypsum Board	1/2" (13)	0.092 (0.16)	R-0.5 (0.08 RSI)	50 (800)
3 5/8" x 1 5/8" Steel Studs with Top and Bottom Tracks	18 Gauge	35.825 (62)	-	489 (7830)
Fibreglass Batt Insulation	3 5/8" (92)	0.024 (0.042)	R-12 (2.1 RSI)	0.9 (14)
Exterior Sheathing	½" (13)	0.092 (0.16)	R-0.5 (0.08 RSI)	50 (800)
Horizontal Z-girts with 1 ½" Flange	18 Gauge	430 (62)		489 (7830)
Exterior Insulation	3 1/8" (80)	0.2 (0.029)	R-15.7 (2.77 RSI)	1.8 (28)
Metal Cladding with ½" vented airspace incorporated into exterior heat-transfer coefficient				
Steel Beam W14x26 (W360x39)	-	347 (50)	-	489 (7830)
Steel Bearing Plates	1 3/16" (30)	347 (50)	-	489 (7830)
Steel Deck	1/16" (1.6)	347 (50)	-	489 (7830)
Concrete Topping	6" (152)	6.3 (0.9)	-	120 (1920)
Isokorb type S modulules	3 1/8" (80)			
Exterior Film (left side	-	-	R-0.2 (0.03 RSI)	-

Best-Case Thermal Loss Reduction: 0.820 Btu/h • ft • °F

Proposed retrofit

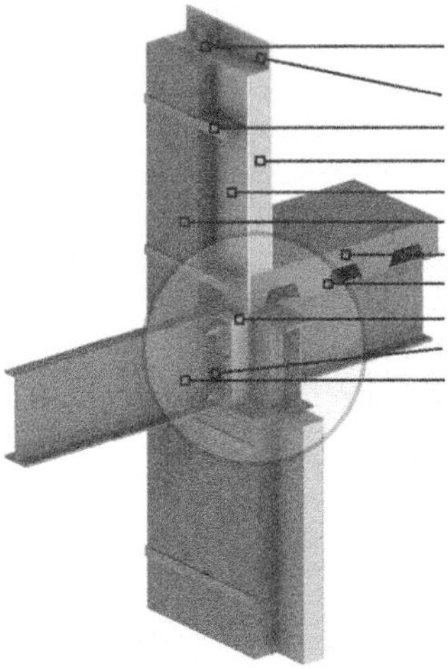

Detail from inset

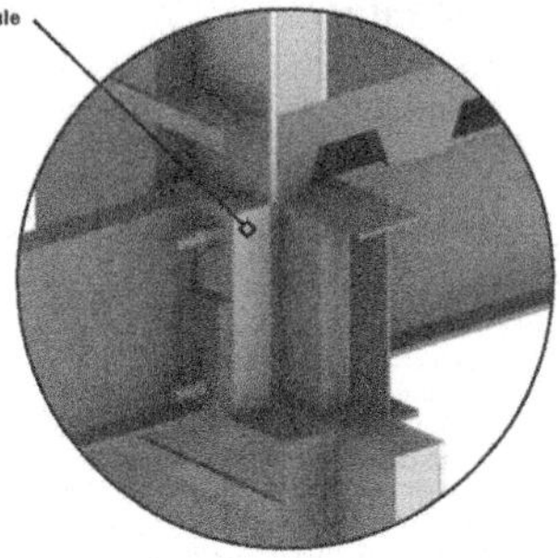

3f Thermal Performance Simulation Results		
Condition	**Clear Wall R-Value (W/m²K)**	**Linear Transmittance (Ψ) Btu/h • ft • °F (W/mK)**
Continuous Beam	R-17 (3.04)	1.73 (0.92)
With Schöck Isokorb Type S	R-17 (3.04)	0.91 (0.48)
As-built transmittance – Lowest retrofit transmittance = Best-case thermal-loss reduction		
1.73-0.910 = 0.820 Btu/h-ft-°F		

Q/C and sequencing

Detail

1. Replace each single, continuous beam with two beams flanged at their respective ends and connected by a structural-grade thermal break such as Isokorb® Type S modules or similar.

Notes

1. This detail must be designed in accordance with all the applicable structural codes and standards for new construction.
2. A sound understanding of the mechanical properties of structural thermal break technologies and materials is a prerequisite for executing this proposed solution.

Important Clear Wall Details

6a Steel Studs with Brick Ties

Commentary

This detail involves 6 in. steel studs constructed 16 in. on center tied to exterior brick veneer. The application of insulation requires care to prevent excessive compression and voids.

For proper suppression of convection loops and reduction of air intrusion to interior linings (gypsum wallboard or equivalent), gaps between the insulation and existing walls should be eliminated, especially at complex floor penetrations, partition walls, etc.

The joints between the barrier elements (air barrier, moisture barrier, thermal barrier, etc.) should almost always be drained joints, in the form of two-stage sealant joints or similar. Upgraded rain-penetration performance will often be required given that a thermal upgrade will require a reduction in wetting to balance the reduction in drying that results from reduced heat flow (either by insulating or reducing thermal bridging or increasing air tightness).

Existing stud

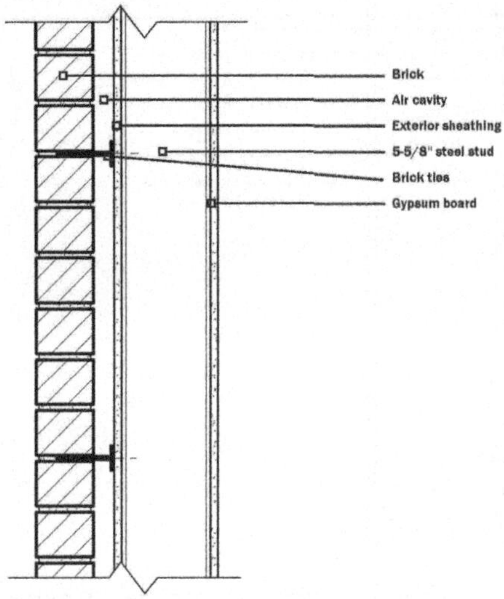

6a Modeling Values for Steel Studs with Brick Ties

Component	Thickness Inches (mm)	Conductivity Btu/h • ft • °F (W/m K)	Nominal Resisitance hr-ft²·°F/Btu (m²K/W)	Density lb/ft³ (kg/m³)
Interior Film (right side)	-	-	R-0.6 (0.11 RSI) to	-
R-0.7 (0.12 RSI)	-	0.578 (1)	R-0.523 (0.092 RSI)	110 (1800)
Gypsum Board	1/2" (13)	0.092 (0.16)	R-0.5 (0.08 RSI)	50 (800)
5 5/8" Steel Studs	18 gauge	35.825 (62)	-	489 (7830)
Air Gap in Stud Cavity	5 5/8" (143)	-	R-0.9 (0.16 RSI)	0.075 (1.2)
Exterior Sheathing	1/2" (13)	0.092 (0.16)	R-0.5 (0.08 RSI)	50 (800)
Air Gap	1 1/4" (32)	-	R-0.9 (0.16 RSI)	0.075 (1.2)
Brick Veneer	3 5/8" (92)	0.451 (0.78)	-	120 (1920)
Batt Insulation in Stud Cavity	5 5/8" (143)	0.02 (.034)	R-21 (3.70)	0.9 (14)
Mineral Fiber Insulation	2" (51)	0.017 (0.029)	R-8.4 (1.48 RSI)	1.8 (28)
Exterior Film (left side)	-	-	R-0.2 (0.03 RSI)	-

Best-Case Thermal Loss Reduction: 19.4 Btu/h • ft • °F

Proposed retrofit

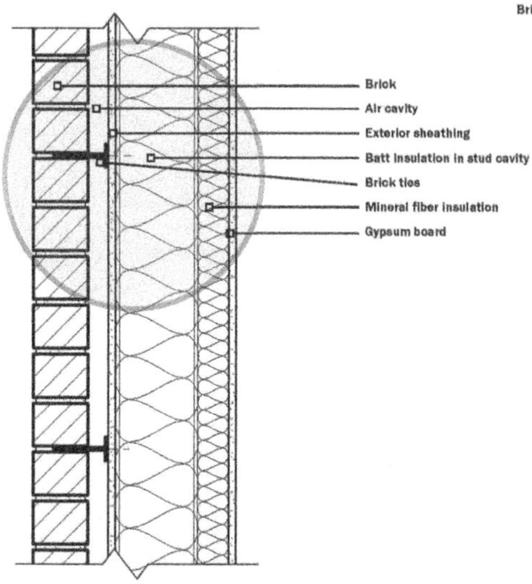

Detail from inset

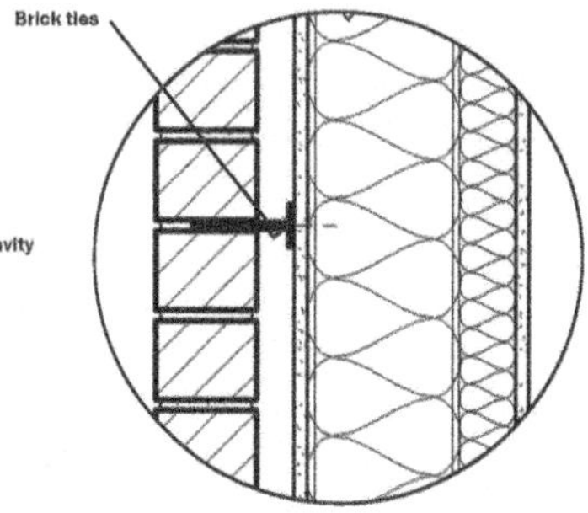

6a Thermal Performance Simulation Results		
Condition	**Clear Wall R-Value (W/m^2K)**	**Linear Transmittance (Ψ) Btu/h • ft • °F (W/mK)**
Wall Clear Field	R-4.3 (0.76)	-
Retrofit with...		
Batt Insulation in Stud Cavity (x)	R-12.7 (2.24)	-
Mineral Fiber Insulation (y)	R-13.3 (2.34)	-
Batt Insulation in Stud Cavity (x) and Mineral Fiber Insulation (y)	R-23.7 (4.18)	-
As-built transmittance – Lowest retrofit transmittance = Best-case thermal-loss reduction		
23.7-4.3 = 19.4 Btu/h·ft²·°F		

Q/C and sequencing

1. Remove the interior wall gypsum board.
2. Fill the steel stud cavity with fiberglass batt insulation. Make sure the cavity is completely filled, with no gap between the batt and the steel studs or exterior sheathings.
3. Attach 2 x 2 in. wood studs or furring strips to the exposed interior steel stud flange faces.
4. Attach plastic stick pins to the cavity-facing side of the gypsum board. These pins should be separated no more than 48 in. vertically, starting from the gypsum bottom edge and upward. The stick pins will provide support to the mineral fiber insulation when it is installed in the rest of the wall assembly and in cavities between the wood spacers.
5. Cut the mineral fiber insulation to fit and place in stud cavities.
6. Install new gypsum board over the mineral fiber insulation and attach to the previously installed wood furring.

Important Clear Wall Details

6b Concrete Wall with Interior Steel Stud Assembly

Commentary

Applying spray foam to the steel stud cavity will improve the thermal-control layer and the air-infiltration control.

For effective convection-loop suppression and reduction of air permeability of interior linings (gypsum wallboard or similar), gaps between the insulation and existing walls should be eliminated, especially at complex floor penetrations, partition walls, etc.

It is critical to apply the insulation outboard of the interior sheathing and framing, as it ensures that these will remain much warmer, and therefore prevent condensation problems from occurring.

The joints between the barrier elements (air and vapor barrier, etc.) should almost always be drained, two-stage sealant joints or similar.

The spray-foamed air gap will improve clear wall thermal performance by approximately R-6, assuming that the foam insulation used has a thermal conductivity of 0.014 h•ft•°F (0.024 W/m K), but thermal performance will be reduced at floors and load-bearing partition areas. A structural component may bypass the foam insulation at these locations, causing thermal bridge heat losses due to the lack of insulation continuity.

Existing stud

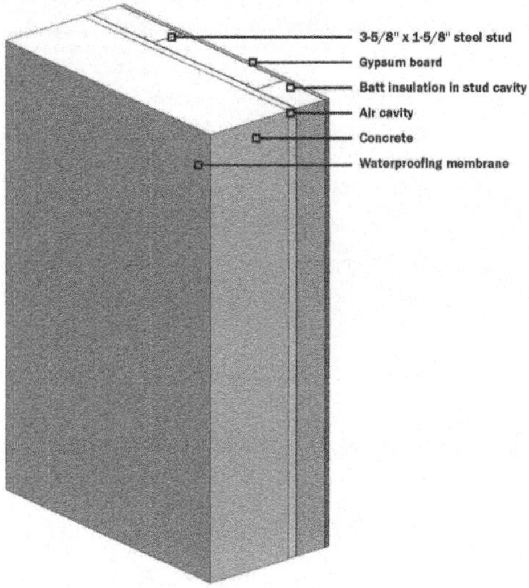

6b Modeling Values for Concrete Wall With Interior Steel Stud Assembly

Component	Thickness Inches (mm)	Conductivity Btu/h • ft • °F (W/m K)	Nominal Resisitance hr·ft²·°F/Btu (m²K/W)	Density lb/ft³ (kg/m³)
Interior Film (right side)	-	-	R-0.6 (0.11 RSI) to	-
R-0.7 (0.12 RSI)	-	0.578 (1)	R-0.523 (0.092 RSI)	110 (1800)
Gypsum Board	1/2" (13)	0.092 (0.16)	R-0.5 (0.08 RSI)	50 (800)
3 5/8" Steel Studs	18 gauge	35.825 (62)	-	489 (7830)
Batt Insulation in Stud Cavity	3 5/8" (143)	0.025 (0.044)	R-12 (2.11 RSI)	0.9 (14)
Air Gap	1" (25)	-	R-0.9 (0.16 RSI)	0.075 (1.2)
Spray Foam Insulation	1" (25)	0.014 (0.024)	R-6 (1.06 RSI)	0.35 (1470)
Concrete	10" (254)	1.04 (1.8)	-	0.20 (850)
Exterior Film (left side)	-	-	R-0.2 (0.03 RSI)	-

Best-Case Thermal Loss Reduction: 6.3 Btu/h • ft • °F

Proposed retrofit

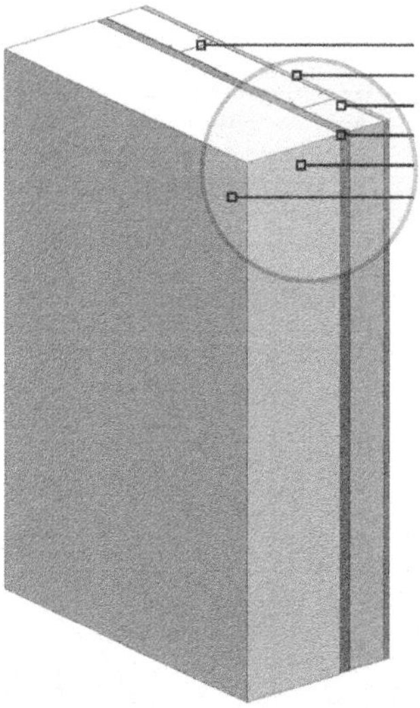

Detail from inset

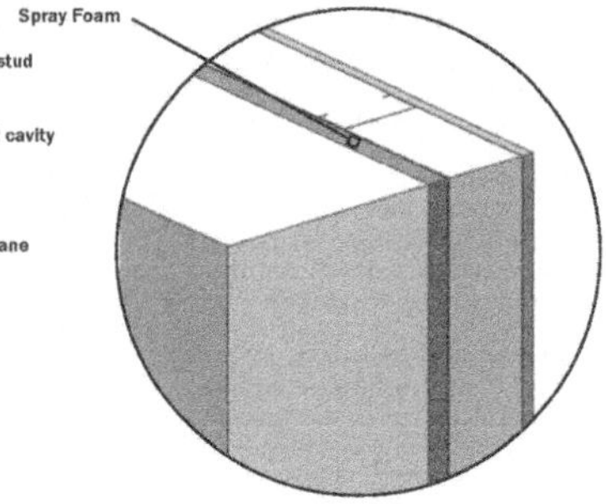

6b Thermal Performance Simulation Results		
Condition	Clear Wall R-Value (W/m²K)	Linear Transmittance (Ψ) Btu/h • ft • °F (W/mK)
Wall Clear Field	R-9.5 (1.67)	-
Retrofit with...		
1" spray foam insulation	R-15.8 (2.78)	-
As-built transmittance - Lowest retrofit transmittance = Best-case thermal-loss reduction		
15.8-9.5 = 6.3 Btu/h·ft²·°F		

Q/C and sequencing

1. Remove the interior finish (gypsum board).
2. Remove the R-12 insulation batt from the stud cavity to expose the gap layer between the concrete exterior wall and the steel stud assembly.
3. Apply the spray foam insulation in the stud cavity. Make sure the entire concrete wall is covered by approximately 1 in. of foam.
4. Replace the R-12 insulation batt removed from the stud cavity.
5. Reinstall the gypsum board interior finish.

7a

Historical Details with Interior Insulation

Window Sill in Solid Brick Masonry

Commentary

The objective of the thermal bridge mitigation approach shown here is to maintain original facade aesthetics while improving the section's thermal performance. However, all plans and designs related to the modification of historical architectural details should first be coordinated with the pertinent authorities before work bgins.

Rigid insulation may be used in place of spray foam at the window opening without significantly affecting the section's thermal performance.

The hollow open space in window frames will promote natural convective heat flow between building interior and exterior. This undesired heat transfer can be reduced by filling voids with factory-installed, custom-formed foam plastic or rigid stone wool insulation sections.

It is critical that the head flashing direct water to the exterior and maintain air tightness at this location.

The thermal performance can be further improved by specifying thicker insulation layers and thinner wood blocking. Insulating the exterior of the rough window opening, behind the precast sill, will also improve performance.

Existing sill

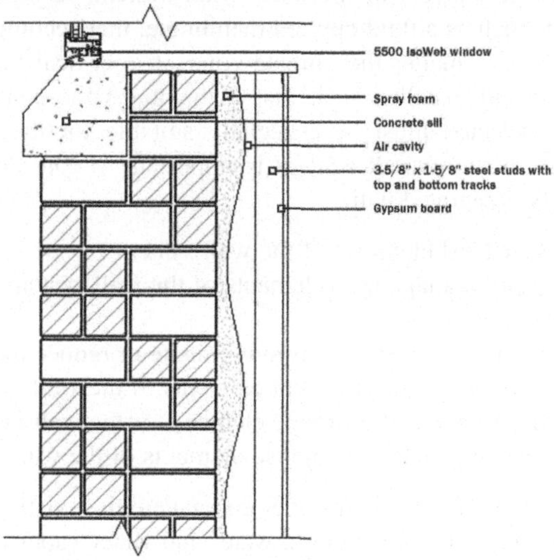

7a Modeling Values for Window Sill in Solid Brick Masonry

Component	Thickness Inches (mm)	Conductivity Btu/h • ft • °F (W/m K)	Nominal Resistance hr·ft²·°F/Btu (m²K/W)	Density lb/ft³ (kg/m³)
Interior Film	-	-	R-0.74 (0.13 RSI)	-
Brick	15 5/8" (397)	0.5778 (1)	R-2.25 (0.397 RSI)	110 (1800)
Spray Foam	2" (51)	0.0139 (0.024)	R-12 (02.11 RSI)	2 (32)
Air Gap	1 1/2" (38)	0.1192 (0.206)	R-1 (0.185 RSI)	
Air Gap with Vertical Steel Studs	1 1/2" (38)	0.806 (1.395)	R-0.16 (0.027 RSI)	
Gypsum Board	1/2" (13)	0.092 (0.16)	R-0.5 (0.08 RSI)	50 (800)
Steel Top and Bottom Track	-	27.7 (48)	-	480 (7800)
5500 IsoWeb Window	-	-	-	-
Exterior Film (Wall)	-	-	R-0.23 (0.04 RSI)	-

Appendix E: Window Installation Guidance

Successful window installation depends on successful integration with the wall's overall water and air management strategies. It must integrate with air, moisture, and vapor control layers of the given wall system. In some cases, there may be one element, such as a fluid-applied membrane, that accomplishes one or more of these functions. No matter the climate zone, it is critical that the specified window be designed and installed such that it maintains the continuity of these layers. The product selected must be especially suitable for the water management system planned for the overall wall. It is important to consider that window installations should be integrated with:

- The structural elements of the wall system
- The water management elements of the wall system
- Flashing systems
- The thermal barrier system to eliminate or reduce thermal bridges
- The air leakage management elements of the wall system
- The decorative and cosmetic elements of the wall system
- The cleaning and maintenance elements of the wall system

It is essential that the installer understand the existing water management system of the building to prevent bulk water and water vapor penetration into and through the building envelope. There are two fundamental methods for preventing bulk water intrusion, the drainage and the barrier methods, detailed in the following sections.

Bulk Water Management: Drainage Method

Drainage systems can be identified as those designed to both (1) deflect exterior bulk water, such as rainfall, by employing a first surface water barrier, such as siding, and also (2) employ a drainage system to aid in the removal of any water that penetrates beyond the first surface water barrier (Fig. E.1). Such drainage systems may employ building papers, house wraps, or other systems to create a water-resistant barrier intended to protect the rest of the wall or roof system. Common examples of drainage systems include building envelopes with exterior surfaces that may include siding, veneers, panels, shingles, wood shakes, brick, or other applied exterior surfaces, that have an air space behind these exterior surfaces, and that also have a weather-resistant building paper, fabric, or other material intended to drain away any bulk water that penetrates the exterior surface finish. Building envelopes with wall systems that include a cavity between the exterior cladding system and the wall sheathing are generally included in this definition. In drainage systems, fenestration units are to be integrated with, and sealed to, the water-resistant barrier to promote drainage. The fenestration system must also be sealed to any flashing systems employed. Flashing systems should direct all incidental water to the outermost

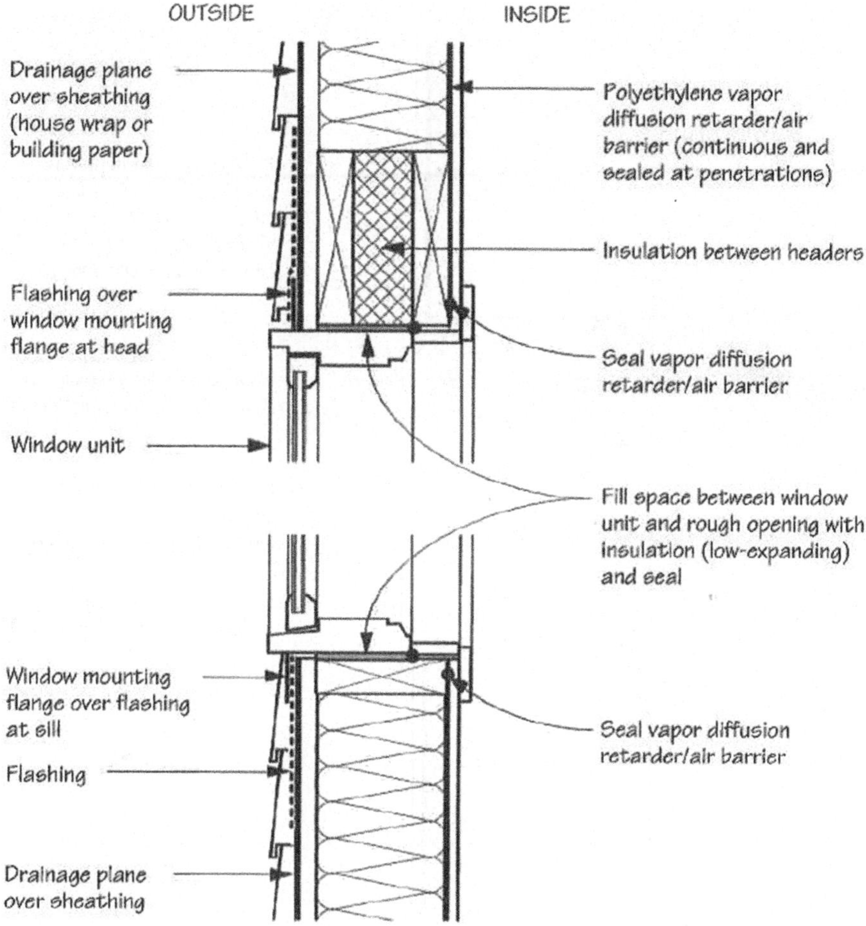

Fig. E.1 Example of window installed as part of drainage wall system. (Illustration taken from *Residential Windows: a Guide to New Technologies and Energy Performance*)

surface of the wall. The design of drainage wall systems may allow the flashing system to route incidental water to the surface of the water-resistant barrier only. Use of pan flashing to direct incidental water onto the outermost wall surface in drainage systems is acceptable and may be preferable.

Bulk Water Management: Barrier Method

Surface barrier systems can be identified as those systems in which the outermost surface of the wall or roof is the sole barrier to intrusion of liquid water (Fig. E.2).

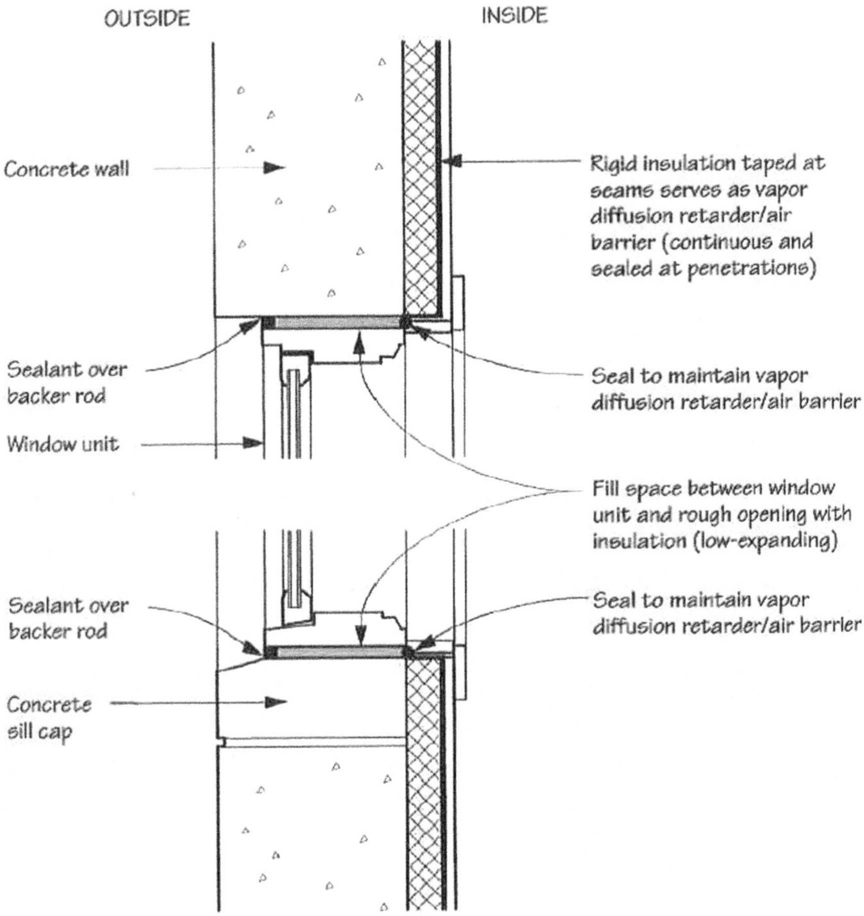

Fig. E.2 Example of windows installed as part of a barrier wall system. (Illustration taken from Residential Windows: a Guide to New Technologies and Energy Performance)

Barrier systems are designed to be sealed at the exterior surface to keep water out. Barrier systems rely on sealants around building penetrations to prevent moisture intrusion. Most systems make no provision for drainage of incidental moisture that may enter the wall or roof system. In addition, they generally include an exterior coating that is relatively impermeable to moisture. The most common surface barrier cladding systems used in buildings are certain forms of EIFS wall cladding and large panel cladding systems installed without water-resistant barrier membranes. In barrier systems, fenestration units are sealed to the outermost wall surface.

Fig. E.3 View of air and weather barrier integrated into rough opening to receive replacement window

Air Leakage Management

Typically, when replacing windows, doors, and skylights, the replacement units are sized to closely fit into the hole(s) made by removing other fenestration products. Proper fenestration replacement requires significant attention to air sealing details and continuity of the surrounding insulation layers, which need to connect with the insulation layers within the window and glass, especially at the gap(s) between the new fenestration and the opening(s). It is essential that the installer understand the air leakage control system employed in the existing building enclosure (if any) to properly integrate the new window, door, or skylight into the building system to prevent air leakage into and through the building envelope (Fig. E.3). Proper sealants, tapes, foams, and other materials must be used to seal all gaps and penetrations. Surface cleanliness – of both the new fenestration product and the surfaces to which it is being attached – helps to ensure good quality and durable air sealing. In some cases, proper air sealing will require surfaces to be primed and cleaned to ensure air sealing material adhesion. Carefully review the material specifications of both the fenestration product manufacturer and those of the air sealing system being employed to ensure material compatibility, proper temperatures for successful application, required surface treatments (if any), and other requirements necessary to achieve proper air sealing.

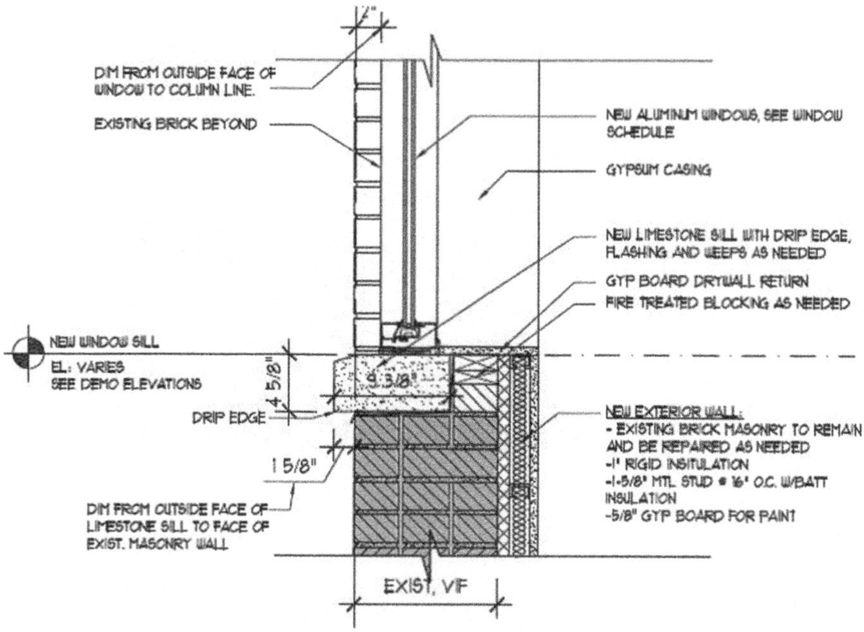

Fig. E.4 Example of thermal bridge in wall construction

Thermal Bridging

Thermal effects of the installed window depend on thermal transmission of its components, the glazing unit and the frame, as well as how it is installed into the wall (effect of thermal bridging and air leakage). Windows should be installed to minimize thermal bridges between their frames and the walls. Windows should be installed as close to the thermal (insulation) plane of the wall as possible. This reduces the possibility of a thermal bridge, poor thermal performance, and formation of condensation. Figure E.4 shows an example of thermal bridging in a masonry wall. The window is installed well outboard of the insulation layer of the wall. The stone sill creates a thermal bridge that bypasses the thermal break of the window and, in cold climates, creates cold surface temperatures of the interior metal. Condensation may form on the window framing when it is exposed to warm, humidified, interior air.

One strategy for minimizing thermal bridges when replacing windows is to move the plane of the window closer to the insulation. However, other elements, including flashing and structural anchorage, must also be considered. The following example (head and sill, Figs. E.5, E.6, E.7 and E.8) shows a scenario in which the replacement window is being installed further inboard than the existing window. Note that it cannot be moved all the way back to the insulation line as the drainage path and

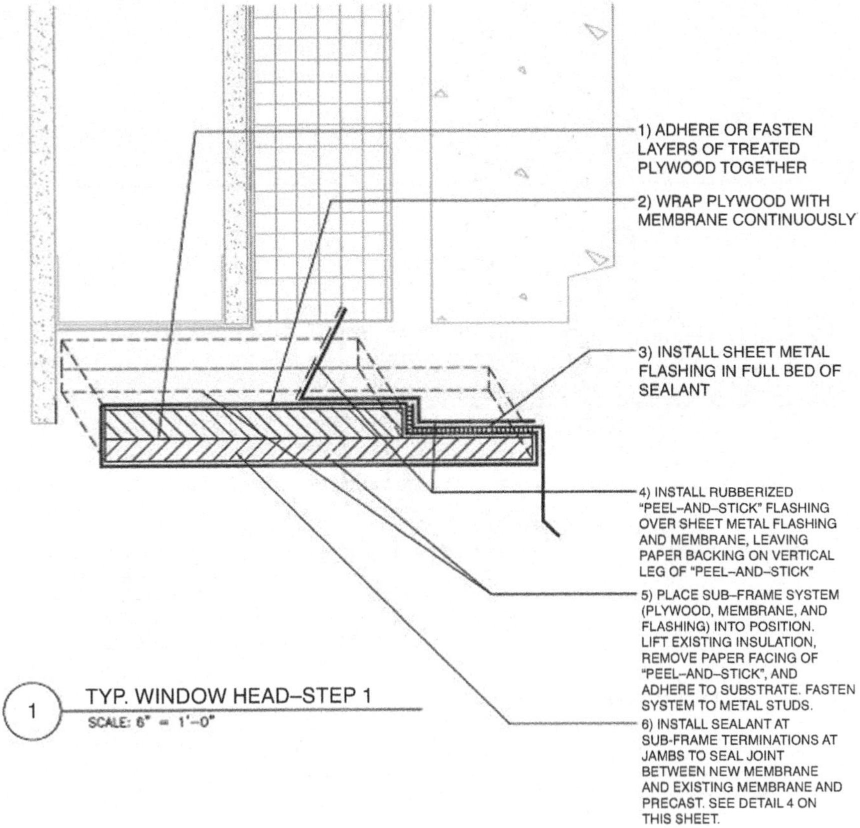

Fig. E.5 Step-by-step illustration of window head installation (Step 1)

flashing and the tie into the air and weather barrier must be addressed as well. Note that there are low-conductivity tensile and compressive structural products and elements that can be used, such as wood, rubber, PVC, high-density polyurethane, plastics with low density, cellular glass, and solutions that reduce the conductive elements within the insulation layer. Some of these can be fixed to or fixed through depending on product application and detailing situation.

Figure E.9 presents an overall view of what is in fact a Passive House compliant detail of window installation. Notice how nothing is materially different from normal construction or detailing; it is easy to construct, and easy to level, and has conventional air sealing that is typical of many countries conscious of energy efficiency. This detail has a completely neutral psi value (no additional heat loss around the window). The window is also rated for performance under arctic conditions, and the glazing can be modified to reduce/increase solar gain. Also the frame is available in many colors and styles that match most common construction designs.

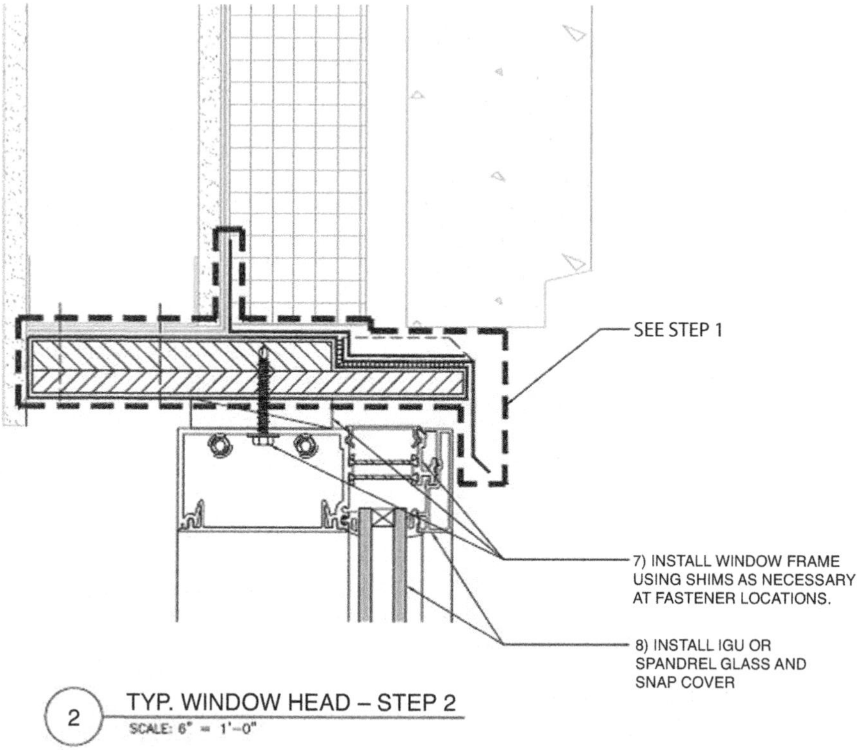

Fig. E.6 Step-by-step illustration of window head installation (Step 2)

This is (intentionally) a very macroscaled detail in terms of drafting, but includes such important aspects as:

- Rubberoid/paper/stucco/intelligent membranes are used in lieu of continuous metal flashings through the line of thermal performance.
- All insulation materials are lining up for the best performance.
- Bulk water is dealt with by heavy flashings that all drain into a sill pan that is returned at the sides, so there are no weaknesses at corners; nevertheless the heavy metal flashings do not need to extend under the entire window; a rubberoid flashing can be used to collect that small amount of water.

An insignificantly lesser performance can be obtained with the same detail by switching the Compaq foam with wood. These components vary in type and form, but are available for this insulation thickness. Figures E.10 and E.11 show examples of replacement of the window installed in a masonry wall.

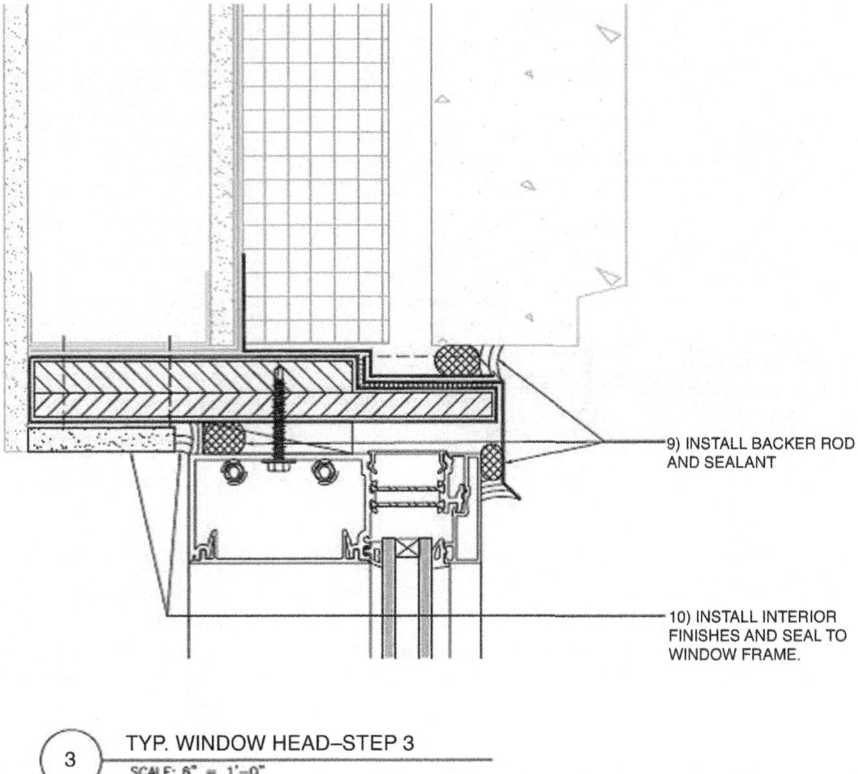

Fig. E.7 Step-by-step illustration of window head installation

When considering the strategies to reduce thermal bridging, it is important to not forget that the window still needs to integrate with the air, moisture, and vapor control layers of the wall system. Figures E.12 and E.13 show a window installed into an EIFS. Note that the foam of the EIFS acts as the insulation layer and that the thermal break (fiberglass pressure plate) is in line with this layer. The window is tied to the air, vapor, and moisture layers with a membrane that glazes into the framing of the window.

Tables E.1, E.2, and E.3 describe the construction process of window installation in a steel stud wall and a masonry wall with exterior insulation and brick façade. Information for these tables is based on studies conducted by the ERDC team and its contractors (Pagan-Vazquez et al. 2015a, b; Lawton Mark et al. 2014).

Tables E.4, E.5, and E.6 show examples of thermal bridges mitigation with building renovation, reproduced with permission from IBO (2016).

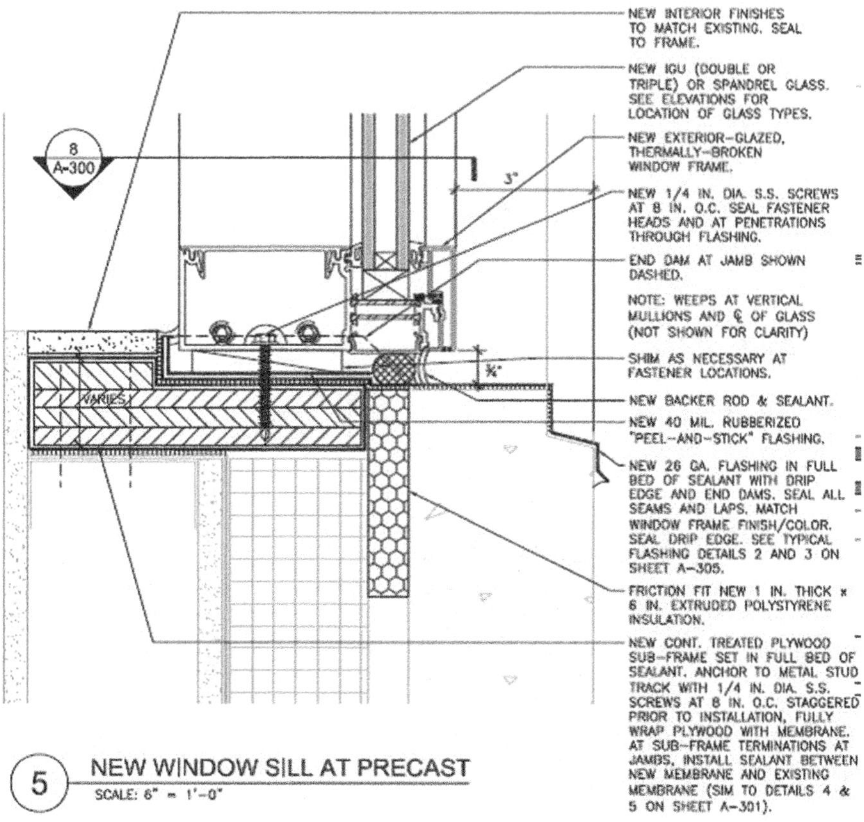

Fig. E.8 Illustration of window sill installation

Post-installation Procedures

After installation, the installer should perform a "punch-list" inspection to verify that windows have been installed correctly. This inspection should include, but may not be limited to:

- The installer should ensure that the fenestration product frame and sash are installed square, plumb, and level within the specified tolerances. If necessary to check all aspects of the installation, the installer should remove all removable sash and screens. Any removed items should be remounted or otherwise reinstalled by the installer before project completion.
- The installer should ensure that all sashes move freely within their frames and that weatherstripping or compressible seals make full contact with mating surfaces.
- The installer should ensure that operable hardware such as locks, cranks, latches, and hinges operate smoothly and that all locking mechanisms engage properly.

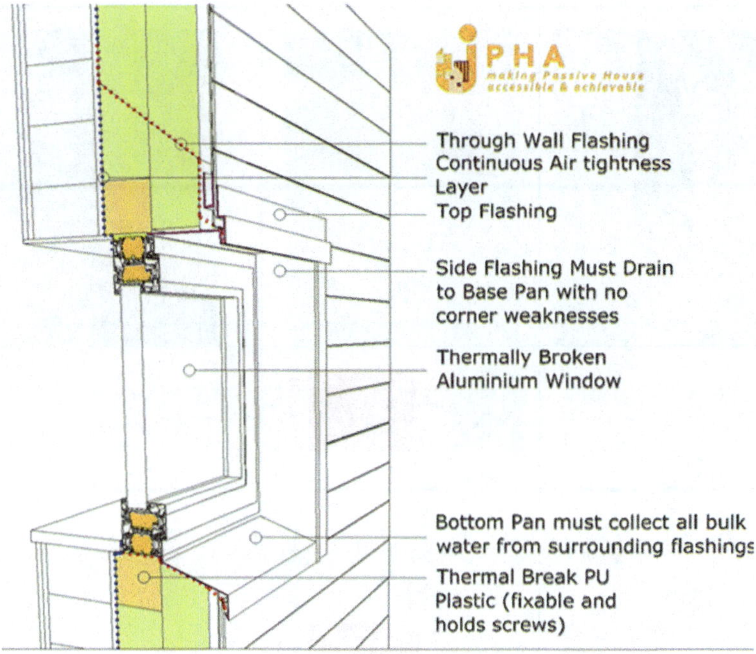

Fig. E.9 Passive House detail of window installation (provided by the Passive House Academy)

- The installer should ensure that all operable elements move freely and function correctly by operating the operable elements of the fenestration product and its hardware.
- The installer should ensure that all accessories and other components of the fenestration product assembly are present, such as screens and hardware.
- The installer should ensure that drainage holes should are inspected for blockage and freed of any obstructions to allow drainage.
- The installer should adjust the hinges so that the seals in a sash or opening section close firmly on surrounding frame.

Post-installation Quality Control

Voluntary performance standards for wood, aluminum, and vinyl windows are established by the American Architectural Manufacturers Association (AAMA). Among others, they set standards for air and water infiltration, thermal efficiency, and structural performance, all of which can have an effect on short- and long-term durability. Standard test methods and procedures are used to evaluate air infiltration, water penetration, and structural and thermal performance. These test methods have

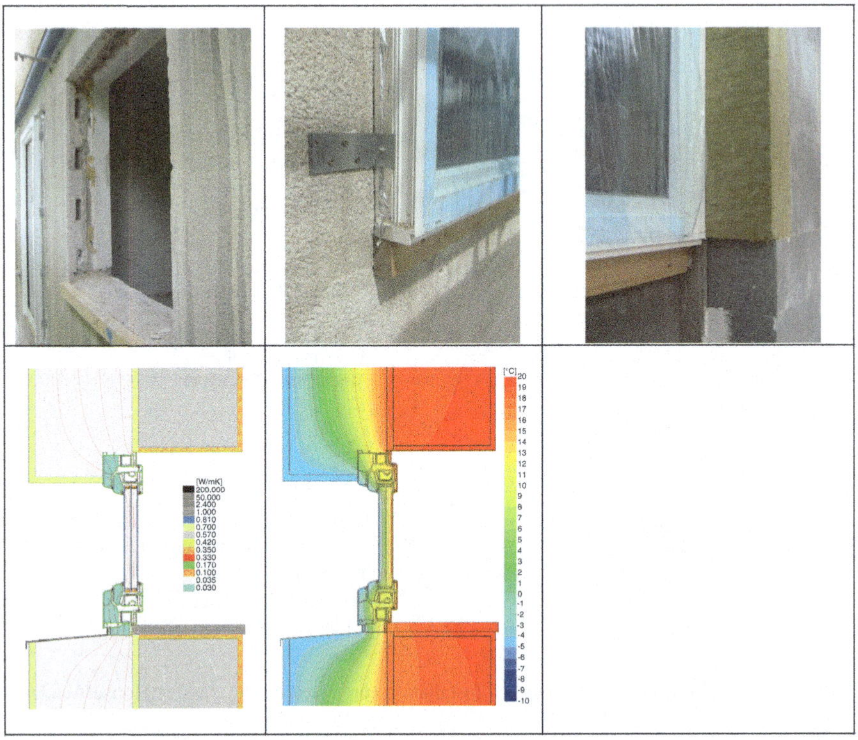

Fig. E.10 Example of old window replacement: (**a**) old Window opening with masonry rabbets removed. (**b**) The addition of a horizontal wooden support facilitates fixture of windows; it is installed and leveled in advance and bears the window's weight. The window is then fixed with stainless steel angle brackets. (**c**) Top and lateral parts of the window frame are covered with the exterior wall insulation as far as possible. (**d**) Schematic of window installation. (**e**) Thermal image of window performance
Boundary conditions: Exterior wall U = 0,107 W/(m^2K) 300 mm 0,035 W/(mK)
Resulting thermal bridge effect:
$\Psi_{fixture}$ (top/lateral) = 0,010 W/(mK)
$\Psi_{fixture}$ (bottom) = 0,058 W/(mK)
Mean: $\Psi_{fixture}$ = 0,021 W/(m^2K)
Overall resulting U-value of window: U_W = 0,86 W/(m^2K); T_{min} = 14,1 °C

been developed for window manufacturers to verify that a specific window system design or model meets the specified performance requirements.

A vital part of a window replacement project is post-installation QC testing. This typically consists of testing for water and air infiltration and will verify that the window is performing as designed and that the installation, including flashing and perimeter sealants, has been installed correctly. The test protocol is outlined in AAMA 502, *Voluntary Specification for Field testing of Newly Installed Fenestration Products.* AAMA 502 specifies that windows should be installed shortly after being installed using ASTM E783, *Standard Test Method for Field Measurement of Air Leakage through Installed Exterior Windows and Doors,* and ASTM E1105,

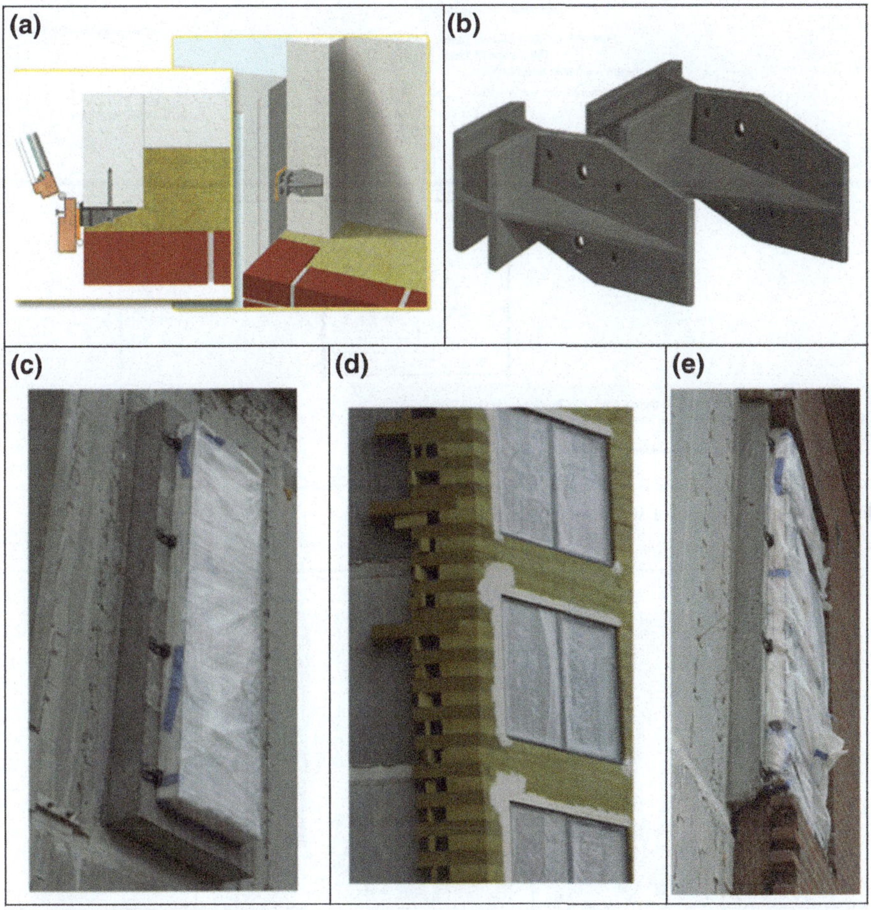

Fig. E.11 Example of window installation within external insulation: (**a**) schematics of window mounting using K-MUR bracket; (**b**) example of brackets for window mounting outside the wall; plastic brackets K-WALL-28/50 reduce thermal bridging and have a load capacity of 120 kg per mounting; (**c**) window is installed outside the wall using brackets; (**d**) mineral wool insulation is installed using adhesive compound; (**e**) insulation is covered with brick; **a** and **b** graphs are provided by Knudsen Killen A/S; **c**, **d** and **e** – pictures taken by Alexander Zhivov at different construction sites in Aalborg, Denmark

Standard Test Method for Field Determination of Water Penetration of Installed Exterior Windows, Skylights, Doors and Curtain Walls by Uniform or Cyclic Static Air Pressure Difference. Both tests involve the use of a chamber to induce a pressure differential on the window in an attempt to draw air or water through the window. The tests are performed in accordance with pressures specified in the project documents or, in the absence of project documents, design values provided by the window manufacturer.

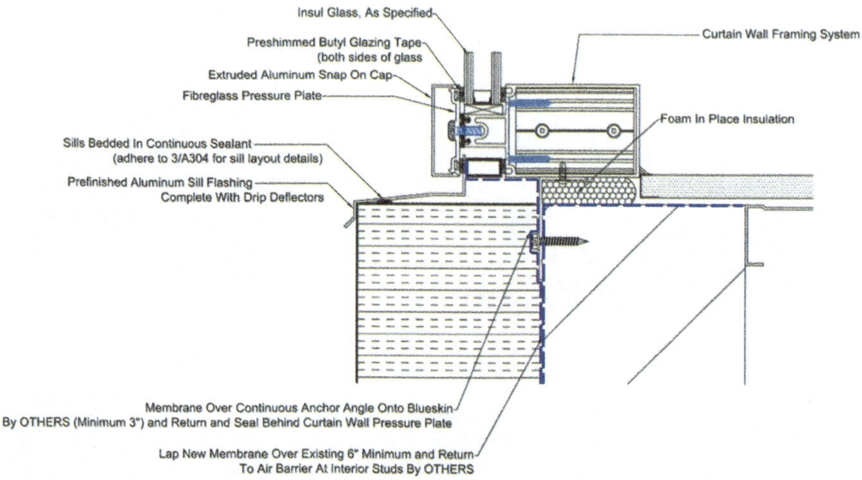

Fig. E.12 Illustration of window installed into EIFS wall

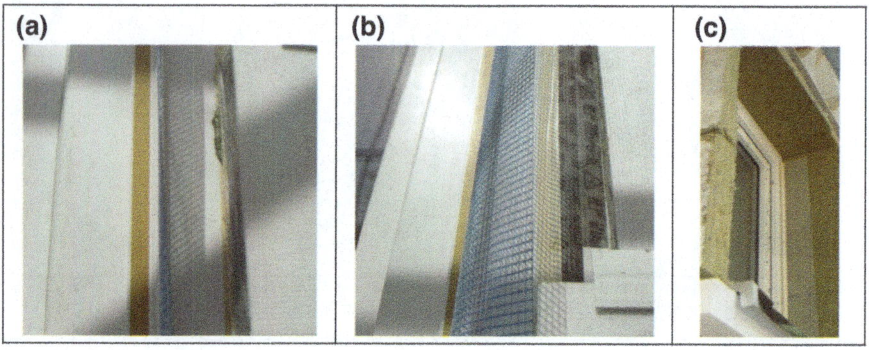

Fig. E.13 Window/wall gap is sealed using mesh glued to the window frame from the outside using adhesive connector (c, d, e). Plaster is laid over the mesh and then painted over

Inspection and Maintenance

No matter the type of window or the material, a post-installation inspection and maintenance program is vital to the longevity of the specified window system. When inspecting windows, the window should not be considered as a separate entity, but as part of the overall wall system. A visual inspection of each window should be done on a regular basis to determine any maintenance or repairs that may be required. Conditions can be determined of all the parts that make up the window including window glazing, frame material, material finishes, hardware, weatherstripping, screens, glass and glazing, and sealant joints. The operability of windows should be checked. The window and surrounding interior finishes should be observed for any signs of water leakage or damage.

Table E.1 Window installation into the externally insulated masonry wall

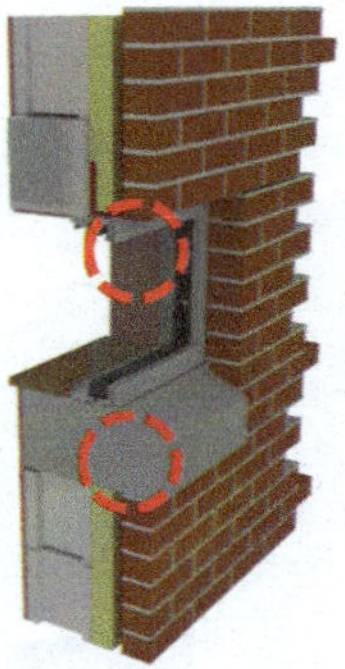		Issues with typical window installation in a masonry wall: no alignment or connection of the thermal break to the insulation layer and highly conductive shelf angle bridging the thermal break
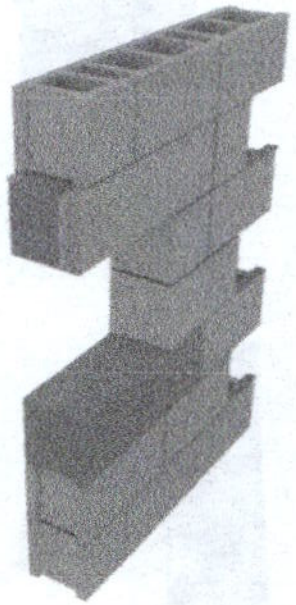		 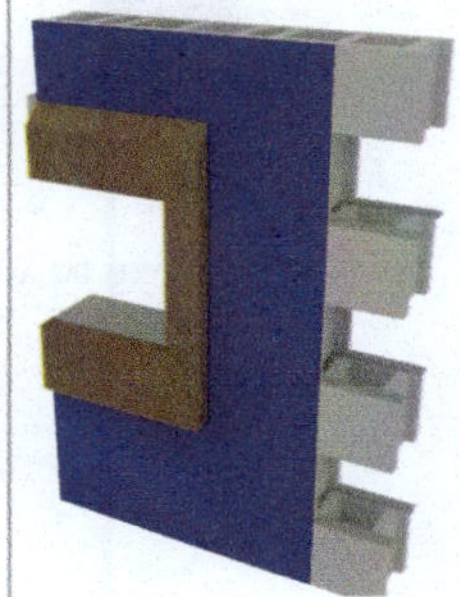
Begin with the CMU wall with lintel cast into U blocks	The wall is covered in a suitable air/water membrane and sealed at all edges with a suitable sealant	A treated timber break is fastened to the CMU

(continued)

Table E.1 (continued)

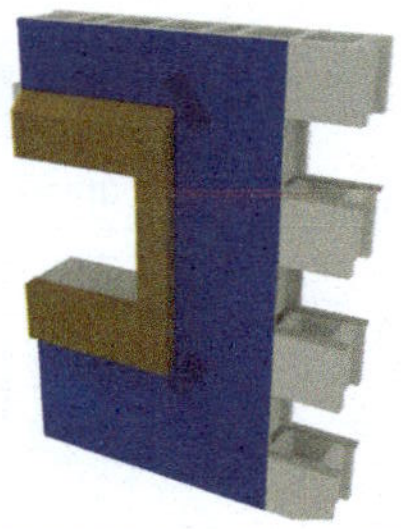	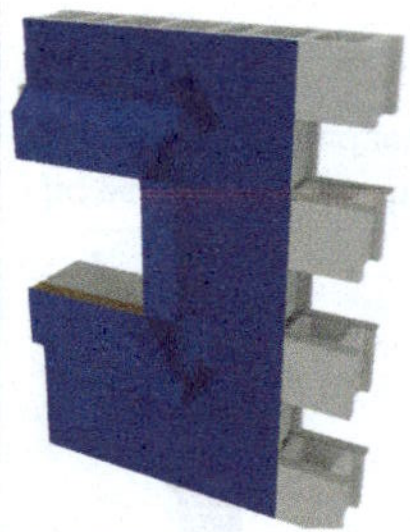	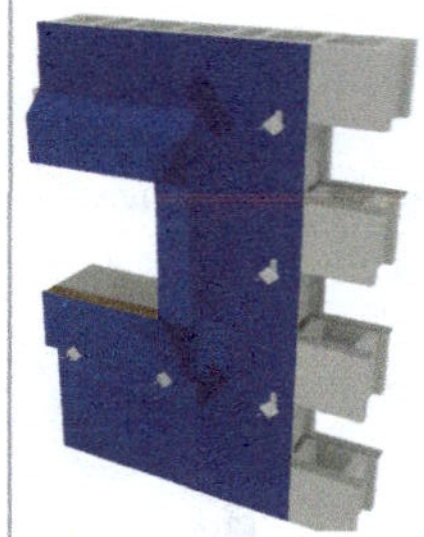
Next the edges of the timber break are taped over with a gusset	This allows a preformed permeable peel-and-stick to be cut and folded covering all edges to complete the airtight seal	Brick anchors can now be fixed
	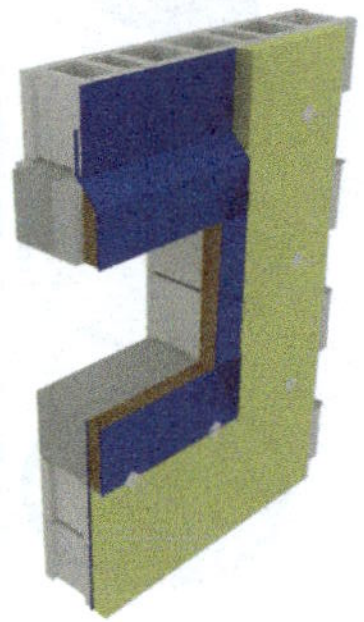	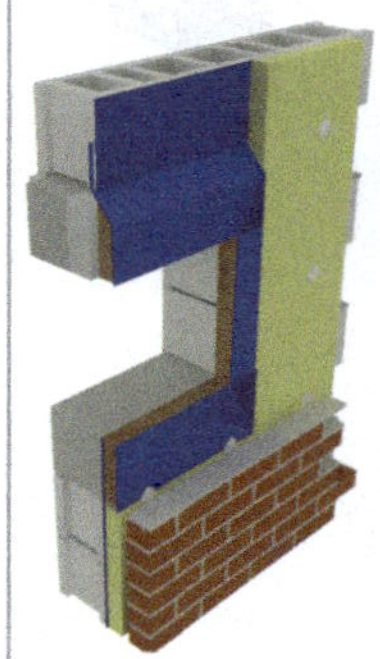 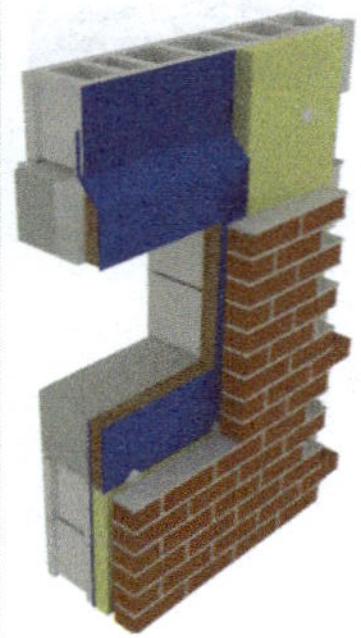
Insulation is added		Brick courses are laid to the level of the sill and head

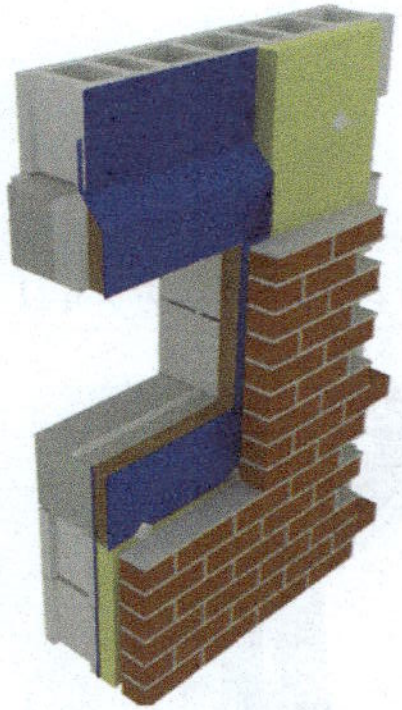	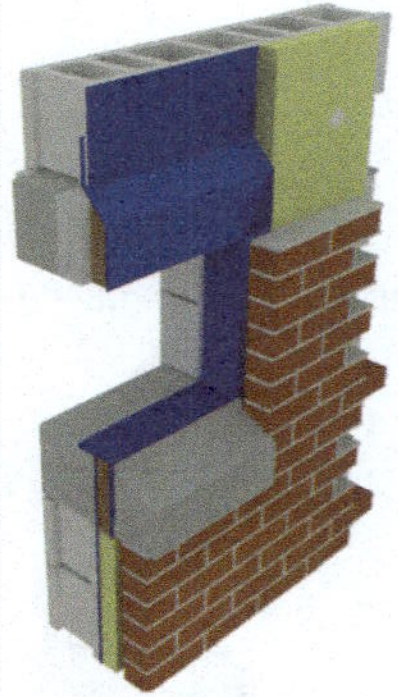	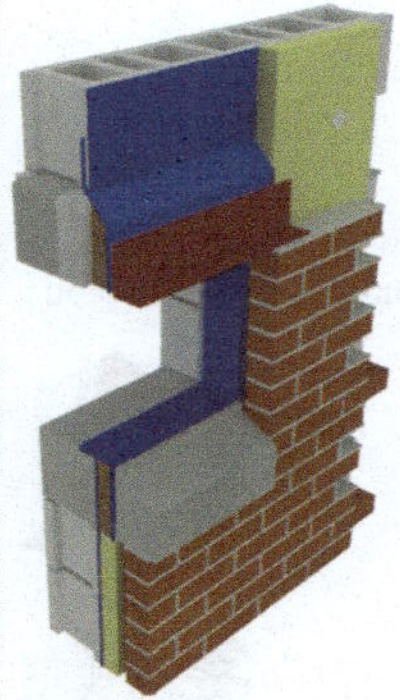
A backdam anchor is added to all four sides of the reveal	A peel-and-stick membrane is added for airtightness and watershed before adding the sill	The loose lintel is also added
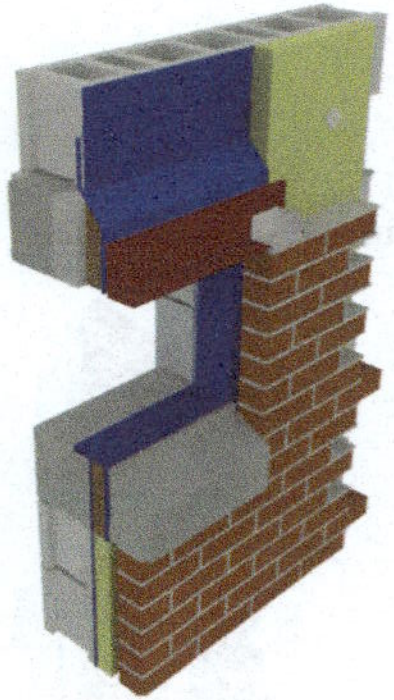	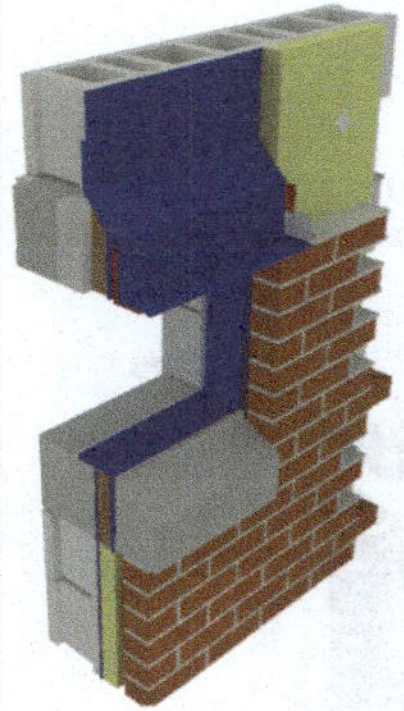	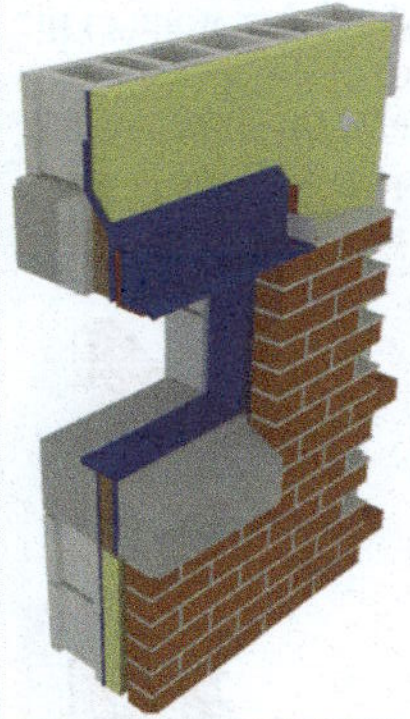
To prevent water ingress to the sides a sidedam is added at the level of the lintel	A through-wall flashing can now be added at the head	The insulation layer and air- and watertightness are now complete

(continued)

Table E.1 (continued)

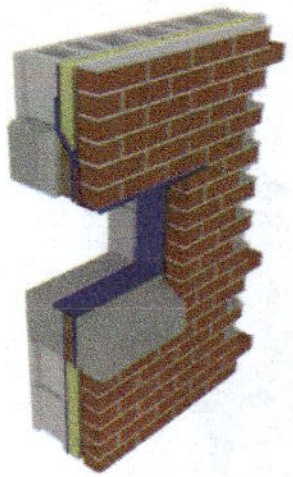		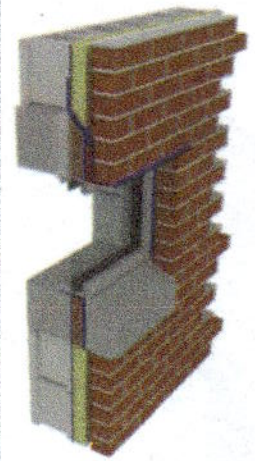
Brick and internal finishes can be completed, so the opening is ready for the window installation	A bead of sealant is added to the inside face of the backdam anchor to form an airtight seal	The window can then be pressed into position bonding with the sealant added to the reveals. The tolerance gaps can be filled at all sides. Care must be taken to leave weep holes at the sill to allow any trapped water to escape
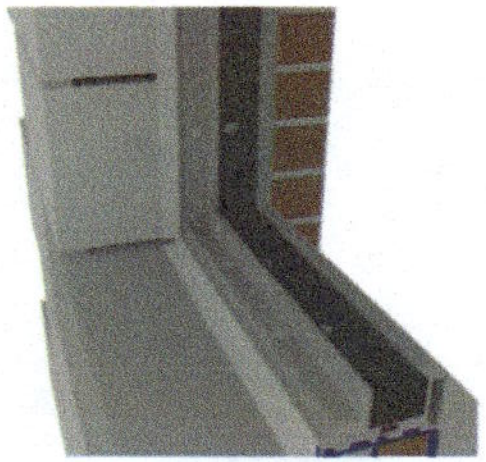	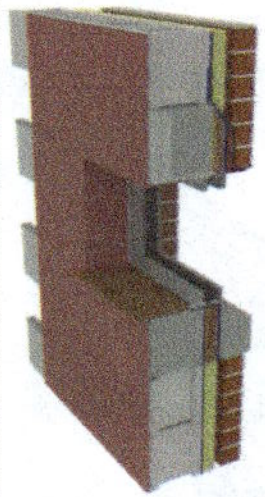	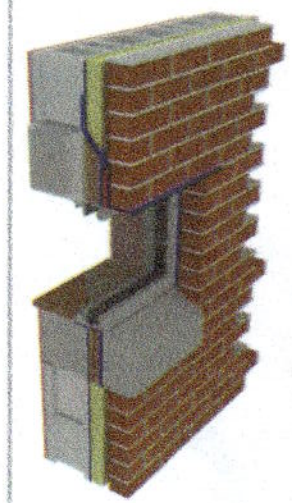
This particular brand of blast-resistant window is also bolted to the structure	Finally internal finishes can be added	The detail is complete

Table E.2 Window installation into the internally insulated masonry wall

	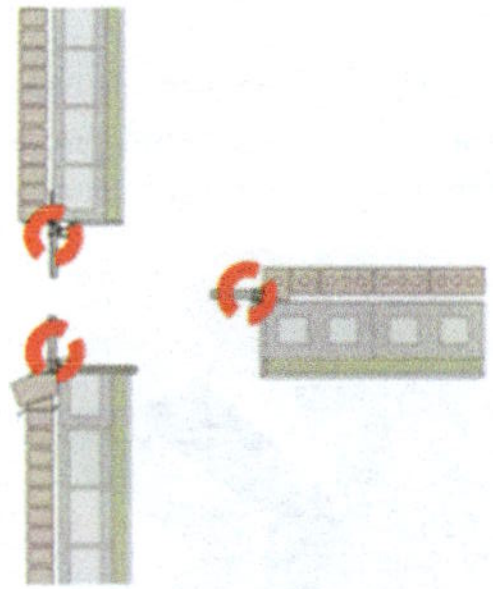 	Problem: there is no continuity of the insulation layer here and no apparent continuity of airtight layer

Retrofit solution: connect the window to the insulation via the reveals. Replace brick sill with a sill with an up-stand to prevent rain penetration. Flash all around and finish with a press-on cover strip	Remove the window and any silicones or seals that may require chiseling out the reveals to remove fixing straps	Remove the window board and if needed the drywall at the reveal

(continued)

Table E.2 (continued)

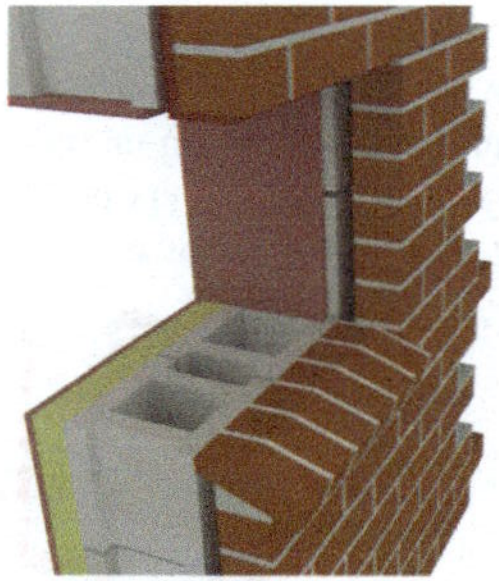	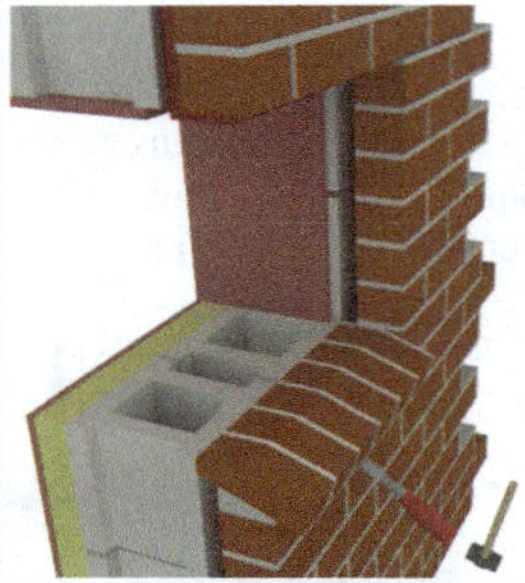	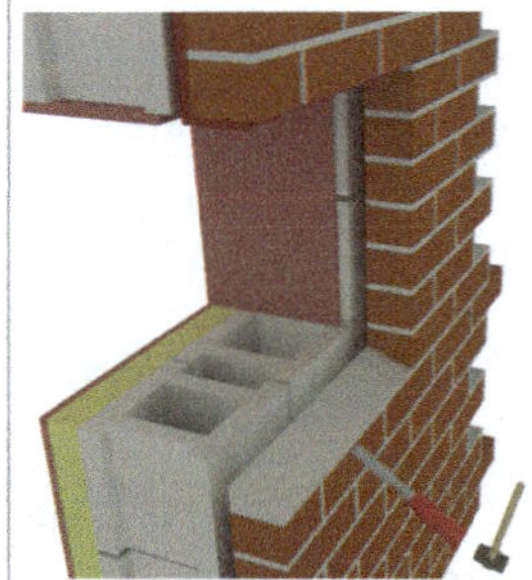
With the window board removed, the internal insulation and CMU blocks are now visible	Chisel out the old brick sill	The mortar bed also needs to be removed
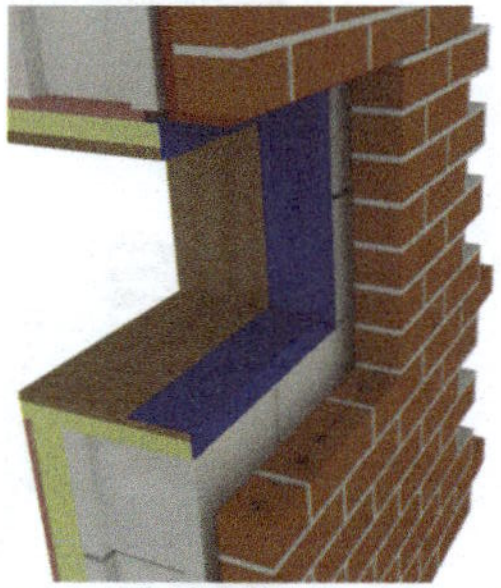	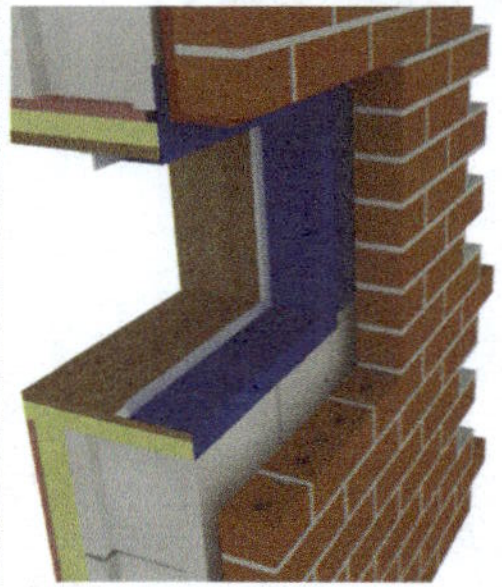	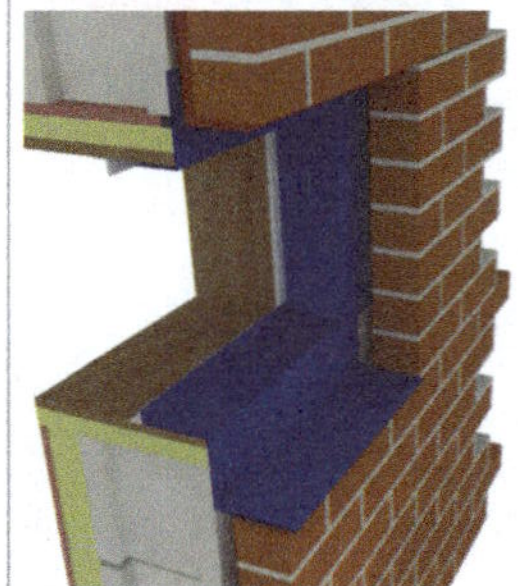
A plywood buck sandwiched by a high R per inch insulation and pre-wrapped in a self-adhesive membrane is added to all four sides of the reveal	A backdam anchor is added to all four sides of the wood buck, and an additional self-adhesive membrane is lapped over the wood buck to the reveals	Apply the self-adhesive watershed membrane at the sill and tape all joints to the reveals

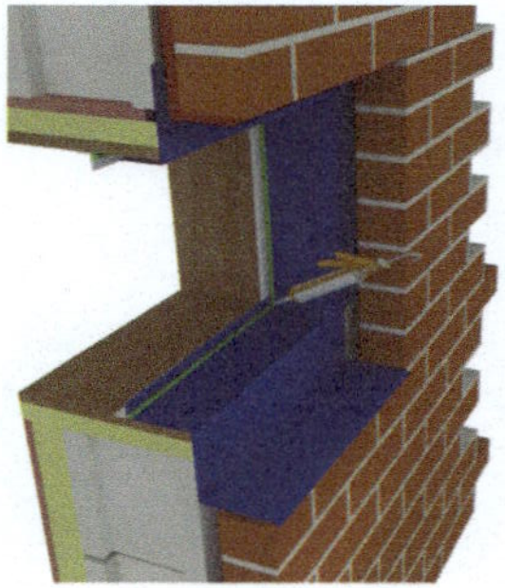	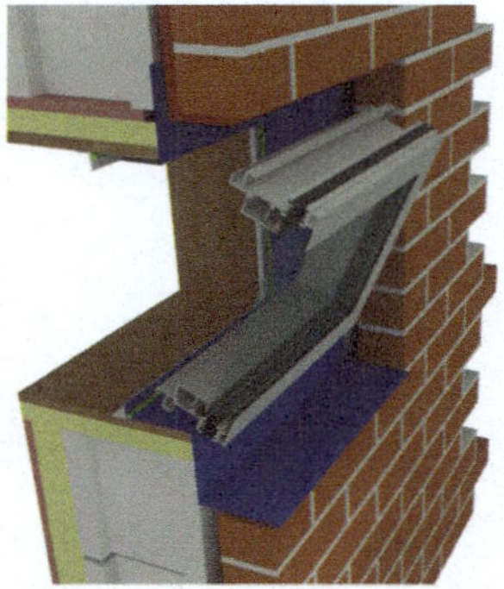	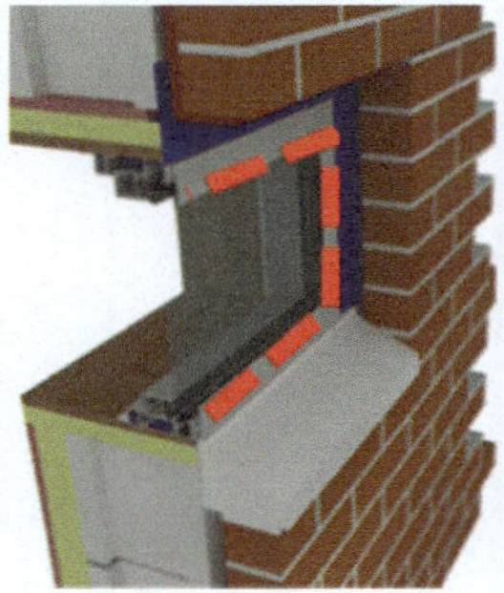
A bead of sealant is added to the backdam at all four sides.	The window can be pressed into position forming an airtight seal against the backdam	The window and air sealing is now complete, but to finish the gaps in the façade, a snap-on trim will be added
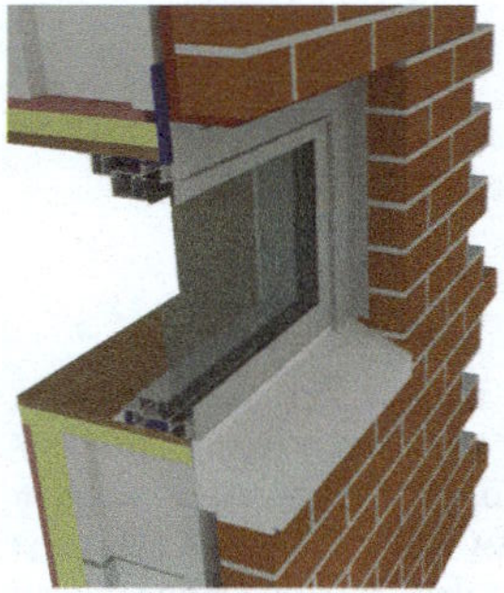	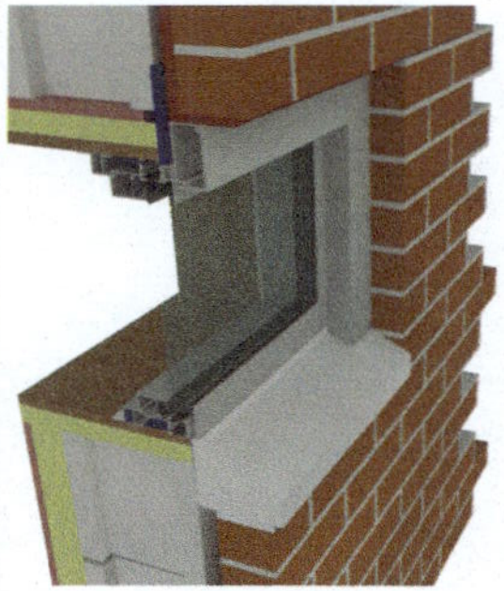	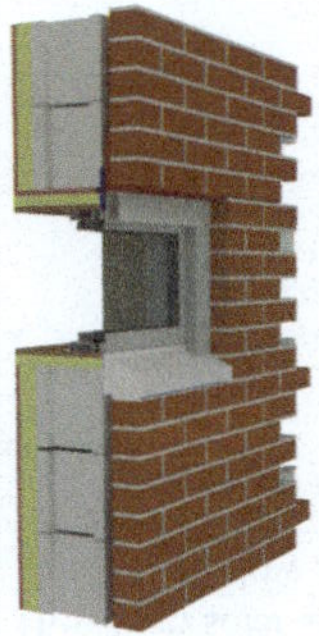

(continued)

Table E.2 (continued)

Apply snap-on trims such as these. A preformed housing is first fixed to the reveal. Most window manufacturers provide snap-on trims that are used to cover the gap between the window frame and the window surround on the two sides and top. The snap-on trims also cover the tape used to seal the window at the sides and top	Lastly, a cover piece snap-on to the preformed housing	This method will reduce thermal bridge effect and reduces risk of surface condensation
		However, solar gains will be reduced due to the smaller window size (1.5 in. on all sides)

Table E.3 Window installation into the steel stud wall

	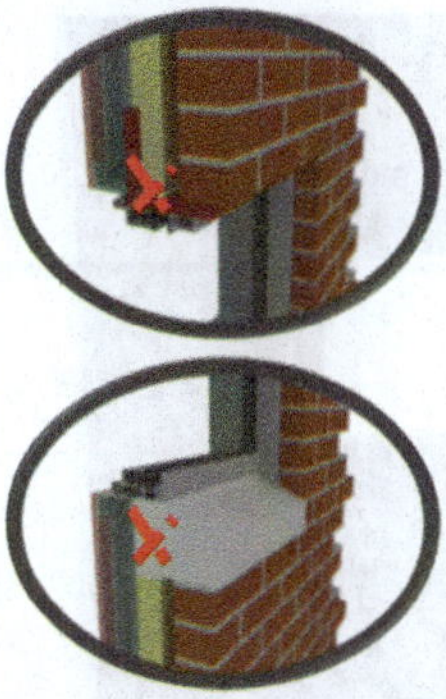	Problem: there is no alignment or connection of the thermal break to the insulation layer There is no thermal break in the existing window; this will cause very high losses As a result, the thermal bridge will likely be ineffective as there are huge losses from the frame itself Condensation may occur on the inside of this frame in cold weather
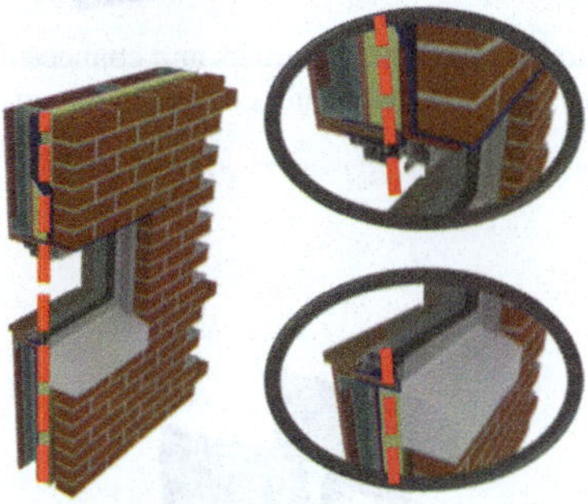		Retrofit solution: A 1.5-in.-thick timber break can be used to bridge the insulation layer to the thermal break in the window

Starting out with the steel studs	Gypsum wall board is added to exterior	An air/water barrier can be placed over the sheeting

(continued)

Table E.3 (continued)

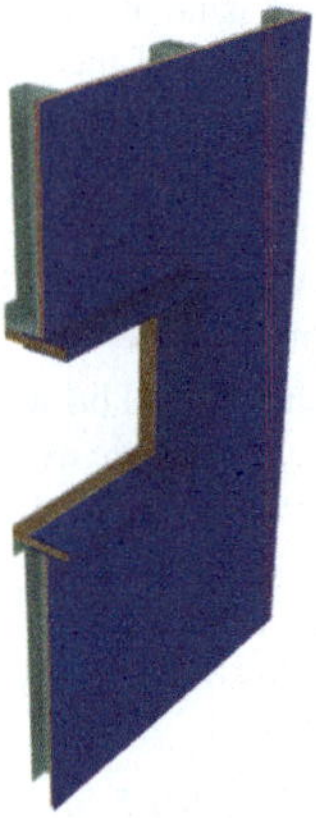		
A pre-wrapped treated timber or plywood buck is added to all four sides of the reveals. This can allow up to 6 in. of insulation in the cavity even though the window may be perfectly aligned within that insulation	The wood buck needs to be sealed at the corners and connected with self-adhesive membrane to the air/water control membrane	
		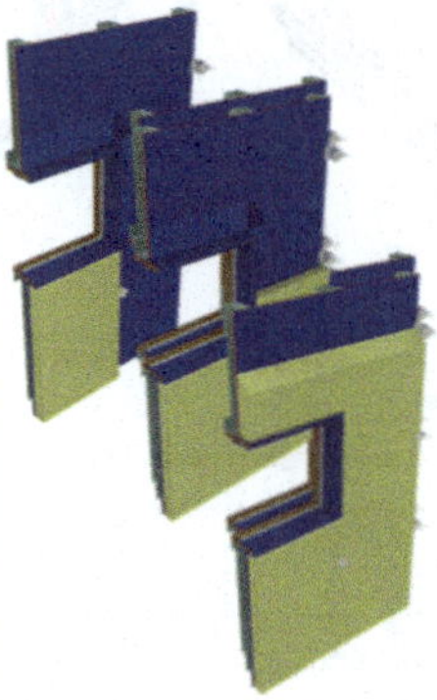
Anchors can be added to hold the insulation and tie the steel studs with the outer brick veneer	A backdam anchor is then added to all four sides of the reveal	Insulation can then be placed between the anchors

(continued)

Table E.3 (continued)

Brick courses are laid to the level of the lintel		The lintel is then place over the window

Before adding the through-wall flashing, a sidedam is added to the sides of the lintel	Once the through-wall membrane at the head is added, then the water and air sealing is complete	Insulation and brick courses can then be continued
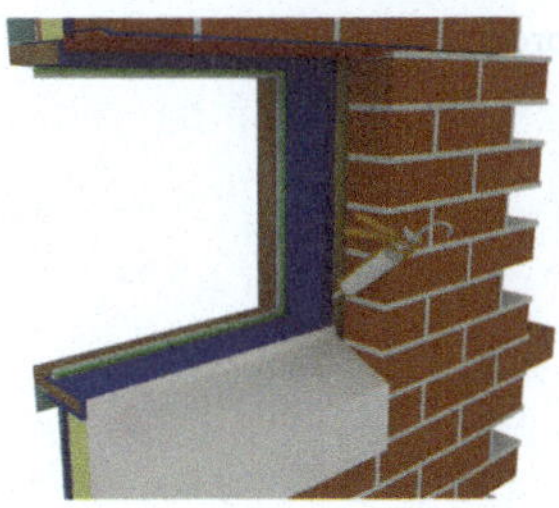		

(continued)

Table E.3 (continued)

Before adding the window, a bead of sealant is applied to all four sides of the reveal onto the backdam anchor to form an air seal	The window is then pushed into position forming an air-tight seal with the sealant	When the window is fully placed, the outer edges can also be filled with sealant. Be sure to cut or leave gaps at the sill allow drainage

The edges of the window can also be covered by proprietary snap-on trims		

Internal finishes and services are then added	The detail is complete	

Some low-energy building standard practices, government control (e.g., in Ireland), and certifiers (including Passive House Academy for thermal verification) require that a photographic protocol of documentation be taken while on-site and be filed with the building control authority to verify what was on-site after the build is complete. This further informs the inspection team if a problem arises.

Table E.4 Building renovation using brick masonry wall insulation with ETICS and window replacement

This insulation concept is not be applicable to historic/listed buildings		
Renovation concept:	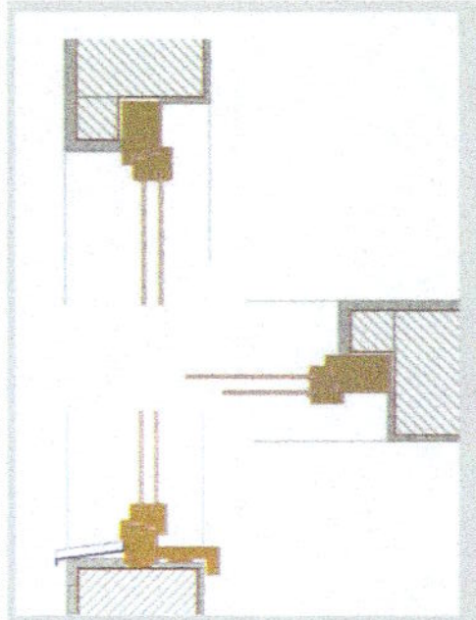	
The external wall is thermally renovated by an external thermal insulation composite system		
The windows are replaced by triple pane windows with thermally broken frames installed within the insulation layer		
Note: shorter aluminum frames and thicker wall insulation reduce thermal losses due to thermal bridging		
Construction process:	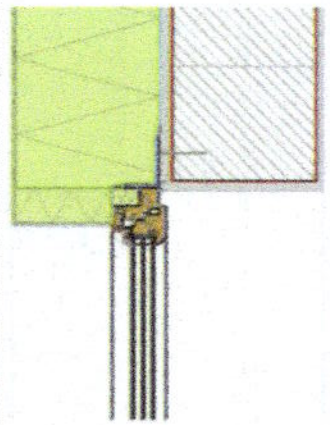	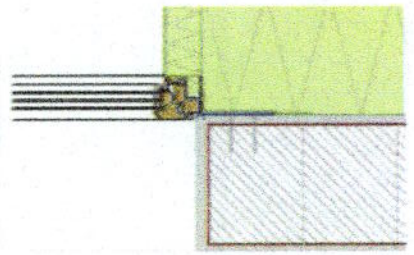
The air barrier is the existing exterior plaster; close the cracks, or, if necessary, fill the entire surface		
Create clean/airtight wall surfaces after existing window is removed by leveling screed/plaster mortar. Bond the existing exterior plaster and interior plaster airtight		
The windows are affixed and sealed with a fleece-laminated butyl rubber strip on all sides on the plaster of the reveal and are then plastered over	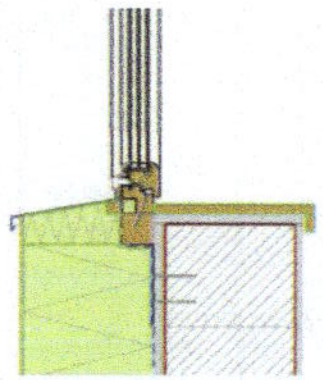	

(continued)

Table E.4 (continued)

Wall structure					
Silicate plaster				0.6	cm
EPS – expanded polystyrene				30	cm
Lime cement plaster				2.5	cm
Crushed brick masonry				30	cm
Lime cement plaster				1.5	cm
U-value	0.12	[W/m²K]	Σ	64.6	cm

Window U-values		
Window	2.6	W/m²K
Glazing	1.04	W/m²K
Window frame	1.04	W/m²K

2-dimensional specific values		
Ψ-value lintel	0.012	W/mK
Ψ-value reveal	0.001	W/mK
Ψ-value parapet	0.024	W/mK
f_{RSi} . lintel	0.79	-
f_{RSi} . reveal	0.79	-
f_{RSi} . parapet	0.75	-

Table E.5 Building renovation using brick masonry wall insulation with ETICS, window replacement, and installation of external blinds. This insulation concept is not applicable to historic/listed buildings (with permission from "Details of Passive Houses Renovation, 2016")

Renovation concept:

The external wall is thermally renovated by an external thermal insulation composite system.

The windows are replaced by triple pane windows with wood-aluminum thermally broken frames installed within the insulation layer

The existing exterior plaster, which is connected airtight to the interior plaster, is the continuous airtight layer. The sash seal of the window is connected to the leveling screed/plaster

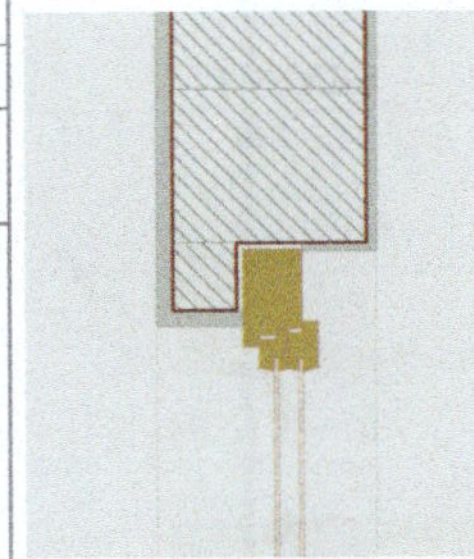

Construction process:

The air barrier is the existing exterior plaster; close the cracks, or, if necessary, fill the entire surface

Create clean/airtight wall surfaces after existing window is removed by leveling screed/plaster mortar. Bond the existing exterior plaster and interior plaster airtight

The windows are replaced by triple pane windows with wood-aluminum frames installed within the wall insulation level

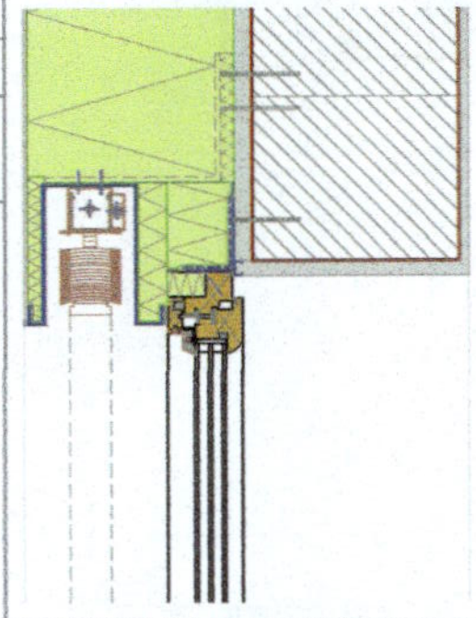

Wall structure		
Silicate plaster	0.6	cm
EPS – expanded polystyrene	30	cm
Lime cement plaster	2.5	cm
Crushed brick masonry	30	cm
Lime cement plaster	1.5	cm
U-value 0.12 [W/m²K]	Σ 64.6	cm

Window U-values		
Window	2.6	W/m²K
Glazing	1.04	W/m²K
Window frame	1.04	W/m²K

2-dimensional specific values		
Ψ-value lintel	0.047	W/mK
f_{RSI}	0.77	-

Table E.6 Building renovation: brick masonry wall is insulated with ETICS, reinforced concrete basement ceiling is insulated on the lower side, and basement windows are replaced

Renovation concept:	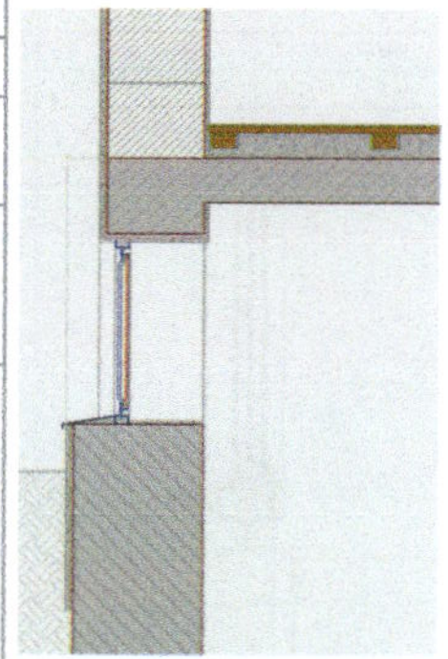
The external wall is thermally renovated by an installing external thermal insulation composite system	
The basement ceiling slab is insulated on the lower side, and the basement walls is insulated externally (horizontal perimeter insulation)	
The windows are replaced by wood-aluminum windows that are installed in the insulation layer. If the distance between the upper edge of the ceiling and the upper frame of the window is small, a frame doubling in the upper window area can be considered	
The temperature of the basement space is expected to be below the temperature of the rest of the building. Due to the lack of moisture source control, the basement will have a higher RH. Therefore basement ventilation is recommended	
Construction process:	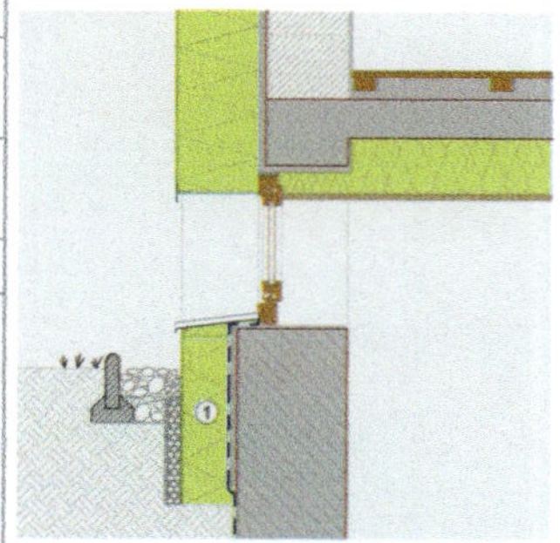
Seal the airtight layer (exterior plaster, fully filled if required) all the way to the lower edge of the ceiling. This also provides a clean surface for applying the seal. Bituminous slurry is to be used in the seal area	
The existing exterior plaster will create a continuous air barrier. This is connected airtight with the interior plaster. Connect the sash seal of the window with the leveling screed/plaster	
Tightly bond the seal on the complete surface up to at least 30 cm above the ground level (splash zone)	
If the external wall in contact with the ground is not suitable for applying the seal, create a clean surface (plaster base coat) before applying the vertical seal. If drainage is required, the ground must be dug up in any case	
Fully bond perimeter insulation panels to the basement wall, insert an expanding foam seal on the upper side, attach a drip edge, and plaster the wall	**1 = XPS**

Wall structure

Silicate plaster	0.6	cm
EPS – expanded polystyrene	30	cm
Lime cement plaster	2.5	cm
Crushed brick masonry	30	cm
Lime cement plaster	1.5	cm
U-value 0.12 [W/m²K]	Σ 64.6	cm

Wall structure, in contact with the ground

Reinforced concrete	40	cm
U-value 3.29 [W/m²K]	Σ 40.0	cm

Ceiling structure

Parquet floor	2.5	cm
Fill, raft battens 5/8	5	cm
Fill	2.5	cm
Concrete slab	14	cm
Glass wool MW-W insulation felt	20	cm
Wood wool lightweight panel	2.5	cm
U-value 0.16 [W/m²K]	Σ 46.5	cm

Window U-values

Window	1.5	W/m²K
Glazing	1.04	W/m²K
Window frame	1.04	W/m²K

2-dimensional specific values

Ψ-value interior/exterior	-0.022	W/mK
Ψ-value interior/exterior with window	0.064	W/mK
Ψ-value interior/basement	0.487	W/mK
Ψ-value interior/basement with window	-0.019	W/mK
f_{RSi}	0.84	-

Appendix F: Air Barrier – Examples of Good Practices

Major Renovation Projects

With major renovation projects, there is no need to conduct assessment of the building enclosure before its renovation. However, it is important to understand the status of the building after it is gutted since the design of the air barrier will be based on the limitations found in the assessment. This section shows some examples of air barrier drawing details for new construction and major renovation projects, developed by:

- The Building Envelope Committee of the Boston Society of Architects (Figs. F.1, F.2, F.3, F.4 and F.5)
- The Association of Airtightness in the Construction Industry (FLiB), Berlin (Figs. F.6, F.7, F.8, F.9, F.10, F.11, F.12, F.13, F.14, F.15, F.16, F.17, F.18, F.19, F.20, F.21, F.22, F.23, F.24, F.25, F.26 and F.27)
- USACE (Figs. F.28, F.29, F.30, F.31, F.32, F.33, F.34 and F.35).

These details were made publicly available to provide examples of how an air barrier can be designed for the building envelope. The details must be reviewed by a design professional before using them in a specific project, including the loads imposed on the air barrier by the differences between the interior and exterior environment of a building.

These details do not separate the designs by their durability; some designs are more durable than others. The cost for each air barrier system will vary depending on the material chosen and the difficulty in the installation. These details may not be appropriate for all climate zones. Any misapplication or misinterpretation of these details is the sole responsibility of the user.

These details are focused on the continuity of the air barrier from foundations to roof, including the sealing of all penetrations. None of the designs have been reviewed by a structural engineer. The structural support of the air barrier to withstand positive and negative air pressures imposed on it by the wind load, the stack effect, and the mechanical effect have been taken into consideration. These details need to be reviewed by a structural engineer as these loads, in many cases, are transferred to the backup wall and super structure of the building. Structural requirements for the systems and anchorages normally designed by specialty engineers for light gauge steel studs, stone, and precast concrete connections have not been engineered in these details.

Figure F.1 details an external air barrier design at roof edge (A), parapet (B), floor slab (C), penetration (D), window head (E), window jamb (F), window sill (G), foundation (H), shingle roof (I), and metal roof (J).

Window crack perimeter sealants should be used that are compatible with polyethylene, such as low- or ultra-low-modulus silicone. For small windows up to 2.5 or 3 m, one-part polyurethane sealant foam may be used. Care must be used when installing the sealant foam as the material, depending on the installation, may not meet the requirements to be an air barrier material. A membrane properly connected with compatible sealants and termination bars to window and membrane may also be used. The same tie-in location is true of louvers, metal door frames, and store fronts. The curtain wall is tied in at the tube face of the glazing pocket.

Advantages and Disadvantages of the Design in Fig. F.1

This wall type has a layer of continuous insulation on the exterior, making it energy efficient. It is easy to build, because the air and vapor barrier and the insulation are easier to make it continuous. By having the insulation and air/vapor barrier on the exterior, the steel stud backup wall is always kept above the dewpoint of the winter indoor air, which reduces the potential for condensation that leads to mold, mildew, corrosion, and rot. Penetration of the air barrier is also avoided by interior electrical

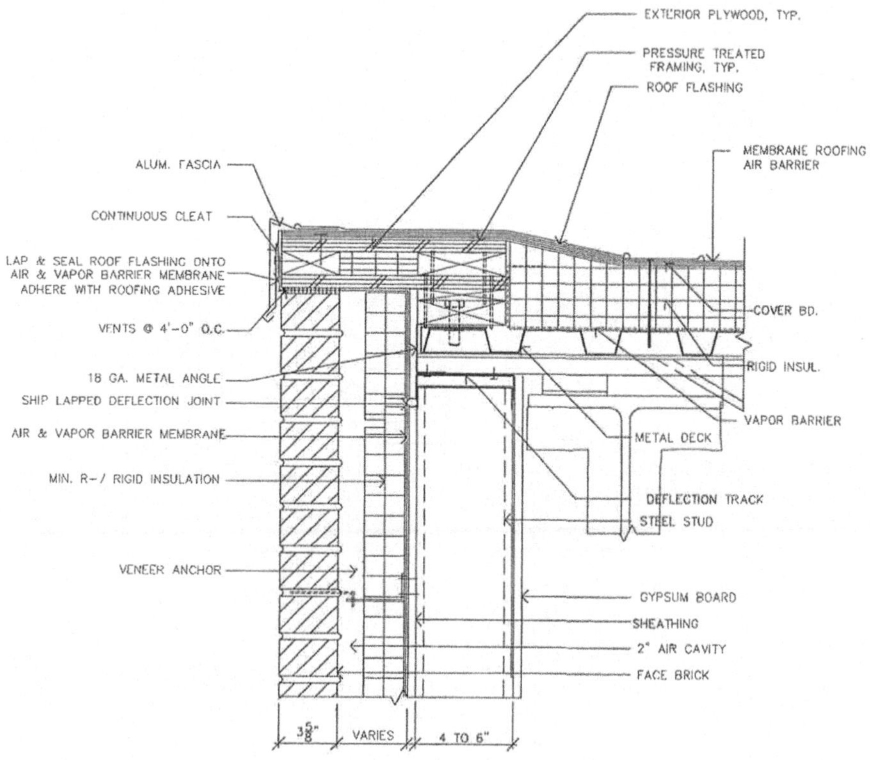

Fig. F.1 Conceptual details of an external air barrier for building with a brick veneer wall, a pitched shingle roof and a metal roof: brick, 2 in. cavity, continuous insulation, air/vapor barrier, sheathing, a steel stud backup wall, and a gypsum drywall interior finish. (**a**) Roof edge; (**b**) parapet; (**c**) floor slab; (**d**) penetration; (**e**) window head; (**f**) window jamb; (**g**) window sill; (**h**) foundation; (**i**) shingle roof; (**j**) metal roof

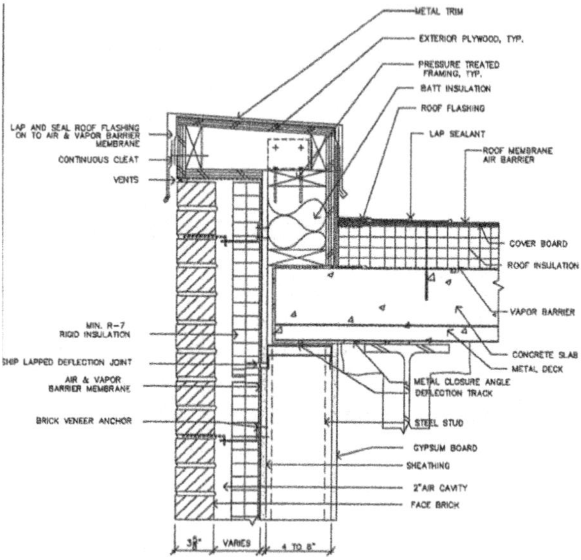

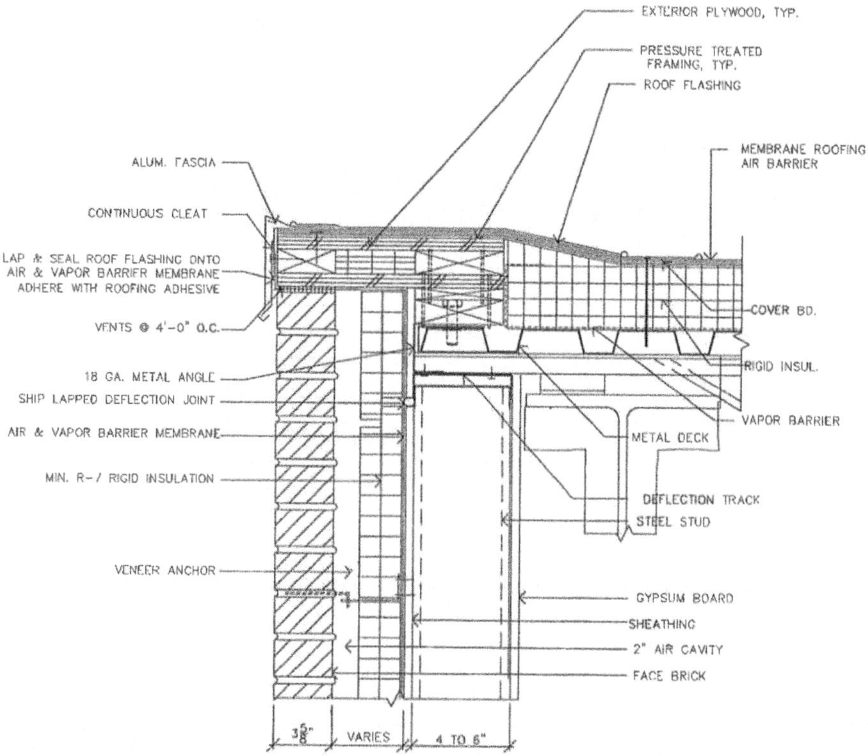

Fig. F.1 (continued)

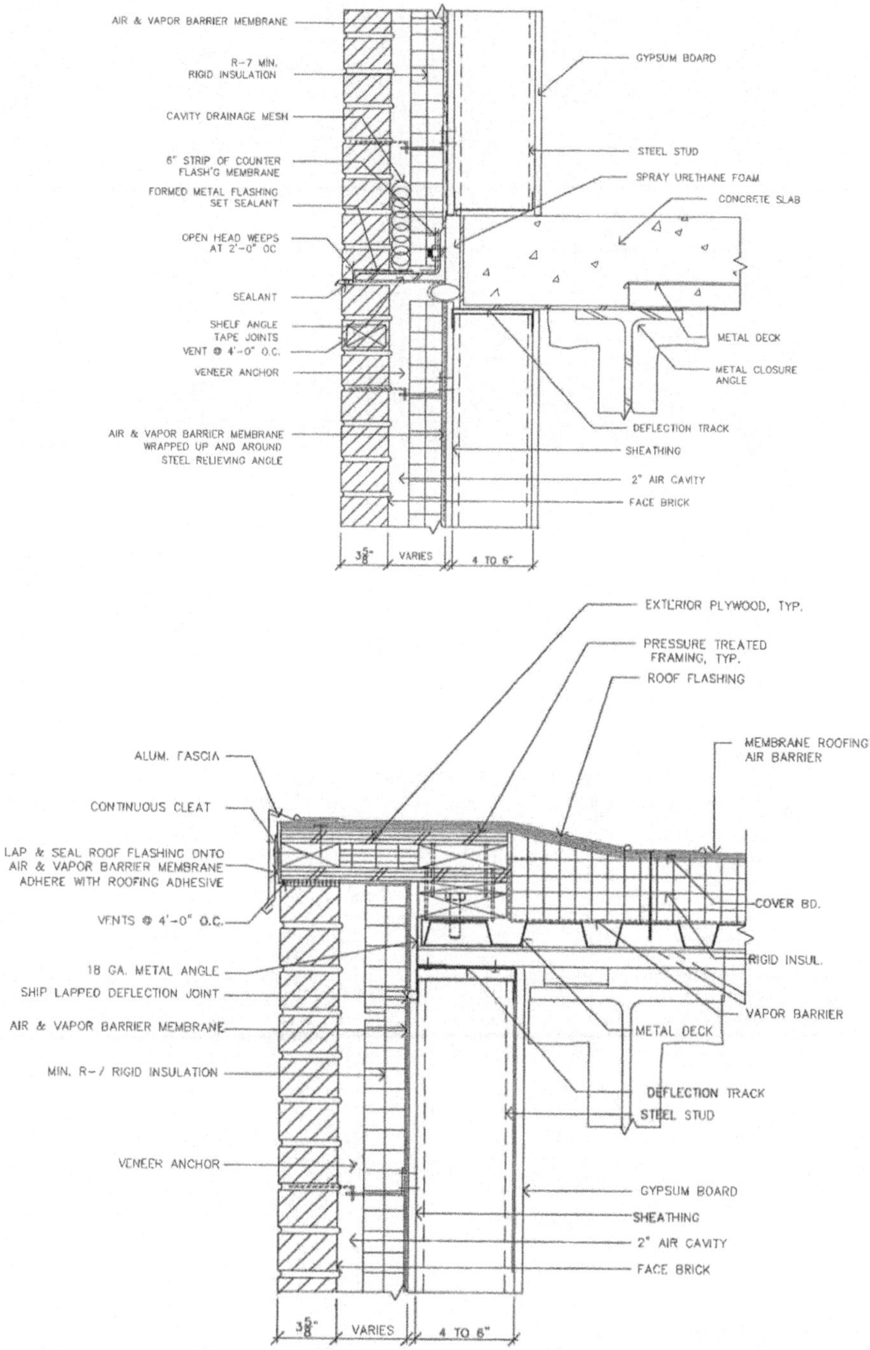

Fig. F.1 (continued)

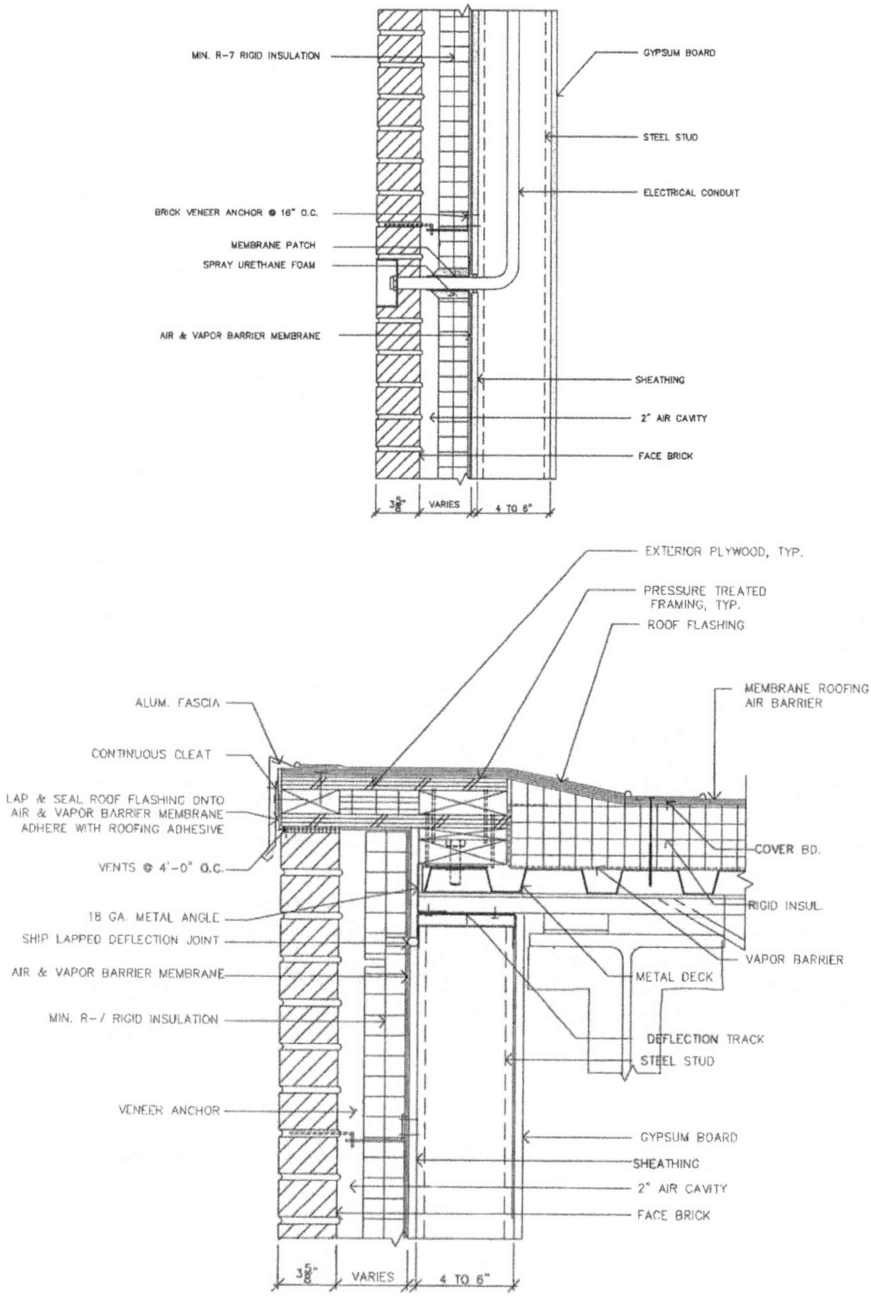

Fig. F.1 (continued)

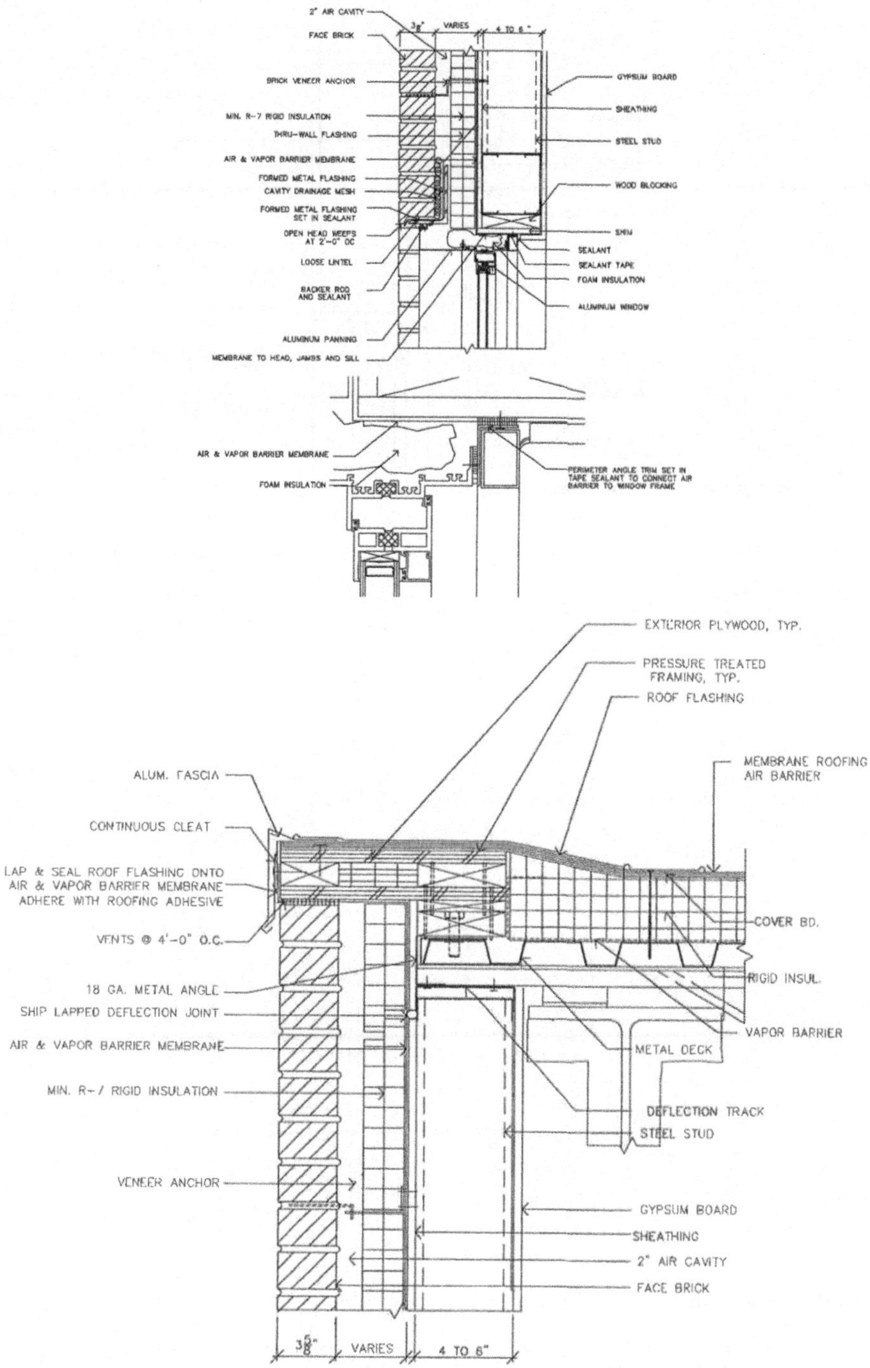

Fig. F.1 (continued)

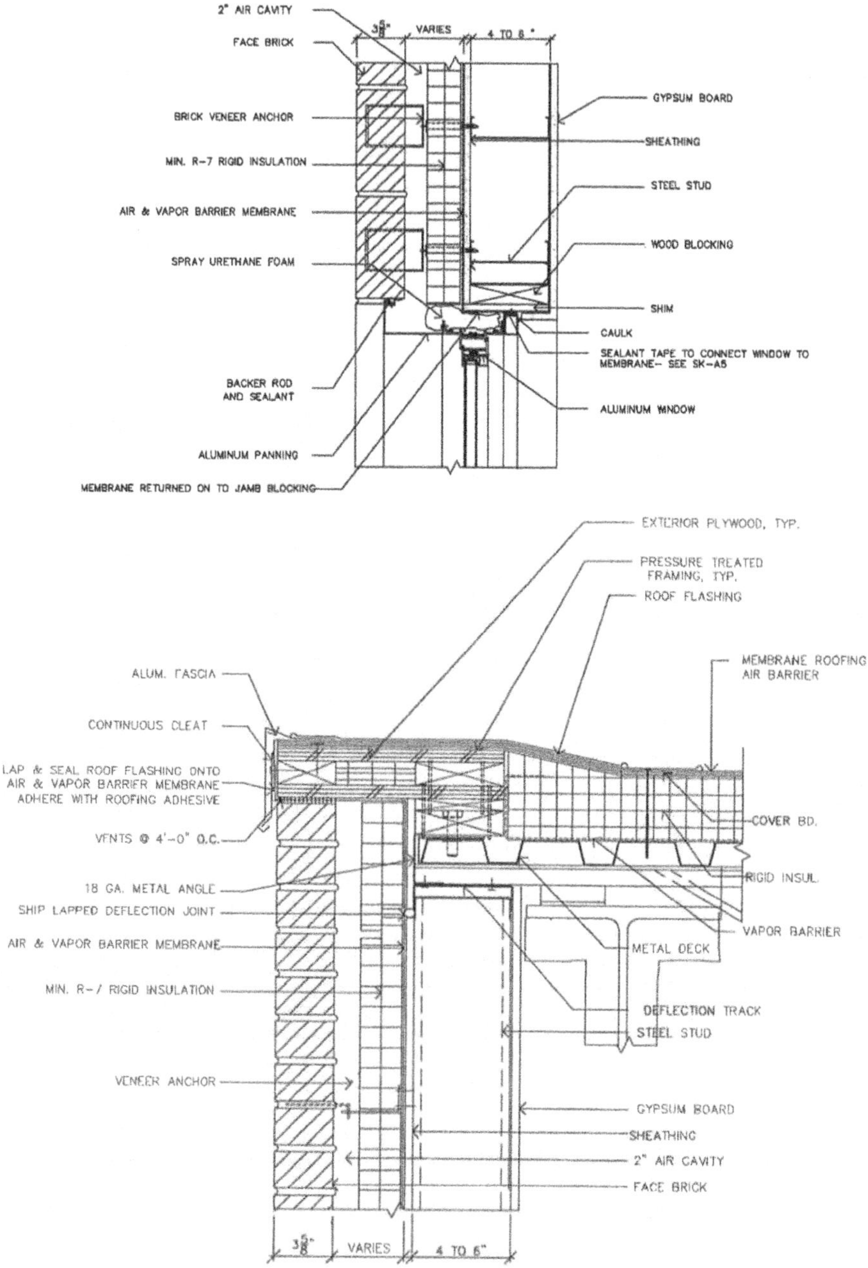

Fig. F.1 (continued)

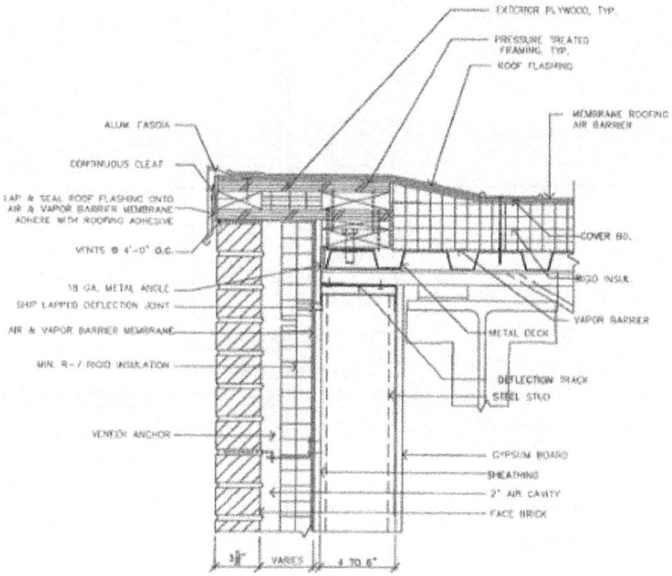

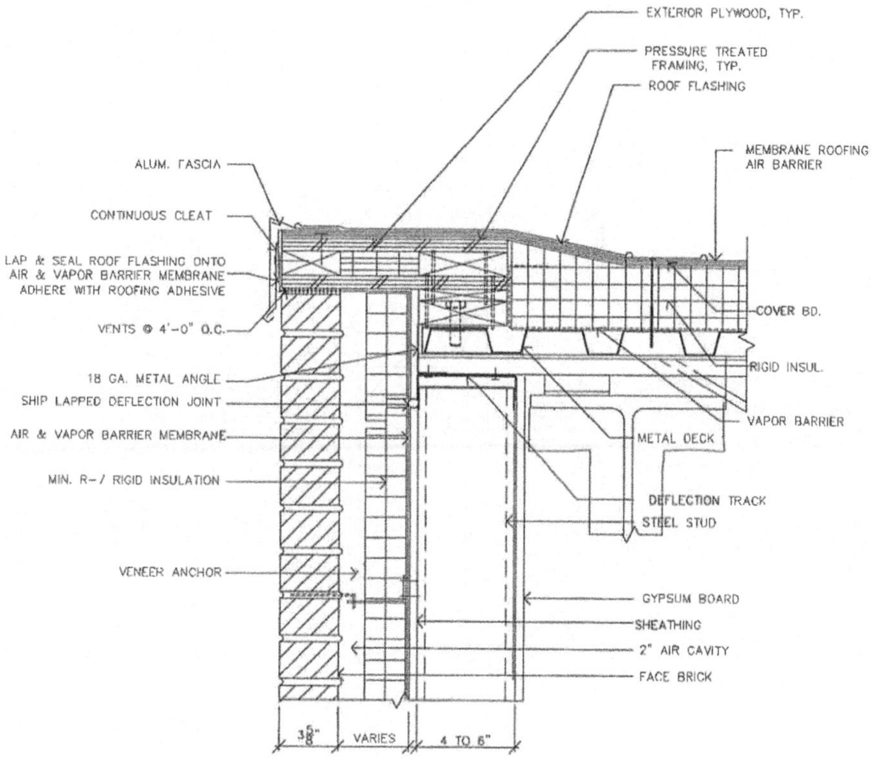

Fig. F.1 (continued)

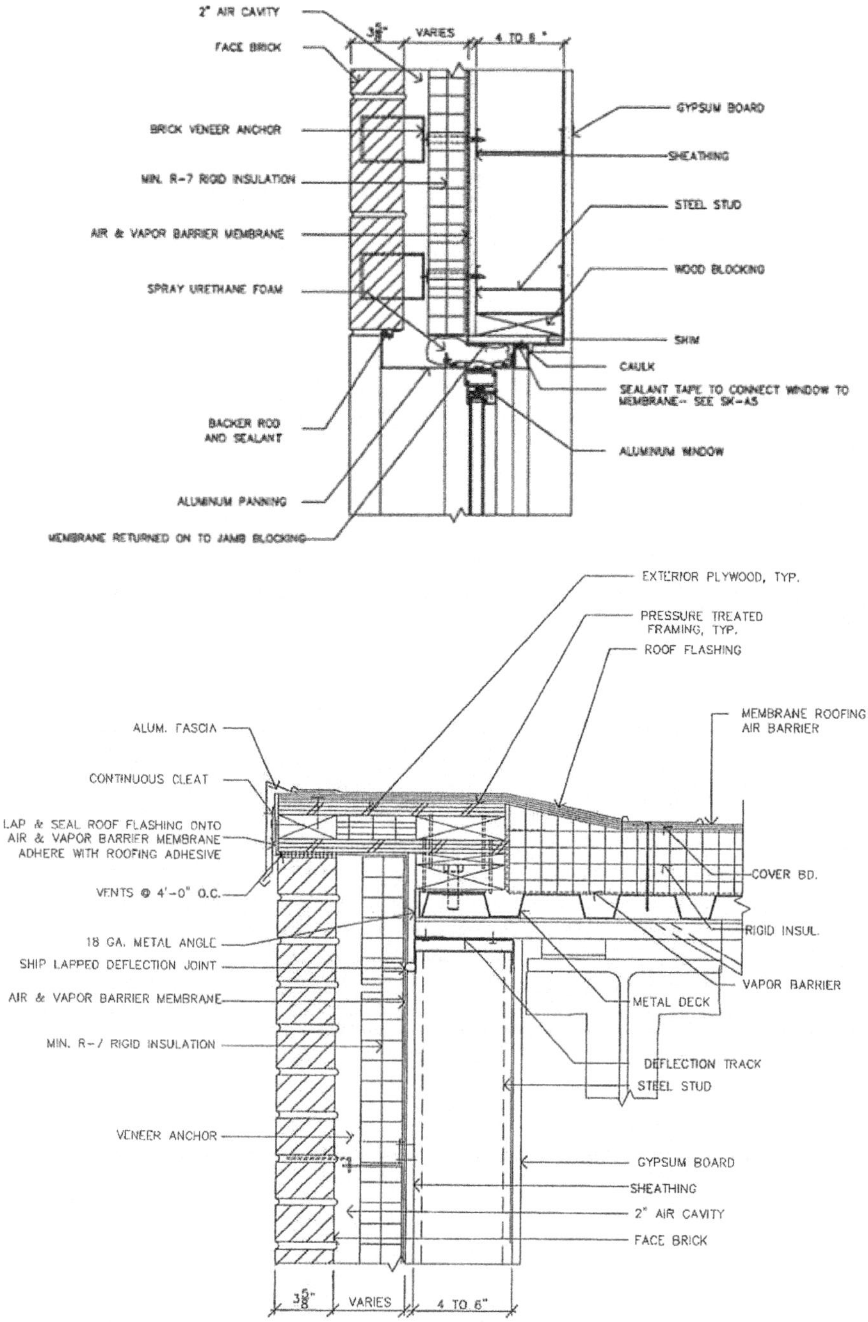

Fig. F.1 (continued)

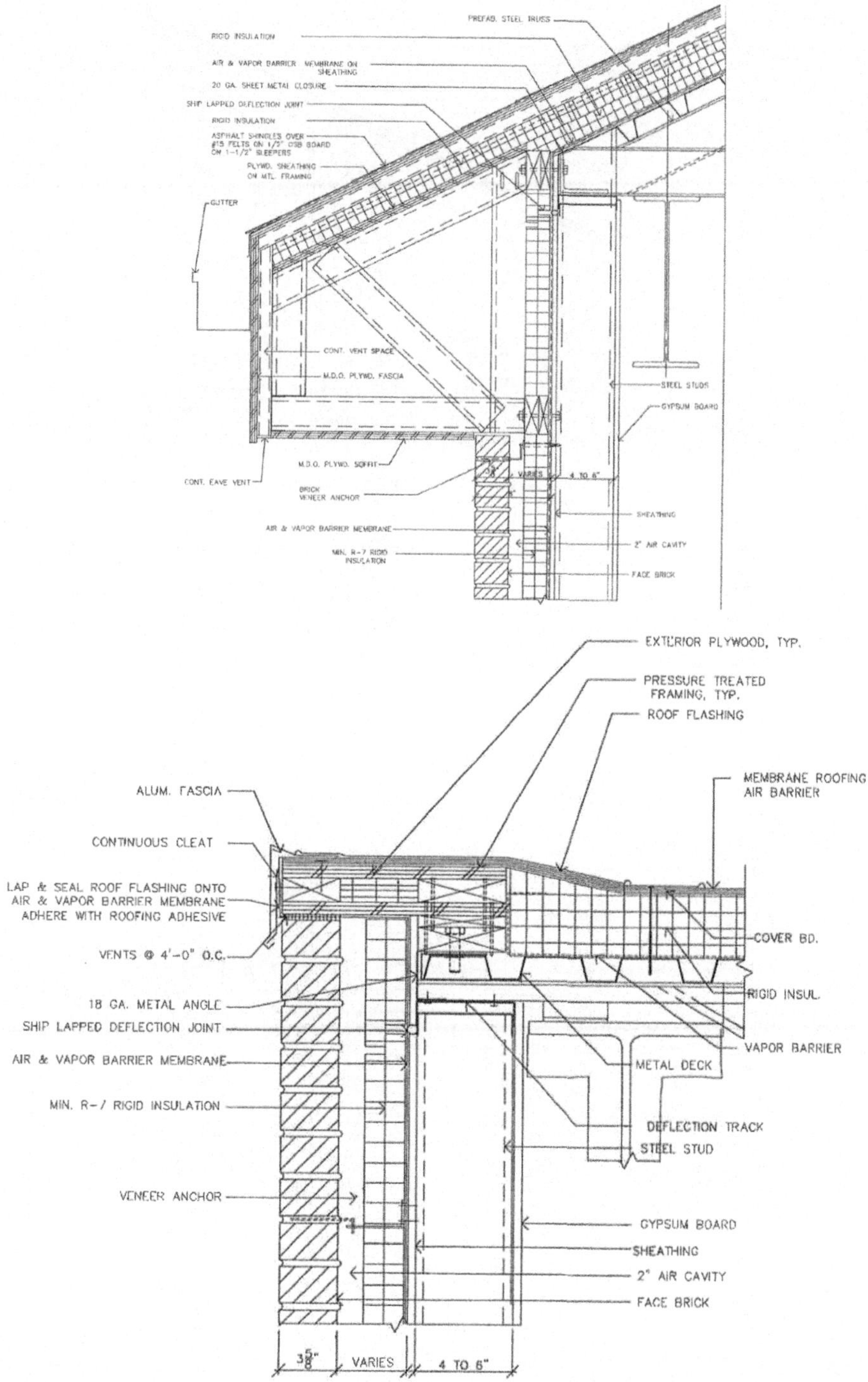

Fig. F.1 (continued)

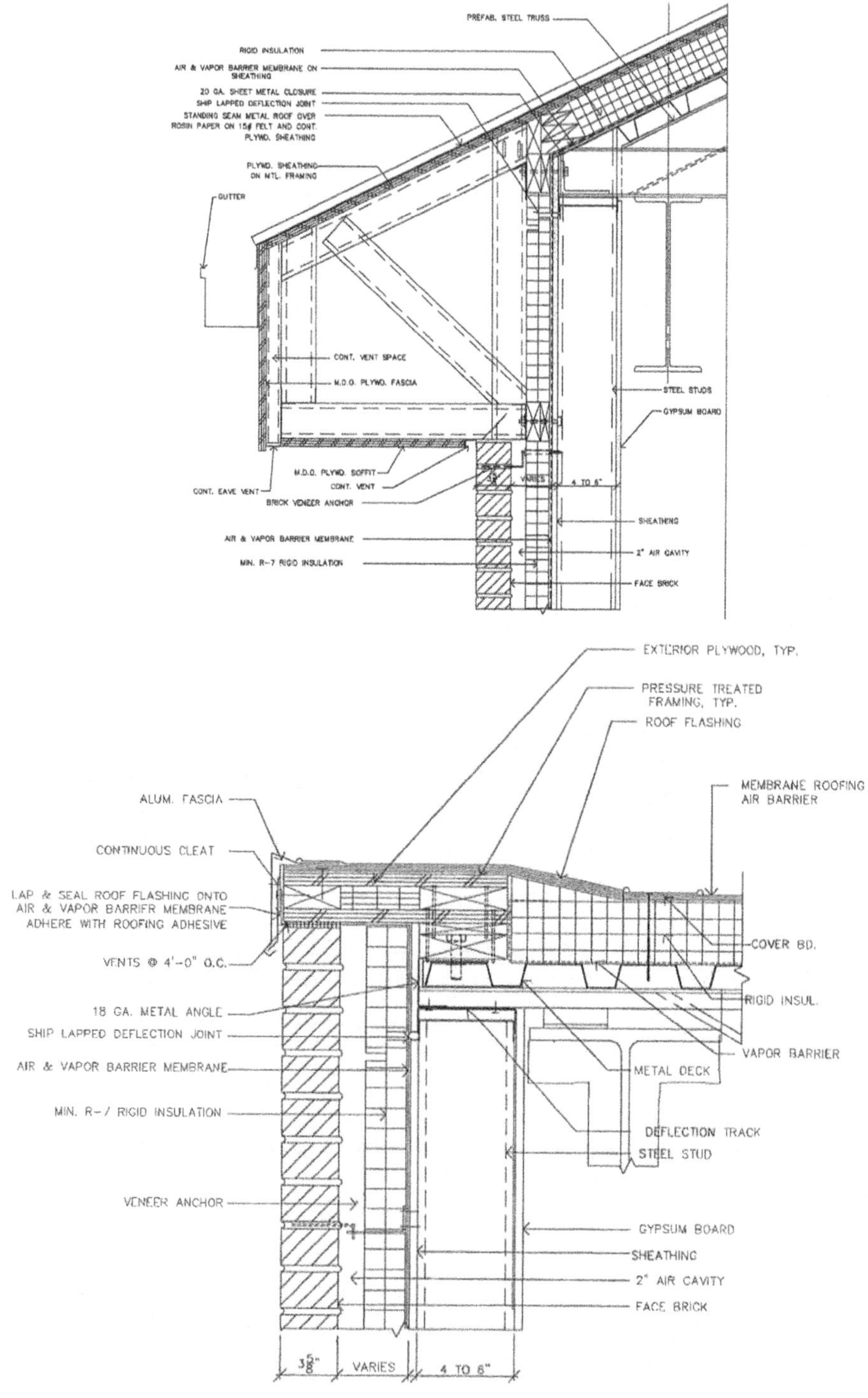

Fig. F.1 (continued)

outlets, switches, and other similar penetrations. The air and vapor barrier on the warm side of the insulation also works as the drainage plane, handling liquid water, water vapor in the winter, and reverse vapor drive in the summer, keeping the backup wall dry. The year-round stable temperature that continuous insulation on the exterior promotes the durability of the air and vapor barrier as it is protected from the weather. The air barrier material is also structurally well supported, especially if the insulation is mechanically fastened rigid insulation and/or if the membrane is self-adhered.

Since this wall (in this climate and interior conditions) can function without a vapor barrier, XPS insulation can be used with a spun-bonded polyolefin air barrier and no vapor barrier. However, the installation requirements for a spun-bonded polyolefin air barrier are completely different from those for installing the same material as a water-resistant barrier. Without the proper fastening pattern and the sealing of the joints and penetrations, the plane of airtightness is not achieved.

The XPS (provided it is the proper thickness to provide the required structural requirements) can be turned into the air barrier by taping the XPS with a durable peel-and-stick tape. The disadvantage of this installation procedure is that the tape is on the cold side of the wall, and subject to the climatic conditions.

Air and vapor barrier alternatives include:

- In the case of closed-cell medium-density spray polyurethane foam insulation, the substrate needs to be prepared by installing flexible transition membranes between different materials and across construction joints.
- Self-adhered modified asphalt/polyethylene membrane.
- Fluid-applied spray-on or trowel-on air/vapor barrier. (Take care that asphalt damp-proofing is not an air barrier material.) Liquid-applied air and vapor barrier membrane may double an insulation adhesive as well.
- If this kind of air and vapor barrier is used, membranes need to be installed at windows and other transitions, usually an asphalt peel-and-stick or a fluid-applied flashing material.
- Reinforced polyethylene sheet. Seal all joints with a compatible sealant and clamp over the seal or use strips of self-adhered membranes.
- Reinforced aluminum foil air/vapor barrier, with joints taped and the transitions detailed with a self-adhered membrane or fluid-applied membrane.
- Spun-bonded polyolefin installed in accordance with the requirements for an air barrier installation (with rigid insulation).

The design shown in Fig. F.2 gives details of external air barrier for brick veneer wall design, with a 2 in. cavity, 1 in. insulating sheathing (air barrier), R-11/R-13 (maximum), unfaced fiberglass batts in stud space and stud backup wall, at roof edge (A), parapet (B), floor slab (C), window head (D), window jamb (E), window sill (F), and foundation (G).

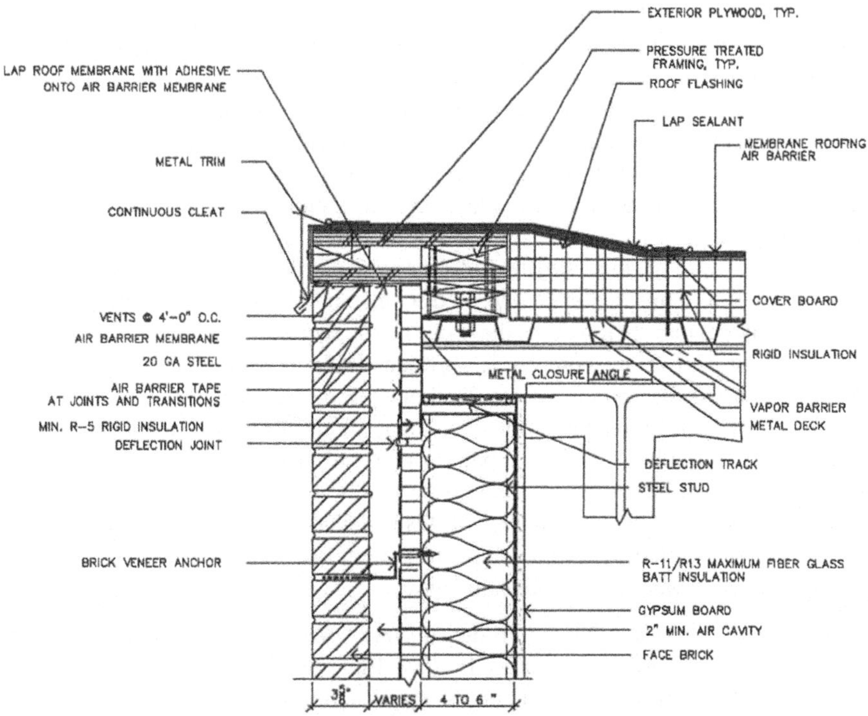

Fig. F.2 Conceptual details of external air barrier for building brick with veneer wall design, with 2 in. cavity, 1 in. insulating sheathing (air barrier), R-11/R-13 (maximum), unfaced fiberglass batts in stud space and stud backup wall: (**a**) roof edge; (**b**) parapet; (**c**) floor slab; (**d**) window head; (**e**) window jamb; (**f**) window sill; (**g**) foundation

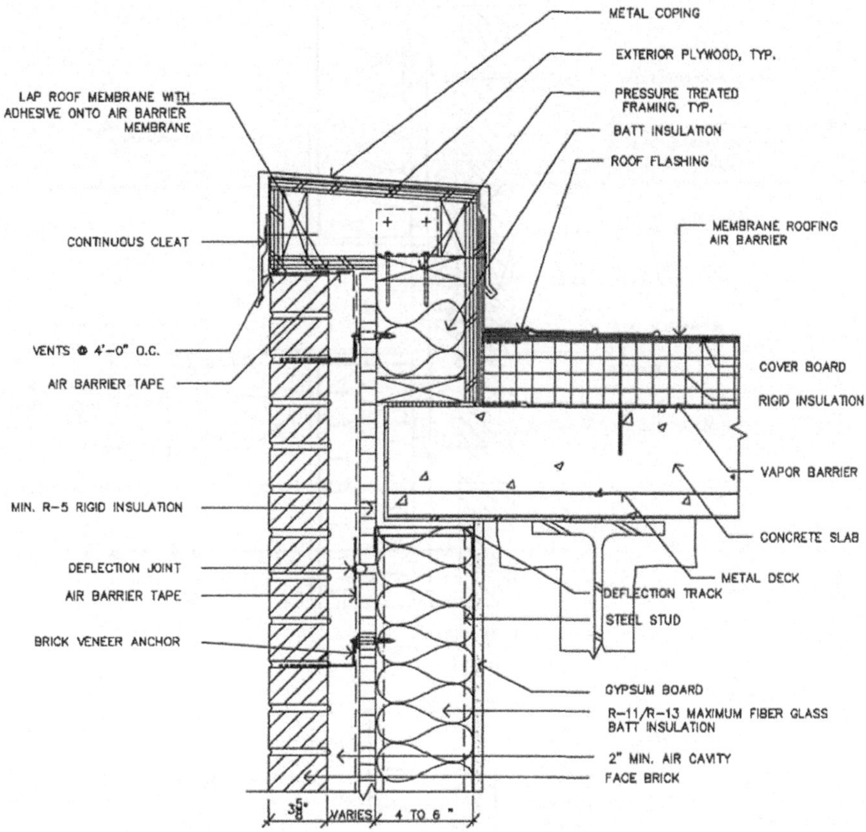

Fig. F.2 (continued)

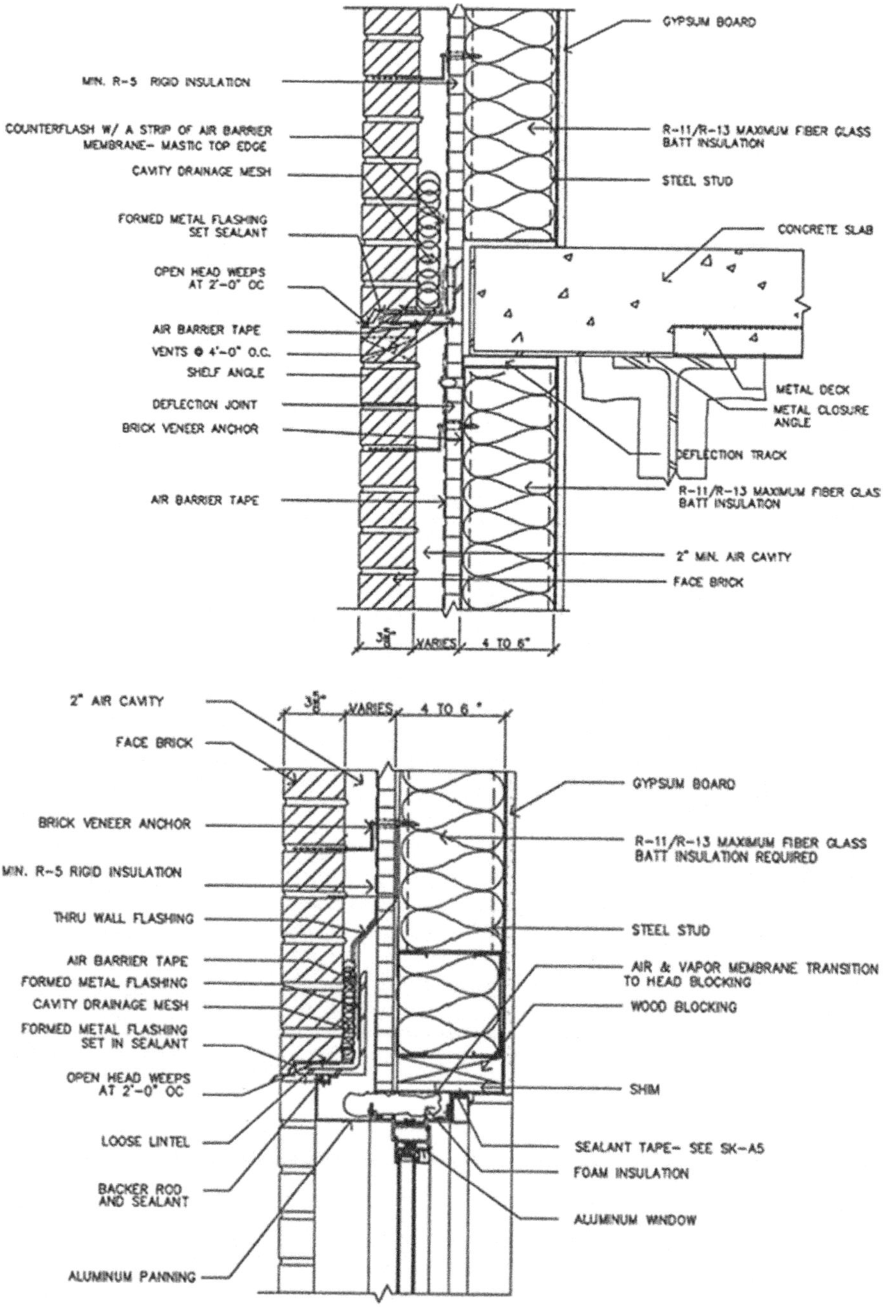

Fig. F.2 (continued)

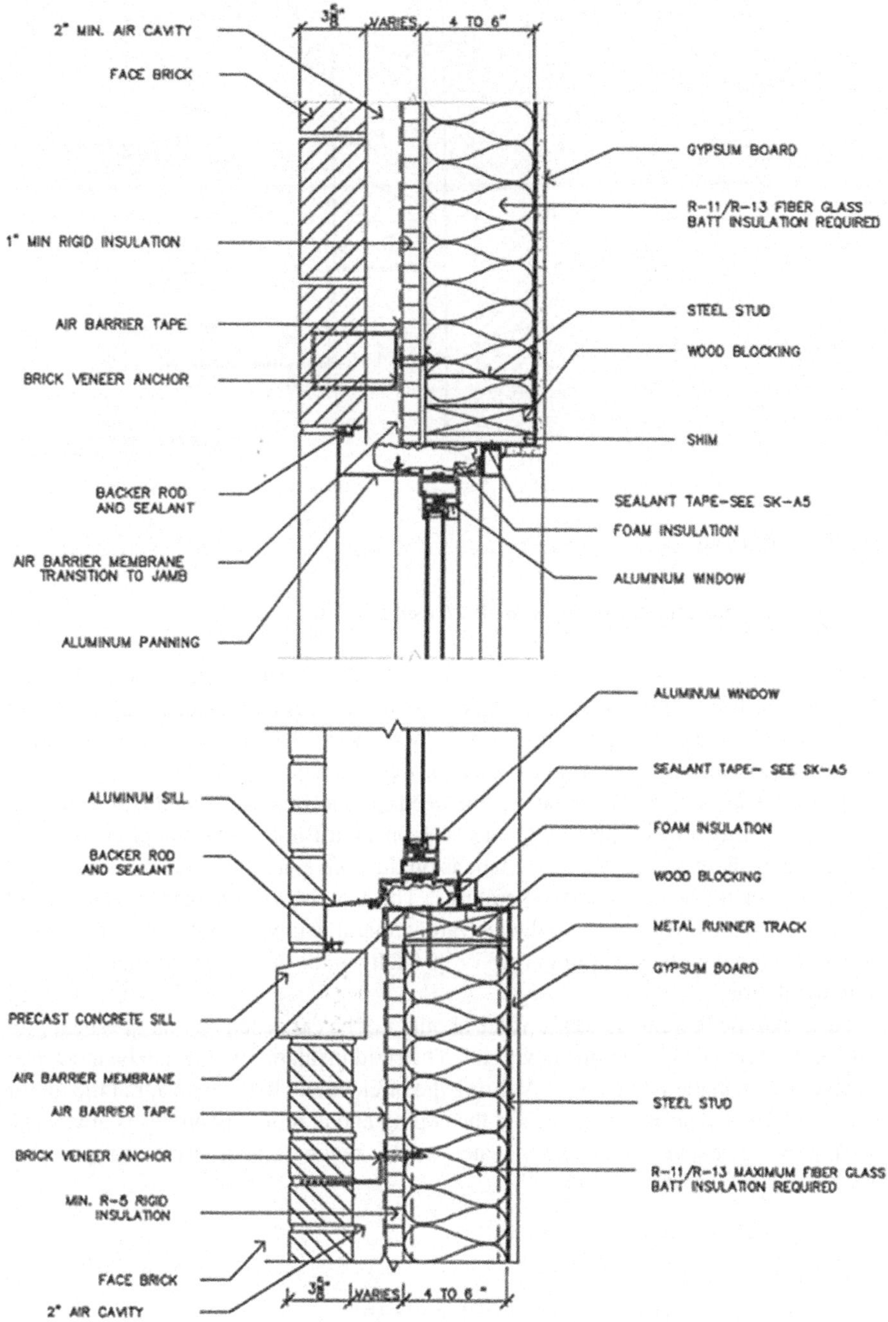

Fig. F.2 (continued)

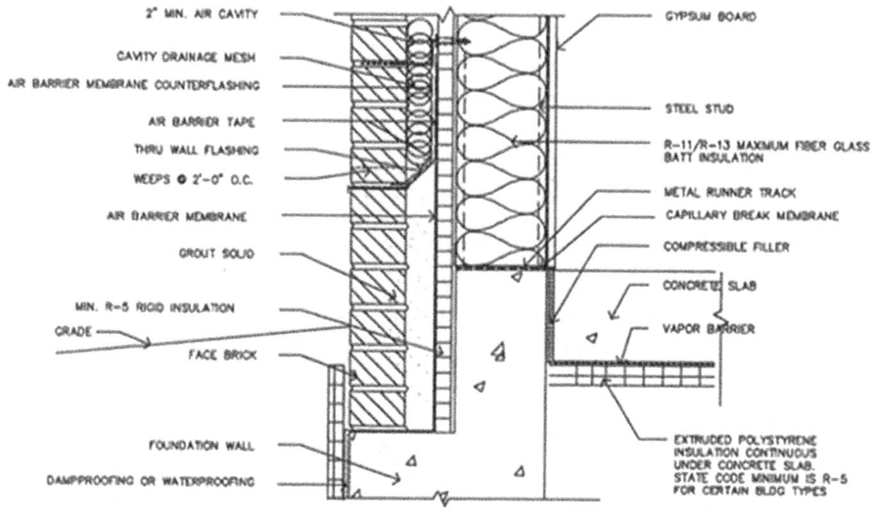

Fig. F.2 (continued)

Advantages and Disadvantages of the Design in Fig. F.2

This wall type uses the rigid cellular plastic insulation boardstock as the air barrier. The steel studs should be separately braced. With R-5 insulating sheathing, this wall design does not need a vapor barrier up to 35% interior RH. If the interior operating conditions are higher, then a separate vapor may be required.

The tape used to seal the sheathing needs to be one specifically designed to provide the function of providing the seal on boardstock. Fluid-applied flashing material can also be used. Since the tape is on the cold side of the wall, it is expected to be subject to the exterior temperatures. The water-resistant barrier is the rigid insulation, so the upper edges of the tape are vulnerable to water intrusion. To further protect the tape on the sheathing, a 15 or 30# felt can be added on the outside of the rigid insulation.

Note that the batt insulation R-value should not be exceeded so that the sheathing temperature remains above the dewpoint. The rigid insulation value can be increased and the batts can then be deleted. Venting the brick cavity to the exterior at the top of any continuous obstruction such as at the coping and at relieving angles is always an advantage, to relieve water vapor generated by reverse vapor drive and promote more rapid drying of the brick.

- Continuous insulation alternatives:
  - XPS
  - Foil-faced polyisocyanurate sheathing board

- Stud cavity R-11/R-13 insulation alternatives:
  - Friction-fit fiberglass batts
  - Rockwool
  - Cellulose
  - Spray foam
- Air barrier alternatives: Taped rigid insulation boards and self-adhered membrane transition membranes. Insulating sheathing shall be mechanically fastened in a pattern that will withstand air pressures loads such as wind, stack, and fan pressures. All penetrations must be made airtight.

The design shown in Fig. F.3 gives details of external air barrier for brick veneer wall design, air cavity, R-3 continuous rigid insulation, 15# felt, sheathing air barrier, stud backup, vapor barrier, and gypsum drywall interior finish: at roof edge (A), at parapet (B), at floor slab (C), at window head (D), at window jamb (E), at window sill (F), and at foundation (G).

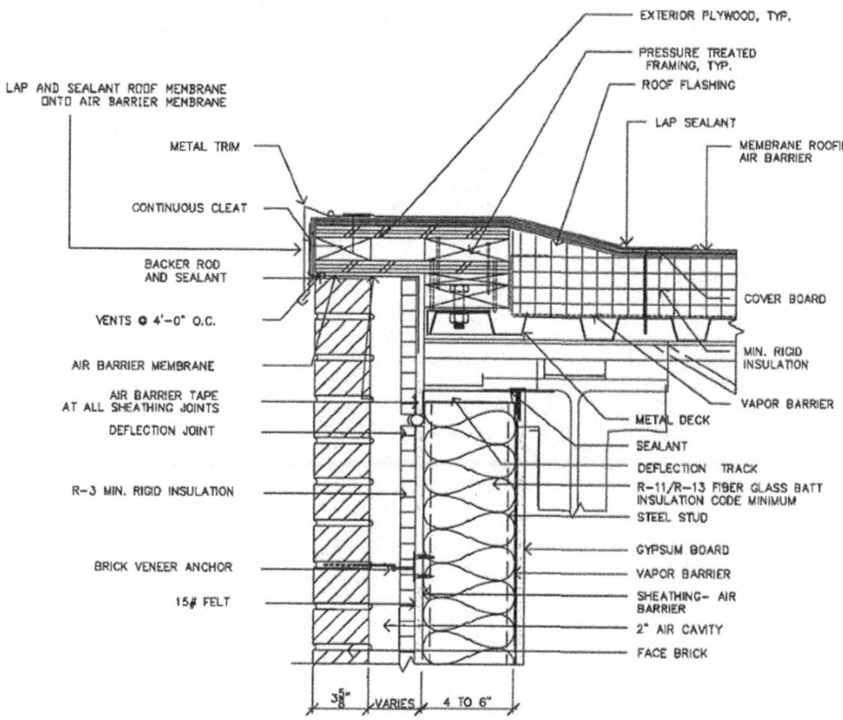

Fig. F.3 Details of external air barrier for brick veneer wall design, air cavity, R-3 continuous rigid insulation, 15# felt, sheathing air barrier, stud backup, vapor barrier, and gypsum drywall interior finish: (**a**) roof edge; (**b**) parapet; (**c**) floor slab; (**d**) window head; (**e**) window jamb; (**f**) window sill; (**g**) foundation

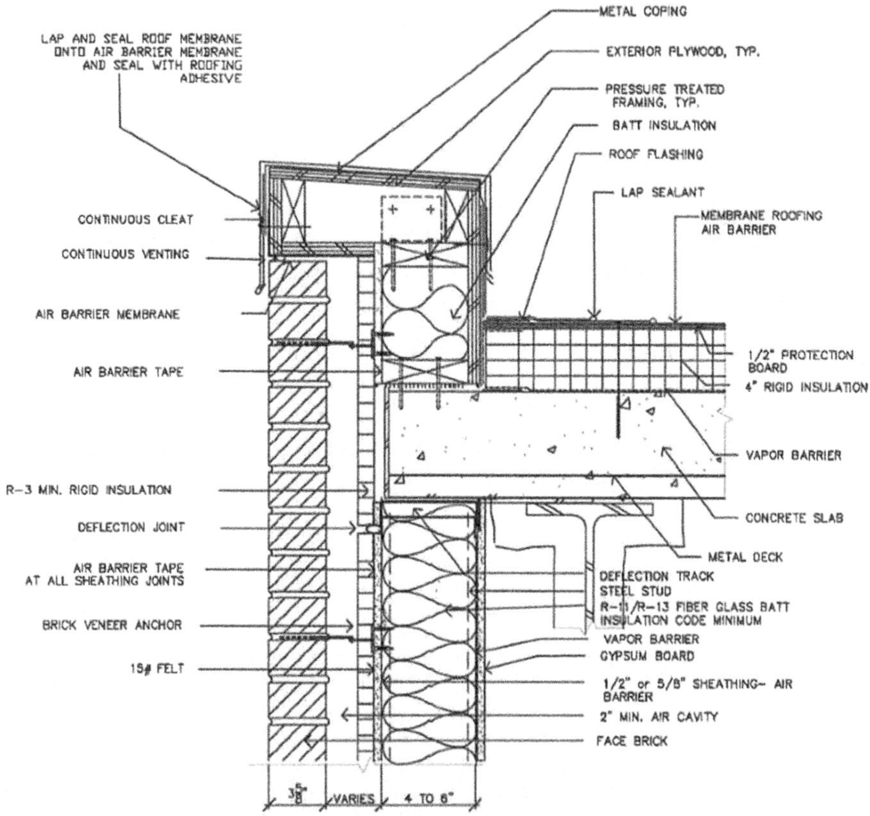

Fig. F.3 (continued)

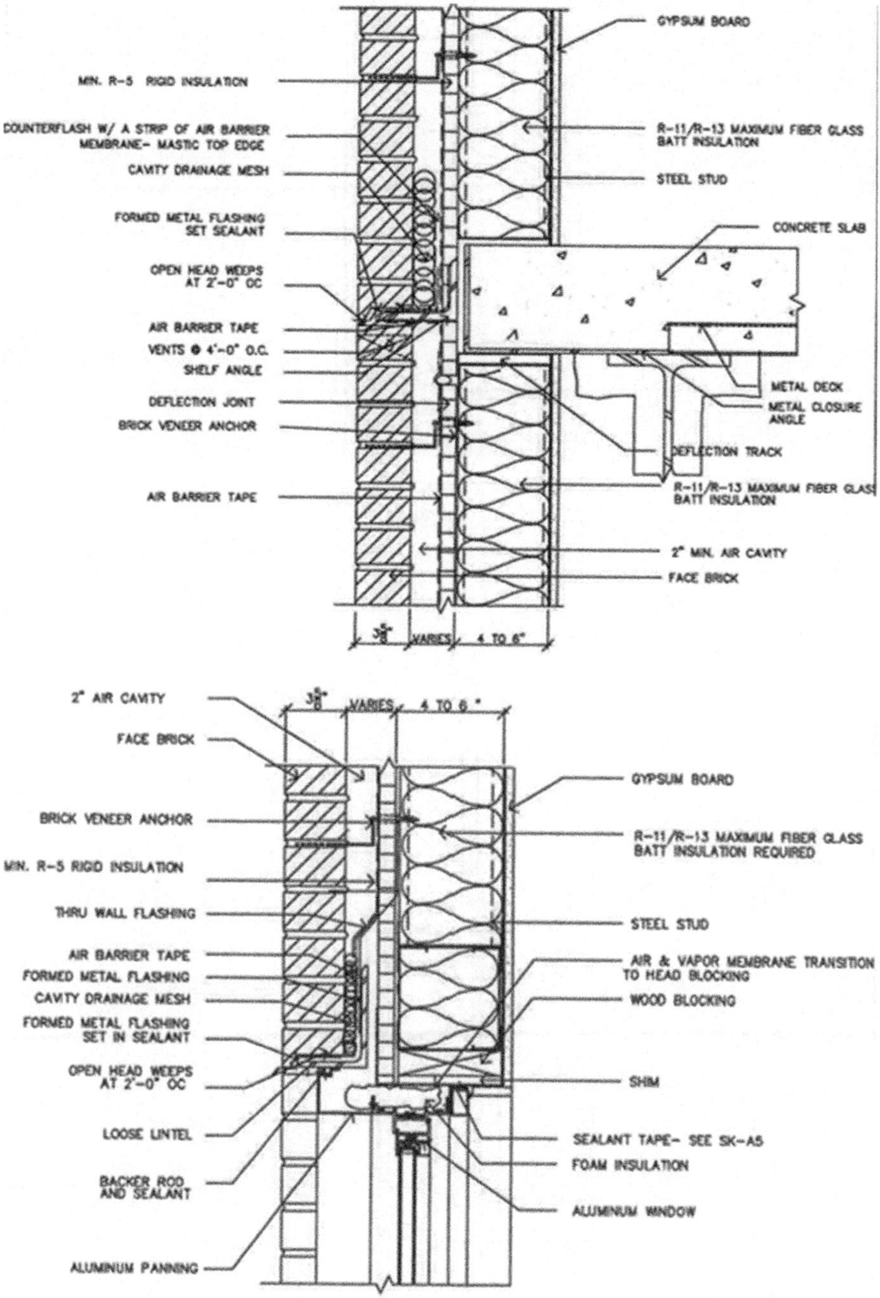

Fig. F.3 (continued)

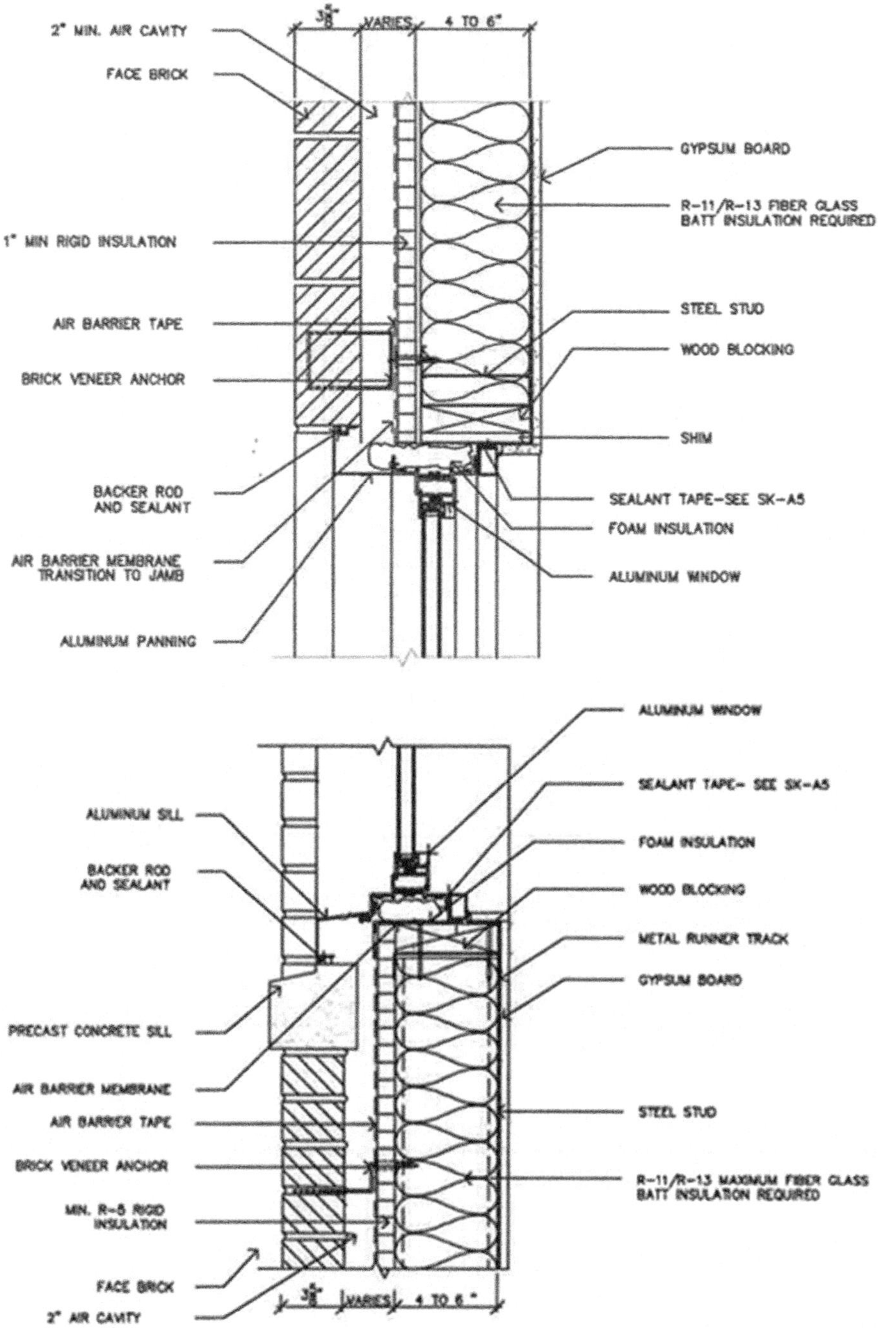

Fig. F.3 (continued)

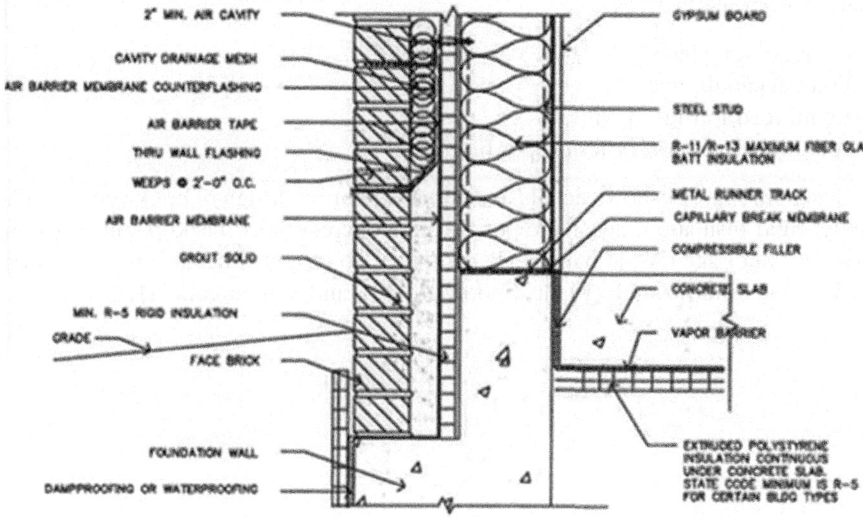

Fig. F.3 (continued)

Advantages and Disadvantages of the Design in Fig. F.3

This design has R-3 rigid continuous insulation and R-11 or R-13 batts in the stud cavity (dependent on climate zone). This wall needs a vapor barrier to work. It is interesting to note that increasing the rigid insulation to R-5 eliminates the need for a vapor barrier up to a maximum interior RH of 35%, which makes for a better wall system, since it can dry out to the inside if incidental wetting occurs. The insulation in the cavity needs to be XPS or polyisocyanurate, since the wall is subject to reverse vapor drive and those kinds of insulation help reduce the effect. It also is important to vent the brick cavity for that same reason. The air barrier is shown to be the gypsum sheathing, but can also be a vapor-permeable liquid-applied spray-on or trowel-on air barrier, which would eliminate the need for the 15# felt.

Continuous insulation alternatives:

- XPS
- Polyisocyanurate
- Medium-density closed-cell spray polyurethane foam

Air barrier alternatives:

- Taped gypsum sheathing and screws.
- Commercial grade spun-bonded polyolefin.
- Liquid-applied vapor-permeable air barrier.
- If medium-density closed-cell polyurethane foam is used on the sheathing, then it should be increased to 25 mm, whereupon it will work as an air barrier material and this design will eliminate the need for a vapor barrier. The 15# felt is also omitted. Transition membranes are required.

Vapor barrier alternatives:

- 6 mil polyethylene
- Foil-backed drywall
- Reinforced foil-faced batts
- Any other 0.1 perm or less vapor barrier

Design in Fig. F.4 shows details of external air barrier design of brick veneer wall, cavity, rigid insulation, air and vapor barrier, concrete block backup, and gypsum board: at roof edge (A), at parapet (B), at floor slab (C), wall detail (D), at window head (E), at window jamb (F), at window sill (G), and at foundation (H).

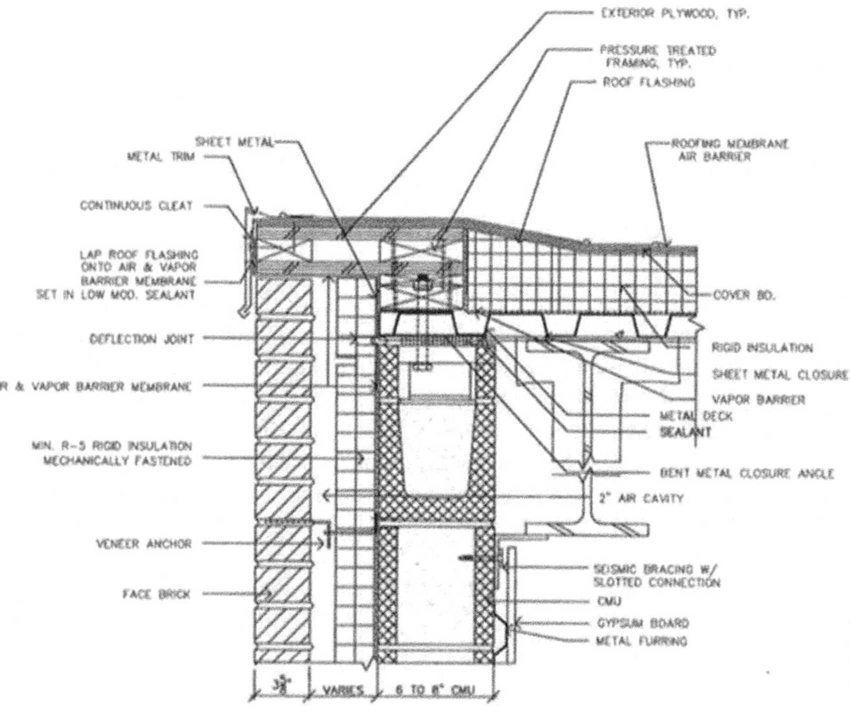

Fig. F.4 Details of external air barrier design of brick veneer wall, cavity, rigid insulation, air and vapor barrier, concrete block backup, gypsum board: (**a**) roof edge; (**b**) parapet; (**c**) floor slab; (**d**) wall detail; (**e**) window head; (**f**) window jamb; (**g**) window sill; (**h**) foundation

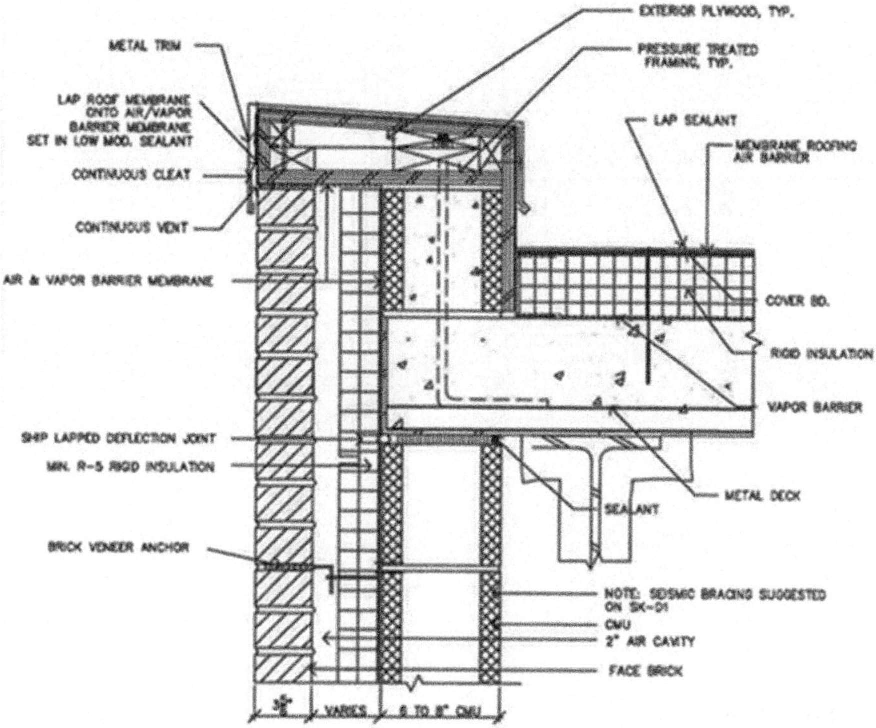

Fig. F.4 (continued)

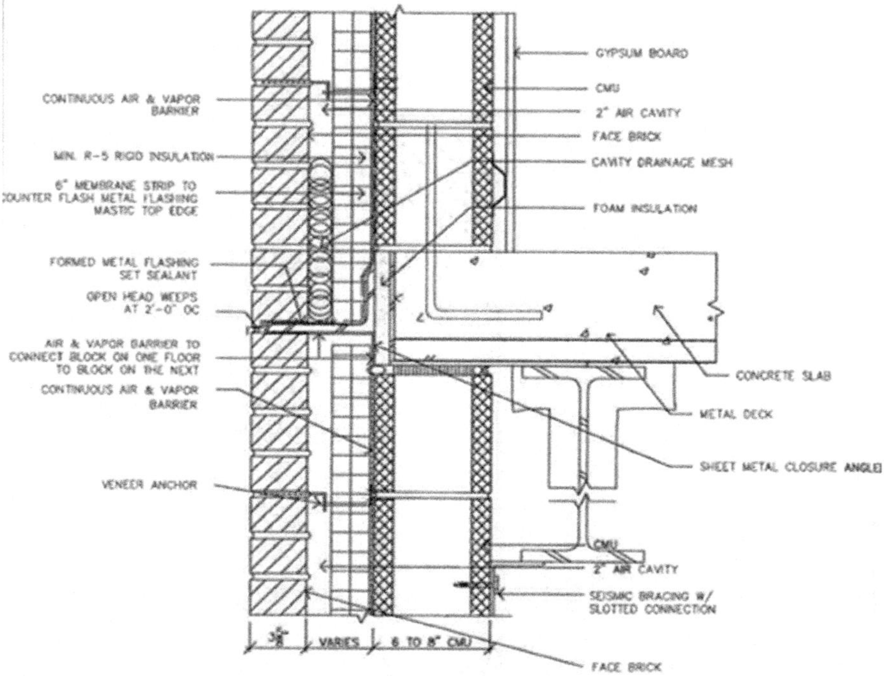

Fig. F.4 (continued)

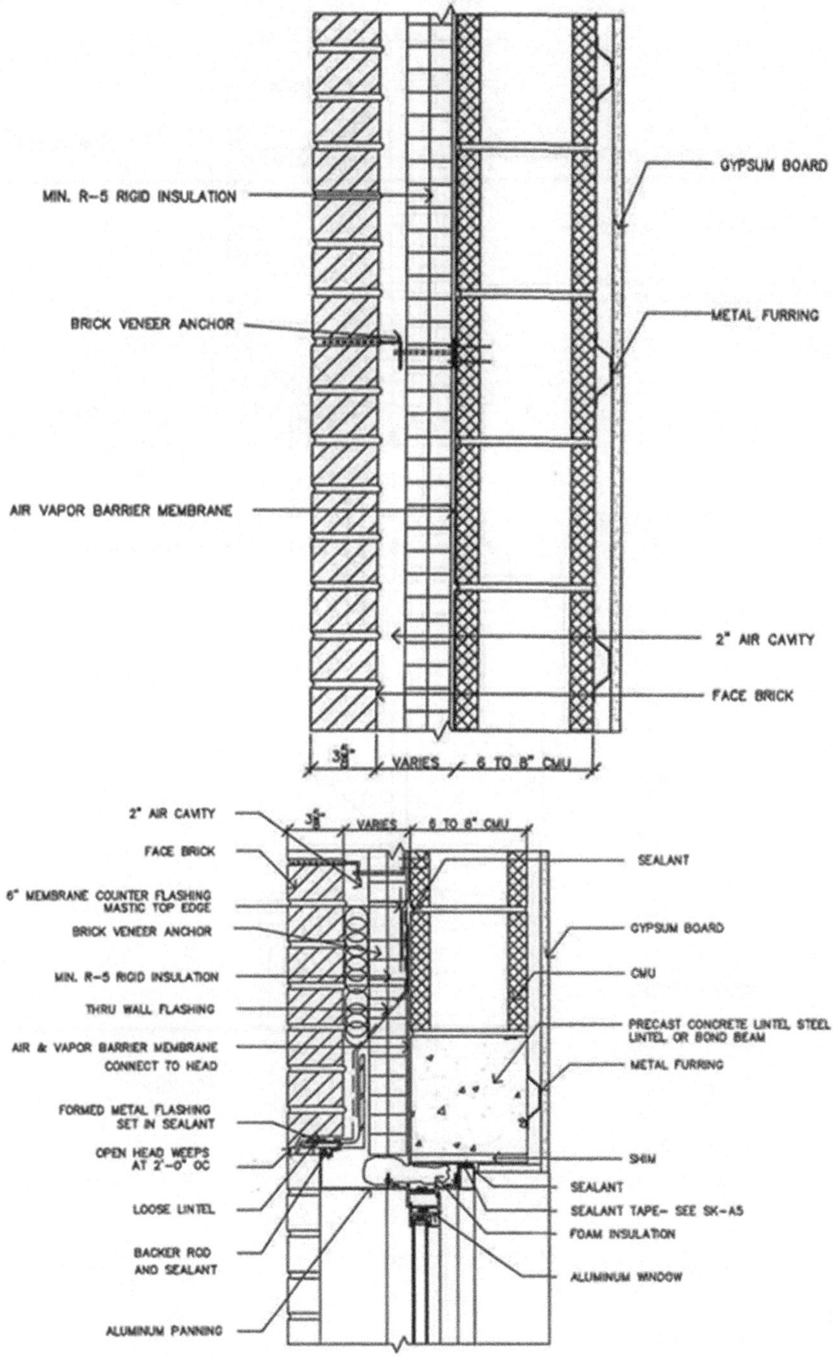

Fig. F.4 (continued)

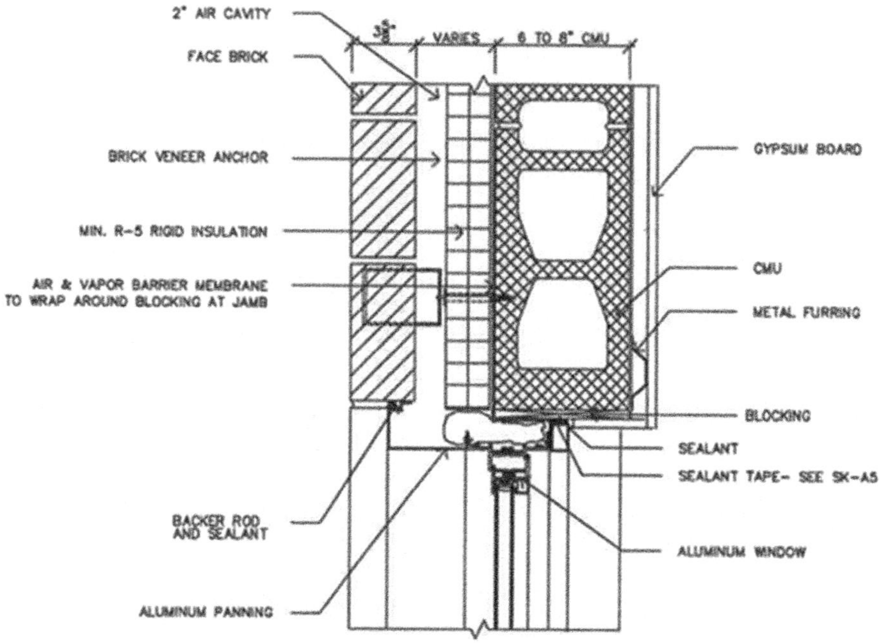

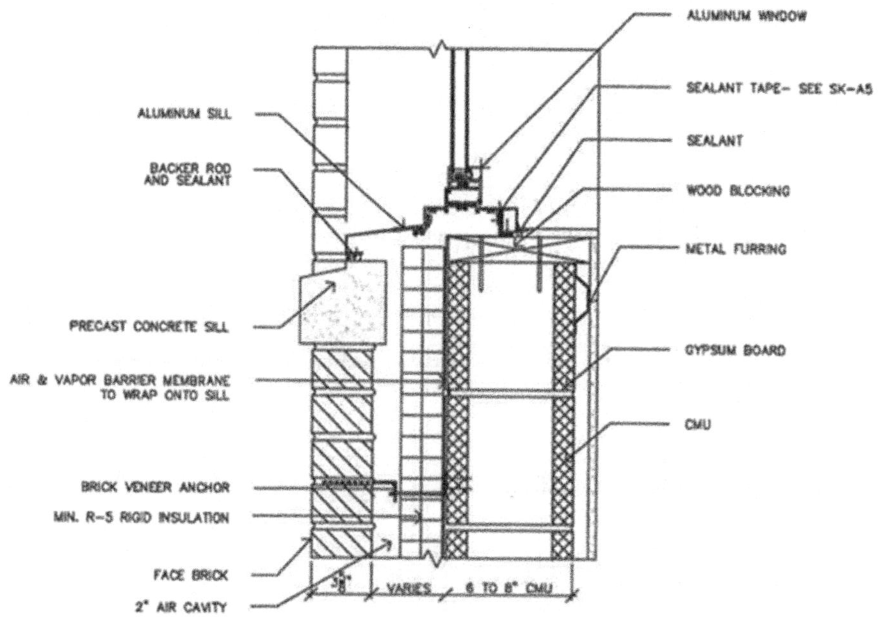

Fig. F.4 (continued)

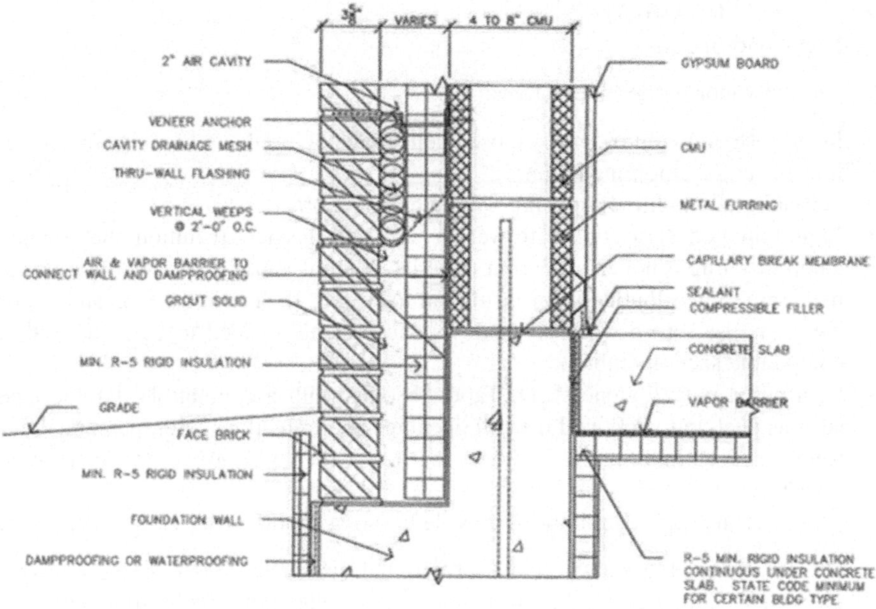

Fig. F.4 (continued)

Advantages and Disadvantages of the Design in Figure F.4

This design is very similar to the design shown in Fig. F.1, except that less insulation (R-5) is used. All the other choices are identical. This wall type has a layer of continuous insulation, making it energy efficient. It is easy to build because the air and vapor barrier and the insulation are continuous. The concrete block backup wall has "thermal mass," which stores/releases energy. The backup wall is always kept above the dewpoint of the winter indoor air.

The air and vapor barrier on the warm side of the insulation also works as the drainage plane, handling liquid water, water vapor in the winter and reverse vapor drive, keeping the backup wall always dry. The year-round stable temperature of the air and vapor barrier promotes its durability. It is also well structurally supported, especially if the insulation is mechanically fastened rigid insulation and/or if the membrane is self-adhered or fluid applied.

With closed-cell board insulation, this wall can function without an additional vapor barrier. Other kinds of insulation need a vapor barrier.

Insulation choices:

- XPS
- Polyisocyanurate
- EPS
- Spray polyurethane foam

- Semi-rigid Rockwool®
- Semi-rigid fiberglass

Air and vapor barrier alternatives:

- In the case of medium-density closed-cell spray polyurethane foam insulation, all it needs is transition membranes.
- Self-adhered modified asphalt/polyethylene membrane.
- Liquid-applied spray-on or trowel-on air/vapor barrier. (Caution that asphalt damp-proofing is not an air barrier material.) Liquid-applied air and vapor barrier membrane may double as an insulation adhesive. If this kind of air and vapor barrier is used, trim work at windows and transitions need to be made with a compatible sheet membrane.
- Reinforced polyethylene sheet. Tape all joints with a compatible durable tape such as peel-and-stick, and trim all openings and transitions with membrane.
- Reinforced aluminum foil air/vapor barrier, with taped joints and membrane transitions.
- Commercial grade spun-bonded polyolefin (not a vapor barrier), with membrane.

Design in Fig. F.5 shows details of external air barrier single wythe concrete block with R-7 interior insulation. Commercial grade spun-bonded polyolefin air

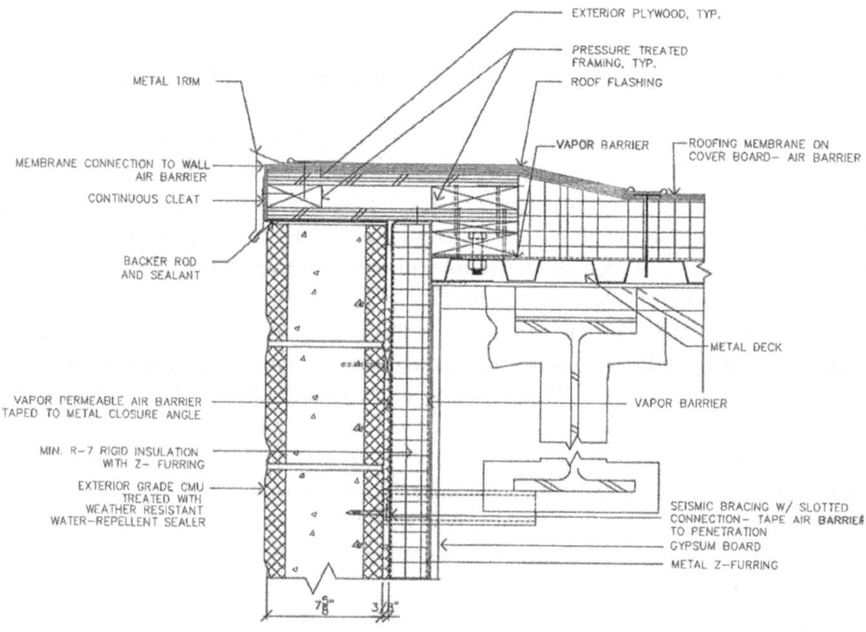

Fig. F.5 Details of external air barrier single wythe concrete block with R-7 interior insulation. Commercial grade spun-bonded polyolefin air barrier, rigid insulation, vapor barrier, interior drywall: (**a**) roof edge; (**b**) parapet; (**c**) floor slab; (**d**) single wythe CMU; (**e**) window jamb; (**f**) window sill; (**g**) foundation

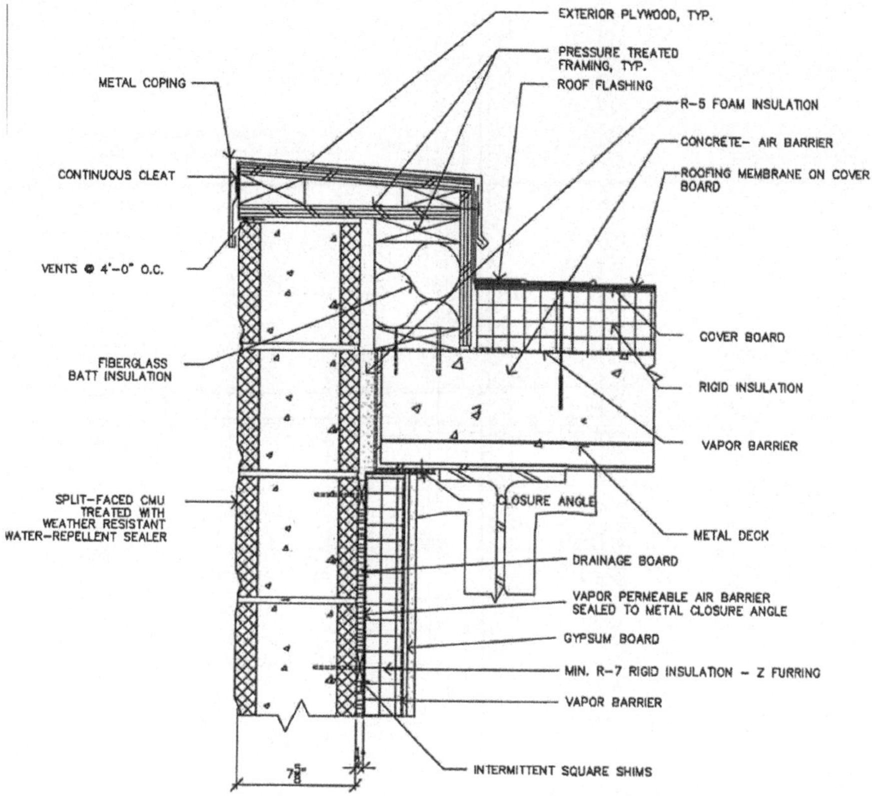

Fig. F.5 (continued)

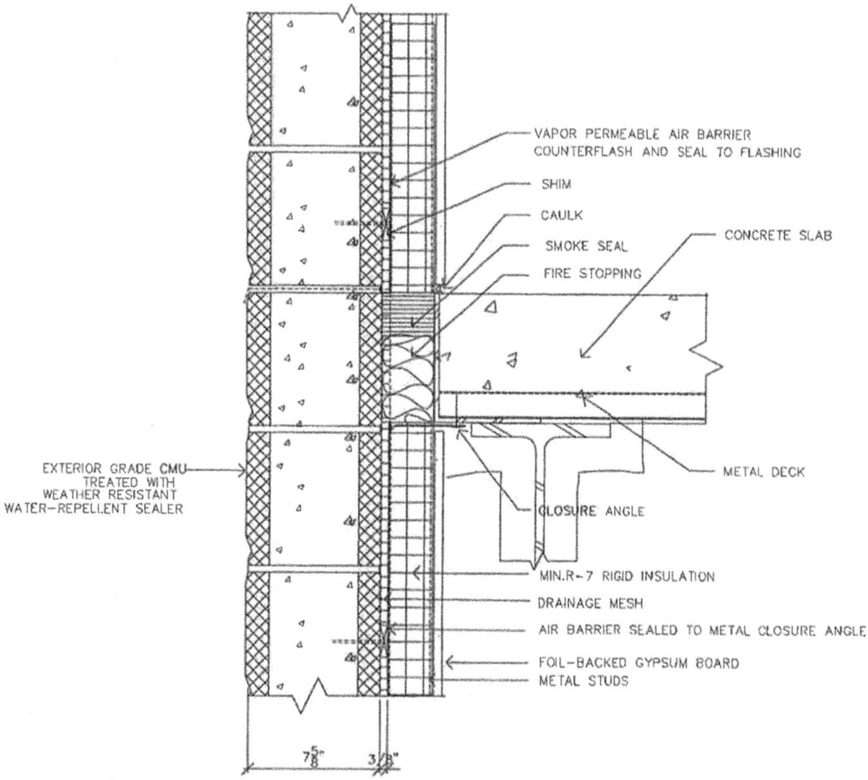

Fig. F.5 (continued)

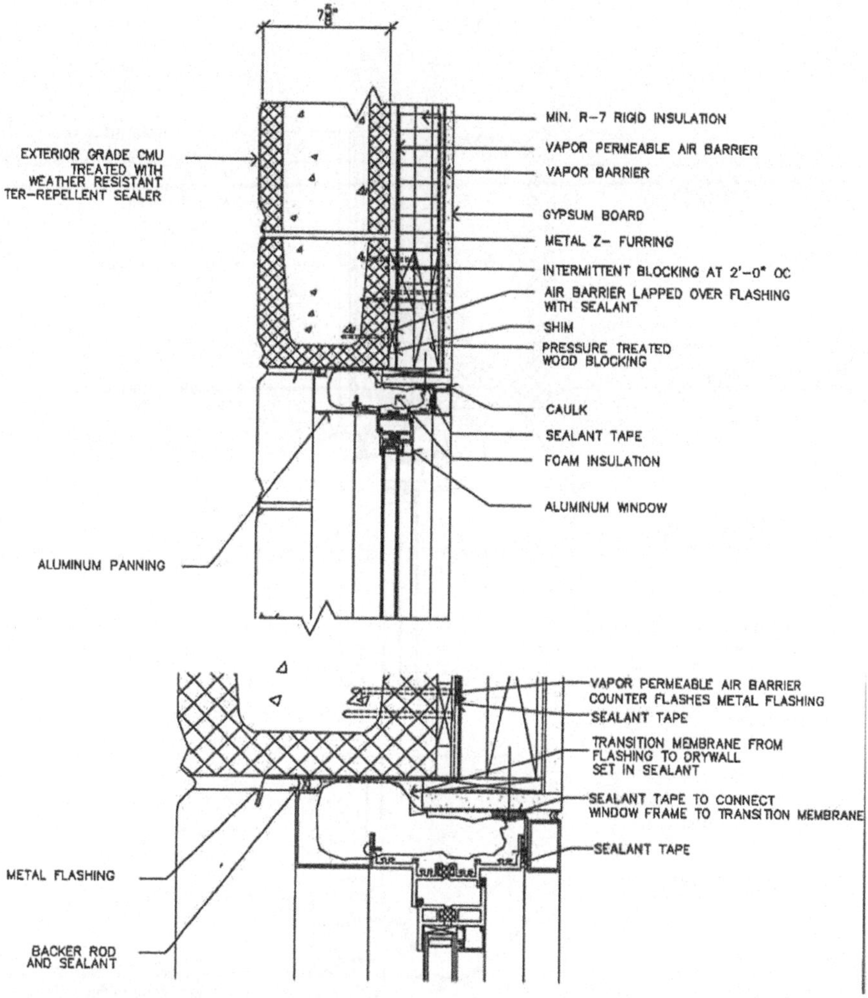

Fig. F.5 (continued)

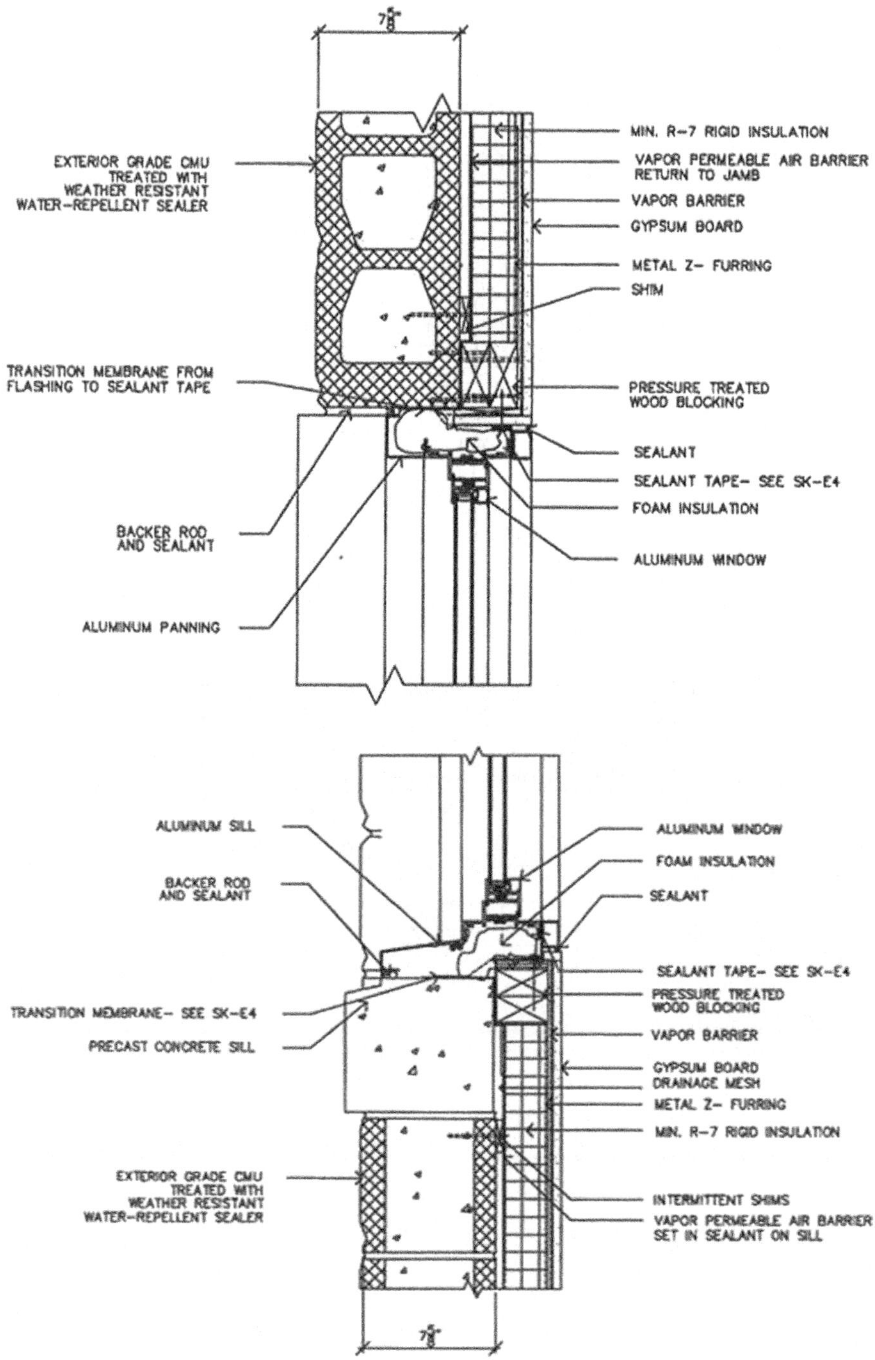

Fig. F.5 (continued)

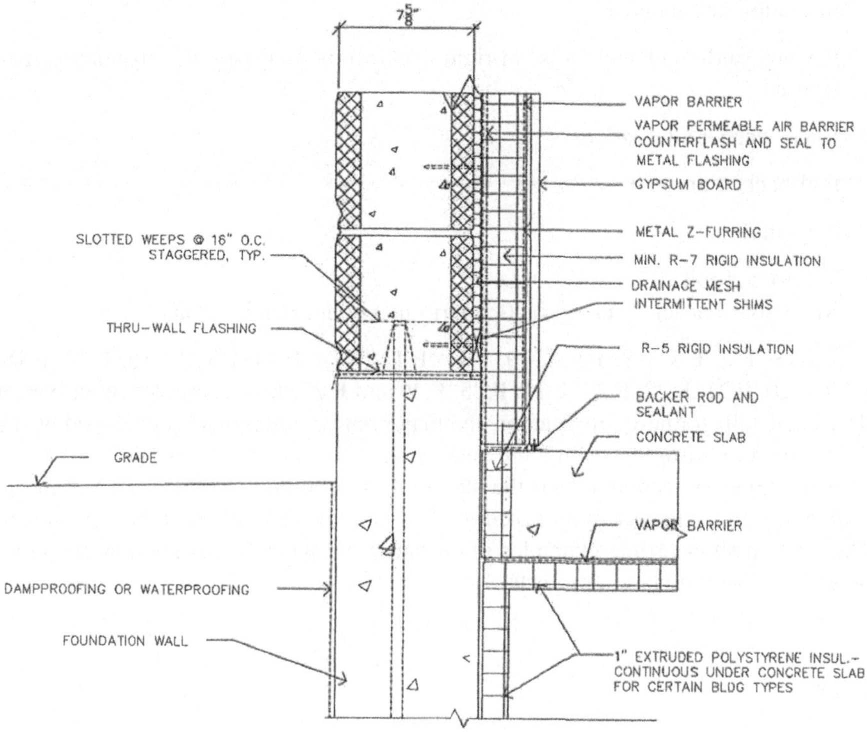

Fig. F.5 (continued)

barrier, rigid insulation, vapor barrier, interior drywall: at roof edge (A), at parapet (B), at floor slab (C), at single wythe CMU (D), at window jamb (E), at window sill (F), and at foundation (G).

Advantages and Disadvantages of the Design in Figure F.5

Single wythe concrete block walls are inexpensive to build, but are susceptible to leaks of both air and water. This design attempts to cure that by creating a drainage cavity on the interior using 8 mm or so thick polymeric drainage and ventilation material (looks like expanded metal lath).

Intermittent shims fastened to the block are used to fasten the air barrier to the wall. These can be lath strips running vertically or intermittently per the spun-bonded polyolefin manufacturer's instructions for air barrier attachment. It is supported in both positive and negative wind directions.

It is important to note that the air barrier is connected with compatible sealants or tape to the metal flashings as well as to metal termination and transition angles.

Air barrier alternatives: None

Insulation alternatives:

- Furring studs and batt or semi-rigid rock wool or glass-fiber insulation may be used.

Vapor barrier alternatives:

- Foil-backed drywall

Polyethylene:

- Reinforced foil
- Any vapor barrier meeting the 0.1 perm maximum requirement

Figures F.6, F.7, F.8, F.9, F.10, F.11, F.12, F.13, F.14, F.15, F.16, F.17, F.18, F.19, F.20, F.21, F.22, F.23, F.24, F.25, F.26 and F.27 show examples of air barrier drawing details for new construction and major renovation projects developed by the Association of Airtightness in the Construction Industry (FLiB), Berlin.

It is important to understand the status of the building after it is gutted and to design the air barrier assembly based on the limitations identified in the assessment. This section shows some examples of air barrier drawing details for new construction and major renovation projects.

Roofs

Slope Roofs

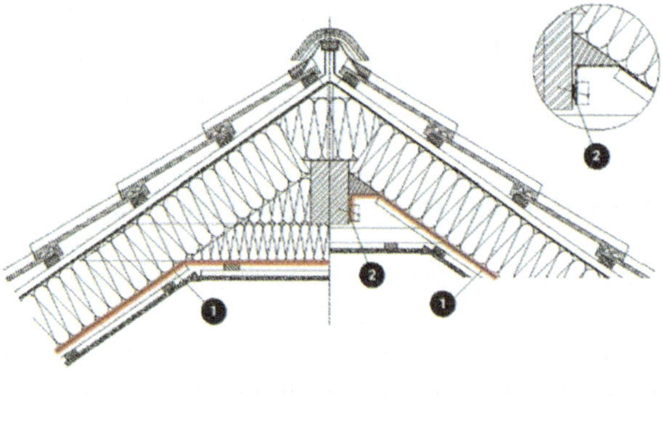

Fig. F.6 Rafter insulation: (1) air barrier, (2) adhesive, tape

Notes:

- The air barrier must be arranged seamlessly on the inside of the wood structure in designs with collar ties or connection of roof rafters.
- In designs without collar ties, the air barrier is connected to the ridge purlin with adhesive or tape.
- The air barrier should be implemented beneath the ridge purlin in designs without collar ties or any cracks in the ridge purlins.

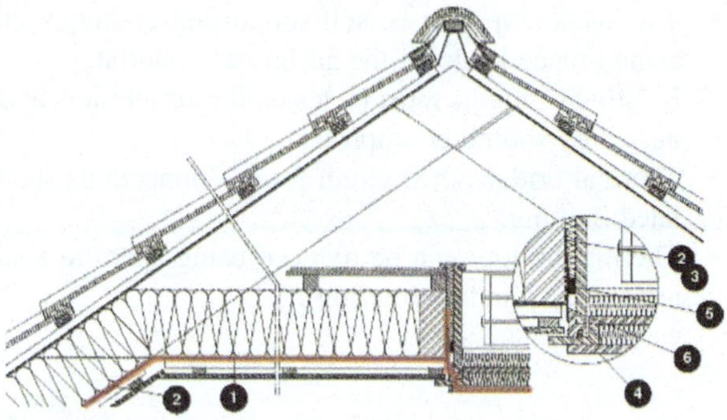

Fig. F.7 Retractable attic: (1) air barrier, (2) tape, (3) adhesive, (4) hatch certified to be airtight, (5) insulation rope, (6) tape with offset opposite adhesive edges

Notes:

- The air barrier will lay under the insulation.
- The air barrier will be attached to the attic ladder.
- The air barrier will be attached with tape, adhesive material, or sealant.
- Existing attic ladders entrance must be tested for airtightness, and if necessary, insulation should be improved.
- The unheated attic must be sufficiently ventilated. Humid air can get into the unheated attic through the functional gaps in the attic stairs; if the moisture cannot escape through vents, it will condense on the cold wood beams causing the wood to rot.

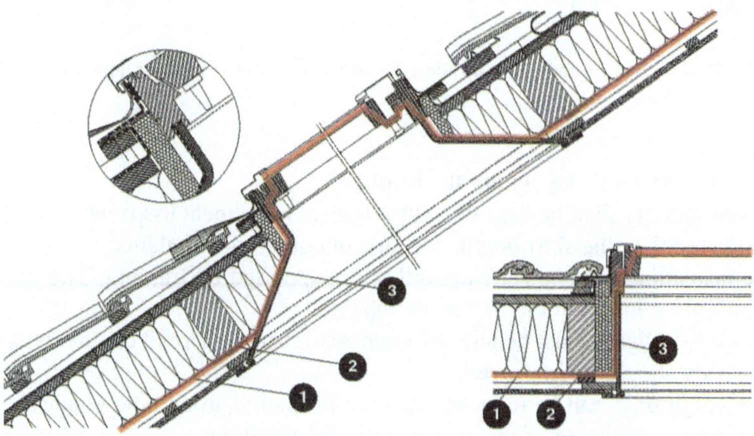

Fig. F.8 Roof with skylights: (1) air barrier, (2) tape, corner tape, (3) skirt board

Notes:

- The air barrier will lay under the insulation.
- The air barriers shall be flexible with sufficient movement loops when attached to the skylights or the skirt boards via tape or corner tape sealants.

- The insulation must be self-supporting or supported by framing. The insulation cannot place loads on the air barrier material.
- Existing skylights must be tested for airtightness and thermal conductivity, and if necessary should be improved.
- Thermal bridges in skylight frame connections should be reduced through insulated framing.
- Thermal bridges can be reduced using aperture insulation (e.g., foam insulation around the window framing).

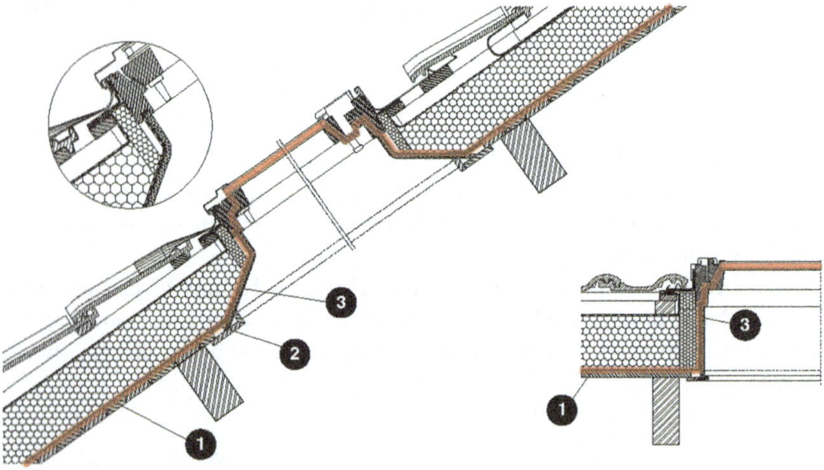

Fig. F.9 Roof with skylights – above rafter: (1) air barrier, (2) tape, corner tape, (4) skirt board

Notes:

- The air barrier will lay within the insulation.
- The air barriers shall be flexible with sufficient movement loops when attached to the skylights or the skirt boards via tape or corner tape sealants.
- The insulation must be self-supporting or supported by framing. The insulation cannot place loads on the air barrier material.
- Existing skylights must be tested for airtightness and thermal conductivity and if necessary should be improved.
- Thermal bridges can be reduced via foam insulation around the window frame.
- Load on the fastening of the air barrier leads to failure.

Flat Roofs

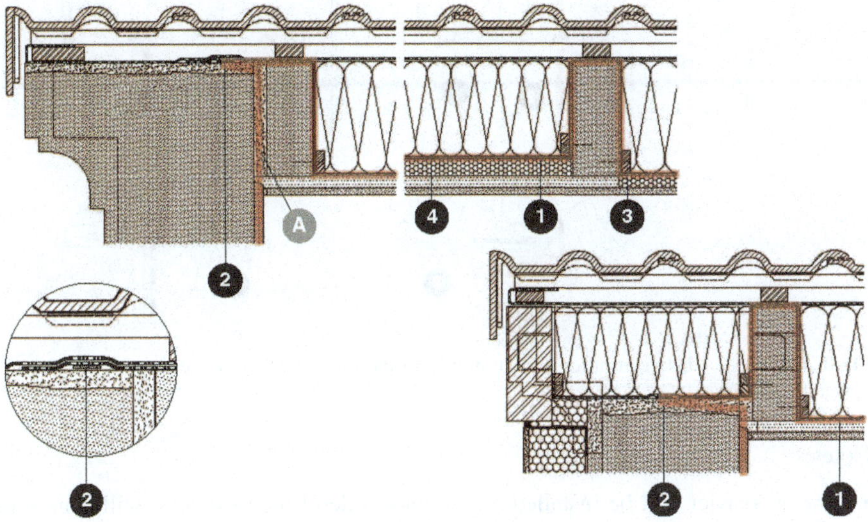

Fig. F.10 Verge – between-rafter insulation sub-top: (1) air barrier, (2) adhesive, (3) supporting lath, (4) protection layer (optional)

Notes:

- Clean out any material between-rafter area, possibly remove nails or put down a protective layer (e.g., soft wood boards or mineral wool insulation mat with adequate compressive strength).
- The air barrier will be laid on the exposed sheathing, will loop around the rafters, and will be fixed on the bottom of the rafters with pressure plates.
- The air barrier will be connected on top of the wall; the surface must be clean, dry, smooth, and sound.
- Ensure that the roof is airtight between the rafter spaces from the gable wall to the interior plaster. If necessary, seal this using another method, e.g., spray foam.
- Protruding internal lining fasteners (nails, clamps, screws, etc.) can perforate the air barrier.

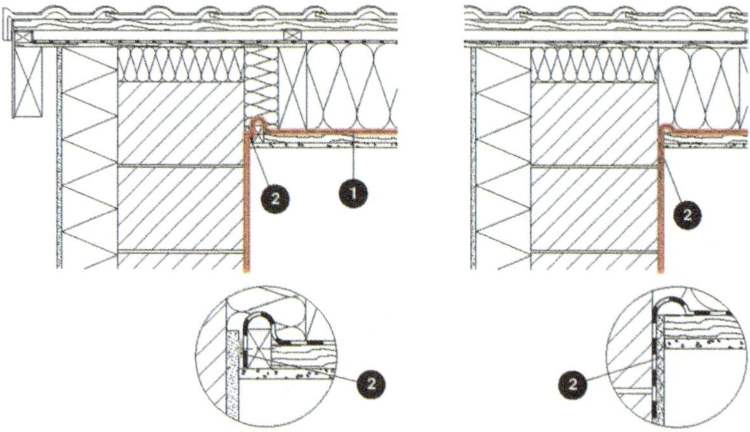

Fig. F.11 Verge – between-rafter insulation: (1) air barrier, (2) adhesive, adhesive tape, pressure plates if necessary

Notes:

- The air barrier will be installed on the underside of the rafters, or will consist of between-rafter insulation.
- Where it connects to the external wall, the air barrier should be well fixed to the gable wall using adhesive so that the connection is airtight.
- The connection can also be achieved by taping the air barrier to the base of the previously plastered surface.

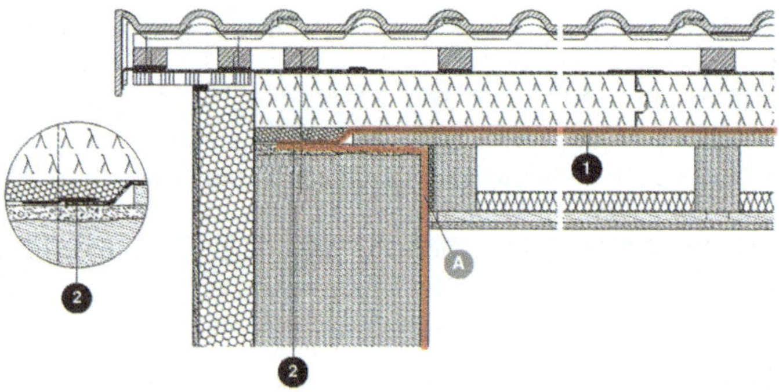

Fig. F.12 Verge – above rafter insulation: (1) air barrier, (2) adhesive

Notes:

- The air barrier is installed on the existing wood structure of the roof truss.
- The air barrier is connected at the top of the wall; the surface of the wall must be clean, dry, smooth, and sound.
- (A) Ensure that the air barrier runs through the rafter space from the gable wall and rafters to the internal plaster. If necessary, use other sealing measures.

- To achieve the airtight connection to the bordering building components, the roof sheathing must be interrupted in the area of the top of the wall to bind the air barrier.
- The transition from the internal plaster to the air barrier must be assessed case-by-case.

Notes:

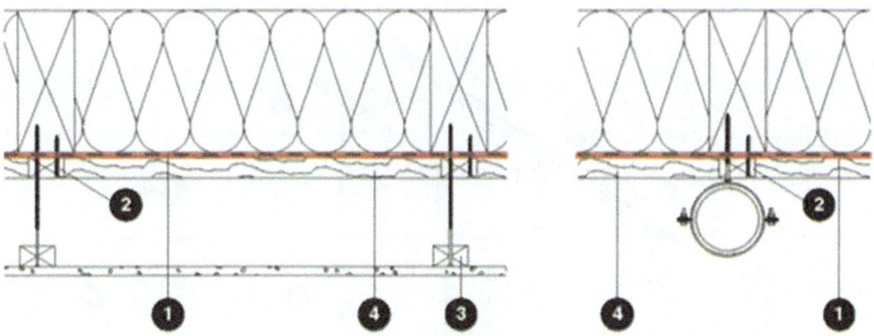

Fig. F.13 Suspension: (1) air barrier, (2) laths, (3) ceiling suspension, (4) loadbearing lath

- The air barrier is attached to the ceiling beams/rafters via laths.
- The fasteners for the suspension are screwed into the ceiling beams/rafters with the laths. This must be done while keeping the penetrations through the barrier as a result of the fastening airtight.
- The air barrier moves during its use. Attachment of the suspended ceiling by screw fasteners only, without additional pressure plates, will cause pressure fluctuations and produce movement that will expand the perforations through the air barrier.

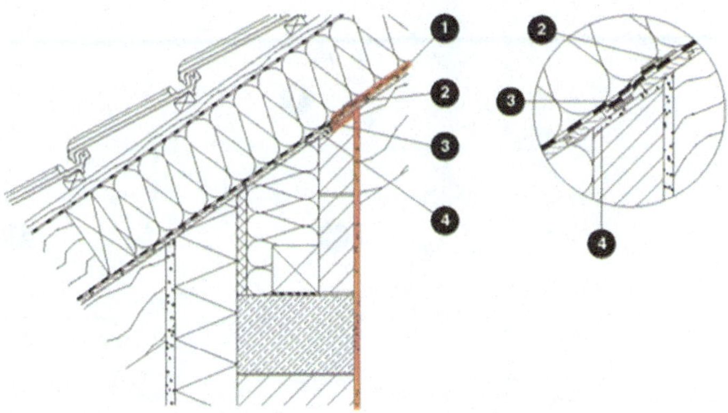

Fig. F.14 Eaves – above rafter insulation, continuous rafters, lined compartments: (1) air barrier, (2) tape, (3) adhesive, (4) smoothed lining at the crown of the wall

Wall-to-Roof Connection

Notes:

- The air barrier is installed on top of the roof boarding.
- The roof boarding is interrupted at the outer wall; the air barrier is installed to the interior.

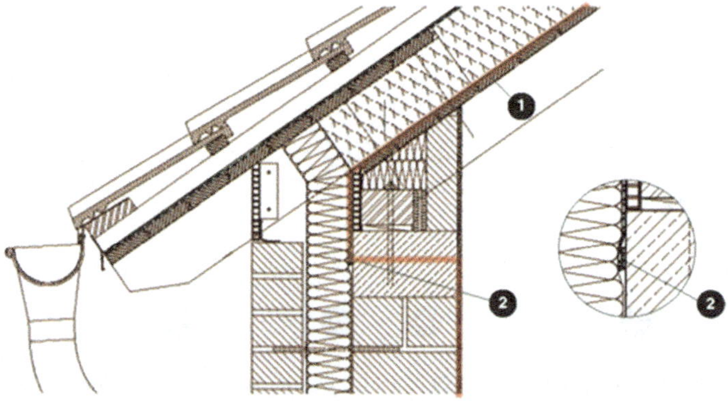

Fig. F.15 Eaves – above rafter insulation, continuous rafters, lined compartments: (1) air barrier, (2) adhesive

Notes:

- The air barrier is installed on the insulating board stock (when present); sealing is done with adhesive on the rim joists.
- The interior plaster shall be sealed to the rim joists.

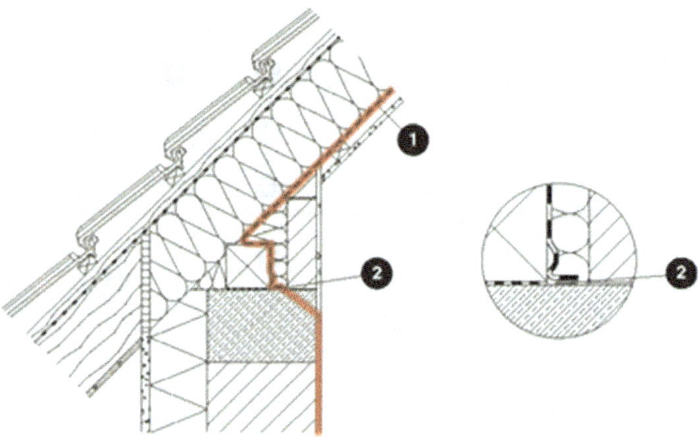

Fig. F.16 Eaves – between-rafter insulation: (1) air barrier, (2) adhesive

Notes:

- The air barrier is installed beneath the insulation.
- The air barrier is connected to the rim joist of the walls.
- The air barrier is adhered with an expansion loop.
- The interior plaster must be sealed to the rim joists.

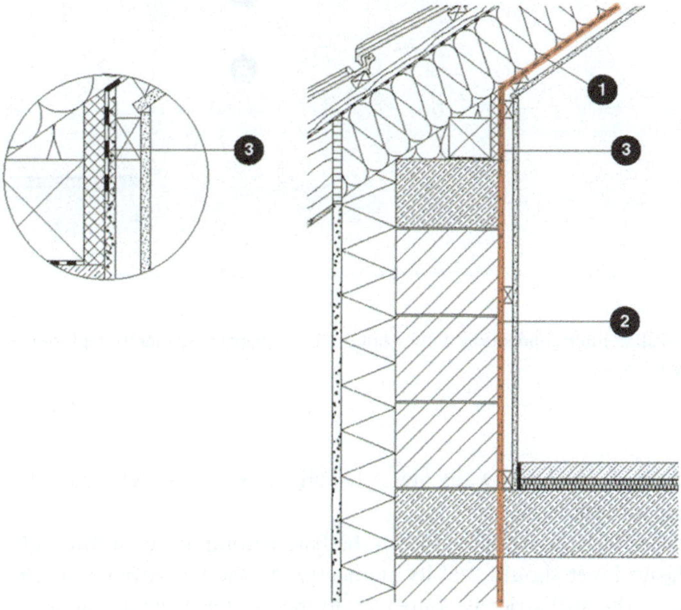

Fig. F.17 Wall/ceiling: (1) air barrier, (2) interior plaster, (3) adhesive, tape, plaster base

Notes:

- The external masonry wall usually has a plaster layer to create an adequate air barrier.
- The plaster layer must extend continuously from the bare ceiling to the unfinished floor.
- The plaster layer should seal its connection to the air barrier in the roof design; otherwise, the air barrier is connected to the plaster using a sealant.
- In areas that will no longer be accessible later in the process, a smooth finish (at least) shall be applied, for example, in the area of wall-mounted installations, suspended ceilings, and jamb walls.
- Where external walls joint the flooring, a plaster layer should be applied to make the external walls airtight.

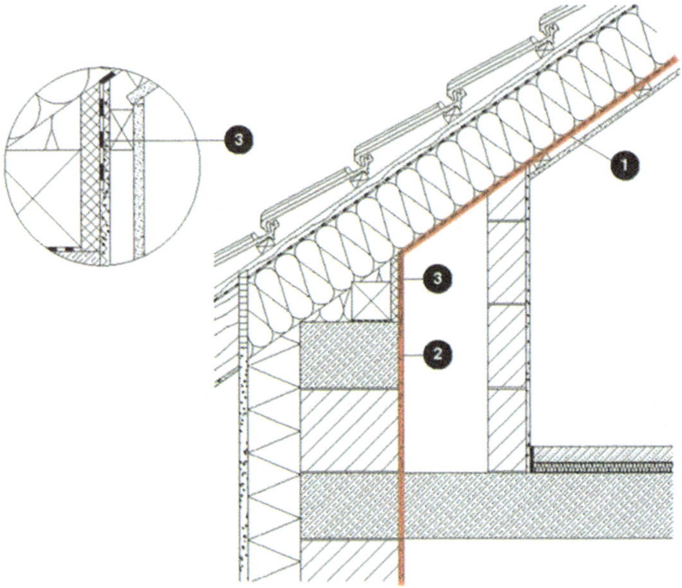

Fig. F.18 Wall/ceiling connection with a jamb: (1) air barrier, (2) interior plaster, (3) adhesive, tape, plaster base

Notes:

- The external masonry walls must usually have a plaster layer that provides adequate airtightness.
- The plaster layer must extend from the bare ceiling to the unfinished floor.
- The plaster layer should seal its connection to the air barrier in the roof design; otherwise, the air barrier is connected to the plaster using a sealant.
- In areas that will no longer be accessible later in the process, a smooth finish (at least) shall be applied, for example, in the area of wall-mounted installations, suspended ceilings, and jamb walls.
- Where external walls joint the flooring, a plaster layer should be applied to make the external walls airtight.

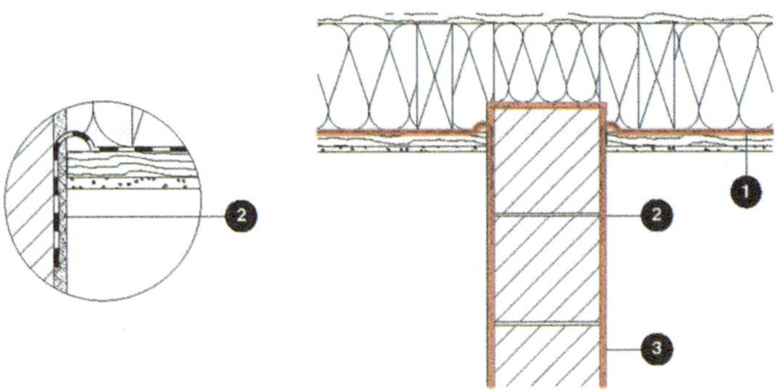

Fig. F.19 Inner wall/roof: (1) air barrier, (2) adhesive or tape (fleece laminated), plaster base, (3) plaster

Notes:

- The air barrier is sealed to the inner wall; the wall must be plastered above the airtight level including the wall crown.
- The connection between the air barrier and the inner wall is sealed with adhesive or a tape. If the foil and or tape is not fleece laminated, the junction must subsequently be sealed using plaster.
- The plastering of the wall coping should take place from the side of the trusses.
- Concrete straps/rim joists on the wall coping shall be airtight.

Wall-to-Window Connection

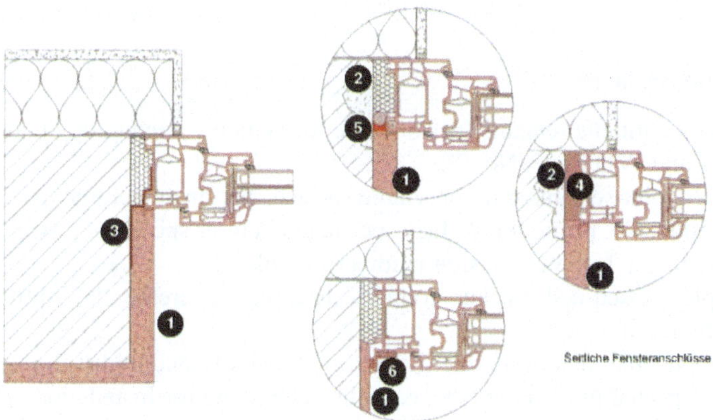

Notes:

- The air barrier is installed around the window, between the window perimeter, and the interior plaster.
- The sealant or sealing system must match the width and condition of the joint.
- Corner areas require special attention.
- If applicable, care should be taken to not over-apply plaster on foil joint seals.
- For waterproof multifunctional sealant strips:
    - Ensure that the sides of the joints are suitably treated (if applicable, smoothed down and/or appropriately covered).
    - The tape compression must be in accordance with the manufacturer specification; if applicable, adapt the size of the tape to the joint width.
    - Install corner areas and shims carefully, in accordance with manufacturer specifications.
    - Shims are not suitable for the underside connection of the window.

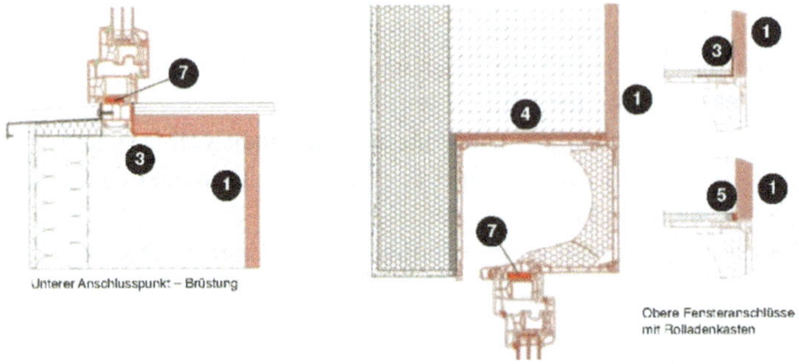

(Load Transfer through Blocking)

Fig. F.20 Components of window installed within the wall: (1) interior plaster, (2) smooth finish. (as needed) alternatives: (3) foil seals; (4) sealant strips, multifunctional sealant strips; (5) spray sealant, (6) tapes with sealant strips; (7) seal window-bank junction profile for windows

Notes:

- For spray sealants:
  - Choose suitable sealants (per Institute for Window Technology Guideline M0 01-1 and IVD Sheet No. 9).
  - Apply joint sealants only on viable surfaces per manufacturer specifications.
  - If applicable, prime the surfaces per manufacturer specifications.
  - Note that a 3-point adhesion is impermissible.
  - Apply closed-cell backer rods as backing to create the optimal cross-section seal.
  - If applicable, determine the movement capability and adjust the joint size to the expected movement (see, e.g., the "Guideline for Installation" [RAL]).
- For tape with sealant strips, take special care in corner areas.
- For connections that will experience a great deal of usage (especially floor-to-ceiling windows and French doors), pay special attention to providing protection against seal damage.
- For renovations that must preserve original windows, pay attention to the windows' suitability with current requirements for air permeability. Check the functionality of the weather seal.

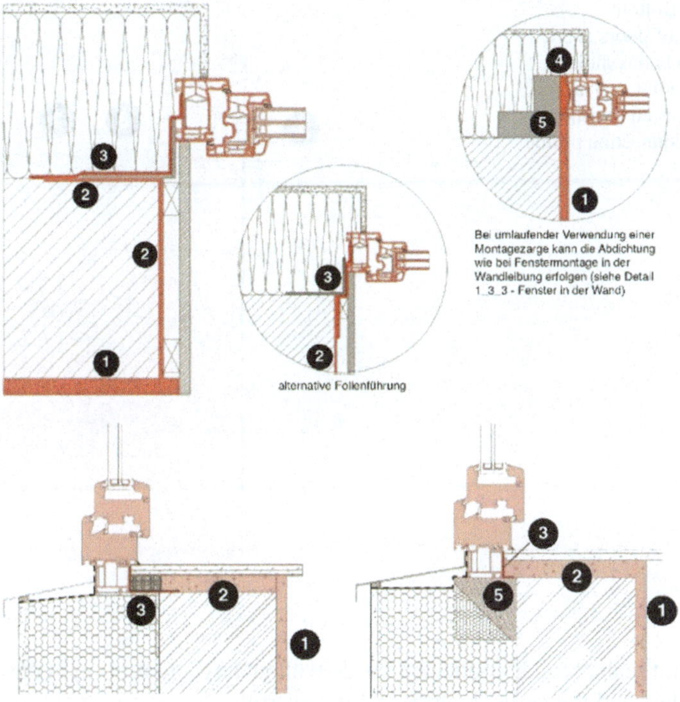

Fig. F.21 Window installed within the wall: (1) interior plaster; (2) smooth finish (if needed); (3) foil joint seals; (4) deviation from the standard installation, e.g., multifunction adhesive strips; (5) subframe

Notes:

- The air barrier is preferably on inside of the window, within the room, as it connects the windows' continuous edge layer. In this context especially, attention must be paid to the connection to the airtight layer of the adjacent building components.
- The sealant material or system is matched to the location of the window, façade construction, and wall conditions.
- The sealant material must be suitable to the application and must meet airtightness requirements.
- All sealant materials must be coordinated with one another (system-suitable).
- Foil joint seals must be adhered free of tension.

Notes:

- The joint sealant foil is connected on the inside without any interruption to the airtight layer of the floor and vertical aperture surface.
- Corner areas require special attention with planning and implementation.
- For doors the airtight layer is ensured against damages by following trades.

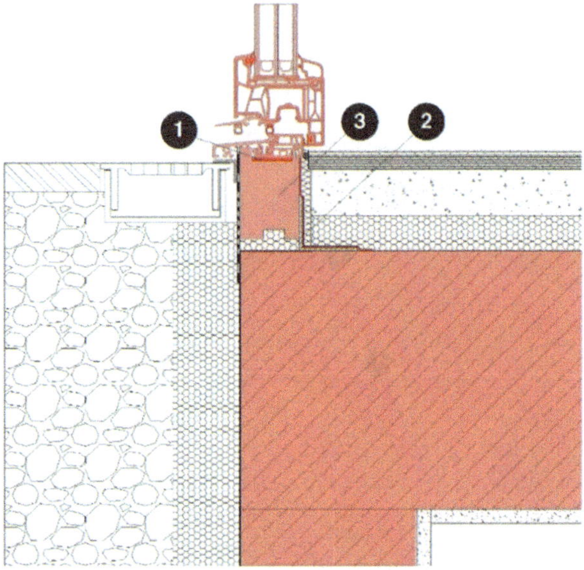

Fig. F.22 Bottom connection of doors and a basement: (1) threshold profile, (2) joint sealant foil, (3) seal between threshold/frame and connection profile

Notes:

- The joint sealant foil is connected inside without any interruption of the seal of the base plate and the vertical surface of the aperture.
- Corner areas require special attention with regard to planning and implementation.

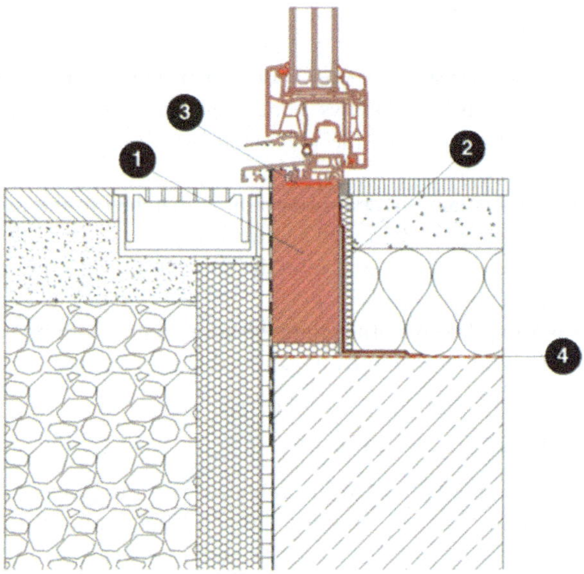

Fig. F.23 Connection of doors at the bottom: (1) threshold profile, (2) joint sealant foil, (3) seal between threshold/frame and connection profile, (4) building sealant

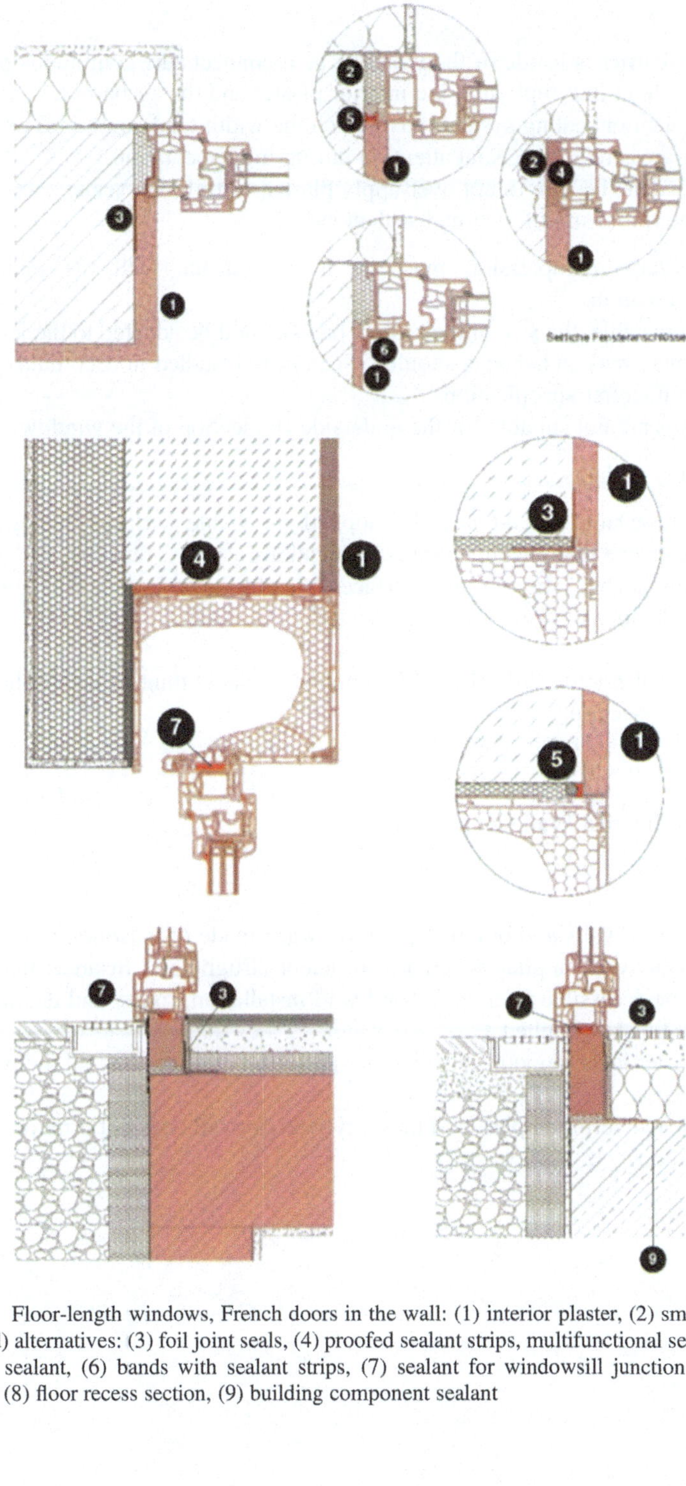

Fig. F.24 Floor-length windows, French doors in the wall: (1) interior plaster, (2) smooth finish. (If needed) alternatives: (3) foil joint seals, (4) proofed sealant strips, multifunctional sealant strips, (5) spray sealant, (6) bands with sealant strips, (7) sealant for windowsill junction profile for windows, (8) floor recess section, (9) building component sealant

Notes:

- The air barrier is inside of the window, as it connects the continuous peripheral layer without interruption to the interior plaster and the air barrier.
- The sealant or sealing system must match the width (and condition) of the joint.
- Corner areas require special attention during implementation.
- If applicable, be sure to not over-apply plaster on foil joint seals.
- For waterproof multifunctional sealant strips:
    - Band/tape compression must be in accordance with the manufacturer specification.
    - If applicable, the size of the sealant band should be adapted to the joint width.
    - Corner areas and shocks should be carefully installed in accordance with the manufacturer specifications.
    - Shims are not suitable for the underside connection of the window.
- For spray sealants:
    - Joint sealants should only be applied on viable surfaces and grounds in accordance with manufacturer specifications.
    - If applicable, the ground surfaces should be primed per manufacturer specifications.

Closed-cell round cords should be applied as backfilling to create the optimal sealant cross-section.

External to Internal Wall Connection

Notes:

- For exterior walls and building partition walls made of masonry, a plaster layer must usually be applied to create sufficient airtightness. In areas that will no longer be accessible later, e.g., front wall installation, risers, and downspouts, a smooth finish is applied.
- The plaster layer/smooth finish is fully and seamlessly applied all the way to the corner area.
- Note that covering unplastered masonry with drywall does not provide sufficient airtightness.

Fig. F.25 Outer wall, building partition wall: (1) interior plaster

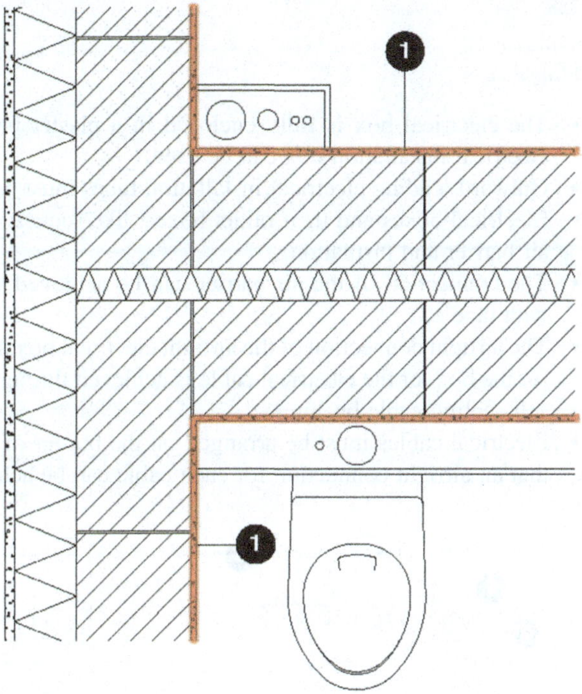

Penetrations

Notes:

- Cables/lines will penetrate the airtight layer.
- The connection of the air barrier to the cable goes through the cable sleeves.
- If done carefully, the connection can also be achieved with the suitable adhesive (e.g., tape with plastic substrates).
- Electrical cables must be arranged under one another and on the border of the building components so that an airtight connection for each cable is possible.
- The use of elastic, self-sealing cable sleeves can allow cables to move and carry electricity without ruining the airtight connection.

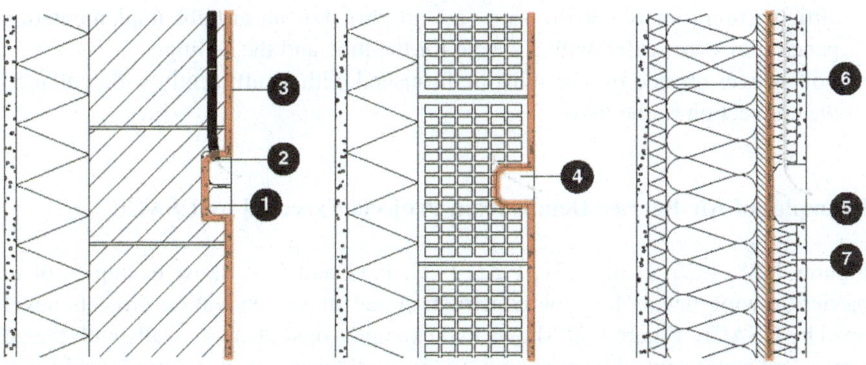

Fig. F.26 Electrical conduits: (1) electrical installation box, (2) sealing plugs, (3) plaster, (4) airtight electrical installation box, (5) installation housing, (6) air barrier, (7) installation level

Notes:

- The electrical box is fully enclosed in a plaster bed. Alternatively, an airtight electrical installation box can be used.
- Outward-leading electrical installation tubes must be sealed.
- Electrical boxes and their outlet boxes, like sockets, may be installed behind the air barrier and insulation.
- The connection of the air barrier can be achieved with self-bonding sleeves or with tape.
- The airtight connection of the airtight electrical box, the airtight electrical housing for the box, or the electrical cable is achieved through the compression of elastic, self-sealing cable bushing.
- Electrical cables must be arranged on the border of the building components so that an airtight connection for each cable can be achieved.

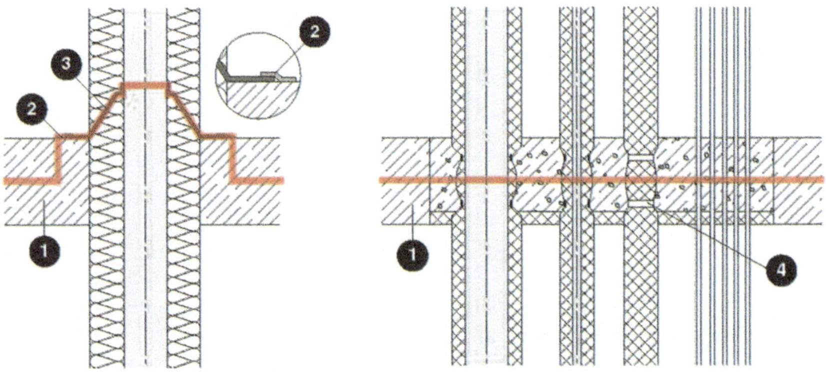

Fig. F.27 Pipe penetration: (1) ceiling, (2) tape, (3) sleeve, molded part, (4) cable tie

Notes:

- Leave sufficient gaps between bordering building components to enable the appropriate connection.
- Mineral wool sleeves for penetrations through the fire compartment are not airtight (fireproof is not the same as airtight). For an airtight implementation, penetrations are sealed with a sleeve for the tube and the ceiling.
- Foil sleeves created on-site must be furnished with a sufficiently wide collar for the connection to the tube.

Examples of Air Barrier Details from Projects Executed by USACE

Figures F.28, F.29, F.30, F.31, F.32, F.33, F.34 and F.35 show examples of air barrier drawing details for new construction and major renovation projects developed by USACE. Figure F.28 shows two examples of stud cavity walls with the air, vapor, and water control layer shown. In this particular case, it is emphasized for air barrier clarity. On the left, a continuous plane for the air barrier runs up the wall,

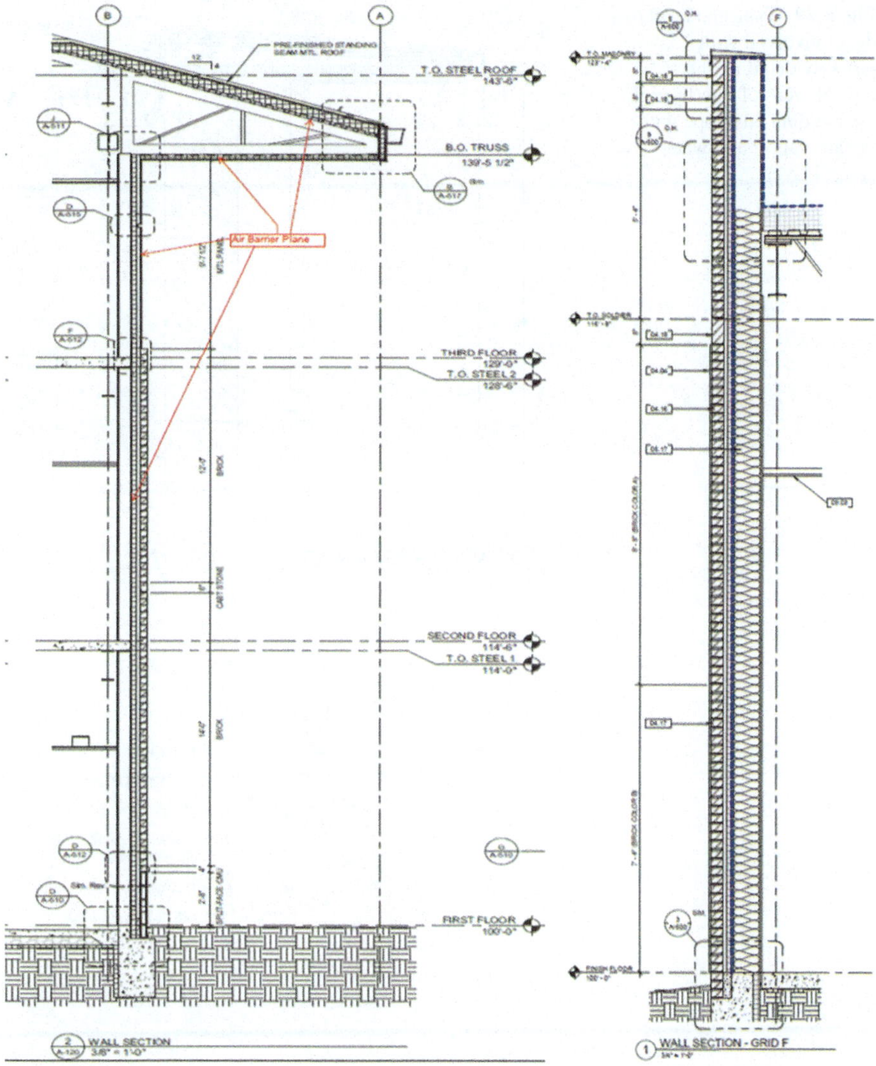

Fig. F.28 Two examples of stud cavity walls showing the air, vapor, and water control layers

across the soffit, and interfaces at the fascia with the roof water proofing layer. On the right, a continuous plane for the air barrier runs up the wall, and interfaces at the top of the parapet with a membrane roof system. Project drawings will identify critical locations of both interfaces, with supplemental pages to provide necessary details.

Figure F.29 shows two examples of the type and level of detail that is involved with properly detailing the critical areas of interface of the building envelope with the air, vapor, and water control layer.

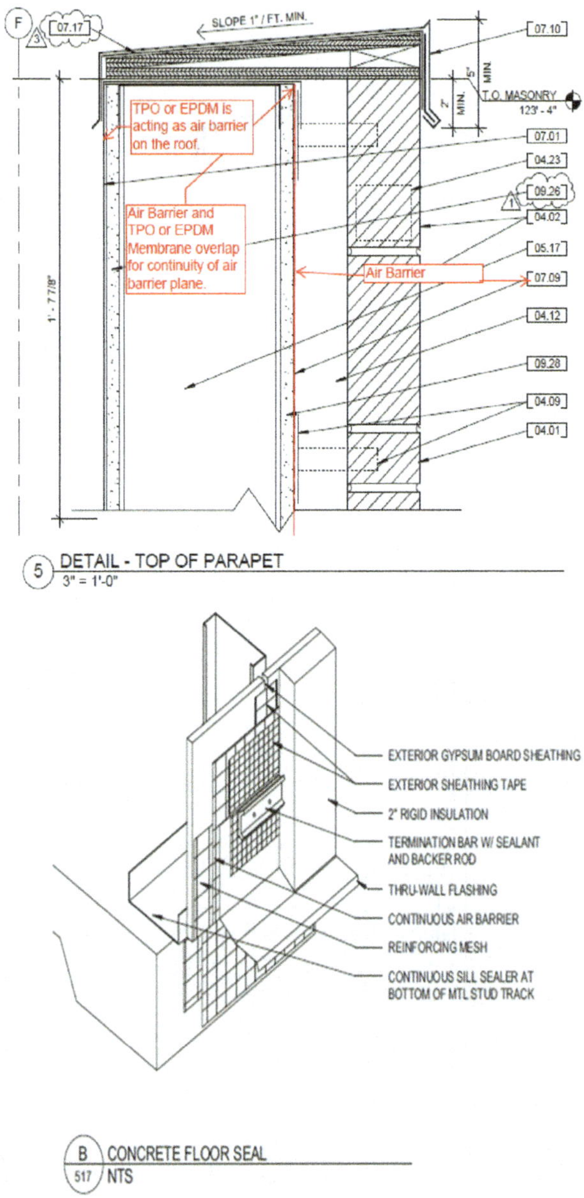

Fig. F.29 Type and level of detail involved with properly detailing the critical areas of interface of the building envelope with the air, vapor, and water control layer

Precast Concrete Walls

Figure F.30 shows two sections for a precast concrete wall assembly. The major advantage of precast concrete walls is that the concrete material provides a complete air, vapor, and water barrier. Details at critical interfaces are illustrated separately.

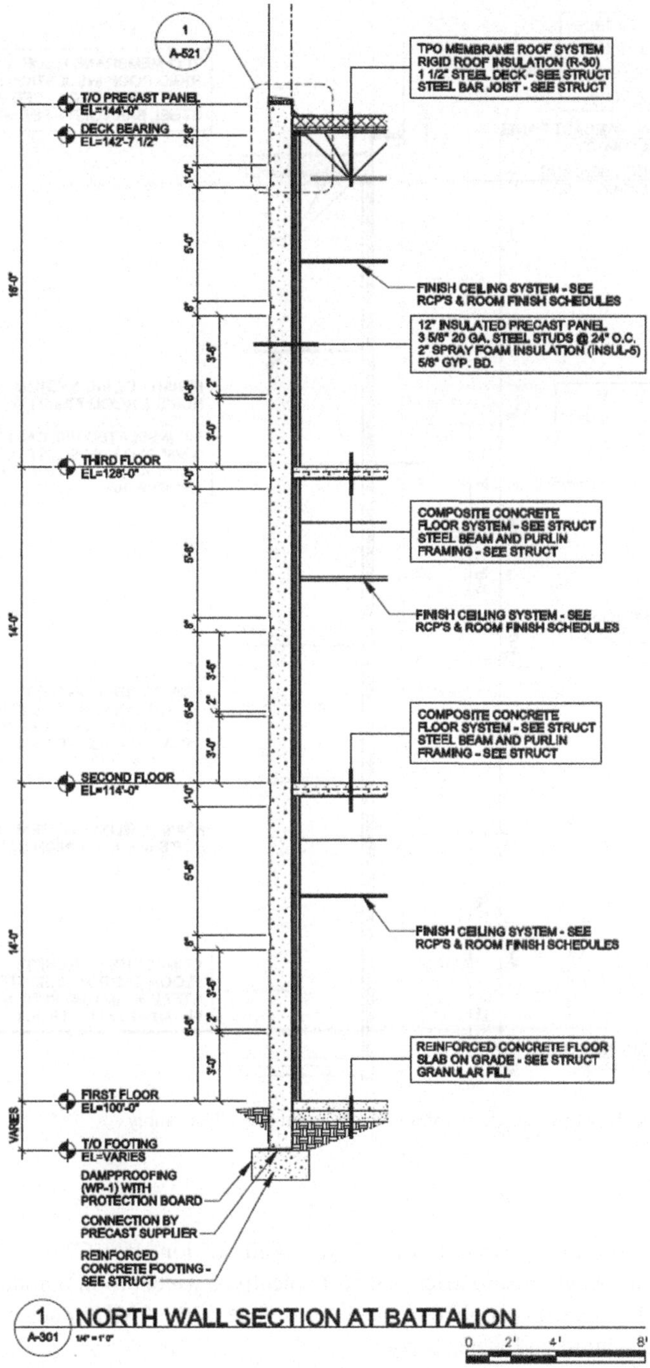

Fig. F.30 First of two sections for a precast concrete wall assembly

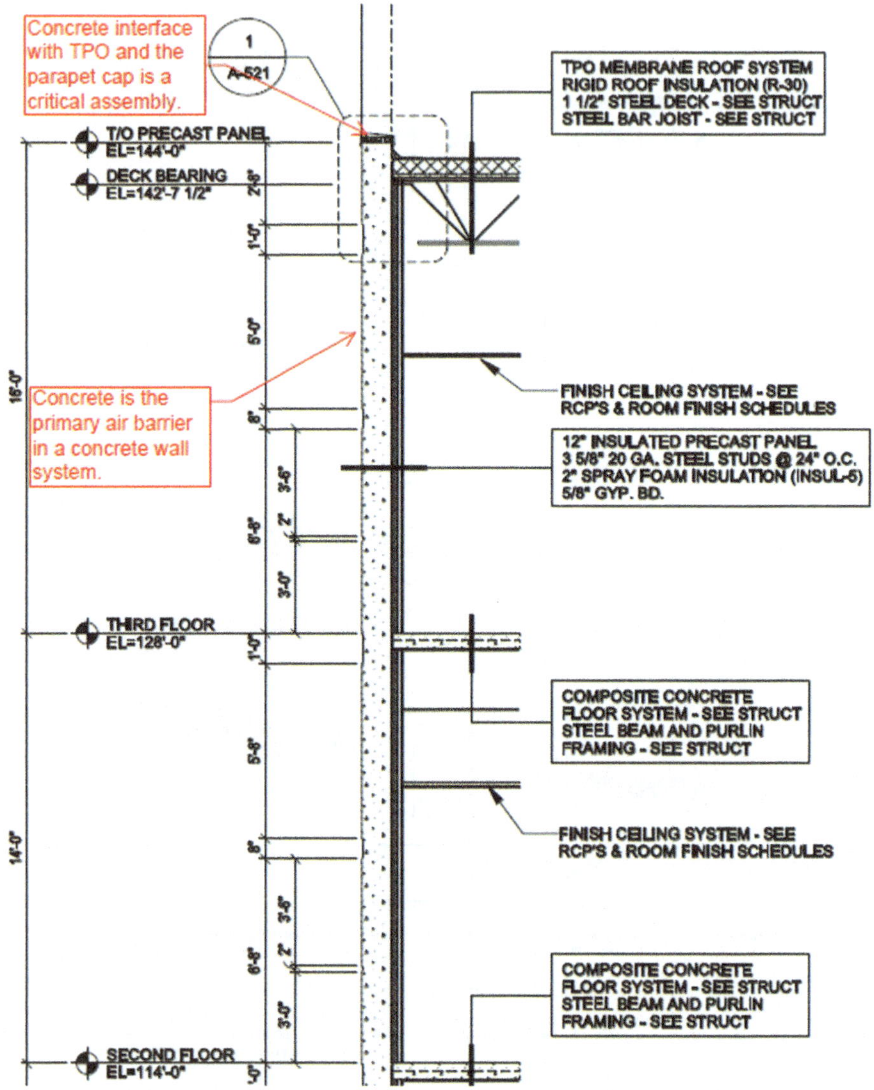

Fig. F.31 Second of two sections for a precast concrete wall assembly

CMU Walls

Figure F.32 shows an example of a CMU wall section. If a CMU wall is fully grouted, it can act as an air barrier as it is. Typically, a water proofing material (an air barrier rated material) is applied to the exterior face of the CMU under the façade to act as an air, vapor, and water barrier.

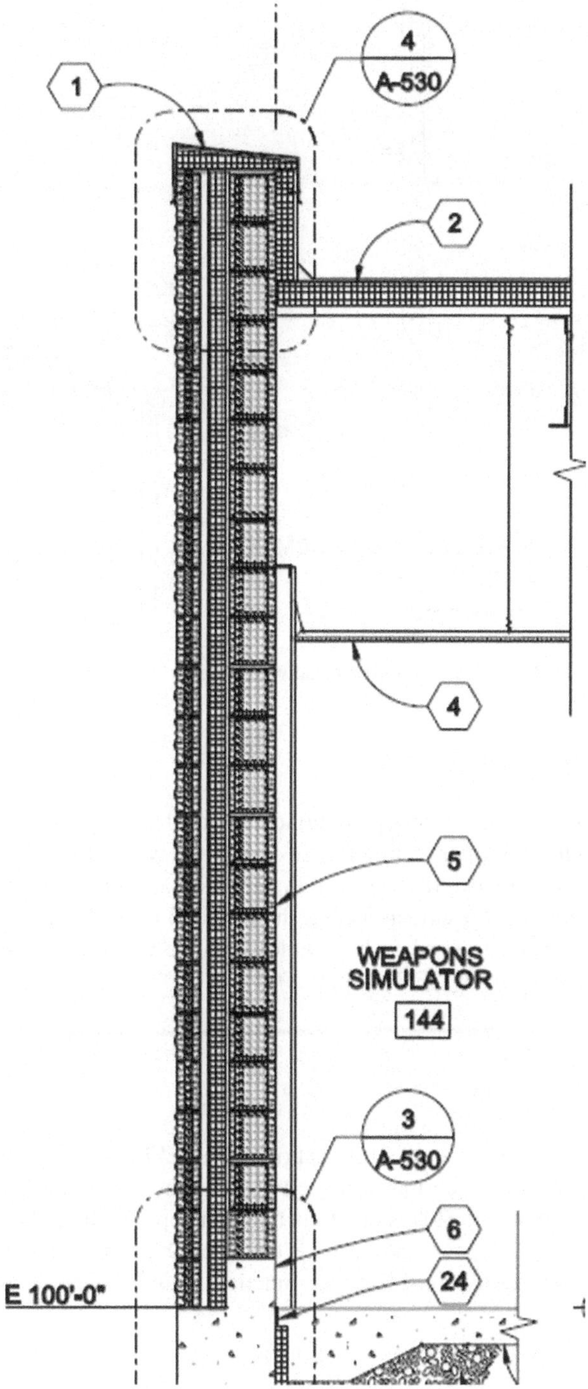

Fig. F.32 CMU wall section

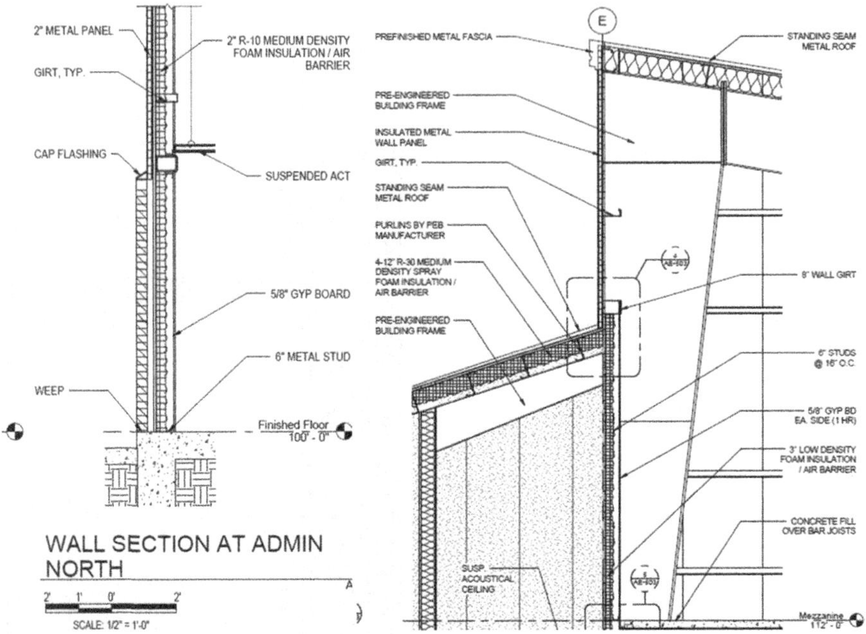

Fig. F.33 Pre-engineered metal wall systems

Pre-engineered Walls

Figure F.33 shows pre-engineered metal wall systems, which are the most difficult to ensure proper sealing for air, vapor, and water. Figure F.33 (left) shows a section of a pre-engineered metal wall system with an SPF application as the air control layer in this specific case as well as the insulation. In these assemblies all of the major points of leakage need to be detailed and the call out details for these critical locations shown in Fig. F.33 (right).

EFIS Walls

Figure F.34 shows the sections of an EFIS wall system. In EFIS wall systems, the most common material used and the air, vapor, and water barrier is a fluid-applied membrane. This is due to is uniform application and the need for back drains in the EFIS system to prevent ice damage. Other materials may be used, but need to be specifically considered for function in an EFIS application.

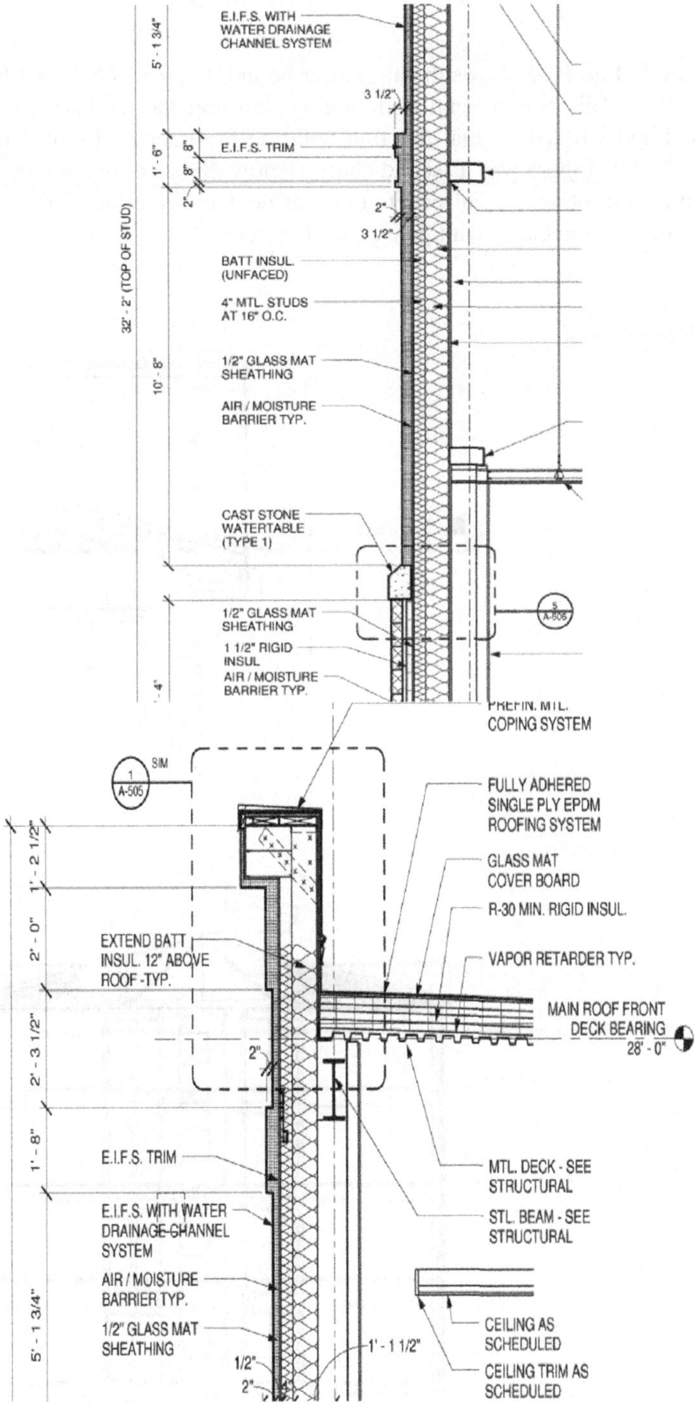

Fig. F.34 EFIS wall system

Roofs

The red line in Fig. F.35 shows the air barrier boundary for this building below the roof assembly. This was a ventilated roof system and the air barrier plane was constructed and worked in dual function with the fire-rated and sealed lid in this building. Note that there was a second ceiling below this layer into which all of the light fixtures and other typical penetrations of devices were built. This is a very functional option especially with precast wall systems.

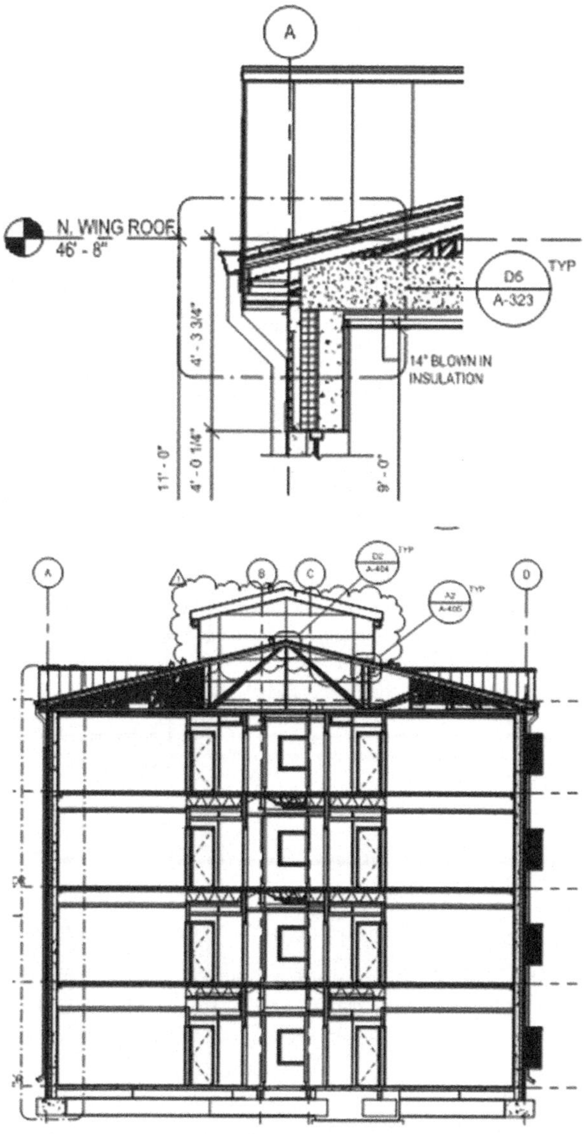

Fig. F.35 Air barrier boundary below the roof assembly

Minor Renovations and Energy-Focused Projects

Before starting a retrofit project, a building enclosure assessment must be performed, to determine the existing paths of air leakage. The results must be documented on drawings of the building and by photographs of existing conditions. Open louvers, conduit penetrations, and elevator or lift machine room vent shafts or like features should be identified and inspected or tested to establish an overall condition of the envelope. A full envelope test is often a good starting point for establishing a scope for a potential renovation or retrofit project, particularly if there is no existing air barrier within or as a component of the building envelope. This test can be performed in accordance with national standards, e.g., ASTM E 779 Standard Method of Determining Air Leakage Rate by Fan Pressurization as well as ASTM C 1060 Standard Practice for Thermographic Inspection of Insulation Installations in Envelope Cavities of Frame Buildings in the United States. Other diagnostics such as using a smoke tracer, either a handheld unit or theatrical smoke fog machine, can be helpful in locating and determining the magnitude of leakage in a particular location. Air leakage testing around openings can also be performed using ASTM E 1886, Standard Practices for Air Leakage Site Detection in Building Envelopes and Air Barrier Systems.

(continued)

Preparing the building for an airtightness test should follow air barrier test standards (e.g., USACE/ Air Barrier Association of America's airtightness test protocol) while using the smoke tracer; infrared thermography can also be helpful in determining existing leakage problems.

The DER of an existing building can include one of three strategies in the retrofit of airtightness:

1. Apply an air barrier to the exterior existing cladding during a re-clad of the building.
2. Apply an air barrier to the interior of the building during a gut rehabilitation while maintain the building exterior.
3. Individually seal existing leakage areas during partial renovation.

This section shows common areas of concern and recommends sequence of building enclosure tightening.

1. Seal the top of the building

Roofs

Low slope roofs are relatively easy to make airtight. During reroofing, a peel-and-stick air/vapor barrier can be installed on the structural deck or on an underlayment board. Peel-and-stick is preferable due to self-sealing characteristics of the membrane to screw penetrations. These areas should be attended to in roofs:

1. Attics
2. Roof/wall intersections and plenum spaces
3. Mechanical penthouse doors and walls
4. HVAC equipment penetrations through the roof
5. Other roof penetrations, such as plumbing vent stack

Component	Roof/wall intersection
Location	Top of building
Examine	Check for passive and direct venting to the exterior in this room. If there is venting, no measures should be taken. If no ventilation is found, check for gap between roof and wall intersection. If the exterior walls are a metal panel system, check the joint between the wall and curb for leakage
Significance	Many buildings use the space between the roof deck and the suspended ceiling as a return air plenum. This makes the infiltration much more significant and causes more energy loss
Test Method	Check for visual indicators (cob webs, dirt streaking). Use a smoke pencil if access to the joint is available. Check curb to wall joint in the same manner (if required). Are there dirt streaks or cob webs by the joint? Is there mortar, insulation, or boarding at the joint? If there are cracks around mortar, dirty insulation, or joints around the boarding, then the joint will be leaky and requires sealing
Solution	Seal roof-to-wall intersection (with foam if no combustion equipment is present; fire-rated caulking if there is combustion equipment). Seal vertical joints in wall panels if present. Seal curb to wall (typically two lines). Seal holes by posts as required
Materials	Small gaps in visible areas such as classrooms with no suspended ceiling will require caulking. Slightly larger gaps will require one-part foam. Gaps approximately 1 to 4 in. will require two-part foam. Larger gaps will require a board

(continued)

Component	Roof/wall intersection
	(drywall or other type of rigid material; fire-rated material may be required). Joints and perimeters will require two-part foam
Quantify	Lineal feet of roof-to-wall, wall to curb, and vertical joints. Specify material to be used
Notes	Note the access to the joint. (Stairwells and duct work will prohibit access.) Note height of joint. If it is over 20 ft, a lift may be required, which is an additional line item cost
Crackage Size	1/8 to 1/16 in

Component	Roof/wall intersection
Location	Top of building
Examine	Above ceiling tiles on top floors of buildings with flat roofs. Visually see and access the underside of the roof deck (metal fluted, tectum, wood, or others) at exterior walls or at roof level changes. This is done from below. Is there a gap between wall and underside of roof deck? Is there a blocking (fiberglass or neoprene pucks or mortar)? Is this open to a soffit beyond the exterior wall?
Significance	This gap is at the top of the "chimney" (in chimney/stack effect) and will be a significant contributor to air leakage
Test method	Check visually and with smoke pencil when access to the actual joint is available. Look for cobwebs, dirt streaking, etc. Check all ages of construction as well as where roof level changes are located. This is all done from below the roof deck. Check that access is available. Are there dirt streaks or cob webs by the joint? Is there mortar, insulation, or boarding at the joint? If there are cracks around mortar, dirty insulation, or joints around the boarding, then the joint will be leaky and require sealing
Solution	Seal the joint between roof deck and wall. (If metal and above suspended ceiling, use foam. If exposed metal such as in a gym, one-part foam or caulking required.) Large gaps may require a backing (fiberglass or some type of boarding such as a Celotex-type product).
Materials	Small gaps in visible areas such as classrooms with no suspended ceiling will require caulking. Slightly larger gaps will require one-part foam. Gaps approximately 1 to 4 in. will require two-part foam. Larger gaps will require a board (drywall or other type of rigid material; fire-rated material may be required). Joints and perimeters will require two-part foam
Quantify	Lineal feet as well as height of joint to be sealed (14 or 24 ft high). Costs vary as well as access requirements (such as lifts). Count number of lines required to seal the location.
Notes	Note the access to the joint. (Stairwells and duct work will prohibit access.) Note height of joint. If it is over 20 ft, a lift may be required, which is an additional line item cost.

Component	Roof/wall intersection
Location	Floors below top floor
Examine	Some buildings have continuous curtain walls or have a slab to exterior wall joints between each floor. This can be identified by lifting ceiling tiles at the exterior walls on a few floors below the top floor

(continued)

Component	Roof/wall intersection
Significance	Stack effect and smoke transfer
Test method	Check for visual indicators such as cobwebs and dirt streaking. On these floors, you may see directly into the floor above. This contributes greatly to stack effect. Smoke pencil test will also demonstrate this effect. Look for visible gaps and air movement with smoke pencil
Solution	Seal between slab and exterior wall or curtain wall. Foam may not always be the best-suited material (if curtain wall or large gap). Foam board or other type of fire-rated boarding may be required to bridge the gap
Materials	Two-part foam, 1-part foam or caulking
Quantify	Note lineal footages. Note size of joint to be blocked and sealed, which type of product to use, and which floors and locations on each floor
Notes	Note accessibility and type of material. Also note the visibility of the joint after completion (aesthetics)
Crackage size	1/8 to 1/16 in. per foot

Roof wall joint in need of sealing

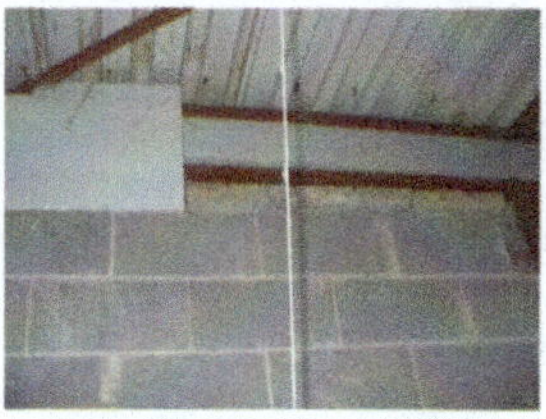
Roof wall boards not effectively sealed

Roof/wall above ceiling tiles (foamed)

Other Roof Wall Before/After Photos

Testing roof/wall intersection

Testing roof/wall once section completed with air sealing foam tape

(continued)

Roof/wall intersection before work	Roof/wall intersection air sealing completed with polyurethane foam

Roof/wall with cementitious wood fiber deck with purlins on CMU wall above ceiling tiles (foamed)	

Seal the Bottom of the Building

Fig. F.36 Loading dock levelers and doors can be a significant source of infiltration

Defined as "the ground floor and anything below grade." It is typically a unique area of the building.

1. Exterior soffits, canopies, and ground floor access doors
2. Underground parking access doors and elevators
3. Exhaust and air intake louvers and vents
4. Pipe, duct, cable, and other service penetrations into core and shafts of building
5. Sprinkler head penetrations, inspection access doors, and hatches
6. Seal core wall to floor slab
7. Residential crawl spaces and penetrations, vented or unvented
8. Loading docks (Fig. F.36)

Seal the Vertical Shafts

1. Stairwell fire doors
2. Fire hose cabinets
3. Plumbing, electrical, cable, and other penetrations within service rooms
4. Elevator rooms: cable holes, door controller cable holes, bus bar openings
5. Garbage chute perimeter and access hatches
6. Hallway pressurization grille perimeters
7. Smoke shaft access doors
8. Elevator shaft smoke control grilles
9. Service shafts

Conduit risers unsealed to floor	Duct and pipe sleeve unsealed and	
		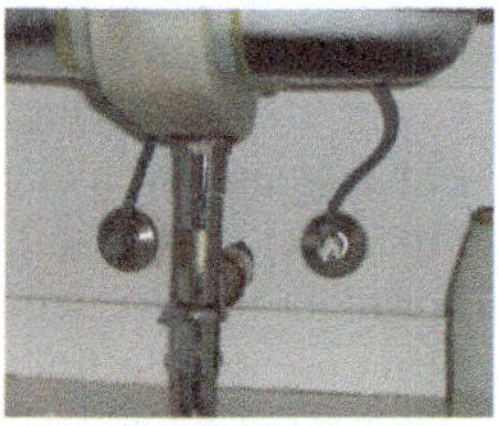
Electric bus bar riser unsealed	Penetrations such as hose/extinguisher cabinets can have penetrations into shafts	Plumbing pipe and plumbing accessory penetrations into plumbing chases

Seal the Exterior Walls

Sealing of the exterior recladding can be an easy retrofit strategy for airtightness.

10. Air barrier applied to exterior of existing walls. Self-adhered membrane being installed on exterior wall area in a retrofit of an existing school. Surfaces must be clean, smooth, and dry and free of any foreign matter that would affect the performance of the installed system.

Fluid-applied air barrier installed over an existing exterior wall area and insulated with medium-density closed-cell spray polyurethane foam.

11. Air barrier applied to interior of existing walls. Medium-density SPF spray polyurethane foam is being installed as an air/vapor and thermal insulation barrier. The interior walls are framed out after the SPF is installed on the interior of the exterior wall surface areas. This greatly reduces thermal bridging and the SPF air seals the exterior walls in the process.

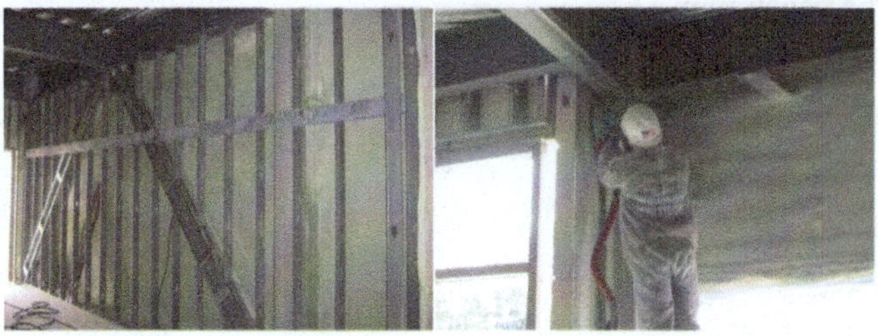

12. Sealing individual leakage pathways with self-adhered membranes.

a. Step 1: Apply the manufacture's approved sealant at the area where the penetration protrudes through the wall.

b. Step 2: Apply manufacture's approved primer/adhesive to the wall in the approved method and allow to flash off or cure according to the climactic conditions for that day. These primers are required on all substrates that are going to have a sheet rubberized asphalt membrane installed. If a butyl membrane is being installed, primers are not required in some cases.

c. Step 3: Sealing round penetrations by cutting self-adhered membrane to fan out around the penetration.

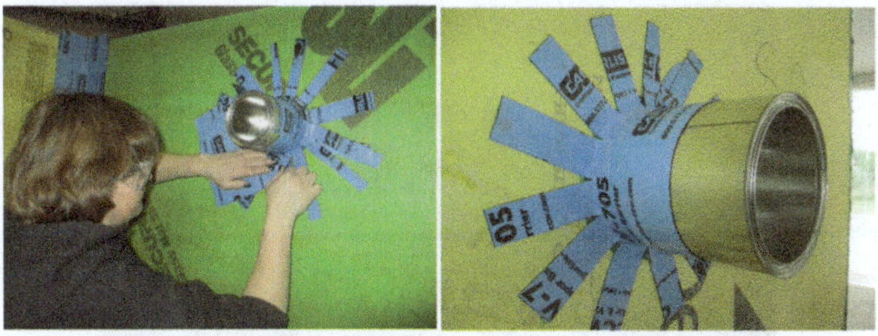

d. Step 4: Install a sheet of self-adhering membrane over the previously installed fanned out membrane. This can be done using two overlapping pieces or one continuous piece over the detailed area.

e. Step 5: When the area has been covered with the self-adhered membrane and rolled with a membrane roller, the installer can then apply a bead of sealant over all non-factory areas that have been cut or where there are any chances of air or water infiltration into the assembly. The sealant then needs to troweled to a smooth finish.

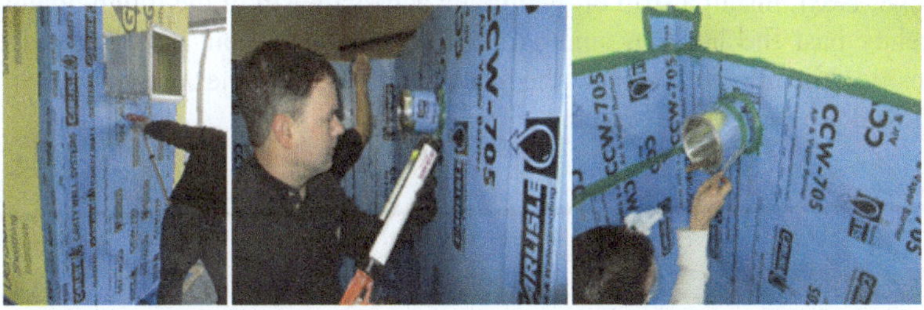

f. Step 6: Detailing around windows and doors. Prime all surfaces as previously mentioned. Then install a piece of membrane into each corner of the rough opening with one half of the membrane folded onto the wall area as shown.

g. Step 7: Then install a bow tie-shaped piece of membrane into each corner of the rough opening. This piece is installed to prevent water or air from entering into the assembly at the small hole that remains in each of the corners of the rough openings.

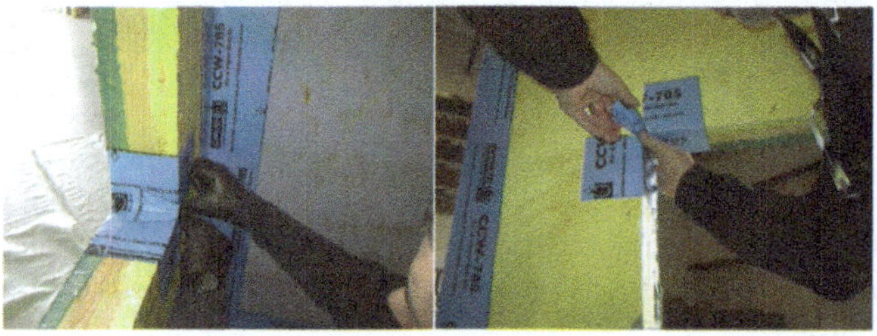

h. Step 8: Next, install a strip of membrane across the sill of the rough opening that extends past the jams by approximately 4 ½ in. if you are using 9-in.-wide transition membrane. Then cut the membrane at each corner of the rough opening and fold the membrane onto the sill.

i. Step 9: Install strips of transition membrane on each of the jams, starting at the top of the sill. Cut horizontally across the membrane from the edge of the rough opening and fold into the rough opening.

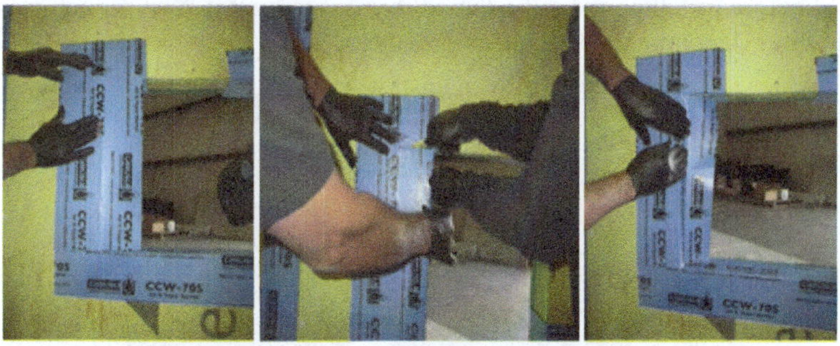

j. Step 10: Now install a piece of transition membrane across the head of the window. Then cut vertically to the inner edge of the rough opening and fold the membrane up into the underside of the head area of the rough opening.

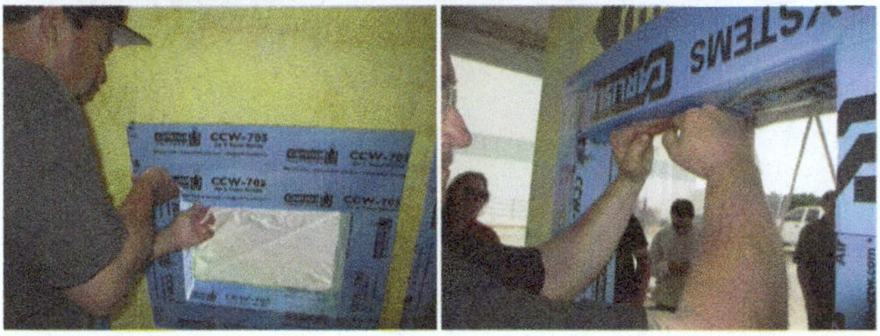

k. Step 11: These installed pieces should now be rolled with a membrane roller to properly adhere them to the substrate.

l. Step 12: All seams should be sealed with the manufacture's approved sealant if the wall is now going to be sprayed with a fluid-applied material. If the wall is going to have a complete self-adhered application do not install any sealants until the entire area has had the self-adhered membrane installed. This would prevent any sealants that may be installed in an area where you would have to wait until it was cured before installing the self-adhered membrane over it. You cannot install any self-adhered membrane over uncured sealants.

Brick masonry that is too deteriorated to repair	Air barrier applied to the existing exterior wall

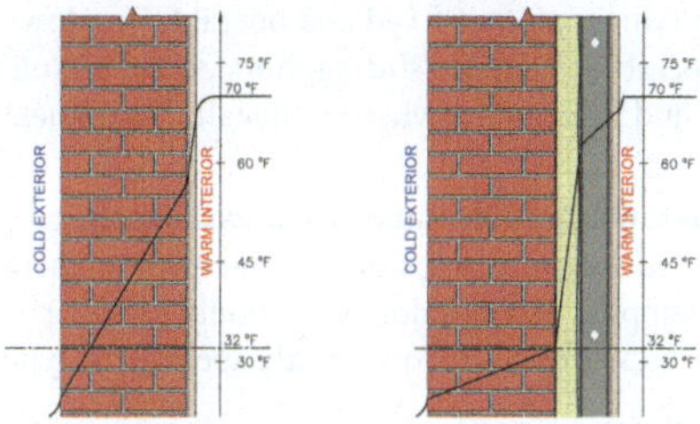	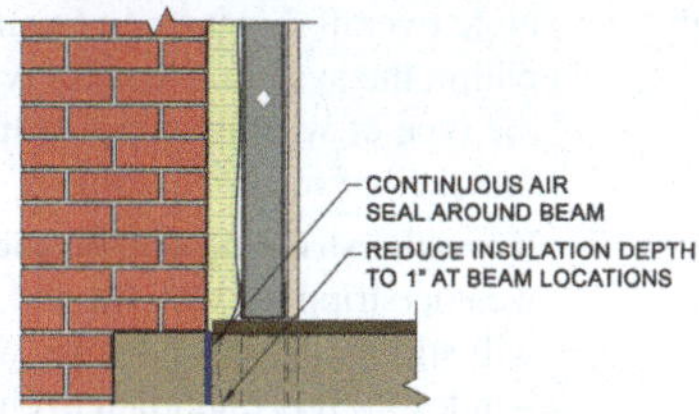
Adding interior air barrier and insulation results in a wall that is colder in winter. Condensation within the brick wall must be controlled so that freeze-thaw cycles do not harm the softer brick that is in the middle of the wall	Structural members penetrating the wall are vulnerable to added wetting. Reduce insulation thickness around existing structural members and seal around the member

Component	**Windows**
Location	Throughout building
Examine	Check windows for perimeter seals, glazing (seal between glass and frames), and if they are operable. There are many different styles (awning, hopper, single/double hung, casement, wood, aluminum, vinyl, steel), and each of these has different points of failures such as weatherstripping, hardware, seals at trim to wall, seals to perimeter, or mullions between each window. In curtain wall and aluminum window systems, the cavities in mullions and frames can be hollow. This creates thermal bridging and can result in ice and condensation buildup.
Significance	There are many windows in most buildings. There are several points on window systems that can be "drafty," which will be a large contributor to comfort concerns
Test method	Check windows for weatherstripping. If there is any, is it a fin seal or something else? Can the existing weatherstripping be replaced? Visually inspect by rattling the operable window to see if there is significant movement between the track and sash. Test using a smoke pencil and look for dirt streaking around trim to wall. Check window perimeters for gaps or cracks (include trim to wall and trim to window joints as well as mullions between windows in banks of windows). Visually inspect and use a smoke pencil. Tap on glass to see if the pane rattles; if it does, then the glazing (the seal between the glass and frame) has failed. Check on window mullions and frames to hear if the cavity is hollow. Check for signs of moisture (stains, etc.). On double hung windows, check the upper window to see if it has dropped

(continued)

Solution	Single/double hung windows can have either fin seal replaced if it is a newer unit. Older-style windows may have sweeps installed on the sash or operable window (but permission from customer is required due to aesthetics). Double hung windows are the same, but if the upper window has dropped, installing a block to hold the upper at the top as well as sealing the upper operable window (sides and top) is required. In sliding, awning, and hopper windows, a fin seal product can usually be used to retrofit (but it is critical to check and note the slot or track where the existing weatherstripping is as T-slots typically come in two different sizes.) Many are only L-slots and existing weatherstripping is glued in place. Some windows require a cast-in-place methodology of weatherstripping (steel casements). This must be noted
Materials	Fin seal for aluminum windows; V-strip, silicone, latex caulking, and other products can be specific to certain window manufacturers
Quantify	Note overall dimension of window and dimension of fixed and operable windows within the system. Note the type of operable window (sliding, hopper, etc.). Note the type of weatherstripping required and location of where to install. Note lineal footages of material required
Notes	If weatherstripping double sliding windows, only the inner windows require weatherstripping with fin seal. (There are weeping holes on the outer set, so there will still be air leakage after weatherstripping the exterior set of double sliding windows.) It is important to check the track accurately to correctly specify the type of material to use
Crackage size	Glazing 1/128 in. perimeter sealing 1/32 to 1/64 in. crack, w/s 1/16 to 1/32 in. crack

Check corner gaskets, which may shrink and leak air.	Air leakage between glass and frame	Double hung upper windows can fall and need to be held up securely and sealed
	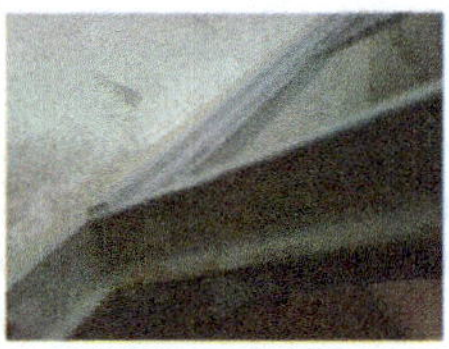	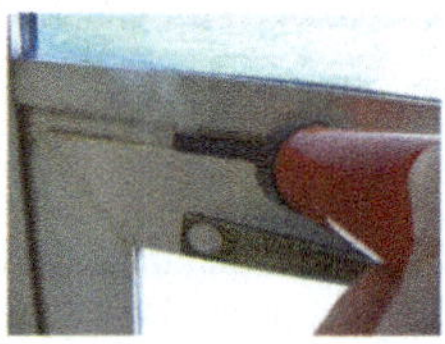
Fin seal will work well in this slot	Rusted lip area is where cast-in-place will go on steel casement window	Aluminum hopper – If no slot or track, do not weatherstrip

Component	Exterior doors (main, sliding)
Location	All exterior entrances/exits
Examine	Check all exterior doors, including stairwell, entrance, maintenance, shipping receiving, courtyards, etc. Check type of doors (commercial, sliding, etc.). Check for existing weatherstripping and whether it is good or bad
Significance	Direct connection to the exterior temperatures
Test method	Visual inspection and smoke pencil test. Do doors require full retrofit weatherstripping or just bottom sweeps or vertical sweeps between double doors? Commercial doors require fin seal replacement. Look for daylight under and around door and dirt accumulating at jam
Solution	Weatherstrip door and seal glass and perimeters as required. If door has handle, DXL may be required due to location of handle and style of handle. A round knob requires DXL where a lever may allow for DX
Materials	DX(L), sweeps (3 and 7 ft), V-strip, caulking
Quantify	Note the number of doors. (A double set of doors is counted as two doors). Note the dimension of oversized doors. Note the type of material to be used (include color if required). Note locations of doors on floor plans
Notes	Note that certain type doors are offset hinges. Sliding doors can also require specific materials along with centers between double doors
Crackage size	1/16 in

Before

After

Component	Soffits (if accessible)
Location	Soffits are typically found at the first floor, but can often be located near balconies on upper floors
Examine	Between the exterior wall and the roof deck, there can be a gap where the overhang of the building is located. This allows conditioned air to enter the area and exit as well as allow unconditioned air to enter via the cracks and gaps found in typical soffits. Check above the vestibule and anywhere if there is an overhang at an exterior wall of the building. There may be pipes and services running through these, so investigate by asking questions to ensure that if this soffit is blocked and treated as unconditioned space; we do not inadvertently cause pipes to freeze. This generally is an issue when there is occupied space over the soffit.
Significance	Connection to the exterior; sometimes ceiling spaces are under negative pressure, which will pull OA into the building
Test method	Visually inspect to see if there is an effective air barrier between the soffit and the conditioned wall. Check to see if the exterior joints (top and bottom of soffit at outer wall) have not been sealed. Check to see if exterior light fixtures are affected. Look for visible daylight at soffit edges, lack of insulation on soffit floor, and cob webs around light fixtures
Solution	Ideally, boarding and sealing off the soffit directly above the exterior wall is recommended. Insulated boarding may be required depending on the height of the gap. If there is an existing blocking system (paper-faced batts or polystyrene), seal the joints
Materials	Rigid insulation, two-part foam, silicone
Quantify	Count lineal footages. Note height of gap to be blocked or amount of lineal footage required to seal perimeter of existing blocking system. Note locations. Inquire if fire-rated material will be required to bridge gap
Notes	Inquire if fire-rated materials are required. Note if it is more practical to seal the joints in the soffit from the exterior (underside) and to seal light fixtures
Crackage size	1/8 in. of lineal footage of soffit

Soffits are often blocked, but seldom sealed. Often boards have fallen down. This shows both	Seal all joints in this soffit using a Dymonic® color-matched product. Need outside quotes for this type of work	Typical soffit: either block at exterior wall and seal or seal joints from exterior

(continued)

Images of open soffits to the conditioned space

Before

After. Typical soffit: usually can be blocked vertically from interior of building and sealed when access is not possible for sealing joints from exterior

Component	**Service penetrations**
Location	Exterior walls
Examine	Check exterior walls for penetrations such as fire suppression systems. Check for seals around these. When found, look from the interior side of the building for these penetrations.
Significance	Connections from exterior or vented rooms (unconditioned spaces) can contribute to stack effect
Test method	Use a smoke pencil and check for daylight, dirt streaking, and visible gaps around penetrations
Solution	Seal perimeters of these penetrations from interior. Exterior side can be sealed for water reflection, but energy savings are seldom the reasons for sealing from the exterior
Materials	Fire-rated caulking, polyurethane foam, silicone
Quantify	Note dimension and number of penetrations. Note locations
Notes	High-temperature pipes will require specific caulking or materials. Note accessibility for the workers to work
Crackage size	0.015 sq ft per penetration or 1/16 in. (Use diameter of penetrations.)

	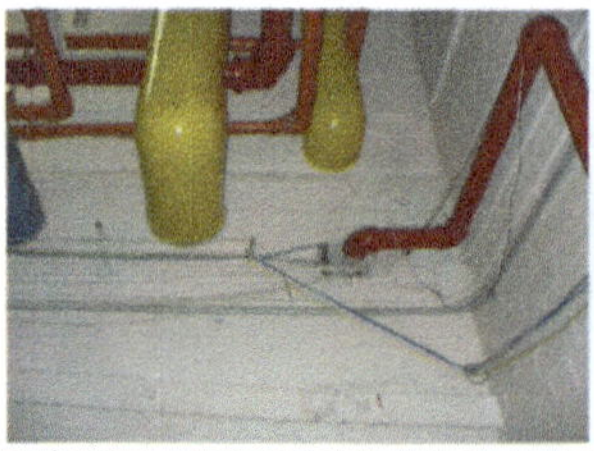	
Seal around service penetrations with fire-rated materials	Typical service penetrations	

		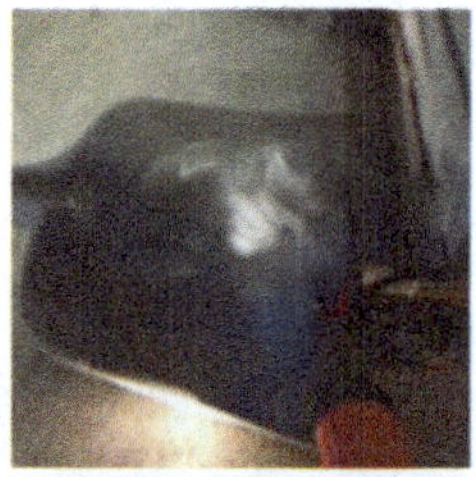
Before		
		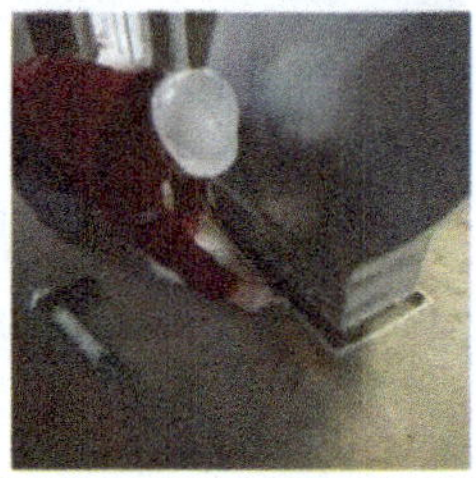
After		

Examples of Details of Possible Options to Seal the Various Penetration that a Building May Have

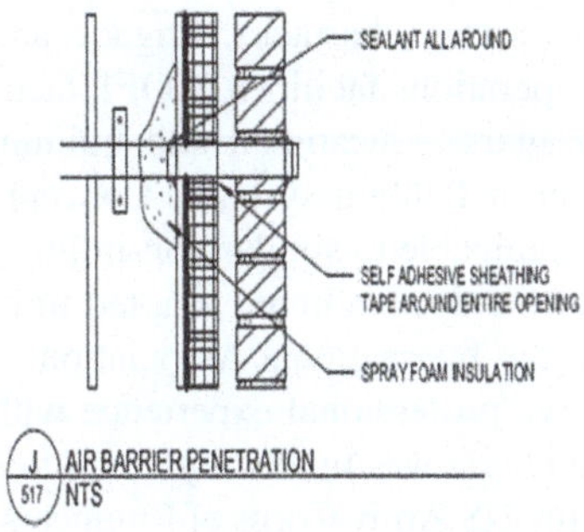

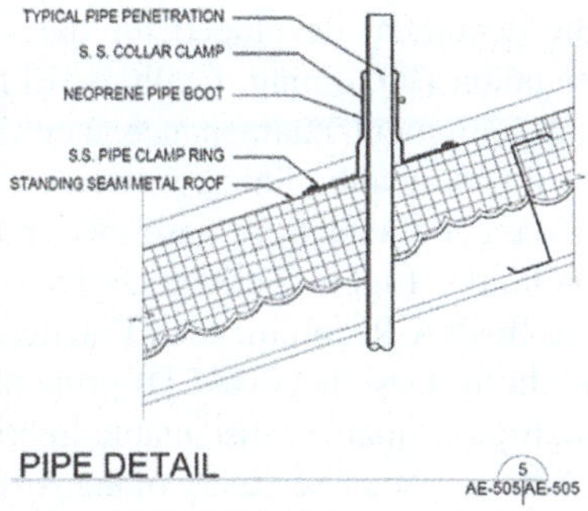

Compartmentalize the Building

Appendix G: Lighting Design Guide

The comprehensive "Lighting Design Guide" presented in this appendix is based on the document developed for the US Army facilities (e.g., barracks, brigade and battalion HQ, dining, facilities [EPDF], company operations facilities [COF], tactical equipment maintenance facilities [TEMF], hangars, educational and training facilities, health clinics, warehouses) and their exterior lighting systems. Concepts and recommendations described in this Guide are applicable to similar non-military buildings. Lighting criteria and design strategies for the document are adapted from the IESNA Recommended Practices, 10th Edition, the International Association of Lighting Designers (IALD) principles, and collective professional experience with high-performance, sustainable lighting design. The Guide was funded by the Office of the Assistant Secretary of the Army (IE&E) and the US Army Corps of Engineers.

Sustainable Lighting Design Approach

The lighting strategies outlined in this Lighting Design Guide are drawn from sustainable design principles. Providing sufficient functional lighting for visual tasks while maintaining comfortable visual environment is required for a lighting design to be considered successful. The goal of sustainable lighting design, and this document, is to satisfy these lighting design criteria using the fewest materials and least energy.

The three cornerstone strategies for sustainable lighting design that were used to create these guidelines listed in the box should be considered in every design.

Cornerstone Design Strategies

- Provide appropriate illuminance levels without over-lighting
- Use efficient lamps, ballasts and luminaires
- Reduce electric lighting usage with controls

More specific energy-saving lighting design tactics that help create visually comfortable, effective, and efficient lighted environments applicable to many space types in addition to those outlined in this document are also listed on the right.

Specific Design Tactics

1. Optimize architecture to provide daylight in frequently occupied spaces
2. Apply light-colored (high-reflectance) surface finishes as appropriate. Finish colors shall be functional, easily maintained and help hide soiling.

(continued)

3. Cluster similar tasks to improve lighting system energy efficiency
4. Locate luminaires close to tasks that require higher illuminance
5. Use linear fluorescent and LED luminaires predominately
6. Use high efficiency ballasts with appropriate ballast factors if using fluorescent sources
7. Use high-efficacy versions of lamps (e.g., 3100 lumen fluorescent T8)
8. Illuminate walls and ceilings to increase perception of brightness
9. Use daylight responsive lighting controls in frequently occupied spaces with daylight access
10. Use vacancy sensors in spaces with daylight access especially in private offices
11. Use luminaire integrated occupancy & daylight sensors with wireless grouping & granular dimming functionality for maximized energy saving strategy by offering light only when it is needed and where it is needed.
12. Use controls that offer luminaire wattage reductions or allow occupants to adjust max trim levels.
13. Use occupancy sensors in spaces without daylight access control lighting with astronomic time-clocks for building-wide energy conservation

Retrofitting

Considerations

This guide is intended as a Lighting Design Guide for major renovation projects, not just a lighting a retrofit guide. All of the strategies contained within are applicable both to new and existing buildings, but may not always be the most cost-effective lighting relighting strategy use with minor renovation.

A one for one technology swap to keep first cost low is often not the most effective improvement for energy savings and simple payback. As with all technological advancements, some are more cost-effective than others, and it is important to perform a financial analysis to determine if there will be sufficient savings to justify the project. Initial simple calculations that consider differences in technology wattage as estimates of energy savings compared to cost are useful. However, the variables of possible changes in operational hours, direct control savings, and any possible maintenance savings should also be incorporated into the analysis for final decisions. It is often control applications and related operational changes that provide the majority of savings.

Ensure effective lighting quality – It is critical to ensure that the proposed new lighting technology offer an effective match to the needs of the space or area. This includes maintaining appropriate light levels as well as ensuring user satisfaction of the lighting environment.

In some cases spaces may be over-lit, and reductions to proper levels should be part of the proposed change when energy savings are important. When considering changes, verify the expected use of the space as far into the future as possible to ensure that chosen project light levels will be appropriate.

Present and promote retrofit projects positively as new and more energy efficient. Lighting education is important so that occupants will understand that reducing energy consumption is important and they will be more accepting of any changes in the case of retrofits. The human visual system does not actively notice changes of up to 20% reductions in light levels when done gradually or when absent. Clean refreshed systems will appear brighter even if illumination and energy are reduced.

This guide provides best practice design guidance for lighting systems for deployment on military installations.

For more detailed information on lighting retrofits, please refer to Annex G-1 to this appendix.

LAMPS

Considerations

A variety of commercially lamp types are available, each with their own unique characteristics, which make some lamp types better suited for some applications. Important lamp performance properties include light output, power consumption, luminous efficacy, rated lamp life, lamp lumen depreciation, color temperature, color rendering, dimmability, and sensitivity to voltage and temperature.

Below is a list of key lamp types used in energy-efficient lighting systems along with recommendations for use.

Linear Fluorescent

These discharge lamps require a ballast to operate and are typically the most efficacious (in lumens per Watt) lamp available for general interior lighting requirements. Linear fluorescent lamps are available today (2010) that operate more efficiently, more quietly, and provide better color performance than those available 15 years ago or more. Linear fluorescent lamps have high efficacy (80–100 lumens per Watt), turn on and warm up virtually instantly, can be dimmed, have rated long lamp life (30,000–46,000 hours), and are relatively inexpensive. For these reasons, linear fluorescent lamps should be considered for the general lighting needs of nearly any interior space or application. Applications that call for small luminaires or precise optics, such as an accent light, often require the use of a different lamp type. With the products available today, 4 ft high-performance T8 lamps coupled with NEMA premium ballasts are the most energy-efficient and cost-effective type of linear fluorescent lamps – exceeding T5/T5HO lamp and ballast combinations. However, the smaller-diameter T5 lamps provide the advantage of greater luminaire efficiency in many cases.

Compact Fluorescent

These smaller, bent versions of linear fluorescent lamps provide many of the benefits of linear fluorescent lamps in a smaller form factor. Compact fluorescent lamps come in many shapes, sizes, and wattages. These lamps typically have good efficacy (50–60 lumens per Watt), turn on and warm up quickly, can be dimmed, and have a reasonably long rated life (10,000–16,000 hours). The efficiency of luminaires that use compact fluorescent is typically low compared to linear fluorescent and ceramic metal halide, due to the size and shape of compact fluorescent lamps. LED is replacing CFL in most applications.

Ceramic Metal Halide

Ceramic metal halide lamps are discharge lamps that are small enough to be considered a "point-source" for most architectural lighting applications, but typically cannot be dimmed and take several minutes to fully ignite. The size, shape, and efficacy of ceramic metal halide lamps make them excellent source for accent lighting or downlighting when instant starting or dimming are not required. Low-wattage versions of these lamps typically have good efficacy (55–70 lumens per Watt), cannot be dimmed, have excellent color rendering (80–95 CRI), and reasonably long rated life (10,000–12,000 hours).

Light-Emitting Diodes (LED)

LEDs are a solid-state electronic light source. The performance of LED equipment varies greatly, and the rapidly changing state of the technology can make this lamp type challenging to specify. Generally, LEDs have a long rated life (50,000+ hours) and are energy efficient, producing 50 to 100 lumens per Watt in many cases, and are dimmable. LEDs are often a good choice for applications where the luminaire will be placed close to the lighted surface (such as task lighting) or where dimming is needed. Designers should specify LED luminaires from only the top five manufacturers that have been photometrically tested in accordance with IESNA LM-79 and IESNA LM-80 methods.

Designers should also prefer the top five manufacturers that offer luminaire integrated controls with wireless grouping and granular dimming.

Technology Schedule

L01 Fluorescent 32WT8

Four-foot linear fluorescent lamp, nominal wattage not to exceed 32W, initial lumen output of 3100 lm or greater, color rendering index of 80 or greater, rated lumen maintenance of 94% or greater.

L02 Fluorescent 32WT8U

"U"-shaped fluorescent lamp, nominal wattage not to exceed 32W, initial lumen output of 2800 lm or greater, color rendering index of 80 or greater, rated lumen maintenance of 90% or greater.

L03 Compact Fluorescent Long Twin-Tube

Long twin-tube compact fluorescent lamp with double bi-pin base, nominal wattage of 36W or 40W, initial lumen output of 2900 lm or 3300 lm, respectively, color rendering index of 80 or greater, rated lumen maintenance of 90% or greater.

L04 Compact Fluorescent Triple-Tube

Compact fluorescent lamp in a triple-tube configuration; double bi-pin base; nominal wattage of 18W, 26W, or 32W; color rendering index of 80 or greater; lumen maintenance of 86% or greater.

L05 Compact Fluorescent GU24 Integrated Ballast

Compact fluorescent lamp with integral ballast and GU24 base, medium screw base is not acceptable, nominal lamp wattage of 13W, 18W, or 26W.

L06 20/39W Ceramic Metal Halide

Ceramic metal halide lamps with nominal wattage of 20W or 39W, color rendering index of 80 or greater.

L07 Fluorescent 17WT8

Two-foot linear fluorescent lamp with a nominal wattage not to exceed 17W and initial lumen output of 1400 lm or greater, color rendering index of 80 or greater, rated lumen maintenance of 94% or greater.

L08 Fluorescent 39WT5HO

Four-foot linear fluorescent lamp, nominal wattage not to exceed 39W, initial lumen output of 3500 lm or greater, color rendering index of 80 or greater, rated lumen maintenance of 94% or greater.

L09 ceramic metal halide 150wmh, ceramic metal halide lamp with nominal wattage of 150w, with mogul base and color rendering index of 80 or greater.

L10 Fluorescent 54WT5HO

Four-foot linear fluorescent lamp, nominal wattage not to exceed 54W, initial lumen output of 5000 lm or greater, color rendering index of 80 or greater, rated lumen maintenance of 90% or greater.

LED and LED Linear Replacement Tubes

Minimum 80% efficacy (80lm/W) and rated life of 50,000hrs or greater, based on IESNA LM-80 -2008 at 70% lumen maintenance, with wattage and lumen output as specified in luminaire schedule.

Recommended Lighting Power Density and Illuminance Values

Space type	Target illuminance	Target LPD
Common spaces		
• Conference room	40 fc	0.80 W/ft^2
• Corridor	10 fc	0.50 W/ft^2
• Dining	20 fc	0.60 W/ft^2
• Dishwashing/tray return	50 fc	0.65 W/ft^2
• Kitchen/food prep/drive thru	50 fc	0.65 W/ft^2
• Living quarters	5–30 fc	0.60 W/ft^2
• Mechanical/electrical	30 fc	0.70 W/ft^2
• Office (open)	30–50 fc	0.70 W/ft^2

(continued)

• Office (enclosed)	30–50 fc	0.80 W/ft^2
• Reception/waiting	15–30 fc	0.50 W/ft^2
• Restroom/shower	20 fc	0.80 W/ft^2
• Server room	30 fc	0.85 W/ft^2
• Serving area	50 fc	0.70 W/ft^2
• Stair	10 fc	0.50 W/ft^2
• Storage (general)	10 fc	0.50 W/ft^2
• Storage (dry food)	10 fc	0.70 W/ft^2
• Telecom/SIPRNET	50 fc	1.20 W/ft^2
• Vault	40 fc	0.70 W/ft^2
Training		
• Readiness bay	40 fc	0.75 W/ft^2
• Training room (small)	15–30 fc	0.70 W/ft^2
Vehicle maintenance		
• Consolidated bench repair	50 fc	0.60 W/ft^2
• Repair bay/vehicle corridor	50 fc	0.85 W/ft^2
Exterior		
• Exterior secured spaces (storage)	1–2 fc	0.07 W/ft^2
• Parking	20 fc	0.80 W/ft^2
• Roadway	30 fc	0.70 W/ft^2
Education		
• Kindergarten classroom	15–30 fc	0.70 W/ft^2
• Active playroom	30–50 fc	0.50 W/ft^2
• Staff lounge	15–30 fc	0.50 W/ft^2
Warehouse/receiving/issue bay	40–60 fc	0.80 W/ft^2
• Hangar		
• Maintenance bays	50–70 fc	1.0 W/ft^2
• Production control	25–30 fc	0.60 W/ft^2
Athletic		
• Gym	40–50 fc	0.70 W/ft^2
• Racquetball	50 fc	1.10 W/ft^2
• Combatives	50 fc	0.90 W/ft^2
Health		
• Resident work area	30–45 fc	0.40 W/ft^2
• Exam room	30–50 fc	0.60 W/ft^2

Ballasts/Drivers

Considerations

All discharge lamps, including fluorescent and metal halide, require a ballast to operate. Specifying an efficient ballast with the appropriate ballast factor, start method and controllability is critical for minimizing the connected lighting load, maintenance and energy use.

Start Methods

Fluorescent lamps are started using instant start or programmed start ballasts. Instant start ballasts use the least power to operate lamps, but abruptly start the lamp, which can shorten lamp life when the lamps are frequently switched. Programmed start ballasts start lamps more gently so that lamp life is not as adversely affected by frequent starting, but these ballasts typically require more power than instant start ballasts. The lamp life and energy effects are typically minor, but the ballast starting method should be chosen carefully in applications where lamp switching cycles are less than 15 minutes or greater than 3 hours. Luminaires that are left on for long periods of time should generally use instant start ballasts for energy savings, and those that are switched on and off many times per day should use programmed start for extended lamp life. Both instant start and programmed start ballasts are widely available for linear fluorescent lamps. However, compact fluorescent ballasts are typically only available as program start.

Metal halide lamps are manufactured to be started in one of two ways, probe start or pulse start. Specify electronic pulse-start ballasts and lamps whenever possible.

Controllability

Static output ballasts drive a given number of lamps at one specified ballast factor. These are fixed output devices and are less expensive than ballasts with variable ballast factors.

Multi-step ballasts can operate lamps at two or more ballast factors. This is often referred to as stepped dimming as the lamp light output can be dimmed by incremental steps. These ballasts provide a more visually comfortable environment compared to multilevel switching, which essentially turns off a number of lamps to save energy. Bi-level and tri-level are the most common and can be controlled by multiple power feeds to a single ballast or by control wires.

Dimming ballasts have the ability to operate lamps with a variable ballast factor so the lamps may be brightened or dimmed in a smooth fashion. The load of the ballast is reduced when lamps are dimmed, providing energy savings. With LED technology the cost of dimming drivers has come down substantially. Dimming drivers should be the default choice when upgrading any lighting design plan to LEDs. Dimming drivers also provide multilevel controls, which makes it easier to meet ASHRAE 90.1 code requirements.

Efficiency

The efficiency of ballasts can vary considerably. To specify the most efficient ballast, choose the ballast with the lowest input wattage for the desired start method and ballast factor. Programs are in place to distinguish efficient ballasts from less efficient ones. The National Electrical Manufacturers Association (NEMA) has an efficiency standard for ballasts designed to operate T8 fluorescent lamps. In this program, ballasts that meet minimum performance criteria may be labeled as NEMA premium. Specifying NEMA premium ballasts where applicable is a convenient and effective way to guarantee that only the most efficient ballasts are used.

Ballast Factor (BF)

Ballast Factor is the ratio of a lamp's lumen output on a particular ballast to the lamp's rated lumen output. By specifying the appropriate BF, the light output and power usage of a luminaire can be carefully chosen to meet the needs of specific applications. Ballasts for T8 linear fluorescent lamps are available in a range of BFs from 0.71 up to 1.37 or higher, while ballasts for many lamp types are only available with a 1.0 BF. Specifying ballasts with a higher ballast factor may allow for the use of fewer lamps per luminaire, which often improves luminaire efficiency and also reduces the number of lamps to be maintained in a building.

Ballasts/Drivers

Technology Schedule

B01 Multilevel

Electronic ballast capable of operating lamps with two or more ballast factors, controlled by low-voltage control signal or multiple circuit feeds, ballasts for 4 ft T8 fluorescent lamps should be labeled as NEMA premium (available for most linear fluorescent lamps, limited availability for other lamp types)

B02 Dimming

Electronic ballast capable of smoothly dimming lamps with a range of ballast factors with a minimum range of at least 10% to 100% of maximum light output, ballasts for 4 ft T8 fluorescent lamps should be labeled as NEMA premium, control signal may be digital, 0-10V, or carried over the power wires.

B03 Instant Start

Electronic ballast capable of operating lamps at one ballast factor ignites lamp using the instant start method; typical ballast factors include 0.77, 0.88, 1.00, 1.18, or higher for linear fluorescent lamps; ballasts for 4 ft T8 fluorescent lamps should be labeled as NEMA premium.

B04 Program Start

Electronic ballast capable of operating lamps at one ballast factor ignites lamp using the program rapid start method, typical ballast factors include 0.71, 0.88, 0.99, or 1.15 for linear fluorescent lamps, ballasts for 4 ft T8 fluorescent lamps should be labeled as NEMA premium.

B05 Electronic CMH

Electronic pulse-start ballast for low-wattage ceramic metal halide lamps.

B06 LED Driver

A LED driver is a self-contained power supply that has outputs matched to the electrical characteristics of an LED array. Typical drivers are static though drivers may also offer dimming through several different means of control. Ensure compatibility of drivers with any dimming controls that may be present or proposed for the system.

B07 Smart Luminaire

A smart luminaire is a self-contained luminaire with dimmable LED driver and integrated controls for occupancy sensing and daylight harvesting. In some cases, a smart luminaire may also have a controller that allows for wireless connectivity with other smart luminaires in the area for a synchronized group/zoned behavior. Smart luminaires also offer additional features like wattage augmentation, automatic daylight calibration, etc. and deliver aggressive energy savings without sacrificing occupancy comfort levels.

Luminaires

Considerations

Luminaires are a critical element of a lighting system as they determine how the light from lamps is directed into a space. Luminaires include housing, mounting hardware, and one or more lamps and may contain any or all of the following: ballast, reflector, lens, and shielding media. Many factors play into the selection of the optimal luminaire for a space or task. The luminaire categories listed below represent the majority of luminaires installed in DoD building projects. Understanding the advantages and disadvantages of these luminaires and the options with which they are provided is critical for specifying high-performance lighting systems.

Recessed Troffers

Recessed troffers with flat prismatic lenses or high-performance non-planar lenses are the most efficient troffers available. Specifying troffers with fluorescent T8 or LED lamping offers the greatest flexibility for providing the appropriate light level in a space with the thoughtful selection of lamp quantity and ballast factor. LED luminaires with smart lighting technologies tend to offer more energy savings and advanced functionalities. Higher luminaire efficiencies may be obtained by specifying luminaires designed specifically around the lamp type specified. Specifying that interior reflectors are painted after fabrication (PAF) often improves luminaire efficiency by 10% or more. Typically, the fewer lamps in a troffer, the more efficiently it operates. Two-foot by 4 ft (2x4) troffers are typically the most efficient and least expensive to install; however, the form factor and size of 1x4 or 2x2 troffers may make them better suited for many applications. In general, 4-ft-long troffers offer the best performance as they use 4 ft lamps, which are the most efficacious and cost-effective.

Linear Pendants

Linear fluorescent pendants can offer an excellent combination of high-efficiency, glare control, and light distribution and allow the location of the light source to be closer to the task area. Pendants with one lamp in cross-section almost always demonstrate the greatest luminaire efficiencies and the best optical performance.

Specifying T8 lamps with high ballast factor ballasts or T5HO lamps are effective ways to use a one lamp cross-section without sacrificing light output.

Recessed Downlights

Recessed downlights are one of the least efficient luminaire types available. These luminaire are often specified when low light levels are required or when the size and shape of downlight luminaire makes it easy to incorporate into a design. When specifying a downlight, consider those with open optics as they offer the highest efficiency. In many cases, led downlights are good alternatives to compact fluorescent downlights as the LED versions often provide greater light output using fewer watts. Larger aperture downlights are typically more efficient than smaller aperture downlights. High bays – The most efficient and effective high bay luminaires use fluorescent lamps. Luminaires that use T5 or T5HO are often more optically efficient than those that use T8 lamps; however T8 lamp and ballast combinations offer greater flexibility and system efficiency than T5/T5HO systems. In general, T8 high bays are the best choice except for when the increased light output provided by T5HO lamps is required. Designers should carefully consider the effect of the ambient air temperature at high bay mounting locations as fluorescent lamps are sensitive to temperature. Consult with the luminaire manufacturer to determine the best specification for the environment in which the luminaire will be used. Specifying high ballast factors will reduce the number of luminaires or lamps per luminaire, which will provide easier system maintenance.

Task Lights

Task lights are an extremely effective way to deliver high illuminance to a work area with the minimal energy use. As the luminaire is typically within 1 to 2 ft from the task, it can be easy to “over-light” a task area. When specifying a task light, choose the lowest wattage option that provides the appropriate illuminance at the task. LED task lights often provide better optics and lower wattage options than fluorescent task lights. Consider task lights with integrated occupancy sensors or wiring task lights though occupancy sensor-controlled power strips for maximum energy savings.

Technology Schedule

F01	Lensed Troffer		Fluorescent Recessed 2'x4' or 1'x4' or 2'x2' high efficiency troffer with planar prismatic lens, designed for T8 lamping, painted after fabrication, min. 75% efficiency
F01-LED	Lensed Troffer		LED Recessed 2'x4' (4000 lm, min. 90lm/W) or 2'x2' (3745lm, min 89lm/W) high efficiency troffer with planar prismatic lens, integrated LED lamping
F02	Wallbracket Direct/Indirect		Fluorescent wall mounted with shielding or lensing to reduce glare and direct light in the desired direction
F03	Non-Planar Lensed Troffer		Fluorescent recessed 2'x4', 1'x4', or 2'x2' high efficiency troffer with non-planar lens, designed for T8 lamping, painted after fabrication, min. 80% efficiency
F03-LED	Non-Planar Lensed Troffer		LED recessed 2'x4' (5840 lm, min. 90lm/W.) or 2'x2' (3125 lm, min. 85lm/W) high efficiency troffer with non-planar lens, integrated LED lamping
F04	Suspended Direct/Indirect		Fluorescent suspended luminaire with downlight and uplight component, shielding to prevent glare at angles between 45° and 90°, wide distribution uplight component
F05	Furniture Integrated		Fluorescent furniture integrated luminaire, downlight and uplight component, shielding or baffling to minimize glare from uplight component
F06	Perimeter		Fluorescent recessed perimeter luminaire, baffling or shielding to prevent direct view of lamps from normal viewing angles
F07	Narrow Lensed Wrap		Fluorescent linear luminaire with lamp completely shielded by prismatic acrylic lens, one lamp in cross-section
F08	Wide Lensed Wrap		Fluorescent wide linear luminaire with lamps completely shielded by prismatic lens, two or three lamps in cross-section
F09	Undercabinet		Fluorescent low-profile under-cabinet luminaire with low ballast factor and integral switch
F10	Suspended Indirect		Fluorescent suspended linear luminaire with 100% uplight, wide light distribution, optional clear dust cover
F11	High Bay		Fluorescent suspended or surface mounted high efficiency high bay luminaire, min. 90% efficiency, optional wire guard, clear lens, and uplight component

Controls

Occupancy Considerations (Some of These May Be Mandatory to Meet Energy Codes Such As ANSI/ASHRAE/IESNA 90.1 and the International Energy Conservation Code)

Automatically reducing electric lighting when spaces are unoccupied provides significant energy savings. There are several factors to consider when specifying automatic occupancy-based lighting controls for user acceptance and maximum energy savings.

Occupancy Sensors

These devices use infrared or ultrasonic sensing or both to determine if the area within its sensing zone is occupied by humans. The sensors automatically power on electric loads when occupancy is detected and power off electric loads after a set timeout period after vacancy is detected. These sensors offered with smart luminaires provide manual override and wireless grouping functionality for additional savings. Most areas only have intermittent occupancies, and therefore with a wireless grouping scenario, these sensors force only those luminaires that detect local occupancy at bright levels while other unoccupied areas operate at a lower background light level.

Vacancy Sensors

These devices are specialized occupancy sensors. When vacancy is detected, these sensor units automatically power off electric loads after a set time. Once the load has been switched off, it must be manually powered back on by switch or other devices. In daylighted spaces, this technology increases energy savings by enabling users to leave lights off when daylight is sufficient. Most occupancy/vacancy sensor units have an internal setting to allow facility management to select either occupancy or vacancy mode.

Mounting

Sensors should be located and aimed to provide the best coverage for a given area. For smaller spaces, such as a private office or storage closet, a sensor integrated into a standard wallbox provides sufficient coverage and is typically the most cost-effective approach. For larger spaces such as corridors or open offices, one or more ceiling-mounted sensors are required to ensure enough coverage to all areas. Such spaces should use smart luminaires with integrated sensors that offer wireless grouping and granular dimming functionality since these fixtures offer maximum coverage area and the flexibility to provide group lighting suitable to various occupancy layouts. Less labor is required to install smart luminaires because the installer need only replace the existing lighting with smart wireless LED luminaires; there is no need to rewire circuits.

Timeout

The time period between when a sensor detects vacancy in a space and when the load is switched is adjustable in the field. For areas with intermittent occupancy of short durations, a short timeout period, such as 5 minutes, should be used. For areas that a false vacancy reading and subsequent darkness may create a safety or security concern, longer timeout periods of 20 minutes or more should be considered. Typical timeout periods between 10 and 20 minutes are suitable for most applications. Smart luminaires offer two types of timeout – one when the lights dim down from a high level to a background level after detecting vacancy and the other timeout from background level to off. The two-step approach allows for additional energy savings within a lighting group without disturbing the occupants through abrupt on/off cycles and also provides a uniform light distribution in the space without dark spots due to luminaires that randomly turn off within the space.

Sensing Technology

Occupancy and vacancy sensors typically use either ultrasonic or infrared sensing or both to determine if a space is occupied by people. Infrared sensors detect occupancy by changes in infrared signals created by people moving within the coverage area. These sensors are best used in spaces where there is a direct line of sight to the sensor from all areas where people will be within a space. Ultrasonic sensors detect changing ultrasonic frequencies created by movement within a space. Ultrasonic sensors are best used in spaces where obstructions prevent a direct line of sight to the sensor occupied areas. Dual-technology sensors may be used in spaces where using only infrared or ultrasonic sensors may not provide adequate coverage or for increased assurance that the sensors will sense occupancy.

Daylight Considerations

Considerable energy savings can be achieved by reducing electric lighting loads when there is sufficient daylight in a space. Daylight responsive lighting controls must be designed and commissioned with care to ensure that the system operates effectively and without disruption of the occupants within the space. Shading devices, skylights, and glazing should be specified and sized to provide usable daylight without excessive glare or heat gain.

Automatic Dimming

Photosensors can be specified to dim electric lighting according to available daylight. Dimming systems provide the most seamless integration of electric light daylight. Dimming systems typically do not turn luminaire power completely off when ample daylight is available; instead lamps are reduced to 5% or 10% of full light output, which requires 20% or more of the ballast rated power.

Automatic Switching

Photosensors can be specified to switch off lighting when enough daylight is available. Switching systems open the circuit, so that luminaire power and light output are completely off. Photosensor setpoints for switching systems should be set to a higher value than with dimming systems to minimize the perception of the abrupt change in the lighting condition.

Manual Switching

Many spaces are not regularly occupied during daylight hours or may have intermittent occupancy patterns. When occupancy sensors are used in intermittently occupied spaces, daylight responsive controls may not be needed as the electric lighting can be turned off by a vacancy sensor for much of the day. If ample daylight is available, an occupant entering a space will be less likely to manually power on the electric lighting. If the occupant does choose to turn on the lighting, it will stay on until they leave the room or manually turn the lighting off. This strategy works especially well with multilevel ballasts, as users will often elect to turn the electric lighting to a lower level when daylight is available.

Sensors and Daylight Controlled Luminaires

Daylight sensors and daylight-controlled luminaires should be located approximately one window height from the window wall and should be unobstructed by ceiling surface-mounted equipment or structural columns. Because the optimal location and configuration of daylight responsive lighting systems is dependent on variables such as climate zone, ceiling height, orientation of the fenestration, as well as glazing properties, the location of daylight sensing equipment should be determined on a case-specific basis. In case of smart luminaires, every luminaire is a daylight zone by default, so no additional work is required for zone planning. If smart luminaires do not detect ambient light levels they do not dim down, but if they do detect it, then additional savings can be achieved because of daylight dimming.

Technology Schedule

C01 Infrared Occupancy/Vacancy Sensor

Sensor capable of detecting motion by changes in the infrared signals may be recessed or surface-mounted, single or bi-pole; occupancy sensors may be low-voltage or line voltage; low-voltage sensors and low-voltage wall switches must be used for vacancy sensor mode.

C02 Ultrasonic Occupancy/Vacancy Sensor

Sensor capable of detecting motion by changes in the ultrasonic signals may be recessed or surface-mounted, single or bi-pole; occupancy sensors may be low-voltage or line voltage; low-voltage sensors and low-voltage wall switches must be used for vacancy sensor mode.

C03 Dual Tech Occupancy/Vacancy Sensor

Sensor capable of detecting motion by changes in the ultrasonic and infrared signals may be recessed or surface-mounted, single or bi-pole; occupancy sensors may be low-voltage or line voltage; low-voltage sensors and low-voltage wall switches must be used for vacancy sensor mode.

C04 Wallbox Occupancy/Vacancy Sensor

Sensor mounted in a wallbox control station capable of detecting occupancy by either infrared or ultrasonic signals.

C05 Luminaire Integrated Occupancy and Daylight Sensor (Smart Luminaire)

Occupancy sensor mounted into a luminaire is capable of detecting motion by changes in the infrared signal it receives; the sensor can also be configured to work in vacancy mode. Daylight sensor responds to ambient light levels and dims the lighting accordingly. The daylight sensor can also be configured for daylight switching functionality.

C07 Dimming Photosensor

Sensor that responds to incident light to determine the quantity of daylight available, capable of sending low-voltage signal to dim electric lighting.

C08 Switching Photosensor

Sensor that responds to incident light to determine if the quantity of daylight present meets a determined setpoint, capable for sending a signal to switch off the electric lighting.

C09 Astronomic Timeclock

Device that controls building systems based on the time of day based on astronomic events such as sunset or sunrise accounting for geographic location and time of year.

Space-Specific Recommendations

Description of Typical Design Guide Page

Lighting technologies		Target illuminance	Target LPD
Lamp	**Luminaire**		
L01 Fluor 32WT8			
Ballast/driver	**Controls**	30 fc	0.80 W/ft^2

Target Illuminance

The target average maintained horizontal illuminance measured at the task area. For most spaces, a variance of 10% is acceptable.

Target LPD

Target LPD for the space type. The actual LPD will vary for actual designs based on conditions of actual spaces, but the LPD for the space type shown here is achievable in most cases.

Lighting Technologies

Lighting technologies that are specifically applicable to the space type. The use of all technologies shown will not be required for every space. More detailed descriptions of these technologies are found in the previous section.

Space Description

General description of the space type.

Considerations

Summary of space-specific considerations that inform the lighting design.

Lighting Approach

A short description of a lighting design approach, which may be used to minimize LPD while maintaining a quality visual environment.

Controls

A brief summary of the type of automatic lighting controls that should be implemented in the space to reduce lighting energy use.

Conference Room

Lighting technologies		Target illuminance	Target LPD
Lamp	**Luminaire**		
L01 Fluor 32WT8	F03 Non-planar Lensed Troffer		
	F04 Suspended Direct/ Indirect		
	F12 Wallwash		
	Downlights		
Ballast/driver	**Controls**	40 fc	0.80 W/ft^2
B06 Dimming/LED driver	F03 Non-planar Lensed Troffer		
B07 Smart Luminaires (scenes)	F04 Suspended Direct/ Indirect		
	F12 Wallwash		
	Downlights		

Space Description

Conference rooms may host audio/visual (AV) presentations, meetings, videoconferences, and teleconferences. These spaces typically have a large table in the center of the room and a presentation wall.

Considerations

Due to varying space functions, conference rooms should be equipped with lighting systems that can provide multiple lighting scenes. For reading and writing tasks, the lighting system should be able to provide 40 fc or more on the work plane. For AV presentation mode, the general work plane illuminance should be able to be reduced to between 10 and 25 fc, with no more than 20 fc on the presentation wall. Harsh shadows on occupants' faces seated at the table should be avoided, especially if the space is equipped with videoconferencing equipment.

Lighting Approach

Use highly efficient recessed fluorescent luminaires with non-planar lenses in spaces with ceiling heights below 9 ft, 6 in.; otherwise use suspended direct/indirect fluorescent pendants. To provide for multiple uses of conference rooms, the lighting system should be divided into two or more zones. Each luminaire in the general lighting zones should be capable of two or more light output modes to achieve recommended illuminances for reading and AV mode. There should be two zones at a minimum – one for general lighting and one for the presentation wall. An additional dedicated lighting zone for lighting videoconferencing participants and the wall immediately behind them may be required.

Controls

Use smart luminaires to set wireless groups and subzones for scenes that can be recalled from a compatible wall station. Any scene or manual override action should take precedence and should be maintained as long as the space is occupied.

Occupancy sensors that automatically set the general lighting to low output mode are acceptable for rooms with no daylight access. If the space has access to daylight, then daylight dimming sensors are recommended. In the case of a networked system, integration with AV and shades is also recommended.

Sample layouts

Sample schematic lighting layouts (not to scale)

Comments	**Type key**
Average LPD = 0.71 W/ft^2	F04: (2) lamp 32WT8, direct/indirect, 0.88 BF, bi-level ballast
Maintained illuminance on conference table = 40–60 fc	F12: (1) lamp 32WT8, 1x4 ft wallwasher, 0.88 BF
F04 mounted at 8 ft, 6 in. AFF	
Ceiling height = 9 ft, 6 in. AFF	

Corridor

Lighting technologies		Target illuminance	Target LPD
Lamp	**Luminaire**		
L01 Fluor 32WT8	F01 Lensed Troffer		
L02 Fluor 32WT8U	F01-LED Lensed Troffer		
L03 CFL 40W Long Twin-Tube	F03 Non-planar Lensed Troffer		
	F03-LED Lensed Troffer		
Ballast/driver	**Controls**	10 fc	0.50 W/ft^2
B01 Multilevel	C01 IR Occ/Vac Sensor		
B06 LED dimming driver	CR 05 Integrated occ./day sensors		
B07 Smart luminaires			

Considerations

Corridors experience intermittent occupancy; therefore, any controls should minimize the disturbances caused to occupants near the corridors. Abrupt switching on/off action due to passing motion should be avoided, but at the same time the system should not force occupants to pass in a dark environment.

Lighting Approach

Use highly efficient recessed LED luminaires with non-planar lenses in spaces with ceiling heights below 9 ft, 6 in.; otherwise use suspended direct/indirect LED pendants.

Controls

Smart luminaires with integrated occ./day sensors that run the lights at lower background level, but ramp up when stable occupancy is detected (e.g., discussions in corridor), are recommended. Any passing by motion should be ignored, but sensors only react when there is stable motion. Lower occupancy timeouts are also recommended to maximize energy savings potential.

Alternately, ceiling-mounted external occupancy sensors that can achieve the same intent as smart luminaires are acceptable.

Dining

Lighting technologies		Target illuminance	Target LPD
Lamp	**Luminaire**		
L01 Fluor 32WT8	F10 Suspended Indirect		
L02 Fluor 32WT8U	F16 Suspended Direct		
L03 CFL Long Twin-Tube	F31 or F52 Downlight		
L04 CFL Triple-Tube	F32 Performance Round		
Ballast/driver	**Controls**	20 fc	0.60 W/ft^2
B01 Multilevel	C01 IR Occ/Vac Sensor		
B06 LED luminaires	C07 Dimming Photosensor		
B07 Smart luminaires	C08 Switching Photosensor		
	C05 Integrated occ./day sensors		

Space Description

Dining areas area primarily used for personnel to eat meals. Seating is provided for counter tops as well as for small, medium, and large table dining. Large dining areas will seldom be used for special events; however smaller (or private) dining areas may occasionally be used for such events.

Considerations

High color rendering lighting is recommended for appealing food appearance. Lighting on the tables should be 20 fc in dining areas per TB MED 530 (for reference, the IESNA level is 10 fc). Furniture arrangement in these areas is subject to change, so lighting should be independent of furniture layout. Dining areas are often expansive with relatively low ceilings with architectural features to break up the space into smaller sections. Lighting uniformity is not a priority on the table surfaces, but because of the furniture layout changes, the zoning of the lighting system should be easily reconfigurable without any wiring changes.

Lighting Approach

Use high color rendering (85+ CRI) sources distributed relatively evenly throughout the space. Use some semi-decorative high-efficiency luminaires to add visual interest. Use linear fluorescent or LED luminaires for general lighting; use LED lamps when linear lamps are not appropriate. Minimize the use of recessed downlights; consider LEDs when recessed downlights must be used. An independent zone of lighting in the daylight zone (approximately two times the window head height) may be switched or dimmed based on available daylight. Select luminaires, lamps, and ballast factor combinations appropriately to provide sufficient illuminance without over-lighting.

Controls

Lighting should be controlled with occupancy sensors and building timeclock. Large sections of lighting should be zoned together and controlled by occupancy sensors. Critical zones, those near primary entrances and exits, can be left on to encourage seating in these areas, with other zones left off – only turning on when occupants move into that zone; smart luminaires with wireless grouping and sub-zoning can deliver the necessary flexibility. Daylight sensors (switching or multilevel) should be used in areas in the daylight zone.

Sample Layouts (Fig. G.1)

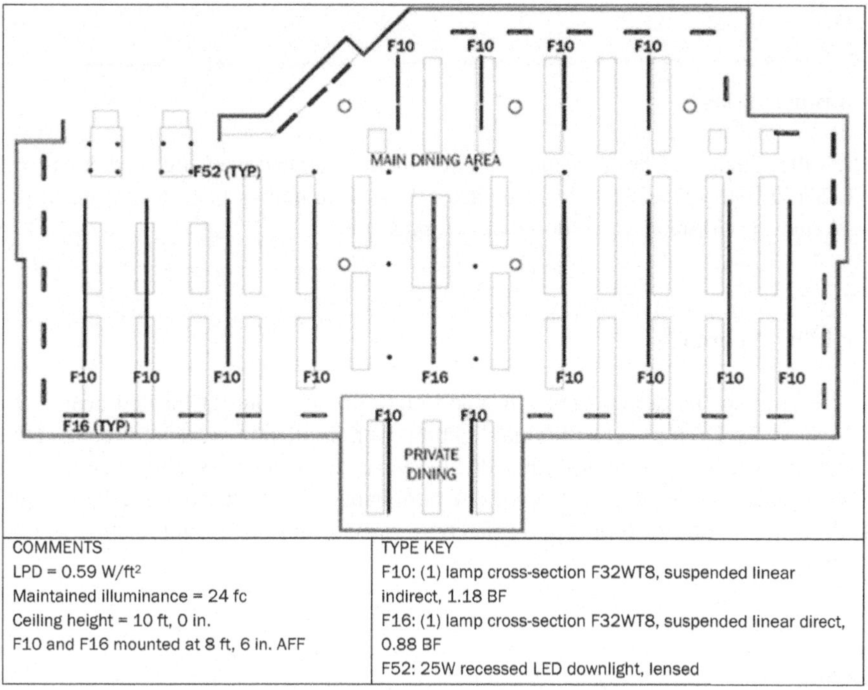

Fig. G.1 Sample schematic lighting layout (not to scale)

Dishwashing/Tray Return

Lighting technologies		Target illuminance	Target LPD
Lamp	**Luminaire**		
L01 Fluor 32WT8	F01 Lensed Troffer		
	F01-LED Lensed Troffer		
Ballast/driver	**Controls**	50 fc	0.65 W/ft^2
B03 Instant Start	C01 IR Occ/Vac Sensor		
B06 LED dimming driver	C04 Wallbox occ./vac. sensor		

Space Description

The primary task in dishwashing areas is loading and unloading industrial dishwashers with dishes and utensils as well as inspecting items for cleanliness.

Considerations

High illuminance of 50 fc or more at equipment or utensil washing work areas (per TB MED 530, the IESNA level is also 50 fc). Luminaires need to be sealed and gasketed to withstand spray down at low pressure.

Lighting Approach

Use high color rendering (85+ CRI) and high-efficacy sources located near/above task surfaces. Do not add additional lighting over circulation areas or directly above large dishwashing machines (that do not have work surfaces above them). Use sealed and gasketed linear fluorescent luminaires. Select luminaires, lamps, and ballast factor combinations appropriately to provide sufficient illuminance without over-lighting.

Controls

Lighting should be controlled primarily via building timeclock. All lighting, including equipment integrated lighting, should be switched off at closing time (with a warning flash for to allow the occupant to override). During operating hours, occupancy sensors can be used to turn the luminaires to a reduced power level when vacancy is detected.

Kitchen/Food Prep/Drive Thru

Lighting technologies		Target illuminance	Target LPD
Lamp	**Luminaire**		
L01 Fluor 32WT8	F01 Lensed Troffer		
Ballast/driver	**Controls**	50 fc	0.65 W/ft^2
B03 Instant Start	C01 IR Occ/Vac Sensor		
B06 LED dimming driver			

Space Description

Low contrast and potentially dangerous tasks are performed in kitchen and food preparation area. Kitchen staff must be able to properly inspect food items and use sharp kitchen utensils and equipment accurately at a fast pace. Equipment is typically washed down with a low-pressure wash.

Considerations

High color rendering sources are strongly desired for food appearance. High illuminance of 50 fc or more on the food preparation surfaces is required by TB MED 530 (for reference the IESNA level is also 50 fc). Lamps must be shielded, coated, or otherwise shatter resistant (TB MED 530). Luminaires need to be sealed and gasketed to withstand spray down at low pressure. Many of the kitchen appliances will come with integrated lighting. Circulation areas do not require high light levels.

Lighting Approach

Use high color rendering (85+ CRI) sources located near/above task surfaces. Do not add additional lighting over circulation areas. Use sealed and gasketed linear fluorescent luminaires. Linear fluorescent lamps should have a long rated life to minimize maintenance over food service areas. Select luminaires, lamps, and ballast factor combinations appropriately to provide sufficient illuminance without overlighting.

Controls

Lighting should be controlled via building timeclock. All lighting, including equipment integrated lighting, should be switched off at closing time (with warning flash for override). Using a long occupancy sensor timeout period, such as 30 minutes or more, will help prevent luminaires being turned off when the space is occupied.

Living Quarters

Lighting technologies		Target illuminance	Target LPD
Lamp	**Luminaire**	5–30 fc	0.60 W/ft^2
L01 Fluor 32WT8	F02 Wall bracket Direct/Indirect		
L02 Fluor 32WT8U	F05 Furniture Integrated		
L04 CFL 26W/42W Triple-Tube	F07 Narrow Lensed Wrap		
L05 CFL GU24 Integ. Ballast	F08 Wide Lensed Wrap		
L07 Fluor 17WT8	F09 or F51 Task		
	F12 Wallwash		
	F13 Strip		
	F14 Vanity		
	F30		
	Lamp		
Ballast/driver	**Controls**		
B04 Program Start	C04 Wallbox Occ/Vac Sensor		
B06 LED dimming driver	C05 Integrated Occ Sensor		

Space Description

Living quarters are similar to a small apartment or dorm suite. Most contain a bathroom, lavatory, an eat-in kitchen, and two private bedrooms each with a bed, night stand, desk, dresser, and a closet. Bedrooms typically are against an exterior wall with a window to outdoors. Kitchens have a refrigerator, oven and range, counter top space, and may be equipped with a microwave and toaster.

Considerations

Reading and food preparation areas require higher illuminance than other areas in the space. Low illuminance is sufficient away from counter and desk tops; 5 to 10 fc is sufficient for most areas. Visually comfortable, effective, permanently installed lighting should be used to prevent the use of uncontrolled plug load lighting. Luminaires may be unintentionally left on when spaces are unoccupied if they are not automatically controlled. Spaces within the living quarters along the core wall can become extremely dim at night time, which may result in some occupants leaving lights on as night lights for wayfinding purposes.

Lighting Approach

Use highly efficient surface-mounted luminaires with appropriate lamps and ballasts to provide sufficient illuminance for localized tasks without over-lighting. For areas that have low light level requirements, this can be achieved by using one lamp luminaires or low ballast factors or both. Provide task lighting at desk with a wall-mounted or furniture integrated luminaire. Provide a reading light on the night stand with a socket specific for compact fluorescent lamps. Low-energy LED nightlights should be provided in rooms without exterior windows.

Controls

Vacancy sensors should be used in each sub-space including bedrooms, bathrooms, kitchens, and other areas. Luminaire integrated occupancy sensors may be appropriate in closets. Some electrical outlets in bedrooms, vanities, and bathrooms should be controlled by vacancy sensor switch.

Sample Layouts (Fig. G.2)

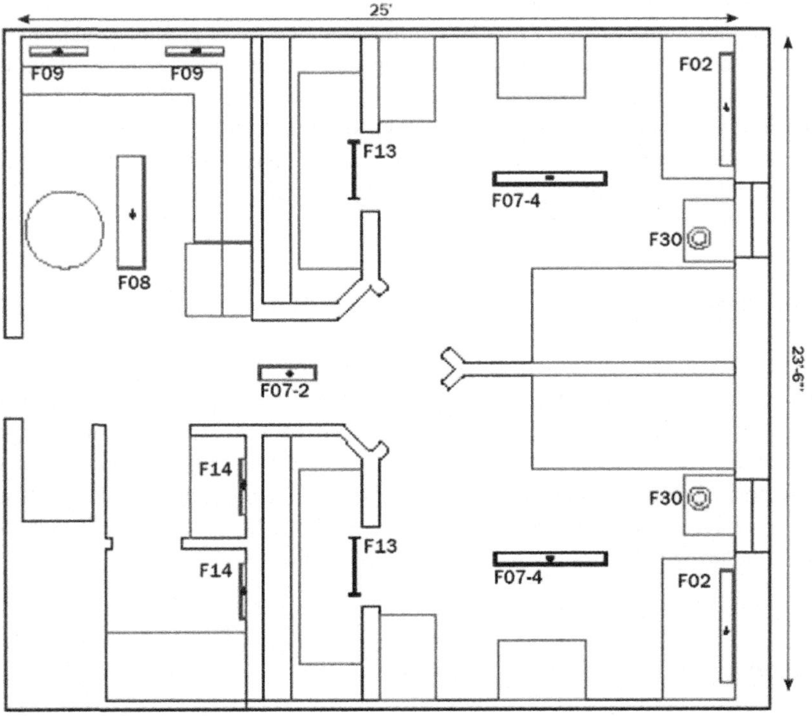

COMMENTS	TYPE KEY
LPD = 0.50 W/ft²	F02: (1) lamp F32WT8, 4' wall bracket direct/indirect, 0.88 BF, wall mount
Maintained illuminance at work-surfaces = 30 - 40fc	F07-2: (1) lamp F32WT8. 2' narrow lensed wrap, 0.77 BF
	F07-4: (1) lamp F32WT8, 4' narrow lensed wrap, 0.73 BF
	F08: (2) lamp F32WT8, 4' wide lensed wrap, 0.88 BF
	F09: (1) lamp F17WT8, 2' undercabinet task, 0.77 BF
	F13: (1) lamp F17WT8, 2' strip light, 1.00 BF, wall mount above door
	F14: (2) lamp F17WT8, 2' vanity wrap, 0.73 BF
	F30: (1) lamp CF13W/GU24, table lamp with dedicated CFL socket

Fig. G.2 Sample schematic lighting layout (not to scale)

Mechanical/Electrical

Lighting technologies		Target illuminance	Target LPD
Lamp	**Luminaire**	30 fc	0.70 W/ft^2
L01 Fluor 32WT8	F07 Narrow Lensed Wrap		
	F13 Strip		
Ballast/driver	**Controls**		
B04 Program Start	C02 Ultra Sonic Occ/Vac Sensor		

Space Description

Mechanical and electrical rooms contain equipment that may include motors, pumps, air handlers, boilers, transformers, lighting controls, circuit breakers, and other similar devices required for building operation. These spaces are typically only occupied by maintenance staff in time of repair or routine maintenance.

Considerations

Facilities personnel must be able to read small print on the surfaces of equipment and perform routine maintenance tasks. Lighting should be provided primarily for the working surfaces of the equipment. Task areas such as the face of a circuit breaker or name plates should be lighted to 30 fc; spaces in between tasks areas should be lighted to 5 fc or more as in corridor spaces.

Lighting Approach

Use high-efficacy sources located near/above task surfaces arranged to light critical task surfaces. Do not add additional lighting over circulation areas. Use linear fluorescent strip or wrap luminaires on chain mounts. Select luminaires, lamps, and ballast factor combinations appropriately to provide sufficient illuminance without over-lighting.

Controls

Ultrasonic occupancy sensors may be used in these areas, but should be placed and commissioned with care to prevent false vacancy readings. Due to obstructions and the typical lack of daylight in these spaces, using a long timeout period, such as 60 minutes, will also help prevent luminaires being turned off when the space is occupied.

Office (Open)

Lighting technologies		Target illuminance	Target LPD
Lamp	**Luminaire**	30–50 fc	0.70 W/ft^2
L01 Fluor 32WT8	F03 Non-planar Lensed Troffer		
LED	F04 Suspended Direct/Indirect		
	F05 Furniture Integrated		
	F09 or F51 Task		
	F12 Wallwash		
	F40 or F50 Adjustable Accent		
Ballast/driver	**Controls**		
B01 Multilevel	C05 Integrated occ./day sensors		
B02 Dimming			
B06 LED dimming driver			
B07 Smart luminaires			

Space Description

Open offices are designed to accommodate multiple individual work areas, typically separated by movable partitions and circulation areas. Individual work areas typically contain a computer, telephone, personal storage, and desk space for reading and writing. Furniture locations are not permanent and may change with needs and staffing. Open offices typically have one or more perimeter window walls that can provide views to the outdoors and usable daylight.

Considerations

Users' age, job function, and occupancy vary in each open office area. Work plane illuminance, as suggested by the IESNA, ranges from 30 fc to 50 fc for most office reading tasks. The visual needs of an older occupant in one work area may be different than that of a younger occupant. In most cases, the circulation space between work areas requires little if any lighting in addition to that provided for work areas. It is typical to find some work areas occupied and some vacant throughout the work day. Direct and reflected glare should be considered. Direct sunlight on work surfaces can contribute to glare and make it difficult to perform work. Lighting in the daylight zone (approximately twice the window head height) can often be turned off or reduced to a low power setting during the day.

Lighting Approach

The lighting system in these spaces should be easily adaptable to suit the needs of occupants. Providing a task/ambient solution is an effective way to minimize energy use while providing sufficient lighting for the occupants. This can be achieved by providing a 15 to 25 fc of ambient (furniture-mounted or overhead) lighting – enough for computer use, facial recognition, and circulation – supplemented by individually controlled task lighting at each workstation. The lack of full height partitions and low-level ambient lighting may leave the space feeling dim in some cases. To avoid this perception, interior full height partitions should be lighted with wallwashers or similar. This is especially effective when lighting the wall opposite the window wall in deeper spaces to balance vertical brightness in the field of view.

Controls

Because of the intermittent occupancy patterns, a lot of energy is wasted in open offices' traditional controls approach, either the entire zone is ON or OFF irrespective of where actual occupancy is. Smart luminaires that offer wireless grouping and granular dimming functionality (bright light levels when there is local occupancy, but dim levels where there is vacancy within a zone/group). Without integrated approach, multiple ceiling-mounted sensors are required to achieve the desired sensing coverage, and the controllability is limited to how lights are wired; there is no room for space churn and easy adaptability, but with smart luminaires these limitations could be overcome. Smart luminaires also offer daylight sensing with automatic daylight calibration thereby making the system immune to space orientation and churn.

Sample Layouts

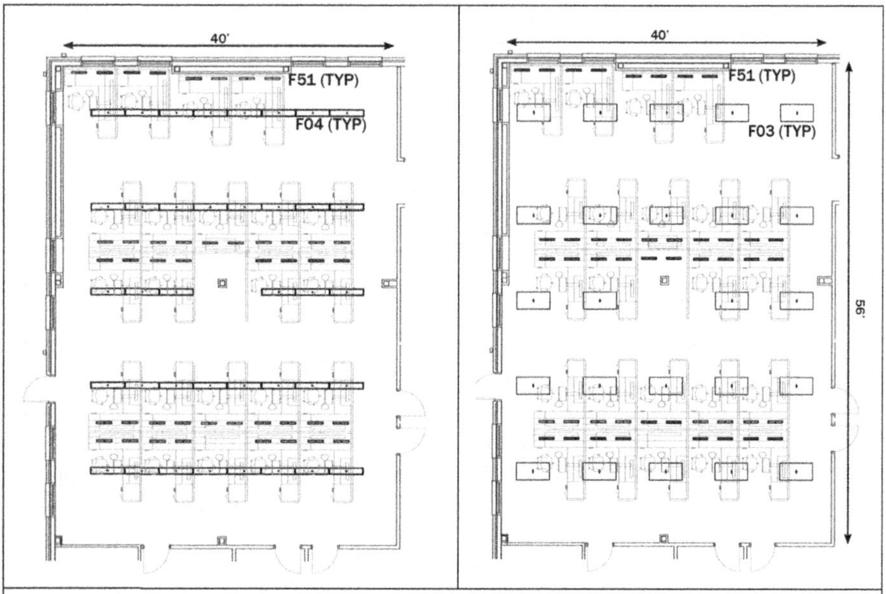

A: SAMPLE SCHEMATIC LIGHTING LAYOUTFOR CEILING HEIGHTS AT OR BELOW 9 FT, 0 IN. (NOT TO SCALE)

B: SAMPLE SCHEMATIC LIGHTING LAYOUT FOR CEILING HEIGHTS ABOVE 9 FT, 0 IN. (NOT TO SCALE)

COMMENTS	COMMENTS
LPD with task lights = 0.77 W/ft^2	LPD with task lights = 0.72 W/ft^2
LPD without task lights = 0.61 W/ft^2	LPD without task lights = 0.56 W/ft^2
Maintained task illuminance = 40 - 54 fc	Maintained task illuminance = 37 - 53 fc
Maintained ambient illuminance = 27 - 50 fc	Maintained ambient illuminance = 20 - 45 fc
TYPE KEY	TYPE KEY
F03: (1) lamp F32WT8. non-planar lensed troffer, 1.15 BF	F04: (1) lamp F32WT8, suspended direct/indirect, 1.00 BF
F51: 6W Undercabinet LED task light (furniture mounted)	F51: 6W undercabinet LED task light (furniture mounted)

Office (Enclosed)

Lighting technologies		Target illuminance	Target LPD
Lamp	**Luminaire**	30–50 fc	0.80 W/ft^2
L01 Fluor 32WT8	F03 Non-planar Lensed Troffer		
LED	F04 Suspended Direct/Indirect		
	F05 Furniture Integrated		
	F51 Task		
	LED luminaires		
Ballast/driver	**Controls**		
B01 Multilevel	C01 IR Occ/Vac Sensor		
B06 LED dimming driver	C05 Integrated occ./day sensor		
B07 Smart luminaires			

Space Description

Enclosed offices are typically intended for use by one person, though some enclosed offices may be used by two people. A typical enclosed office contains a computer, telephone, personal storage, and desk space for reading and writing and may contain an additional seating area.

Considerations

Users of private offices vary by age and job function. Work plane illuminance, as suggested by the IESNA, ranges from 30 fc to 50 fc for most office reading tasks. The visual needs of an older occupant in one office may be much different than that of a younger occupant in another. Private offices often have a window to the outdoors and are frequently unoccupied.

Lighting Approach

The lighting system in these spaces should be easily adaptable to suit the needs of occupants. Providing a task/ambient solution is an effective way to minimize energy use while providing sufficient lighting for the occupants. This is achieved by providing 15 to 25 fc ambient lighting – enough for computer use, facial recognition, and circulation – supplemented by individually controlled task lighting at each workstation. Alternatively, a bi-level or dimmable overhead lighting system can provide similar savings as many occupants may choose the lower lighting level.

Controls

Vacancy sensors should be used in offices with daylight access. For interior private offices, occupancy sensors that activate a low light output should be used. All task lighting should be controlled by independent vacancy sensor. Dimmers that can control the overhead lighting throughout its dimming range are recommended. Smart luminaires that are wirelessly linked to a compatible wireless dimmer switch offer the most flexibility.

Lighting technologies		Target illuminance	Target LPD
Lamp	**Luminaire**	30–50 fc	0.80 W/ft^2
L01 Fluor 32WT8	F03 Non-planar Lensed Troffer		
LED	F04 Suspended Direct/Indirect		
	F05 Furniture Integrated		
	F51 Task		
	LED luminaires		
Ballast/driver	**Controls**		
B01 Multilevel	C01 IR Occ/Vac Sensor		
B06 LED dimming driver	C05 Integrated occ./day sensor		
B07 Smart luminaires			

Space Description

Enclosed offices are typically intended for use by one person, though some enclosed offices may be used by two people. A typical enclosed office contains a computer, telephone, personal storage, desk space for reading and writing and may contain an additional seating area.

Considerations

Users of private offices vary by age and job function. Work plane illuminance, as suggested by the IESNA, ranges from 30 fc to 50 fc for most office reading tasks. The visual needs of an older occupant in one office may be much different than that of a younger occupant in another. Private offices often have a window to the outdoors and are frequently unoccupied.

Lighting Approach

The lighting system in these spaces should be easily adaptable to suit the needs of occupants. Providing a task/ambient solution is an effective way to minimize energy use, while providing sufficient lighting for the occupants. This is achieved by providing 15 to 25 fc ambient lighting – enough for computer use, facial recognition, and circulation – supplemented by individually controlled task lighting at each workstation. Alternatively, a bi-level or dimmable overhead lighting system can provide similar savings as many occupants may choose the lower lighting level.

Controls

Vacancy sensors should be used in offices with daylight access. For interior private offices, occupancy sensors that activate a low light output should be used. All task lighting should be controlled by independent vacancy sensor. Dimmers that can control the overhead lighting throughout its dimming range are recommended. Smart luminaires that are wirelessly linked to a compatible wireless dimmer switch offer the most flexibility.

Sample Layouts

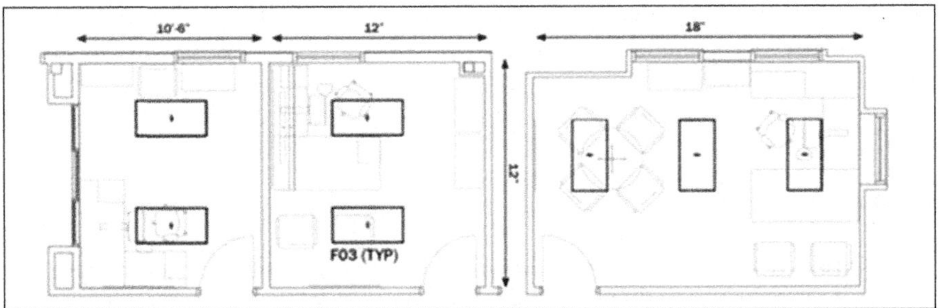

A: SAMPLE SCHEMATIC LIGHTING LAYOUTFOR CEILING HEIGHTS AT OR BELOW 9 FT, 0 IN. (NOT TO SCALE)

COMMENTS	TYPE KEY
LPD with task lights = 0.77 W/ft^2 LPD without task lights = 0.61 W/ft^2 Maintained task illuminance = 40 - 54 fc Maintained ambient illuminance = 27 - 50 fc	F03: (1) lamp F32WT8. non-planar lensed troffer, 1.15 BF F51: 6W Undercabinet LED task light (furniture mounted)

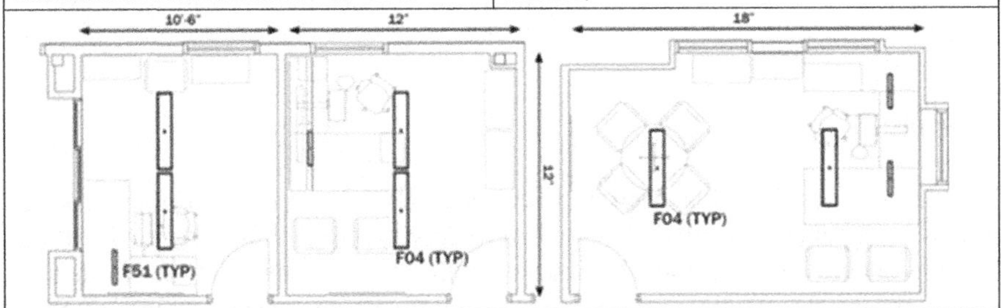

B: SAMPLE SCHEMATIC LIGHTING LAYOUT FOR CEILING HEIGHTS ABOVE 9 FT, 0 IN. (NOT TO SCALE)

COMMENTS	TYPE KEY
LPD with task lights = 0.77 W/ft^2 LPD without task lights = 0.70 W/ft^2 Maintained task illuminance = 37 - 65 fc Maintained ambient illuminance = 20 - 40 fc	F04: (1) lamp F32WT8, suspended direct/indirect, 1.00 BF F51: 6W Undercabinet LED task light (furniture mounted)

Reception/Waiting

Lighting technologies		Target illuminance	Target LPD
Lamp	**Luminaire**	15–30 fc	0.60 W/ft^2
L05 Compact Fluor	F03 Non-planar Lensed Troffer		
L07 Fluor 17WT8	F30 Task		
	LED luminaires		
Ballast/driver	**Controls**		
B06 LED dimming driver	C02 Ultra Sonic Occ/Vac Sensor		
B07 Smart luminaires	C07 Dimming Photosensor		
	C05 Integrated occ./day sensor		

Space Description

Reception area typically serves as a transitional space between the exterior and the rest of the interior. Activities expected in this space include egress, reading, and writing at the reception desk task plane. This space also serves as an initial impression of the function and value of the facility.

Considerations

As an initial impression of the function and value of the facility, the reception area should be relatively congruent in light color (expressed in °k) and illuminance to the corridors throughout the facility. Reading and writing activities can be reasonably expected by users from 25 years to >65 in the reception desk area.

Lighting Approach

In waiting areas and circulation areas 15 to 20 fc ambient lighting facilitates facial recognition, and circulation and short reading tasks. The lighting system at the reception desk area should be easily adaptable to suit the needs of occupants. Providing a task/ambient solution is an effective way to minimize energy use while providing sufficient lighting for the occupants. This is achieved by providing 15 to 30 fc ambient lighting – enough for computer use, facial recognition, and circulation – supplemented by individually controlled task lighting at the workstation. Alternatively, a bi-level or dimmable overhead lighting system can provide similar savings since many occupants may choose the lower lighting level.

Controls

Because of intermittent occupancy patterns in this space and overall dynamics more closely associated with an open office, smart luminaires with wireless grouping and granular dimming are recommended.

Lighting controlled by occupancy set at a predetermined delay of 10–20 minutes, with auto-off when unoccupied. All task lighting should be controlled by independent vacancy sensors at each workstation. Daylight responsive lighting controls should be used in areas with sufficient access to daylight.

Sample layout

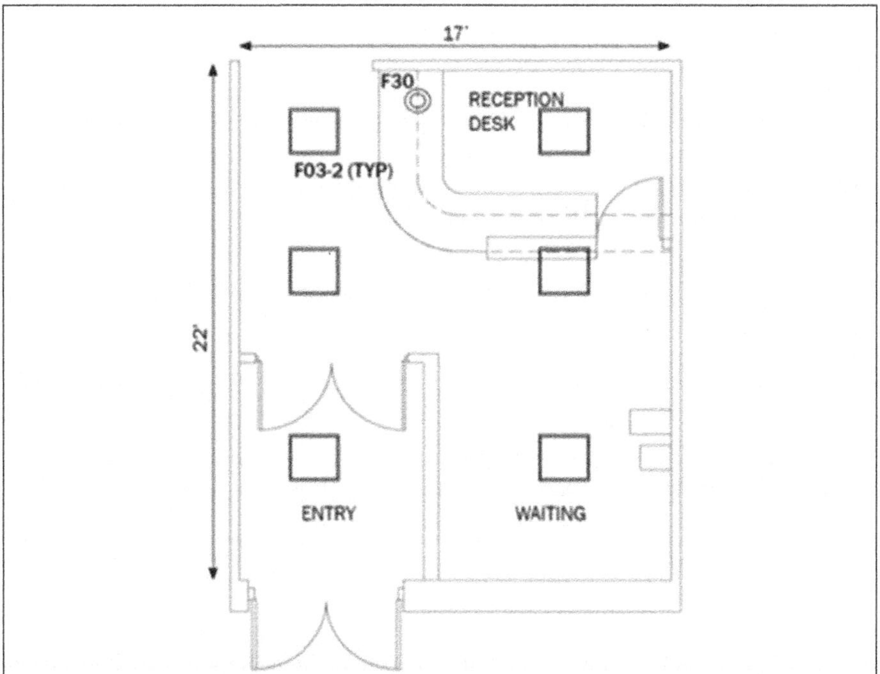

Sample schematic lighting layout (not to scale)
Ceiling = 2'x2' acoustical ceiling tile, 8 ft, 8 in. AFF Typical

COMMENTS	TYPE KEY
LPD = -2: 0.54 W/ft^2 -LED: 0.58 W/ft^2 Illuminance on work surface = -2: 18-30fc -LED: 25-35 fc Maintained average illuminance in circulation areas = 10-30 fc Calculated with 80%/70%/20% ceiling/wall/floor reflectance and 95% Luminaire Dirt Depreciation	F03-2: (2) Lamp 17WT8, 2'x2' Non-Planar Lensed Troffer, 0.88 BF F03-2: 2'x2' Non-Planar LED Lensed Troffer, 0.75 BF F30: (1) 18WCFL Desktop Task

Restroom/Shower

Lighting technologies		Target illuminance	Target LPD
Lamp	**Luminaire**	20 fc	0.80 W/ft^2
L01 Fluor 32WT8	F01 Lensed Troffer		
	F06 Perimeter		
	F14 Vanity		
	F31 or F52 Downlight		
Ballast/driver	**Controls**		
B04 Program Start	C02 Ultra Sonic Occ/Vac Sensor		
	C04 Wallbox Occ/Vac Sensor		

Space Description

Restrooms include one or more toilets and lavatories with a mirror for grooming. Some restrooms include showers and changing facilities.

Considerations

Restrooms are intermittently occupied and used for short durations. The primary visual task is hand washing and grooming. The IESNA illuminance for this space is 5 fc, though lighting the walls and providing illuminances in the range of 10 fc to 20 fc near the sinks and toilets can help make the space feel brighter, which may promote cleanliness. The illuminance provided between the sink and toilets is non-critical and may be as low as 5 fc. Luminaires in the shower areas need to be rated for wet locations.

Lighting Approach

Provide overhead perimeter lighting above the sinks and along the toilet wall. Some restrooms may require an additional luminaire near the room entry, though often the perimeter luminaires are all that is necessary. Luminaires with one lamp may and a low ballast factor should be considered. Overhead lighting provided above the showers and near the locker areas is often all that is required in these areas.

Controls

Ultrasonic occupancy sensors should be used in these spaces to prevent false vacancy readings that may occur with infrared sensors due to interior partitions. These systems should be automatic-on, automatic-off.

Server Room

Lighting technologies		Target illuminance	Target LPD
Lamp	**Luminaire**	30 fc	0.85 W/ft^2
L01 Fluor 32WT8	F01 Lensed Troffer		
	F07 Narrow Lensed Wrap		
	LED luminaires		
Ballast/driver	**Controls**		
B07 Smart luminaires	C02 Ultra Sonic Occ/Vac Sensor		
	C05 Integrated occ./day sensor		

Space Description

Server rooms house semi-permanently installed computer equipment in racks, typically arranged in rows to allow for easy access to the front and back of server equipment.

Considerations

IT personnel must be able to identify computer equipment, data connections, and cables and install components with small fasteners. Lighting should be provided to illuminate the vertical faces of each server rack. These spaces are frequently unoccupied and seldom have access to daylight.

Lighting Approach

Locate and specify highly efficient luminaires between server racks to light vertical surfaces to 30 fc.

Controls

Server rooms have very rare movements, but in some cases for safety reasons, the lights need to remain on. With smart LED luminaire, it is possible to deliver high light output when the space is occupied and low light levels during periods of vacancy. Alternately, ultrasonic occupancy sensors should be used in these areas. Multiple sensors and zones may be required for larger spaces with several server racks. Sensors integrated into wall switches should be used in small areas.

Serving Area

Lighting technologies		Target illuminance	Target LPD
Lamp	**Luminaire**	50 fc	0.70 W/ft^2
L01 Fluor 32WT8	F01 Lensed Troffer		
L06 20/39W CMH	F16 Suspended Direct		
	F40 Adjustable Accent		
	F31 or F52 Downlight		
	LED luminaires		
Ballast/driver	**Controls**		
B03 Instant Start	C01 Occ/Vac Sensor		
B05 Electronic CMH	C05 Integrated occ./day sensor		
B06 LED dimming drivers	C08 Switching Photosensor		

Space Description

The serving area is in buffet format with food serving areas arranged around the perimeter of the space with a central island with circulation space in between. The space behind the food serving areas is used for light food preparation and storage.

Considerations

High color rendering lighting is recommended for appealing food appearance. Lighting on work surface and food presentation areas should be 50 fc. Spill light from lighting the food displays will often be enough to light the circulation areas. Lensed luminaires are required.

Lighting Approach

Use high color rendering (85+ CRI) sources. Use linear fluorescent lensed troffers for lighting larger food displays and work areas. Use ceramic metal halide accent luminaires to highlight special food display areas or those in a center island, for example. Minimize the use of recessed downlights; consider LEDs when recessed downlights are to be used.

Controls

Lighting should be controlled with a building timeclock. All luminaires, including equipment integrated lighting, should be switched off at closing time. Occupancy sensors may also be used in these areas for further energy savings.

Sample layout

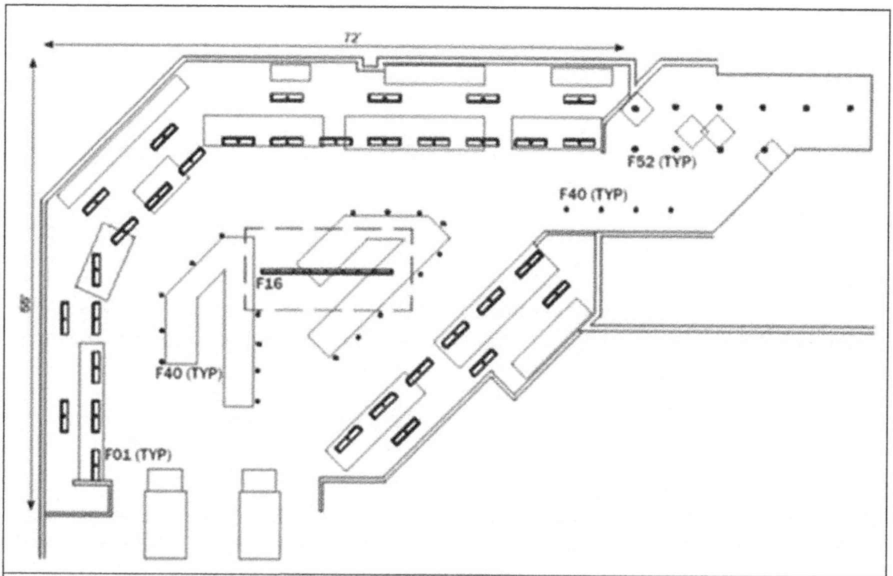

SAMPLE SCHEMATIC LIGHTING LAYOUT (NOT TO SCALE)

COMMENTS	TYPE KEY
LPD = 0.58 W/ft²	F01: (1) lamp F32WT8, 1x4 ft recessed lensed troffer, 1.18 BF
Illuminance on food displays = 39-54 fc	F16: (2) lamp F32WT8, suspended linear direct, 0.88 BF
Illuminance in circulation area = 11-33 fc	F40: (1) lamp 39W CMH, adjustable accent, 1.0 BF
	F52: 25W recessed LED downlight, lensed

Stair

Lighting technologies		Target illuminance	Target LPD
Lamp	**Luminaire**	5 fc	0.50 W/ft^2
L01 Fluor 32WT8	F07 Narrow Lensed Wrap		
	F15 Bi-level Occ Sens Wrap		
Ballast/driver	**Controls**		
B01 Multilevel	C05 Integrated Occ Sensor		
B06 and BO7 LED dimming drivers			

Space Description

Stairwells are typically enclosed and often contain a landing between two building floors. These spaces are used for daily circulation as well as for egress in the case of emergency.

Considerations

Stairways are typically critical exit paths in the case of an emergency. The IESNA recommended target illuminance for stairs is measured at 5 ft, 0 in. above the finished floor. These spaces are intermittently occupied, and some stairwells are used only in case of an emergency. There are conflicting code requirements that stipulate minimum illuminance on stair treads. Building codes typically require a minimum illuminance level even when occupied. Refer to applicable building and safety codes in determining compliance of stair lighting.

Lighting Approach

Locate luminaires at each landing, mounted on wall or ceiling at door locations while meeting applicable safety code requirements.

Controls

Use luminaire integrated occupancy sensors to reduce luminaire to low light output, low power setting when vacancy is detected.

Storage (General)

Lighting technologies		Target illuminance	Target LPD
Lamp	**Luminaire**	10 fc	0.50 W/ft^2
L01 Fluor 32WT8	F01 Lensed Troffer		
	F13 Strip		
Ballast/driver	**Controls**		
B04 Program Start	C01 IR Occ/Vac Sensor		
	C02 Ultra Sonic Occ/Vac Sensor		
	C04 Wallbox Occ/Vac Sensor		

Space Description

Storage areas typically contain shelving to store items vertically. These spaces are typically occupied for brief periods to store or retrieve items.

Considerations

Occupants must be able to identify stored objects and read labels. Lighting should be provided to light the faces of stored items. Lighting the top of shelves is not useful, while light on the front face of shelving will allow for quick identification of items.

Lighting Approach

Use high-efficacy sources located near/above task surfaces arranged to light critical task surfaces to 10 fc. Do not add additional lighting over circulation areas or directly above shelving units. Use linear fluorescent strip or wrap luminaires. Select luminaires, lamps, and ballast factor combinations appropriately to provide sufficient illuminance without over-lighting.

Controls

IR and/or ultrasonic occupancy sensors should be used in these areas depending on shelving and other obstructions. Sensors integrated into wall switches should be used in small storage areas.

Storage (Dry Food)

Lighting technologies		Target illuminance	Target LPD
Lamp	**Luminaire**	10 fc	0.70 W/ft^2
L01 Fluor 32WT8	F01 Lensed Troffer		
Ballast/driver	**Controls**		
B04 Program Start	C02 Ultra Sonic Occ/Vac Sensor		

Space Description

Dry food storage area used to store food and contain movable shelving. Kitchen staff must be able to identify labels and textures of stored food items.

Considerations

Relatively low light levels are required, 10 fc (per TB MED 530, the IESNA level is 5 fc). Lighting on vertical surfaces is critical as items are typically stored on shelving. If movable shelving units are used, lighting should be sufficient on shelf surfaces for all typical storage positions. Lamps need to be shielded if storage areas will contain open packages or other exposed food.

Lighting Approach

Use high-efficacy sources located near/above task surfaces arranged to accommodate movable storage units if in use. Do not add additional lighting over circulation areas. Use sealed and gasketed linear fluorescent luminaires. Select luminaires, lamps, and ballast factor combinations appropriately to provide sufficient illuminance without over-lighting.

Controls

Lighting should be controlled with an occupancy sensor.

Telecom/SIPRNET

Lighting technologies		Target illuminance	Target LPD
Lamp	**Luminaire**	50 fc	1.20 W/ft^2
L01 Fluor 32WT8	F01 Lensed Troffer		
	F07 Narrow Lensed Wrap		
Ballast/driver	**Controls**		
B04 Program Start	C02 Ultra Sonic Occ/Vac Sensor		
B06 and B07 LED dimming drivers	C04 Wallbox Occ/Vac Sensor		
	C05 Integrated Occ sensors		

Space Description

Telecom and SIPRNET rooms house semi-permanently installed computer and telecommunications equipment arranged in racks or against walls.

Considerations

Personnel must be able to identify computer equipment, data connections and cables and install components with small fasteners. Lighting should be provided to illuminate the vertical faces of equipment. These spaces are frequently unoccupied and seldom have access to daylight.

Lighting Approach

Locate and specify highly efficient luminaires to light critical surfaces to 50 fc, avoid locating luminaires directly above or behind equipment to maximize the light falling on critical surfaces.

Controls

Ultrasonic occupancy sensors should be used in these areas. Multiple sensors and zones may be required for larger spaces, for such scenarios to get optimum coverage areas smart luminaires with integrated sensors are recommended. Sensors integrated into wall switches should be used in small areas.

Vault

Lighting technologies		Target illuminance	Target LPD
Lamp	**Luminaire**	40 fc	0.70 W/ft^2
L01 Fluor 32WT8	F01 Lensed Troffer		
	F09 or F51 Task		
Ballast/driver	**Controls**		
B04 Program Start	C01 IR Occ/Vac Sensor		

Space Description

Vaults are secure spaces used to temporarily store sensitive items, such as weapons and ammunition, non-sensitive items with high value, serial numbered items, or secure telecommunication items. Typically there is an administrative workstation and in some cases a workbench area in each vault.

Considerations

Personnel must be able to identify stored objects and read serial numbers with fine print. Lighting should be provided to light the faces of stored items and the workbench and desk area.

Lighting Approach

Use high-efficacy sources located near/above task surfaces arranged to light critical task surfaces to 40 fc. Do not add additional lighting over circulation areas. Use linear fluorescent strip or wrap luminaires. Select luminaires, lamps, and ballast factor combinations appropriately to provide sufficient illuminance without over-lighting.

Controls

IR occupancy sensors should be used in these areas.

Training Room (Small)

Lighting Technologies		Target illuminance	Target LPD
Lamp	**Luminaire**	30–35 fc	0.80 W/ft^2
L01 Fluor 32WT8	F03: Non-planar Troffer		
L07 Fluor 17WT8	F03-LED: Non-planar Troffer		
L10 54WT5HO	F04 Suspended Direct/ Indirect		
LED			
Ballast/driver	**Controls**		
B02 Dimming	C05 Integrated occ./day sensors		
B04 Program Start	C01 Dual Tech Occ/Vac Sensor		
B06 and B07 LED dimming drivers	C07 Dimming Photosensor		

Space Description

Training rooms are multi-use spaces accommodating lecture, discussion, interactive/ collaborative exercises, AV presentations, reading, and note taking.

Considerations

These spaces are small areas where highly interactive and visual activities often take place. Smart boards in these areas act as projection surfaces as well as writing surfaces. Glare control is imperative for clarity during AV presentations. Due to the adaptive nature of this room, training rooms should be equipped with lighting systems that can provide different lighting scenes (multi-scene controllers or manual switches/dimmers). For reading and writing tasks, the general lighting system should be able to provide an average of 35fc on the work plane. When a smart board is in use, general lighting should be able to dim to preset levels, or manually dimmed to the presenter's preference. Lighting on the teaching wall should be separately controlled by a wall control located adjacent to the smart board.

Lighting Approach

Primarily indirect lighting is preferred in spaces where AV presentations are frequent. In spaces with ceilings over 9 ft, 0 in., dimmable, 70–75% indirect distribution pendants provide ergonomic, low-glare lighting from a maximum of 25–35fc with levels variable according the preference of the presenter to accommodate reading, note taking, and facial recognition. In ceilings below 9 ft, 0 in., non-planar lensed troffers provide similar high-efficiency, indirect lighting.

Controls

These spaces are similar to conference rooms where a minimum of two zones are required in these spaces, one for the teaching wall, one for the general lighting, and another may be required for spaces with access to daylighting to allow for daylighting dimming of luminaires adjacent to windows. Wall-mounted dimming interfaces should be provided at room entries and additional scene recall controls at the teaching wall are desirable. Smart luminaires with wireless grouping functionality are recommended to allow for flexibility in zone creation as per user preferences. The controls should also allow for light levels to be stored at different levels as scene.

Sample Layout

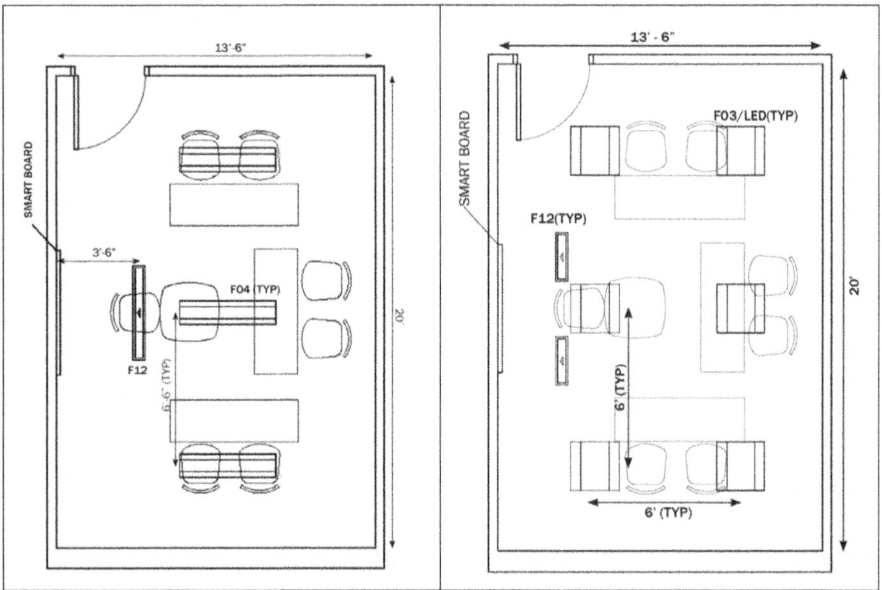

A: SAMPLE SCHEMATIC LIGHTING LAYOUT FOR CEILING HEIGHTS BELOW 9' AFF (NOT TO SCALE)

B: SAMPLE SCHEMATIC LIGHTING LAYOUT FOR CEILING HEIGHTS ABOVE 9' AFF (NOT TO SCALE)

COMMENTS	TYPE KEY
LPD = 0.8 W/ft^2 Illuminance on work surface = 30-35 fc -LED: 25-40 fc Maintained average illuminance in circulation areas = 30-50 fc Vertical illuminance teaching wall: 15-19 fc wallwash off/ 30-34 fc wallwash on Calculated with 80%/70%/20% ceiling/wall/floor reflectance	F03: (2) lamp F17WT8. non-planar troffer, 0.88 BF Bi-level ballast F03-LED: 2'x2' LED non-planar troffer, 0.75 BF F04: (1) lamp F54WT5HO, suspended direct/indirect, 1.0 BF, mounted 8'AFF F12: (1) Lamp 32WT8 1x4 ft/ (1) 17WT8, 1'x2' Wallwasher, 0.88 BF

Parking

Lighting technologies		Target illuminance	Target LPD
Lamp	**Luminaire**	4.0 fc	0.10 W/ft^2
L09: 150W CMH	B01 Multilevel		
139W LED	B04 Program Start		
Ballast/driver	**Controls**		
B01 Multilevel	C05 Integrated Occ Sensor		
B04 Program Start	C09 Timeclock		

Space Description

Open parking facilities accommodating vehicular traffic and pedestrians.

Considerations

Safe operation of vehicles day or night as well as the safety of pedestrians in the area is of primary concern. Energy usage during times when the lot is not in use should be kept at a minimum. For the safety and comfort, the uniformity of illuminance levels should also be taken into consideration with the ratio of the average level of illuminance to the minimum should be less than 4:1.

Lighting Approach

Type II distribution ceramic metal halide or LED roadway lighting on 24 in. arm mounted on 22 ft poles, respectively, provide maximum spacing with a maintained average of 4 fc.

Cut off fixtures prevent light pollution by not directing any light above 90 °F (32 °C).

Controls

LED lighting should be specified with integrated occupancy sensors and programmed to reduce to 50% when area unoccupied and come up to 100% when pedestrian or vehicular traffic is detected.

Metal halide roadway fixture should be controlled by timeclock to switch off when ambient natural lighting can predictably exceed target.

Maximum energy savings are achieved through LED roadway/sight lighting due to low-level setting during periods of inactivity.

Sample Layout

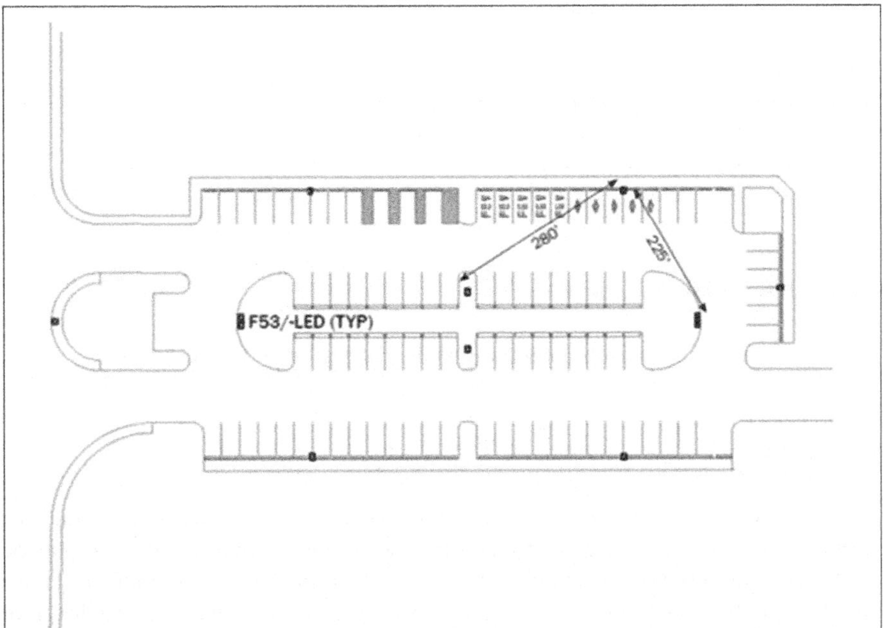

SAMPLE SCHEMATIC LIGHTING LAYOUT (NOT TO SCALE)

COMMENTS	TYPE KEY
LPD = F53: 0.08 W/ft² F53-LED: 0.04 W/ft² Maintained average illuminance = F53: 3.2 fc F53-LED: 1.9 fc Minimum illuminance: F53: 0.5 fc F53-LED: 0.4 fc Maximum/Minimum Ratio = F53: 19.6:1 F53-LED: 19.5:1 Calculated with ground = 0.10% reflectance	F53: Type II distribution Ceramic Metal Halide (CMH) roadway luminaire BF 1.0 Mounting = 24' pole F53-LED: Type II distribution LED roadway luminaire Mounting = 22' Pole

Roadway Lighting

Local with Medium Pedestrian Activity

Lighting technologies		Target illuminance	Target LPD
Lamp	**Luminaire**	1 fc	0.50 W/ft^2
L09: 150W CMH	F53 CMH Roadway		
	F53-LED LED Roadway		
Ballast/driver	**Controls**		
B01 Multilevel	C05 Integrated Occ Sensor		
B04 Program Start	C09 Timeclock		

Space Description

Local roadways with medium pedestrian conflict as defined by the IES in roadway lighting RP-8-00 are generally used for direct access to residential or commercial or industrial facilities, tend to carry light vehicular traffic. Few pedestrians (11–100/hour in a typical 200 meter (656 ft.) section) are expected to frequent the area during the night hours.

Considerations

Pedestrian safety and comfort is a primary concern for local roadways. Roadways should be useful and welcoming during night hours as well as the daytime.

Vehicular operation should also be facilitated and encouraged with proper levels of visibility and contrast to minimize night accidents and maximize the utility of a roadway investment.

Lighting Approach

Type II distribution CMH or LED roadway lighting on 24 in. arm mounted on 24 ft poles provide maximum horizontal beamspread with minimal light behind the luminaire directed toward pedestrian ways, thereby minimizing the number of fixtures required to light a designated span of roadway. Cutoff fixtures prevent light pollution by not directing any light above 90 °F (32 °C).

Controls

LED lighting should be specified with integrated occupancy sensors and programmed to reduce to 15% when area is unoccupied and come up to 100% when pedestrian or vehicular traffic is detected.

Metal halide roadway fixture should be controlled by timeclock to switch off when ambient natural lighting can predictably exceed target.

Local with Medium Pedestrian Activity: Sample Layout

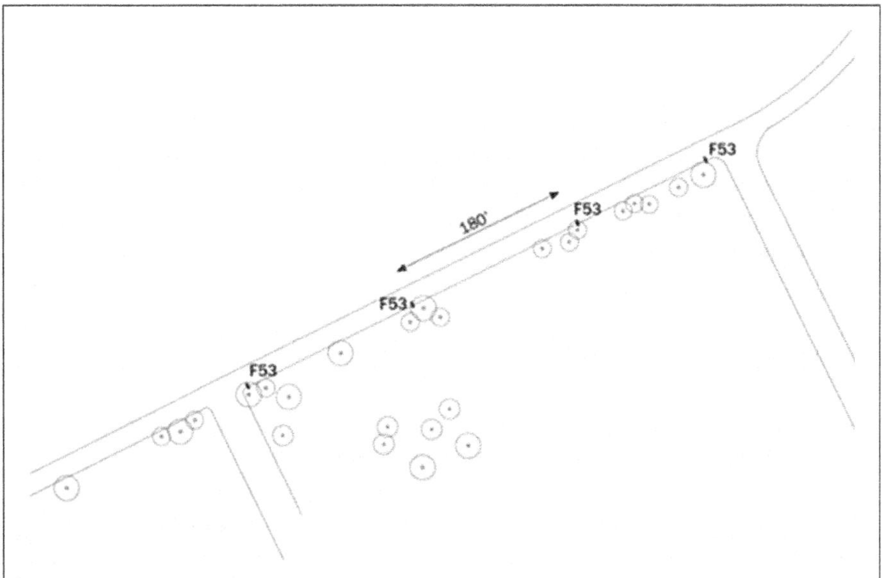

SAMPLE SCHEMATIC LIGHTING LAYOUT (NOT TO SCALE)

COMMENTS	TYPE KEY
LPD = F53: 0.03 W/ft^2	F53: Type II distribution CMH roadway luminaire BF 1.0
F53-LED: 0.02 W/ft^2	Mounting = 24-ft pole with 24-in. arm
Maintained illuminance in circulation areas =	F53-LED: Type II distribution LED roadway luminaire BF 0.75
F53: 0.5-4 fc	Mounting = 24-ft pole with 24-in. arm
F53-LED: 0.6-3.5 fc	
Calculated with ground = 0.10% reflectance	

Education: Kindergarten/Preschool Classroom

Lighting technologies		Target illuminance	Target LPD
Lamp	**Luminaire**	15–30 fc	0.70 W/ft^2
L01 Fluor 32WT8	F01-LED: LED Troffer		
	F01-2: Fluorescent Troffer		
	F12 Wallwash		
Ballast/driver	**Controls**		
B01 Multilevel	C05 Integrated occ./day sensor		
B06&B07 LED dimming drivers			

Space Description

Classrooms may be used for any or all of the following activities: teaching, AV presentations, team exercises, reading, and fine motor activities.

Considerations

Due to space activities, classrooms should be equipped with lighting systems that can provide different lighting scenes (multi-scene controllers or manual switches/ dimmers). For reading and fine motor tasks, the lighting system should be able to provide 40 fc or more on the work plane. For AV presentation mode, the general classroom work plane illuminance should be able to be reduced to between 10 and 20 fc, with no more than 20 fc on the presentation wall. Lighting on teaching walls should be independent from general lighting system.

Lighting Approach

Use highly efficient lensed recessed fluorescent luminaires for general lighting. To provide for multiple uses of classrooms, the lighting system should be divided into two or more zones. Each luminaire in the general lighting zones should be capable of two or more light output modes to achieve recommended illuminances for reading and AV mode. At a minimum, this should include one for general lighting and one for the teaching wall. For larger perimeter classrooms with daylight access, three or more zones may be required, minimally including a general lighting zone, a general lighting daylight zone, and a teaching wall zone.

Controls

Wall-mounted lighting controls should be provided at room entries as well as additional scene recall controls at the teaching wall are desirable for larger classrooms. Smart luminaires with integrated occ./day sensor and wireless grouping functionality are preferred for maximum savings and for providing flexibility to teachers to define various lighting zones and scenes based on their preference. All luminaires must be able to dim to accommodate AV presentations.

In some instances modern technology like Tunable White could be used to help teachers' condition student activities and their mood settings by simply altering the color temperature and intensity of the lighting systems.

Education: Kindergarten Classroom

Sample Layout

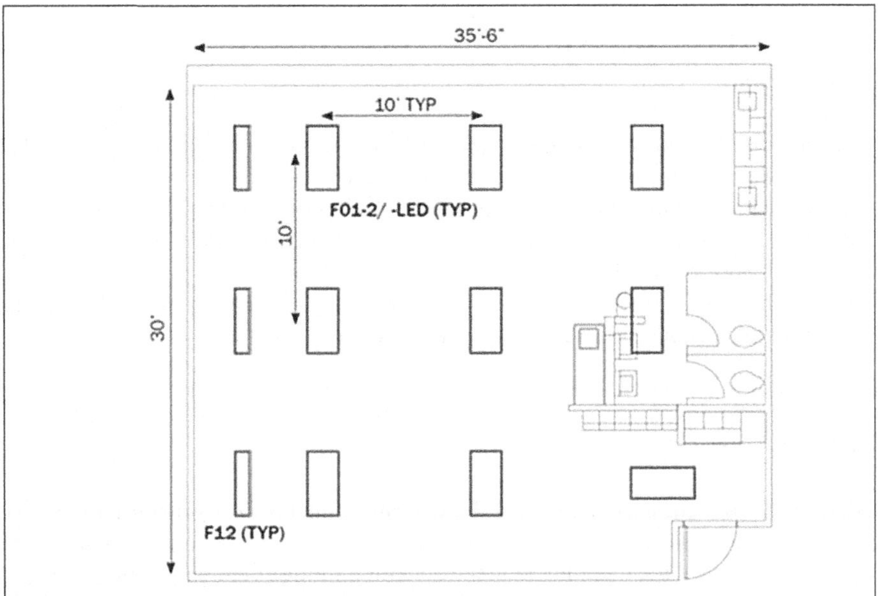

SAMPLE SCHEMATIC LIGHTING LAYOUT (NOT TO SCALE)
CEILING = 2X2 FT ACOUSTICAL CEILING TILE, 8 FT, 8 IN. AFF TYPICAL

COMMENTS	TYPE KEY
LPD = F01-2: 0.60 W/ft^2 F01-LED: 0.50 W/ft^2 Maintained illuminance on work surface = F01-2: 30-50 fc F01-LED: 25-40 fc Maintained average illuminance in circulation areas = 30-50 fc Vertical illuminance teaching wall: 45 fc	F01-2: (2) Lamp 32WT8, 2x4 ft Lensed Troffer, 0.88 BF F01-LED: 2x4 ft LED Lensed Troffer, 0.75 BF F12: (1) Lamp 32WT8, 1x4 ft Wallwasher, 0.88 BF

Education: High School Classroom

Lighting technologies		Target illuminance	Target LPD
Lamp	**Luminaire**	15–30 fc	0.70 W/ft^2
L01 Fluor 32WT8	F01-LED: LED Troffer		
	F01-2: Fluorescent Troffer		
	F12 Wallwash		
Ballast/driver	**Controls**		
B01 Multilevel	C01 IR Occ/Vac Sensor		
B06&B07 LED dimming drivers	C07 Dimming Photosensor		
	C05 Integrated occ./day sensor		

Space Description

Classrooms may be used for any or all of the following activities: teaching, AV presentations, team exercises, reading, and note taking.

Considerations

Due to space activities, classrooms should be equipped with lighting systems that can provide different lighting scenes (multi-scene controllers or manual switches/dimmers). For reading and writing tasks, the lighting system should be able to provide 40 fc or more on the work plane. For AV presentation mode, the general classroom work plane illuminance should be able to be reduced to between 10 and 20 fc, with no more than 20 fc on the presentation wall. Lighting on teaching walls should be independent from general lighting system.

Lighting Approach

Use highly efficient lensed recessed fluorescent luminaires for general lighting. To provide for multiple uses of classrooms, the lighting system should be divided into two or more zones. Each luminaire in the general lighting zones should be capable of two or more light output modes to achieve recommended illuminances for reading and AV mode. At a minimum, this should include one for general lighting and one for the teaching wall. For larger perimeter classrooms with daylight access, three or more zones may be required, minimally including a general lighting zone, a general lighting daylight zone, and a teaching wall zone.

Controls

Wall-mounted lighting controls should be provided at room entries as well as additional scene recall controls at the teaching wall are desirable for larger classrooms. Smart luminaires with integrated occ./day sensor and wireless grouping functionality are preferred for maximum savings and for providing flexibility to teachers to define various lighting zones and scenes based on their preference. All luminaires must be able to dim to accommodate AV presentations.

In some instances, modern technologies like Tunable White could be used to improve the mood in office or classroom settings by simply altering the color temperature and intensity of the lighting systems.

Sample Layout

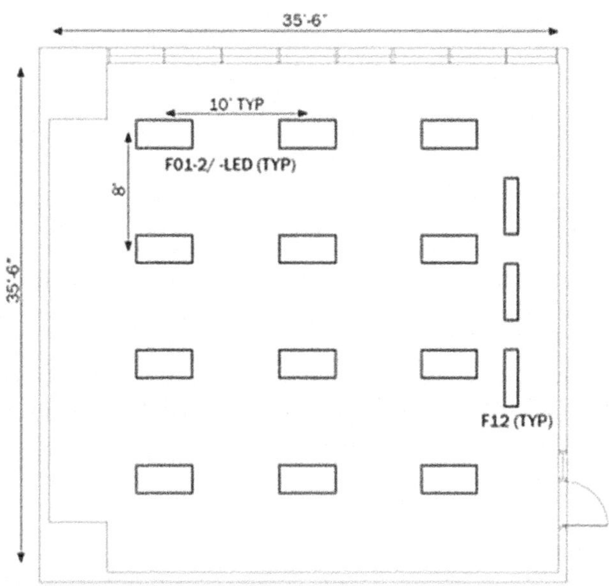

SAMPLE SCHEMATIC LIGHTING LAYOUT (NOT TO SCALE)
CEILING = 2'X2' ACOUSTICAL CEILING TILE, 8 FT, 8 IN. AFF TYPICAL

COMMENTS	TYPE KEY
LPD = -2: 0.63 W/ft² -LED: 0.50 W/ft² Illuminance on work surface = -2: 30-45 fc -LED: 25-40 fc Maintained average illuminance in circulation areas = 30-50 fc Vertical illuminance teaching wall: 45fc Calculated with 80%/70%/20% ceiling/wall/floor reflectance	F01-2: (2) Lamp 32WT8, 2x4 ft Lensed Troffer, 0.88 BF F01-LED: 2x4 ft LED Lensed Troffer, 0.75 BF F12: (1) Lamp 32WT8, 1x4 ft Wallwasher, 0.88 BF

Education: Active Play Area

Lighting technologies		Target illuminance	Target LPD
Lamp	**Luminaire**	30–50 fc	0.50 W/ft^2
L01 Fluor 32WT8	F01-2: Fluorescent Troffer		
	F01-LED: LED Troffer		
Ballast/driver	**Controls**		
B01 Multilevel	C01 IR Occ/Vac Sensor		
B06&B07 LED dimming drivers	C07 Dimming Photosensor		
	C08 Switching Photosensor		
	C05 Integrated occ./day sensors		

Space Description

Multifunctional space that accommodates indoor physical education, as well as various unrelated activities including team exercises, after-school recreation, dining, and small-scale dramatic presentations.

Considerations

Furniture arrangement in these areas is subject to change so lighting should be independent of furniture layout. The most visually demanding task expected – reading – should be accommodated with 20–30 fc for reading. Low illuminance is sufficient for most other activities with 10 to 15 fc being sufficient for most.

Lighting Approach

High-efficiency recessed fluorescent or LED troffers provide 30–50 fc diffuse general lighting for facial recognition, short reading, and writing tasks, and various physical activity. Each luminaire in the general lighting zones should be capable of two or more light output modes to achieve recommended illuminances for reading and AV mode.

Controls

Wall-mounted lighting controls should be provided at room entries. Vacancy sensors are preferred in spaces with access to daylight. Occupancy sensors that automatically set the general lighting to low output mode are acceptable for rooms with no daylight access.

Smart luminaires with integrated occ./day sensor and wireless grouping functionality are preferred for maximum savings and for providing flexibility to teachers to define various lighting zones and scenes based on their preference. All luminaires must be able to dim to accommodate AV presentations.

In some instances modern technology like Tunable White could be used to help teachers' condition, student activities, and their mood settings by simply altering the color temperature and intensity of the lighting systems.

Sample Layout

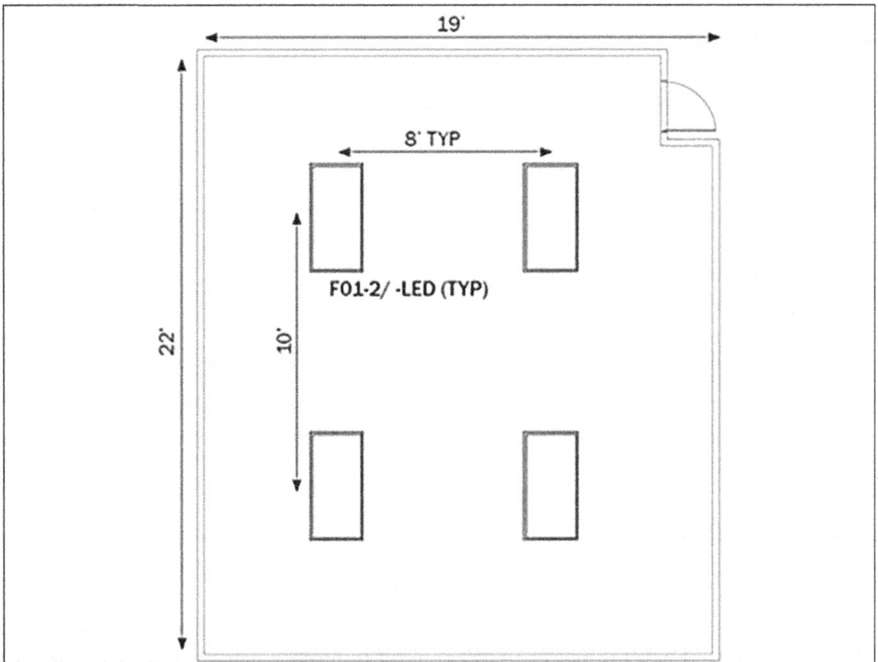

SAMPLE SCHEMATIC LIGHTING LAYOUT (NOT TO SCALE)
CEILING = 2x4 ft GRID ACOUSTICAL CEILING TILE 8 FT, 8 IN. TYPICAL

COMMENTS	TYPE KEY
LPD = F01-2: 0.50W/ft² F01-LED 0.40W/ft² Maintained illuminance on work surface = F01-2: 30-50 fc F01-LED: 20-30 fc Maintained average illuminance in circulation areas = 25 fc	F01-2: (2) Lamp 32WT8, 2x4 ft Lensed Troffer, 0.88 BF F01-LED: 2x4 ft Non-Planar LED Lensed Troffer, 0.75 BF

Education: Staff Lounge

Lighting technologies		Target illuminance	Target LPD
Lamp	**Luminaire**	15–30 fc	0.50 W/ft^2
L01 Fluor 32WT8	F01-LED: LED Troffer		
	F01-2: Fluorescent Troffer		
	F51: LED Task		
Ballast/driver	**Controls**		
B01 Multilevel	C01 IR Occ Sensor		
B04 Program Start	C02 Ultra Sonic Occ/ Vac Sensor		
B06&B07 LED dimming drivers	C07 Dimming Photosensor		

Space Description

The staff lounge includes facilities for minor food preparation, storage (lockers,) and seating. Kitchenettes have a refrigerator, dishwasher, counter top, and storage space and may be equipped with a microwave and toaster.

Considerations

Food preparation areas require higher illuminance than other areas in the space. Low illuminance (5 to 10 fc) is sufficient for all other areas. Luminaires may be unintentionally left on when spaces are unoccupied if they are not automatically controlled.

Lighting Approach

Use highly efficient recessed lensed luminaires with appropriate lamps and ballasts to provide sufficient general illuminance for anticipated tasks without over-lighting. Provide localized integrated under cabinet task lighting at kitchenette.

Controls

Wall-mounted lighting controls should be provided at room entries. Vacancy sensors are preferred in spaces with access to daylight. Occupancy sensors that automatically set the general lighting to low output mode are acceptable for rooms with no daylight access. Luminaire integrated occupancy sensors may be also appropriate in these spaces.

Sample Layout

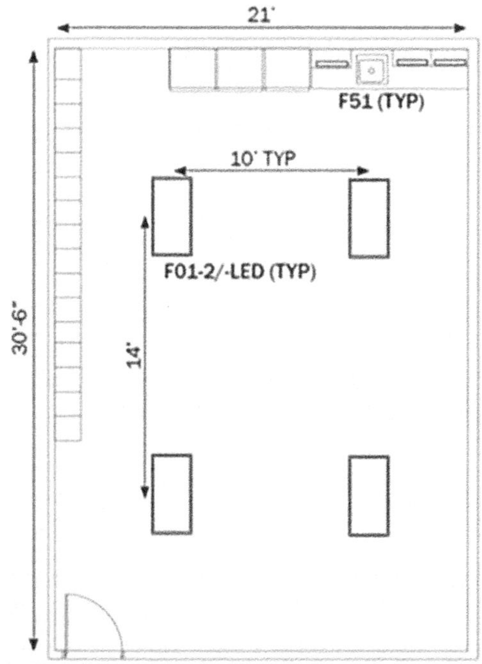

SAMPLE SCHEMATIC LIGHTING LAYOUT (NOT TO SCALE) CEILING = 2'X2' ACOUSTICAL CEILING TILE, 8 FT, 8 IN. AF TYPICAL	
COMMENTS	TYPE KEY
LPD = F01-2: 0.4W/sf^2 F01-LED: 0.36W/ft^2 Maintained illuminance 3' AFF = F01-2: 20-45 fc F01-LED: 10-30 fc Maintained average illuminance in circulation areas = 10-20 fc	F01-2: (2) Lamp 32WT8, 2x4 ft Lensed Troffer, 0.88 BF F01-LED: 2x4 ft LED Lensed Troffer, 0.75 BF F51: 10.5W Undercabinet LED task light (furniture mounted)

Athletic: Gym

Lighting technologies		Target illuminance	Target LPD
Lamp	**Luminaire**	40–50 fc	0.70 W/ft^2
L01: Fluorescent 32WT8	F03 Non-planar Lensed Troffer		
L08: Fluorescent 39WT5HO	F12 Wallwash		
	LED luminaire		
Ballast/driver	**Controls**		
B06&B07 LED dimming driver	C05 Integrated occ./day sensor		
	C07 Dimming Photosensor		
	C09 Timeclock		

Space Description

Fitness areas accommodate basic workout and conditioning and may also accommodate video watching and conversation. Various moving equipment is located throughout the space, and one wall adjacent to free weights may be entirely mirrored.

Considerations

Personal and strength training require a minimum of 40 fc according to the IES standards for lighting of fitness facilities. Harsh, directional sources that cause unflattering shadows and glare experienced by the user while horizontally positioned should be avoided. Screens also present potential for glare from direct sources.

Additional vertical illuminance may be required at a mirrored wall location.

Lighting Approach

Harsh, directional sources that cause unflattering shadows and glare experienced by the user while horizontally positioned should be avoided. Non-planar recessed indirect troffers that provide adequate ambient lighting – 40 to 50 fc – will provide an esthetically pleasing, diffuse lighting that flatters the user and provides an aesthetic form on the ceiling plane.

Controls

In spaces with daylight access, daylight dimming photosensors will ensure steady light levels and maximum energy efficiency. Smart luminaires with integrated occ./ day sensor will provide greatest accuracy in sensing occupancy/ vacancy.

Sample Layout

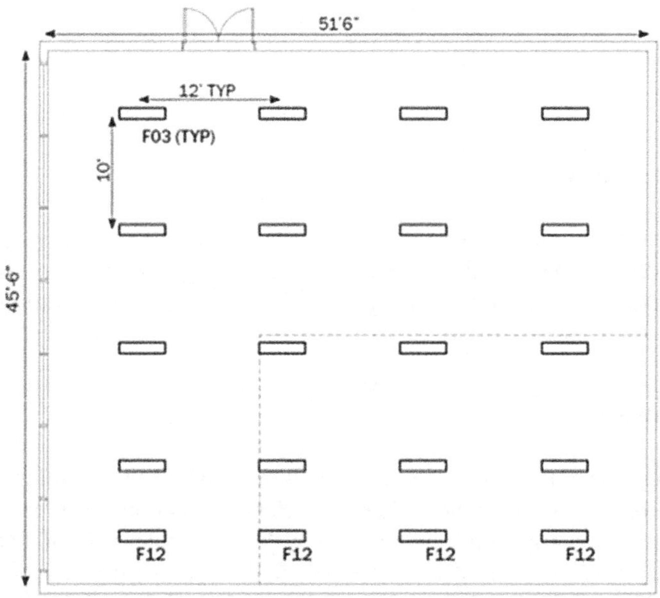

SAMPLE SCHEMATIC LIGHTING LAYOUT (NOT TO SCALE) CEILING HEIGHT = 14' AFF	
COMMENTS	TYPE KEY
LPD = 0.66W/ft^2 Maintained illuminance on 30-in. AFF work surface = 40-50 fc Maintained average illuminance in circulation areas = 35-50 fc Calculated with 80%/50%/40% ceiling/wall/floor reflectance	F03: (2) lamp F39WT5HO. non-planar troffer, 0.88 BF Bi-level ballast F12: (1) Lamp F32WT8, 1x4 ft Wallwasher, 0.88 BF

Athletic: Two-Court Gym With Jogging Track

Lighting technologies		Target illuminance	Target LPD
Lamp	**Luminaire**	50–60 fc	0.80 W/ft^2
L01 Fluor 32WT8	F11 High Bay		
20W LED	F52 LED Downlight		
Ballast/driver	**Controls**		
B06&B07 LED dimming driver	C05 High Bay integrated occ./day sensor		
	C08 Switching Photosensor		

Space Description

Two-court gym with tournament court and jogging track above. Class III facility (Competition play with some spectator accommodations) according to the IESNA Recommended Practice for Sports Lighting. Bleachers located on two walls behind basketball net. Matte finish, 40% reflective floors typical.

Considerations

Basketball is a multi-directional, aerial sport with a minimum of 50 fc required at floor level. Jogging must be accommodated by a minimum of 30 fc. Ceiling height should be kept unobstructed to minimum of 22 ft (6.71m). Suspended fixtures must be able to withstand impact.

Lighting Approach

General lighting provided by suspended fluorescent high bay luminaires with wire guards provide durable, highly efficient general lighting for the courts (40 to 50 fc) and jogging track (35 to 40 fc). Higher level lighting in the circulation areas accommodates for facial recognition and circulation for users acclimation to and from the courts.

Controls

For gyms with access to daylighting, vacancy sensors and daylight dimming can minimize energy usage during hours where daylight can supplement electric sources. Not all courts are used at the same time; with smart high bay luminaires offering wireless grouping functionality, one can create flexible lighting zone per courts, and the granular dimming functionality will deliver high output levels only for occupied courts, while vacant courts operate at low light levels, thereby maximizing energy savings in such spaces.

Sample Layout

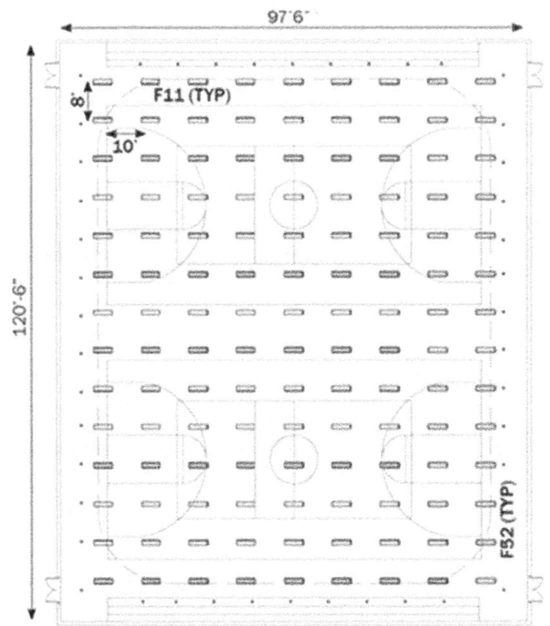

SAMPLE SCHEMATIC LIGHTING LAYOUT (NOT TO SCALE)

CEILING NOTE: EXPOSED STRUCTURAL AND MECHANICAL SYSTEMS TYPICAL; 28' AFF UNOBSTRUCTED

COMMENTS	TYPE KEY
LPD = 0.8W/ft^2 Maintained illuminance on floor plane = 40-50 fc Maintained average illuminance in circulation areas = 30 fc; jogging track = 35-40 fc Calculated with 80%/50%/40% ceiling/wall/floor reflectance	F11: (3) lamp F32WT8, open high bay, 1.0 BF, bi-level circuiting Mounting height = 28'AFF F52: 20W (880lm) LED recessed downlight

Athletic: Racquetball

Lighting technologies		Target illuminance	Target LPD
Lamp	**Luminaire**	50 fc	1.1 W/ft^2
L01 Fluor 32WT8	F01 Fluorescent Lensed Troffer		
Ballast/driver	**Controls**		
B04 Program Start	C01 IR Occ/Vac Sensor		

Space Description

Racquetball Class III facility (competition play with some spectator accommodations) according to the IESNA Recommended Practice for Sports Lighting. Spectators are located at the opposite side of the window wall. Matte finished wall, ceiling and floor panels with of 70–80% reflectance (wall and ceiling) and 50% reflectance typical.

Considerations

Racquetball is a multi-directional, aerial sport that requires a minimum of 50 fc average horizontal illuminance as well as a maximum ratio of maximum illuminance levels to minimum no greater than 2.5. Luminaires should be fully recessed and capable of withstanding impact.

Lighting Approach

General lighting provided by recessed fluorescent 2x4 luminaires with high-impact lenses, providing highly efficient uniform lighting for the court – average of 50 fc. Higher level lighting may be provided at the front playing wall, but where the walls are of a minimum of 70% reflectance, uniform illuminance levels are acceptable.

Controls

Infrared occupancy/vacancy sensors provide appropriate control to ensure that players are never left in darkness and deliver maximum energy efficiency. Controls can be programmed with a 20–30 minute delay and should be directed toward the center and back of room to avoid activation by passers-by.

Sample Layout

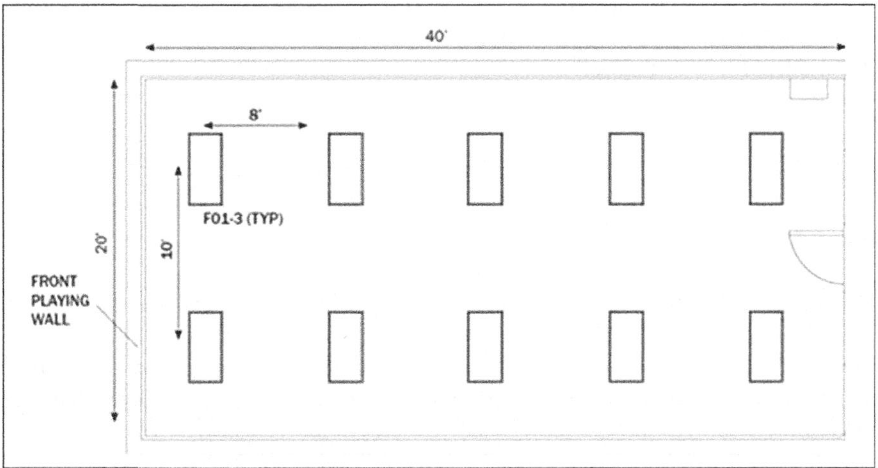

SAMPLE SCHEMATIC LIGHTING LAYOUT (NOT TO SCALE)
CEILING NOTE: PREFORMED HIGH DENSITY RESIN CORE RACQUETBALL COURT PANELS ON T-GRID SYSTEM 20' AFF TYPICAL

COMMENTS	TYPE KEY
LPD = 1.1W/ft^2 Maintained average horizontal illuminance = 54.0 fc Maintained average vertical illuminance = 25 fc Ratio Max/Min = 1.4 Calculated with 80%/70%/50% ceiling/wall/floor reflectance	F01-3: (3) Lamp 32WT8, 2x4 ft Lensed Troffer, 0.99 BF

Athletic: Combatives

Lighting technologies		Target illuminance	Target LPD
Lamp	**Luminaire**	50 fc	0.90 W/ft^2
Fluor 32WT8	F01: Lensed Troffer		
	F03: Non-planar Lensed Troffer		
Ballast/driver	**Controls**		
B01 Multilevel	C03 Dual Tech Occ/Vac Sensor		
B04 Program Start			

Space Description

Combatives Class IV facility (competition play with no spectator accommodations) according to the IESNA Recommended Practice for Sports Lighting. Matte finishes throughout.

Considerations

Combatives encompasses a variety of hand fighting techniques and are a multi-directional, ground-level sport, which requires a minimum of 50 fc average horizontal illuminance, as well as a maximum ratio of maximum illuminance levels to minimum no greater than 2.5. Luminaires should be fully recessed and avoid any fatiguing and unpleasant glare.

Lighting Approach

High-efficiency, recessed non-planar fluorescent luminaires with holographic film coating provide evenly distributed illuminance without offensive or fatiguing glare in the main combatives room. High-efficiency recessed direct fluorescent luminaires provide average illuminance levels in the vestibule and storage rooms of 20 fc.

Controls

Dual tech occupancy/vacancy sensors provide high-sensitivity motion-sensored control to bring general lighting on to 50% when activated; lights must be switched to come on to 100% to provide maximum energy efficiency and ensure the safety and comfort of the user.

Sample Layout

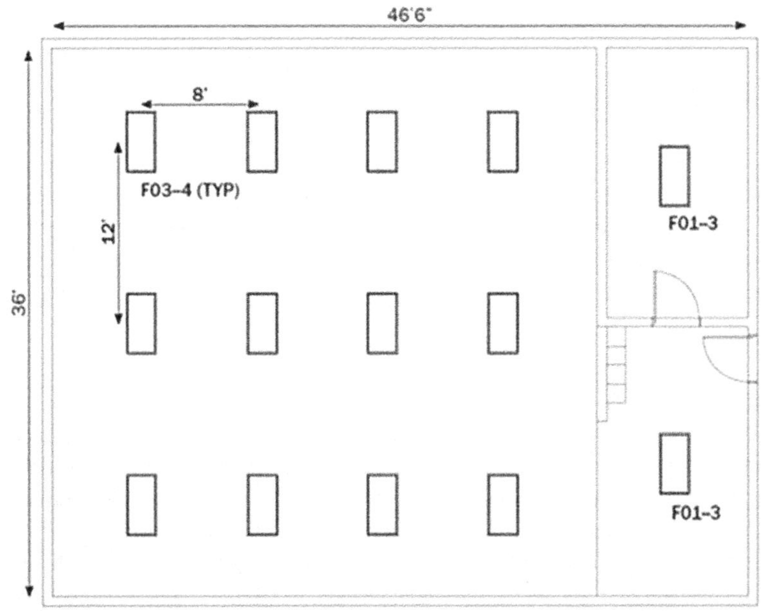

SAMPLE SCHEMATIC LIGHTING LAYOUT (NOT TO SCALE)	
CEILING = 2'X2' ACOUSTICAL CEILING TILE, 10' AFF TYPICAL	
COMMENTS	TYPE KEY
LPD = 0.9W/ft^2	F03-4: (4) Lamp 32WT8, 2x4 ft Lensed Troffer, 0.99 BF
Maintained average horizontal illuminance = 53.5 fc	F01-3: (3) Lamp 32WT8, 2x4 ft Lensed Troffer, 0.88 BF
Ratio Max/Min = 4	
Calculated with 80%/50%/20% ceiling/wall/floor reflectance	

Healthcare: Resident work area

Lighting technologies		Target illuminance	Target LPD
Lamp	**Luminaire**	30–45 fc	0.70 W/ft^2
L01 Fluor 32WT8	F03 Non-planar Lensed Troffer		
20W LED	F51 Task		
	F52 LED Downlight		
Ballast/driver	**Controls**		
L01 Fluor 32WT8	C03 Dual Tech Occ/Vac Sensor		
20W LED	C05 Integrated occ./day sensors		
B06&B07 LED dimming drivers	C07 Dimming Photosensor		

Space Description

Open offices are designed to accommodate multiple individual work areas, typically separated by movable partitions and circulation areas. Individual work areas typically contain a computer, telephone, personal storage, and desk space for reading and writing. Furniture locations are not permanent and may change with needs and staffing. Open offices typically have one or more perimeter window walls that can provide views to the outdoors and usable daylight.

Considerations

Users' age, job function, and occupancy vary in each open office area. Work plane illuminance, as suggested by the IESNA, ranges from 30 fc to 50 fc for most office reading tasks. The visual needs of an older occupant in one work area may be different than that of a younger occupant. In most cases, the circulation space between work areas requires little if any lighting in addition to that provided for work areas. It is typical to find some work areas occupied and some vacant throughout the work day. Direct and reflected glare should be considered. Direct sunlight on work surfaces can contribute to glare and make it difficult to perform work. Lighting in the daylight zone (approximately twice the window head height) can often be turned off or reduced to a low power setting during the day.

Lighting Approach

The lighting system in these spaces should be easily adaptable to suit the needs of occupants. Providing a task/ambient solution is an effective way to minimize energy use while providing sufficient lighting for the occupants. This can be achieved by providing a 15 to 25 fc of ambient (furniture-mounted or overhead) lighting – enough for computer use, facial recognition, and circulation – supplemented by individually controlled task lighting at each workstation. The lack of full height partitions and low-level ambient lighting may leave the space feeling dim in some cases. To avoid this perception, interior full height partitions should be lighted with wallwashers or similar. This is especially effective when lighting the wall opposite the window wall in deeper spaces to balance vertical brightness in the field of view.

Controls

Because of the intermittent occupancy patterns, a lot of energy is wasted in open offices' traditional controls approach, either the entire zone is ON or OFF irrespective of where actual occupancy is. Smart luminaires that offer wireless grouping and granular dimming functionality (bright light levels when there is local occupancy, but dim levels where there is vacancy within a zone/group). Without an integrated approach, multiple ceiling-mounted sensors are required to achieve the desired sensing coverage, and the controllability is limited to how lights are wired; there is no room for space churn and easy adaptability, but with smart luminaires these limitations could be overcome. Smart luminaires also offer daylight sensing with automatic daylight calibration, thereby making the system immune to space orientation and churn.

Sample Layout

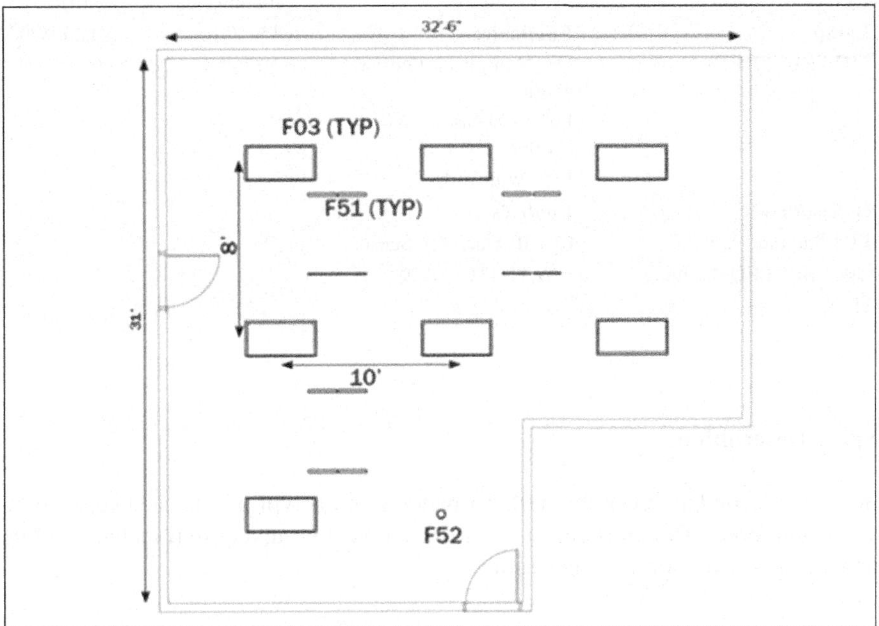

SAMPLE SCHEMATIC LIGHTING LAYOUT (NOT TO SCALE) CEILING = 2'X2' ACOUSTICAL CEILING TILE, 9' AFF TYPICAL	
COMMENTS LPD with task lights = 0.68 W/ft^2 LPD without task lights = 0.4 W/ft^2 Maintained task illuminance = 30-45 fc Maintained ambient illuminance = 27 - 50 fc Calculated with 80%/50%/40% ceiling/wall/floor reflectance and 95% luminaire dirt depreciation (LDD)	TYPE KEY F03: (2) lamp F32WT8 non-planar lensed troffer, 1.00 BF F51: 10.5W Undercabinet LED task light (furniture mounted) F52: 20W Recessed LED downlight

Healthcare: Nurses Station

Lighting technologies		Target illuminance	Target LPD
Lamp	**Luminaire**	15–30 fc	0.80 W/ft^2
L07 Fluor 17WT8	F03 Non-planar Lensed Troffer		
	F03-LED Non-planar Troffer		
	F12: Wallwash		
Ballast/driver	**Controls**		
B04 Program Start	C01 IR Occ/ Vac Sensor		
B06&B07 LED dimming drivers	C09 TIMECLOCK		

Space Description

Serves as secondary reception desk for patients and is typically located centrally to 10+ exam rooms. Contains one or two workstations for nursing staff including filing and storage, work surface, and seating.

Considerations

Facial recognition and color rendering are important considerations in this space since conversation and informal diagnosis can be expected to take place on a regular basis. Reading and writing activities can be reasonably expected by users from 25 years to >65 in the reception desk area.

Lighting Approach

In circulation areas 15 to 20 fc ambient lighting facilitates facial recognition and circulation and short reading tasks. Providing a task/ambient solution is an effective way to minimize energy use, while providing sufficient lighting for the occupants. This is achieved by providing 15 to 30 fc ambient lighting – enough for computer use, facial recognition, and circulation – supplemented by individually controlled task lighting at the workstation. Recessed fluorescent T8 wallwash luminaires highlight the space as a landmark and a destination for patients and provide indirect lighting for short chart and schedule reading tasks.

Controls

The general lighting and the wallwash should be controlled separately; during evening hours the user may choose to use only the wallwash and task lighting since frequency of reading and writing activities are reduced. If the area is not occupied during evening hours, pre-programmed timeclock control may be acceptable.

Sample Schematic Lighting Layout (Not to Scale)

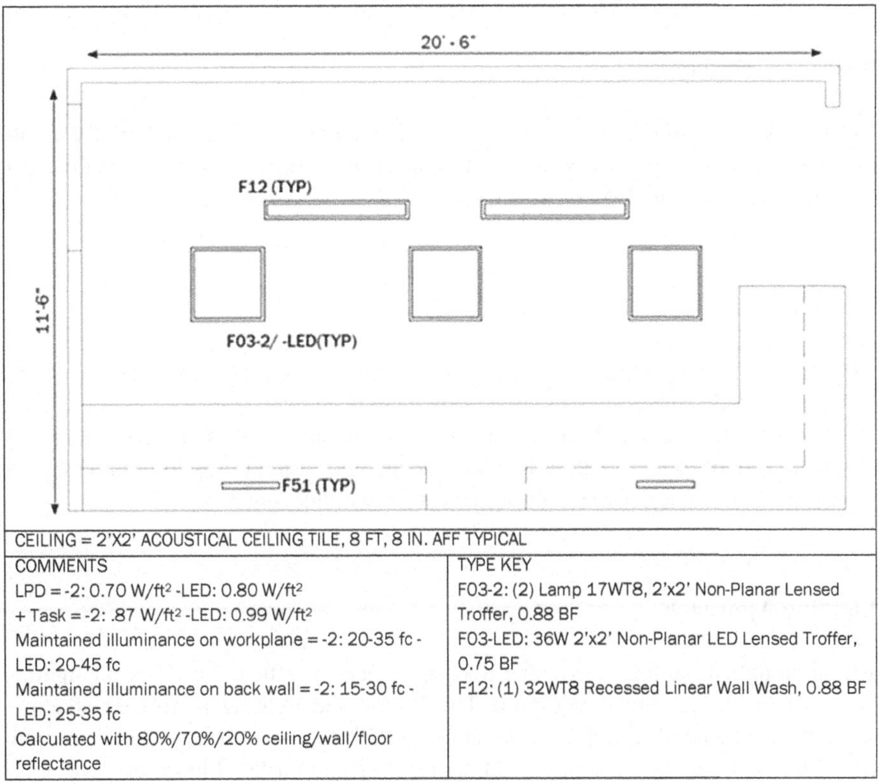

CEILING = 2'X2' ACOUSTICAL CEILING TILE, 8 FT, 8 IN. AFF TYPICAL	
COMMENTS	TYPE KEY
LPD = -2: 0.70 W/ft^2 -LED: 0.80 W/ft^2 + Task = -2: .87 W/ft^2 -LED: 0.99 W/ft^2 Maintained illuminance on workplane = -2: 20-35 fc - LED: 20-45 fc Maintained illuminance on back wall = -2: 15-30 fc - LED: 25-35 fc Calculated with 80%/70%/20% ceiling/wall/floor reflectance	F03-2: (2) Lamp 17WT8, 2'x2' Non-Planar Lensed Troffer, 0.88 BF F03-LED: 36W 2'x2' Non-Planar LED Lensed Troffer, 0.75 BF F12: (1) 32WT8 Recessed Linear Wall Wash, 0.88 BF

Healthcare: Exam Room

Lighting technologies		Target illuminance	Target LPD
Lamp	**Luminaire**	15–250 fc	0.64 W/ft^2
L01 Fluor 32WT8	F02 Direct/Indirect Wall mount		
	F03 Non-planar Lensed Troffer		
	F03-LED Non-planar Troffer		
Ballast/driver	**Controls**		
B01 Multilevel	C03 Dual Tech Occ/Vac Sensor		
B02 Dimming			
B04 Program Start			

Space Description

Exam rooms accommodate (1) patient and (1) medical personnel for brief waiting, examination, and diagnosis. Verbal communication, computer use, and brief reading and writing tasks can also be expected.

Considerations

Accurate color rendering and shadow-free lighting is essential in exam spaces where visual diagnostic screening can be reasonably expected.

Supplementary task lighting for detailed examination may be required and may require fixed or portable task-specific luminaires. Another option that may be desirable is to provide lower lighting levels for waiting rooms.

Lighting Approach

An ambient/task strategy provides maximum energy efficiency. General lighting provided by fluorescent direct/indirect wall-mounted lighting minimizes shadows and provides general illuminance level of 15-25fc to accommodate facial recognition, some computer use, and reading and writing tasks. Fluorescent or LED non-planar troffer provides additional 20–35 fc for up to 60 fc to accommodate visually demanding diagnostic tasks. Minimum CRI of 80 ensures accurate visual diagnostic. Supplemental task lighting whether fixed or portable will vary by application and examination task.

Controls

Occupancy sensors should be used to control general lighting. Task lighting should be manually switched, with minimum of two levels of output.

Sample Layout

F03 EXAM LIGHT
Exam Light Is Not Considered to Be Included in the General Illumination and Provides Additional Illumination up to 60 fc for Medical Use. per ASHRAE Guidelines, It Is Not Part of the LPD Calculation. Exam Light Must Be Operated from Separate Switch.

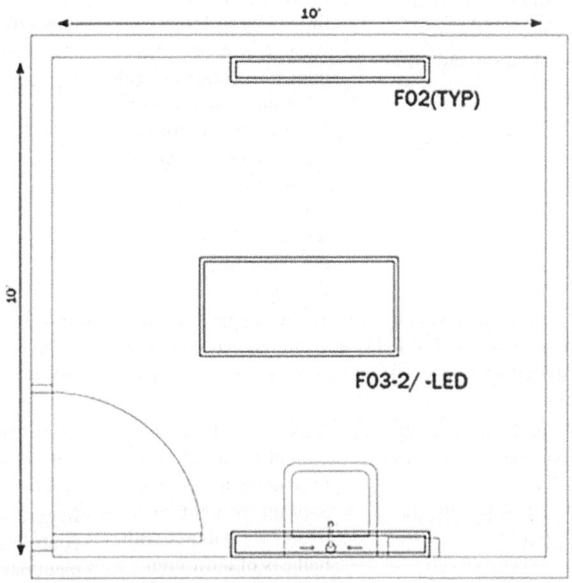

SAMPLE SCHEMATIC LIGHTING LAYOUT (NOT TO SCALE)
CEILING = 2'X2' ACOUSTICAL CEILING TILE, 8' AFF TYPICAL

COMMENTS	TYPE KEY
LPD = -2/-LED: 0.640 W/ft^2 Maintained illuminance on workplane = Ambient only: 15-25fc With Task: -2: 25-50 fc -LED: 20-45 fc Calculated with 80%/70%/20% ceiling/wall/floor reflectance and 95% Luminaire Dirt Depreciation	F02: (1) Lamp F32T8 Fluorescent Direct/Indirect Wall Mount F03-2: (2) Lamp F32WT8, 2x4 ft Non-Planar Lensed Troffer, 0.88 BF F03-LED: 63W 2x4 ft Non-Planar LED Lensed Troffer, 0.75 BF

Table G.1 Interior lighting recommendations

Existing technology	Recommended improvement technology	Recommended (ideal) upgrade	Dos and don'ts (for ideal upgrade)
Incandescent	LED retrofit lamps	Replace standard incandescent applications with LED luminaires with integrated LED drivers. For locations where luminaires cannot be replaced, use screw-based replacement LEDs	Do not use dimming controls unless the LED driver is specified for dimming operation
Note: CFLs are an improvement from incandescent lamps, but are not recommended due to their low service life and low efficacy			
Manual switches	Occupancy-based sensors: Occupancy sensors (auto-on, auto-off) Vacancy sensors (manual-on, auto-off)	Apply PIR for small spaces, ultrasonic where there might be minimal movement or no direct line of sight. Use vacancy sensors where manual-on can eliminate false starts such as conference rooms, private offices with daylight, etc. Ensure that sensors are calibrated to perform as required – test units to eliminate false starts and timeouts	Do not use vacancy sensors where being able to turn lights back on might be an issue if they turn off accidentally
Note: Occupancy sensors are a simple and effective upgrade to the lighting system that can yield great energy savings. Ensuring that lights in conference rooms, restrooms, lunchrooms, copy rooms, and other shared spaces are off when not in use has shown a significant reduction in lighting energy consumption			
Over-lit spaces	T8 or T5 lamps with appropriate electronic ballasts Lower ballast factor ballasts	Redesign lighting system with efficient T5 or T8 luminaires to achieve appropriate light levels or replace ballasts with ballasts of a different ballast factor that will lower lights to provide only what is recommended and necessary (refer to IES guidelines for recommended light levels)	Consider de-lamping only as a last resort option Integrate daylighting controls if there is sufficient daylight
Note: It will be important to educate existing occupants to transition to lower light levels if needed. There are current theories that suggest spectrally enhanced lighting (more bluish light) will allow light levels to be dropped even lower than current IES recommendations. While this idea may have merit, it has not been verified to provide superior lighting environments than standard light reductions in field applications.			

(continued)

Table G.1 (continued)

Existing technology	Recommended improvement technology	Recommended (ideal) upgrade	Dos and don'ts (for ideal upgrade)
Daylit spaces	Daylighting step or continuous dimming controls and photosensor(s)	Integrate a daylight harvesting system with photocontrols and dimmable fluorescent luminaires. Incorporate vacancy sensors where manual-on is practical. Commission the system to reduce light output in response to daylight. Use a ballast with the capability to continuously dim the lights in consistently occupied spaces, or step dim the lights in intermittently occupied spaces. Daylighting controls will only be cost-effective when there is sufficient daylighting available	Ensure that photocells are located to provide accurate sensing of useful daylight
	Vacancy sensors (to achieve further savings during unoccupied hours and allow occupants to leave lights off if desired)		Commission the system to ensure effective operation throughout the day

Note: Daylighting can usually be cost-effective in new buildings where skylights and windows are correctly sized and located and appropriate controls can be integrated. The conditions in existing buildings for retrofits make it more difficult to implement the most effective daylighting strategies, but larger open spaces are still a prime candidate for daylighting control.

Source: Adapted from Energy and Water Conservation Design Requirements for SRM Projects 8.1 Section 8.2.

Glossary

Ballast Factor (BF): The ratio of a lamp's lumen output on a particular ballast to the lamp's rated lumens in the testing environment. Allows for the prediction of actual lamp light output for commercially available lamp-ballast combinations.

Brightness: The attribute used to describe perceived luminous flux per unit area. Cannot be measured or quantified.

Color Rendering Index (CRI): A measurement that describes how similar objects' colors appear under a specific light source as compared to a reference source of like color temperature.

Daylight Zone: An area within a space with enough exposure to daylight that electric lighting may be turned off or dimmed for a portion of the day. For spaces that are primarily sidelighted with daylight, this area extended two times the window height into the spaces.

Efficacy: A measure of a lamp's effectiveness in converting electrical energy into light. Expressed in units of lumens/watt [lm/W].

Energy: The generation or use of electric power over a period of time. Expressed in units of kilowatt-hours.

Glare: An unpleasant or disabling visual sensation stimulated by luminance in the field of view that is significantly higher than the adaptation level of the visual system.

Illuminance: Measures light incident on a point or surface, defined as luminous flux per unit area incident on a point or surface. Expressed in units of foot-candles [fc] or [lm/ft^2] and lux [lm/m^2].

Lamp Lumen Depreciation (LLD): Describes the decrease in lumen output of a lamp during its operable life. Usually expressed as the percentage of initial light output that a lamp emits at 40% of its rated lamp life.

Luminaire Efficiency: The ratio of lumens that exit a luminaire to the lumens that are emitted by the lamp(s) contained within.

Luminance: The magnitude of light energy propagating in a specific direction from an area. Expressed in units of candela/m^2.

Luminous Flux: The time rate flow of light energy. Used to describe the total light output of lamps. Expressed in units of lumens.

Luminous Intensity: The magnitude of light energy propagating in a specific direction from a point. Expressed in units of candelas.

Power: The amount of work done by an electric current in a unit time. Expressed in watts or kilowatts.

Rated Lamp Life: Time, in hours, after which half of a statistically large group of lamps are still in operation, under specific operating conditions.

Reflectance: The ratio of luminous flux reflected off of a surface to the incident luminous flux. Usually expressed as a percentage.

Window Head Height: The distance from the floor to the top of the glazing.

Annex G-1: Strategies for Minor Renovations

Considerations

While many of the strategies of this lighting design guide are applicable to existing buildings, they may not always be the most cost-effective relighting strategy. Opportunities for improved lighting control in existing conditions should always be evaluated since the greatest energy saving benefits are derived from lighting control strategies including occupancy sensors, bi-level ballasts, and daylighting responsive controls.

A dual purpose should be achieved in lighting renovation, improved energy efficiency, and increased user comfort. To this purpose, recommended luminaires generally offer higher efficacies and improved distribution over typical pre-existing luminaires.

Recommendations

For renovation projects, one should consider switching to LED luminaires with integrated sensors because they not only provide the benefit of energy savings from LEDs but also enable the renovation project to be code compliant (ASHRAE 90.1) with respect to additional controls functionality. These smart luminaires also offer additional flexibility and functionality that can increase the overall visual comfort and impression in these spaces.

Because many typical spaces are already wired for two-circuit control, it is recommended that bi-level ballasts are used to provide two levels of illumination. The maximum light level provided by the new system will be within IES recommended levels to provide satisfactory ambient lighting quantity with a reduced connected lighting energy load. Users that prefer lower light levels will be able to reduce their illuminance and energy usage accordingly. The better photometric performance of the new luminaires recommended throughout this guide will improve the balance of horizontal task and vertical wall illuminance to increase overall visual comfort and the impression of brightness in these spaces.

The existing luminaires in offices, corridors, classrooms, training rooms, and reception areas typically consists of a variety of inefficient recessed 2x2 and 2x4 fluorescent troffers with prismatic lenses or deep cell parabolic louvers lamped with 3–4 fluorescent T8 lamps per luminaire. Existing lighting levels should be evaluated before fixture replacement. In some cases, spaces may be over-lit, and reductions to proper levels should be part of the proposed change. The new, lower level of illumination provided by the luminaires recommended in this guide specific to each space type will often seem brighter and more pleasant to the users of the space because the emphasis on indirect lighting reduces glare and emphasizes the illumination of vertical surfaces, thereby increasing the perceived level of general illumination.

In spaces where maintaining luminaires is difficult and time-consuming such as warehouse spaces and maintenance areas with high ceilings, high-performance LED high bay luminaires offer a long rated life (50,000+ hours) and are an energy-efficient alternative to CMH or high-pressure sodium and require far less maintenance than 8-10 lamp fluorescent luminaires.

When considering changes, verify the expected use of the space as far into the future as possible to ensure that chosen project light levels will be appropriate. Refer to Space-Specific Recommendations in this guide for General Target Illumination Levels, as well as task-specific illumination levels described in the Lighting Approach.

In all typical areas, select ceiling tiles will require replacement as 2x4 and 2x2 luminaires are to be removed, leaving voids in the ceiling. Pendant luminaires require a full tile at the ceiling plane with cutouts for the mechanical and electrical connections to pass through.

A key to energy savings throughout renovation is high-performance T8 lamp technology. These lamps, often called super T8 lamps, have a cost premium, but provide more light, better color, and better maintained light output and last longer than the 735 series T8 lamps. By using high lumen lamps coupled with highly efficient luminaires, the proposed changes improve the lighting maintenance requirements by reducing the quantity of lamps that need to be stocked and maintained as well as reduce the relamping frequency. LED non-planar lensed troffers are becoming more cost-effective. As of this writing, the first cost and energy savings have made them almost compatible with high-performance T8 systems (Table G.1).

Analyzing the Cost of a Lighting System (Source: *Energy and Water Conservation Design*, Section 8.6.)

The implementation of advance lighting systems varies in cost and complexity. The application options discussed in this appendix will incrementally reduce lighting energy consumption and have been proven to be cost-effective for most applications. The following general simple payback calculation illustrated below can provide a basic indication of the potential cost-effectiveness of a lighting project.

$$Y_{Payback} = \text{Cost of upgrade}/(T_A * E_{Rate} * W/100)$$

where:

$Y_{Payback}$ = length of time (in years) it takes for energy savings to pay for new lighting system
T_A = annual length of time affected luminaires are operated
E_{Rate} = cost of electricity (per kWh)
W = (difference in) wattage of affected luminaires

The simple payback calculation for lighting systems is highly dependent on the method of control, hours of operation, and complicated utility rates. For a more in-depth cost analysis, be sure to find a more complete method of analysis that characterizes all of the variables specific to the project (the IEA lighting templates are a great place to start understanding the affects specific technologies can have on energy consumption).

Summary

By carefully considering retrofit strategies and integrating task/ambient approaches to spaces that previously have not had task lighting (or were over-lighted), large energy savings can be gained. The tendency to simply replace existing fixtures one for one with new technology often leads to the lowest acceptable savings. Deep retrofits that take into account not only lighting energy savings, but the potential to resize HVAC components also offers the greatest potential for overall energy savings.

Appendix H: Product Delivery Quality Assurance Process (PDQA) for Deep Energy Retrofit (DER)

Introduction

A DER building project must be properly implemented through all phases to accomplish the goals and required performance levels of the Owner. Like all construction projects, there are many steps, decisions, and operations that require an orderly and logical process to be successful. A properly implemented DER will increase a building's value, improve its indoor climate and thermal comfort, and meet Owner's energy and sustainability goals. DER is best accomplished by adopting a project-specific PDQA process. The process as described in this document supplements those procedures addressed in current standards and guides of which specifically address DER, sustainability, and energy conservation in buildings (NIBS 2012, ASTM 2015 and ASHRAE Guideline 0-2013). The PDQA for a specific project must be developed to suit the needs and goals of that specific project.

A solid PDQA for a DER project starts with:

- Formulation of detailed definition for project requirements and criteria, such as the SOW or Owner's Project Requirements Document. These documents establish the basis for the design of the building, against which tenders (i.e., bids) will be made for both design and construction services. Acceptance of these criteria indicate verification of understanding of the criteria by those proposing to provide design or construction services.
- Proper definition in SOW/OPR of areas of major concern to be addressed and checked during the entire design construct, commissioning, and post-occupancy phases.
- Clear delineation of the responsibilities and qualifications of stakeholders in this process.
- In addition to defining the Owner's Project Requirements and criteria through the SOW/OPR process, the methodology for validating that both the design and construction will deliver the desired results must be considered and identified within the project contract documents for each step of the process.
- Funding sources and business model which will be used to implement the project need to be identified early that adequately cover the scope of the project. Well-intended underfunded projects will fail to meet all of the intended goals in some manner.

This PDQA process is applicable to DER projects using both design-bid-build (DBB) and design/build (DB) procurement methods. Select elements of the PDQA process may be adaptable for use with even minor renovation and system replacement projects. Based on whether the enclosure work includes full removal and replacement only or also additions, much of the field PDQA process that has been applied to the retrofits would be applied to enclosure of addition as well.

When properly implemented, a PDQA process will have a significant effect on the building enclosure performance improvement of the completed work.

Annexes H-1, H-2, H-3, and H-4 to Appendix H provide additional support for the user:

- H-1 Selected Energy-Related Parameters and Their Targets for DER Projects to be Included in the SOW/OPR
- H-2 Examples of Accountability and Responsibilities Matrix
- H-3 Mock-ups
- H-4 Mock-Up Process

How It Works

The PDQA process can be considered as an eight-phase operation, one for each of the basic phases for all construction projects.

The eight phases of construction projects can be viewed as:

1. Predesign/Programming
2. Design Procurement
3. Design
4. Construction Procurement
5. Preconstruction
6. Construction
7. Acceptance
8. Post-occupancy Evaluation

In addition to these eight phases, there is the commissioning process, which cuts across all of these phases. Not every construction project will be recognized to fall into these specific defined steps; however nearly every project contains these steps in one form or another. The one phase that is least often accomplished formally is the last post-occupancy evaluation.

The following provides a brief summary of each of the eight phases.

1. **Predesign/Programming** – Development of SOW/OPR must provide a clear and concise documentation of the Owner's goals, expectations, and measurable performance criteria, cost considerations and benchmarks, and success criteria to be obtained by the DER for a building renovation. The SOW or OPR shall be used throughout the project delivery to provide an informed baseline for the renovation and focus for design development and for validating building' energy and environmental performance. Based in part on this document, bidders for both design and construction services will be selected on demonstrating their ability to deliver a DER renovation project that will accomplish the goals of the Owner as defined in the SOW or OPR of the project. Use quality-based sampling for verification of each activity or task to determine how well it meets or relates to the OPR in the Predesign phase. This includes programming documents, defined scope-of-design services, special reports and workshop outcomes, and other activities in the Predesign phase. The determination of the type of procurement process to be used will help frame how the design and construction is selected and

the amount of funding required for each phase of the project. In some cases, the Owner may elect to obtain assistance in writing the SOW and/or OPR, and in evaluating the proposals before future phases of the project are awarded. Owners with little experience in building projects in particular should consider seeking guidance.

2. **Design Procurement** –The purpose of this phase is to procure the services of a designer who the Owner determines will be well suited to provide professional leadership and the design and technical services necessary for the project. The designer must demonstrate a clear understanding of the Owner's Project Requirements, as established by the owners SOW and OPR during the predesign phase of the project. In addition, the designer must demonstrate previous design and analytic experience, ability to successfully coordinate different design disciplines, and the ability to deliver the deep energy renovations that meets the SOW/OPR criteria.
   The procurement method for establishing the project contractual basis may be different from project to project. However all parties should be experienced in the specific project delivery process selected, should demonstrate a clear understanding of the process, and should be comfortable with the approach. Proposals for design services shall be solicited in explicit and clear language. Include a statement regarding design professional commissioning responsibilities and scope in the request for design services. In some cases, the Owner may elect to obtain assistance in writing the SOW and/or OPR, and in evaluating the proposals before award. Owners with little experience in building projects in particular should consider seeking guidance.
3. **Design** – The design phase starts with concept development by the designer and continues through the completion of documents for bidding or negotiating. During the design phase, the following is established: the appearance, configuration, basic system selections, terminations, materials, performance criteria, and interface conditions with other building systems. There is a set of procedures that the designer must follow at this time to make sure the exterior enclosure systems are appropriately considered, designed, specified, and drawn to attain an enclosure that performs properly, and in compliance with the OPR. Assign operations and maintenance personal to participate in the design phase coordination meetings. The designer must identify how any work excluded to meet budget constraints will affect the OPR and adjust evaluation criteria accordingly.
4. **Construction Procurement** – This should include analysis of construction bidder's qualifications, their understanding of the OPR, previous construction and validation of experience and ability to coordinate different construction trades, performance to meet established schedule and budget, and deliver the deep energy renovations in compliance with the SOW/OPR. Bidders should provide evidence of experience installing any type of specialty system that is not commonly used on all types of construction projects.
5. **Preconstruction** – As with the procurement phase, there are provisions the designer can build into the documents, including drawings, technical specifications, and front-end documents for the Preconstruction phase, the exterior enclosure systems, and other DER provisions.

The Preconstruction phase covers the activities between award of the contract and delivery of materials, products, and systems to the building site. This includes final design and engineering, completion of mock-up construction and testing necessary before production, and fabrication and delivery of materials and systems for incorporation into the building.

6. **Construction** –When implementing quality assurance procedures (QAP) during the Construction phase, it may be necessary to explain the intent and process of the QAP to all parties. Many site representatives, manufacturers, and tradesmen may not be familiar with the process and may not understand the goals and objectives. However, if you can successfully explain the program and gain their active support and participation, the results can be impressive. Assign operations and maintenance personal to participate in the construction phase coordination meetings.

   During the construction phase of the project, the previous efforts of the Owner, designers, consultants, fabricators, material suppliers, and contractors are brought together for the true test. Will it work, will it fit, will it look right, will it perform properly, and can the work be completed on time? If the project team has performed their jobs correctly, and if they have communicated properly, and paid attention to the details, the answer will be yes.

   In-progress testing and inspection of the constructed work, as it occurs, is one of the primary tools of the Owner to assure compliance with the project requirements and the SOW/OPR for deep retrofit work.

7. **Acceptance** – Acceptance can be considered an ongoing process applicable to any or all of the phases indicated for a DER, including post DER occupancy and use of the facility. It can be applied in different forms for different projects.

   Acceptance should include a specific and predetermined approach that is included within the SOW/OPR, and the construction contract documents. Testing and inspection of the completed work is often the last opportunity to assure compliance with contract documents and performance levels of the various systems before the building is accepted on behalf of the Owner.

8. **Post-occupancy Evaluation** – Within the warranty period, key DER project performance should be evaluated to determine whether primary project goals related to energy reduction and system operation are being met. The post-occupancy evaluation should be conducted by members of the original PDQA team or a third-party commissioning specialist separately contracted by the building Owner.

   Post-occupancy evaluation can be tailored depending on Owner budget and project goals. The primary deliverable of the post-occupancy phase is an ongoing commissioning report that documents key building performance metrics including extrapolated annual EUI, zone temperature set point deviation profiles, and HVAC system trend data summaries. Secondary deliverable is an updated Issues and Resolutions Log. Require in the SOW/RFP that the contractor perform corrective action for issues determined to be unrelated to deviations from design or unforeseen conditions within the warranty period. Additional optional deliverables include seasonal endurance testing, occupant comfort surveys, Lessons Learned Workshop, and DER project document updates based on the Ongoing Commissioning (OCx) Report results.

Predesign: Phase I

Development of Statement of Work/Owner's Project Requirements (SOW/OPR)

The SOW/OPR are intended to define the scope, goals, and functional performance requirements of the building undergoing DPR renovation. These documents must address both the design and construction phases of the project, and the expectations of the Owner relative to the intended use, occupancy, and service life of the DER aspects of the project. These documents must be prepared in sufficient detail to define criteria for both design and construction services and will provide the basis for verification of understanding and ability to provide services meeting these goals by potential designers and constructors.

The project SOW/OPR should clearly define each goal of the DER project and must be established with realistic and definable goals and criteria. These goals and criteria must be established with the recognition of the limitations and opportunities of the existing building; the goals must be reasonably accomplishable as demonstrated by concept level engineering analysis, and in the recognition of limitations of contemporary available design and analytical tools, and construction materials, systems, and products. SOW/OPR, which are anticipated to require advanced analytic, design/engineering or materials, products, or systems beyond those readily attainable in contemporary terms, may require a different and specialized approach beyond the scope of this Appendix.

It is important that the SOW and bids include specific energy targets, energy security, and system redundancy requirements to be achieved through the DER. Providing the appropriate level of specificity in the SOW/OPR is the first step in assuring that these requirements become contractually binding and will be attained.

There are currently Standards and Guidelines in place through many standards and governmental organizations providing detailed guidance for development of these documents. A current list of such standards and guides and the corresponding agencies is annexed to the end of this Appendix.

Each project and building will be unique and each will require considerable thought in preparing appropriate and attainable SOW/OPR requirements. However, the list below provides a starting point for discussion and perhaps preparing a first draft of an SCW/OPR for many projects. Since each project and building are unique, this list will serve as a guide, but is not exhaustive or fully applicable for all projects.

The SOW/OPR may be prepared by the Owner when they have knowledgeable staff skilled in providing such services. Preparation of the SOW/OPR may also be contracted to consultants specializing in the specific areas of DER work being considered.

- In addition to defining the Owner's Project Requirements and criteria through the SOW/OPR process, the methodology by which the goals and expectations have been established must be cross-checked before finalization to confirm that they are both reasonable and attainable within the project definition of scope, budget, and schedule.
- SOW/OPR, which are not reasonably attainable and do not confirm to other project constraints, will not lead to successful projects.

Building Utility Consumption Data

This information provides the baseline for measuring the energy savings of the completed DER project. To determine proper goals, and compliance with these energy goals, it is necessary to determine/obtain the following before construction:

1. Building use and operating schedule before renovation
2. Number of occupants, occupant schedule, and activity level before the renovation
3. Requirements to indoor air temperature, ventilation rate, and RH before the building renovation
4. Design drawings for the building enclosure showing construction of existing walls, floors, and roofs (as-builts if possible)
5. Design drawings showing building/floor plans and any drawings for modifications from original construction
6. Building orientation
7. Building elevations showing existing opening/fenestration characteristics, including materials, anticipated performance characteristics, sizes, numbers, and locations
8. Types of HVAC system and equipment list (for both cooling and heating systems) currently installed and known modifications from original building construction.
9. List of existing lighting fixtures (types and quantities)
10. Plug loads (types, quantity, and power consumption requirements)
11. Sub-metering data, if available
12. Building envelope data as determined by investigation, including:

    a. Air infiltration as determined by blower door tests
    b. Infrared camera investigation indicating thermal short circuits and air infiltration/exfiltration locations
    c. Building geographical location with the weather file source.

13. Building area breakdown by end-use.
14. HVAC sequences of operation.
15. Interval sub-metering of separate equipment or processes, at least 1 week.
16. Site-source utility factors, from utility source preferred.
17. A pre-renovation building energy model can be calibrated against utility data using one of the following standard approaches:

    a. If the use and building schedules remain unchanged by renovation, the input data listed in items 1 through 16 above is usable for modeling of the post-renovation scenario.
    b. If the use and schedules of the building will change, this information must be provided in the SOW/OPR and used for the post-renovation building model.

Design Procurement: Phase 2

For the bidding process, it is important that the contractor presents a review of the energy requirements for the project to include site and source energy targets, energy calculation and modeling methodologies, and a discussion and resolution of any conflicts or questions to the SOW/OPR. Rather than a restatement of RFP content (which will not be allowed), this review document will demonstrate the contractor's understanding of project and system-level performance goals through description of solutions, challenges, and potential mitigation. The models presented during the bidding and design processes shall provide an initial list of all energy parameters for the project to include operational runtimes, load peaks/schedules, equipment efficiencies/set points/sizes, insulation values, etc. Each entry/value shall be identified as being supplied from the project SOW, a specific design guide, a specific energy standard. These models have to be calibrated against the baseline energy values of the building before DER and should show building schedules before and after renovation. The modeling results should verify that the proposed design meets the Owner's energy targets and shall be reviewed by the Owner to assure assumptions and calculations are valid.

Procurement of design services for DER renovations in buildings is similar to procuring services from a professional for any design service. The following steps are recommended for all DER projects:

- Prequalify those competing for the contract by following a two-step process.
- Submit request for qualifications (RFQ) to establish a list of the most qualified candidates.
- Obtain specific proposals from the list of candidates determined to be most qualified.
- Follow a specific process when issuing the RFQ. The goal is to identify the firms or individuals most likely to provide the required design services most successfully. Projects are sometimes not successful due to the failure to properly vet the qualifications of those invited to perform services in the RFQ.
- Provide clear, unambiguous parameters describing the types of services and expertise to be provided:
  - Obtain third-party assistance when the Owner's organization does not have sufficient levels of expertise to adequately define and evaluate specific aspects of the qualifications required to provide the stated project needs.
  - Provide the same information to all parties invited to submit qualifications. Provide the SOW/OPR for review by those submitting qualifications.
  - Indicate the related design criteria which the designers need to be knowledgeable in applying to the project.
  - Indicate the related design criteria that the designers need to be knowledgeable in applying to the project.
  - Evaluate qualifications based on the criteria issued.

- Meet with the likely finalists to be invited to submit proposals for services before making final determination of bidders list.
- Evaluate the qualifications of each organization on the same basis, the stated goals.
- Evaluation is best performed by a team representing various aspects of the Owner's organization and the technical needs of the project.
- Evaluate bidders on the basis of past performance, technical capacity, and understanding of industry standards and guidelines relating to energy and sustainability practices for design and construction.
- Bidders must/should show their proficiency in executing work as required by the RFP.

- Require bidders for design services contracts to provide the following items as part of their pre-qualification, or as part of their proposal for services:

  - Information reflecting past experience with energy-focused building renovation projects. Including narratives and examples of technical solutions that are relevant to the DER project or similar building type: e.g., how air barriers were selected and designed, tested, and verified; examples of proper detailing of continuous insulation; examples of energy-efficient HVAC and lighting systems designs; adjustments made for unexpected conditions.
  - Qualifications and past experience on energy modeling and calculations. Including narrative description of design approach, distinguishing between site and source energy to comply with energy requirements of RFP. If possible, submit examples of previous energy models/calculations for renovation projects.
  - Demonstration of past proper understanding of energy calculation strategy as outlined in RFP; specifically how site and source energy will be calculated and reviewed.
  - If possible submit experience with renovations of similar building type. If not available, provide examples of other types of successful energy renovation projects.
  - A review of the energy requirements for the project to include site and source energy targets; energy calculation and modeling methodologies.
  - Demonstration of the designer's understanding of project and system-level performance goals through description of solutions, challenges, and potential mitigation strategies.
  - Examples of probable design cost to actual construction cost and percentage of difference related areas outside of design control.
  - At a minimum, the designer may need to be familiar with the following procurement methods to provide adequate service to the Owner during this phase of the project:

    - Design/Bid/Build – "traditional" in the past
    - Design/Build
    - Design Assist

Integrated Project Delivery.

- Evaluation Criteria: Bidders will be evaluated on past performance. Specifically, proficiency in designing similar building type renovation, and in designing a renovation inside a constrained building shell and how well they met the energy objectives.
- During the pre-bidding process, require prospective bidders for design services to attend a site visit with a building walk-through to get an idea of the existing building conditions and space available for mechanical equipment and to validate information for the energy model and clarify issues that can arise during the SOW/OPR review.
- In addition to defining the Owner's Project Requirements and criteria through the SOW/OPR process, the methodology for validating that the design will deliver the desired results must be considered and identified within the project contract documents for design.
- The bidder shall provide a preliminary results of modeling analysis to demonstrate theoretical feasibility of meeting energy targets and understanding what it will take to achieve energy goals using simulation program allowing for monthly analysis along with an initial list of all energy parameters for the project to include operational runtimes, load peaks/schedules, equipment efficiencies/set points/sizes, insulation values, etc. Each entry/value shall be identified as being supplied from the project SOW, a specific design guide, a specific energy standard.
- Follow a specific process when issuing the RFP. The goal is to determine the firm or individual most likely to provide the required design services.
- Provide clear, unambiguous parameters for method of submitting proposals for service.
- Provide a clear understanding of intended method of selection including any scoring or grading parameters that Owner intends to follow in selection process.
- Provide clear, unambiguous, description of SOW/OPR.
- Provide opportunity for all proposing to provide services to meet with the Owners project team before submitting proposals for services, including any consultants participating in establishing SOW/OPR.
- Allow adequate time for response to RFP.
- Allow opportunity for those proposing to provide services to submit questions regarding proposed services and any information provided by Owner.
- Distribute questions and Owner response to all proposing to provide services.
- Answer questions honestly and directly.
- Remove ambiguities whenever possible. Document meeting and agreements made during meeting and distribute to all involved parties.
- Revise SOW/OPR if necessary when questions from designers indicate a problem that may not be able to be overcome.
- Provide questions and answers to all proposing to provide services.
- Require (mandatory) all proposing to provide design services to attend a meeting at the building before proposals.
- Make reports, record documents, and provide other information addressing the project and existing building available to all proposing to provide services.

- Validate scope of work, fee and schedule with apparent successful bidder before award of contract.
- Confirm apparent successful bidder has appropriate staff to perform the services in a timely manner.
- Resolve any outstanding issues or questions before award of contract.
- Do not enter into a contract without clear documentation of any concerns.
- In the event that agreement cannot be reached with first choice bidder, move on to next bidder and perform same validation process.
- Hold a project kick off meeting to confirm protocols, procedures, and schedule as soon as possible after award of contract.

Design: Phase 3

These guidelines apply to the renovation of exterior building enclosure systems for DER projects. During the design phase, attention must be paid to architectural details to be used for the building enclosure renovation; continuity of thermal; water, vapor, and air barriers; enclosure systems coordination; and constructability of the design. Third-party review, or commissioning, can provide a valuable tool in this effort.

It is important to clearly understand at the commencement of the design phase that the design shall not be accepted until it is adequately demonstrated that the energy targets can be met by the proposed design. If the designer cannot produce a design that meets the energy targets and criteria of the SOW/OPR, a review of the problems or constraints shall be presented along with a list of the proposed changes to the requirements.

At a minimum, the following should be considered during the design phase:

- Wind exposure
- Precipitation conditions
- Ambient outdoor conditions of temperature and humidity
- Daylighting improvements
- Solar exposure
- Interior pressurization
- Interior ventilation temperature and humidity conditions
- Energy goals stated in the SOW/OPR
- Constructability
- Review and approval schedules
- Mock-up and testing requirements
- Manufacture/fabrication schedules
- Construction/erection schedules
- Compatibility with adjacent systems
- Appearance criteria
- Durability/service life
- Initial cost
- Life cycle cost

- Code/regulatory approvals
- Maintenance requirements
- Water resistance
- Ability to carry structural loads:
  - Wind
  - Seismic
  - Maintenance
  - Gravity
- Air infiltration/exfiltration resistance
- Fire performance
- Acoustical performance
- Blast/glass shard resistance
- Ballistic resistance
- Security or forced entry resistance
- Thermally induced movement
- Thermal isolation of exterior and interior
- Moisture migration resistance
- Durability of finishes
- Durability of base materials
- Chemical resistance
- Required phasing for occupation of spaces or connection to critical equipment/ systems during renovations

QA efforts by the designer during the design phase should include the following basic steps:

- Evaluate project criteria that establish characteristics of the acceptable end product.
- Make sure the items in the list above have been addressed.
- Review the schedule for completion of the project including consideration for:
  - Owner reviews
  - Design team checking
  - Cost estimating
  - Manufacturers/fabricators review and comment
  - Consultant reviews
  - Review by regulatory or code agencies
  - Peer reviews, redesign
  - Special testing or analysis such as wind tunnel load analysis
- Confirm that good communication procedures between the project team have been established including:
  - Design team
  - Owner/tenant
  - Contractor/construction manager

  - Consultants/testing agencies
  - Cost estimators
  - Manufacturers/fabricators/erectors

- Implement periodic reviews and sign-off of the criteria and design as they are developed. These can be "on-board" reviews that do not stop the project development, or they can be scheduled for a longer length of time with envelope development halted until the review is completed and the review comments are received. All review comments should be addressed and resolved.
- Require final sign-off for the documents before they are released for bidding or negotiations. The documents will often receive a more conscientious review when formalized sign-off procedures are established.
- Models presented during the design phase shall provide an initial list of all energy parameters for the project to include operational runtimes, load peaks/schedules, equipment efficiencies/set points/sizes, insulation values, and other applicable points of information affecting model validity. Each entry/value shall be identified as being supplied from the project SOW, a specific design guide, or a specific energy standard. These models are to be calibrated against the baseline energy values of the building before DER and show building schedules before and as predicted after renovation. The modeling results should verify that the proposed design meets the Owner's energy targets and shall be reviewed by the Owner or their representative to assure assumptions and calculations are valid.
- On completion of 65% design, the designer shall provide results of detailed energy modeling using computer-based program such as DOE-2, Energy Plus or similar per the ASHRAE Standard 90.1, Section G2.2.1, to demonstrate theoretical feasibility of meeting energy targets.

Construction Procurement: Phase 4

The following guidelines apply to all disciplines within specific DER projects.

The procurement phase may include negotiating or bidding, as a means of selecting a contractor to provide the exterior wall construction. In today's construction market, there sometimes seem to be as many methods of contracting for construction as there are projects. The specific method selected may depend on a number of factors including:

- Size of the project
- Cost of the project
- Complexity of the design
- Local or regional market conditions
- Overall project delivery system
- Material availability
- Schedule requirements
- Local or regional practices
- Client purchasing preferences or requirements

The following precautions are designed to assist in attaining an accurate understanding of the project criteria on the part of the bidders, or in the case of award by negotiation, the parties making proposals, and on the part of those considering the bids and proposals (the designer and Owner). All major issues should be understood by both sides before a contract agreement is reached. To this end, follow these guidelines to assist in the success of the procurement phase:

- Allow adequate time for the preparation of bids or proposals.
- Make complete sets of documents available to all bidders.
- Require pre-bid conferences to allow bidders the opportunity to ask questions regarding the documents. If the project is a retrofit or renovation include on-site walk-through.
- Require mandatory attendance at the pre-bid meeting as a condition of bid.
- Require certification by bidders that their bid is based on full contract documents including addenda and that they meet the quality standards of the documents for length and type of prior experience.
- Limit the number of pre-bid addenda to avoid confusion. However, if pertinent and legitimate questions arise during bidding, make sure that they are answered by an addendum communicated to all bidders.
- Do not provide verbal responses to bidders' questions. If a response is required, include it in an addendum or in some logical manner to assure all bidders have the same information available to them.
- Document the results of the pre-bid meeting to all participants by addendum. Do not allow dependence on the spoken word. Require that only the written minutes issued by an addendum be relied on.
- On receipt of bids, allow adequate time for evaluation of each bid. Review each bid individually to ascertain that the bid is responsive and is comprehensive in nature and that the proper bidder certifications are provided.
- Do not base contract award on price alone. Award of contract based on a combination of quality, price, and schedule can provide a more successful selection process.
- Require proposal drawings for performance-based bids. Test reports and calculations should also be submitted for evaluation. Fully evaluate these documents and resolve any concerns before contract award.
- Meet with the bidders for a face to face review of their bids during the bid evaluation process. Perform a detailed review of their bids to confirm compliance with project requirements.
- Allow time for the bidders to consider and respond to any questions that may result from the detailed review. If necessary, allow the bidders to modify their bids in response to these issues.
- For projects with complex phasing, scheduling, or other special schedule-related concerns, provide this information and require the bidders to confirm their ability to comply with these special needs. It is prudent to require the bidders to submit a detailed response indicating their plans on how to handle these conditions.

- In addition to defining the Owner's Project Requirements and criteria through the SOW/OPR process, the methodology for validating both the design and construction will deliver the desired results must be considered and identified within the project contract documents for all parties of the process.

Preconstruction: Phase 5

As with the procurement phase, there are provisions the designer can build into the documents, including drawings, technical specifications, and front-end documents for the Preconstruction phase, that will help to increase the quality of the completed project and the exterior envelope.

The Preconstruction phase covers the activities between award of the contract and delivery of materials, products, and systems to the building site. This includes final design and engineering, completion of mock-up construction and testing necessary before production, and fabrication and delivery of materials and systems for incorporation into the building.

The critical point regarding QA during this phase is to maintain control of the process through communication and attention to detail.

The following guidelines apply to all disciplines within specific DER projects:

- A submittal schedule from the contractors should be the first issue to be considered. This schedule needs to provide a comprehensive listing of all expected and required submittals and should include time for multiple submittals and reviews. It is recommended that a sampling strategy of randomly selecting 5% to 10% of the submittal be used to focus on the quality and ability of the submittal to achieve the OPR. Special attention must be paid to substitutions and proposed deviations from the contract documents and the Basis of Design (BoD) that could adversely impact the OPR.
- Require pre-submittal meetings to review submittal requirements and expectations. Review the technical and procedural requirements of the project as well as any revisions that may be required or that may be found to be helpful as the project develops. Require documentation of these meetings and distribute minutes to those involved.
- Require distribution of approved submittals to contractors of adjacent construction for purposes of coordination review. Conversely, provide submittals of critical construction systems adjacent to the contractors of adjacent systems for coordination. It is essential that coordination problems be resolved during this phase, not on the job site.
- Include a requirement to establish and identify means and methods of communications between the construction team members. Whatever the details are to be, make sure that communication is facilitated not hindered. It is not the designers or Owner's job to manage or control such communication. It is however within their means to encourage, and require contractually, open lines of communication to foster a valuable exchange of information and coordination.

- Require regularly scheduled meetings to review project development. These should be planned to keep the team in tune with the progress of development and identify problems that may require resolution. These meetings serve to notify the team of what level of activity will be required. This will help each team member perform in a timely fashion by providing advance notice of the type and amount of effort required of them. Again, it is not the designer or Owner's role to control or run these meetings. However, it is within their purview to require such meetings and require that it is demonstrated that they occur.
- Encourage separate problem resolution or working meetings outside of the regular progress meetings. These meetings can often be held with fewer people, will be more focused on specific technical or coordination issues, and are more likely to conclude with solutions to specific problems or concerns.
- Establish adequate review, transmittal, and response times for each cycle and each item required by the submittal process. Submittals that are prepared or reviewed in a rush are more likely to contain errors or be inadequately prepared or reviewed. This leads to frustration on the part of the team and too often to unacceptable or inadequate construction.
- Evaluate and resolve problems as they are discovered. At this stage of the project, there is no time available to push problems downstream. Questions that go unanswered at this point are likely to result in delays or unacceptable "field solutions." Additional costs are often the direct result of not adequately addressing issues remaining at this time.
- Require submittal of mock-up, production, fabrication, delivery, and erection/installation schedules. These are necessary to ascertain that adequate time will be available to deliver and install the wall system to meet the overall project schedule. These schedules should be provided in a format common to the overall project to allow integration and coordination. Projects that encounter schedule difficulties are likely to have a higher degree of systems failures.
- Resolve monetary issues as expeditiously as possible. If requests for additional compensation are submitted due to problems discovered during this phase, resolve them. If resolution is not possible, at least document the issue thoroughly and fairly for future resolution. Money problems have a way of disrupting good thinking and common sense, often subsequently leading to envelope performance problems.

The following procedures are suitable for use with projects of varying size, degree of difficulty, and delivery method, with modification to suit the specific project.

- In addition to defining the Owner's Project Requirements and criteria through the SOW/OPR process, the methodology for validating both the design and construction will deliver the desired results must be considered and identified within the project contract documents for all parties of the process.

Construction: Phase 6

The efforts required of the QAP during this phase may vary by the complexity of the building project and systems included. However some general guidelines should be followed for any project. These procedures are suitable to projects of varying size, degree of difficulty, and delivery method, with modification to suit the specific project. In all cases, the goals should be to communicate and pay attention to detail.

During the Design and Construction phases, special attention needs to be paid to details related to building enclosure systems renovation, with regard to continuity of thermal, water, vapor, and air barriers, and their systems coordination and installation. Testing and third-party review or commissioning provide valuable tools to successfully accomplish these tasks.

It is important to clearly understand at the commencement of the Construction phase that the work will not be accepted until it is adequately demonstrated that the energy targets are being met by the completed work. If the contractor cannot produce construction meeting the energy targets and criteria of the SOW/OPR, a review of the problems or constraints shall be presented along with a list of the proposed changes to the requirements.

Fabrication and Delivery Sub-phase. By following the proper QAP procedure during the fabrication and delivery sub-phase of construction, performance problems related to materials and systems can be minimized or avoided entirely. Action items during this phase should include:

- Periodic inspections of fabricated materials to ensure that the required quality is being attained.
  Fabricated products and assemblies should be checked to confirm that they conform to the project requirements and the systems as approved by testing and mock-ups. This will include any revisions made as a result of mock-ups and/or testing.
- Review erection drawings and installation instructions to confirm that any revisions required as a result of the mock-ups and/or testing have been included.
- Monitoring of the procurement of components to be delivered directly to the job site from separate suppliers. Confirm that it will be the correct material and will fit with other products being fabricated.
- Delivery dates and locations should be confirmed.
- Confirmation that the on-site construction schedule has been considered and accommodated by the fabrication and delivery schedule. The order of on-site erection and/or assembly can be critical.
- Review delivery requirements to confirm that the phasing of material delivery will accommodate the amount of available and appropriate on-site storage locations. Materials and fabricated systems that are not properly stored are more subject to problems in the finished construction.
- Review of the method of delivery as related to on-site materials handling capability. Fabricated assemblies that are improperly handled on-site may be subject to damage that can be avoided by planning ahead.
- Monitoring of packaging practices and procedures. Improperly packaged materials can be damaged to the point of being unusable on receipt at the job site. This can result in schedule problems and the attendant risk of errors in fabrication as pressure builds for the damaged material to be replaced or, worse, incorporated into the building.

- Review of the packing lists and manifests to confirm that all of the required components are delivered to the correct location, and on time. Missing items may be replaced with the incorrect material or left out completely in the rush to keep the project moving. This can result in problems.
- Verify that products and assemblies are packaged and shipped in a manner that will prevent damage. This is especially true of large metal windows or frames that can move during handling or transport, thus possibly opening corner joints. Confirm that needed bracing and/or packaging material is provided and that these assemblies will be correctly placed and secured in the shipping vehicles. Finished surfaces are also prone to damage by improper packaging and delivery.
- Require the erecting or installing contractor to check or "shake-out" the delivered material and assemblies for completeness and condition as it arrives. Erection or installation should not be started before this process is completed. It is critical to get the process for replacement of damaged or otherwise unacceptable material, or the delivery of missing material started as soon as possible. In this way, problems can be resolved before they become schedule critical.
- In addition to defining the Owner's Project Requirements and criteria through the SOW/OPR process, the methodology for validating both the design and construction will deliver the desired results must be considered and identified within the project contract documents for all parties of the process.

Mock-Up: Construction and Performance Testing Sub-phase. Many exterior envelope systems will require mock-up or performance testing to confirm that the wall design will conform to the desired appearance and provide performance to meet the criteria of the project. This process is especially important to those systems with appearance, design, or performance criteria that are unusual or that have not been previously constructed. Mock-ups may be performed and tested under laboratory conditions, on the construction site, or both. For more information on mock-up construction, see Annex H-3.

Preconstruction Meeting Before Arrival of Specific Materials and Systems on the Job Site. Issues to be addressed include:

- Phasing and schedule of construction.
- Owner occupancy and ongoing use of the facility for existing occupied buildings being repaired or retrofit.
- Review of previously constructed wall mock-ups where appropriate.
- Review of erection and installation drawings and instructions and systems interface details.
- Communications between designer, consultants, owners, and contractors.
- Review of problem resolution techniques and procedures to be followed.
- Lift schedules where common use of cranes or material hoists is provided.
- Site access for cranes where common use cranes or material hoists are not provided.
- Weather limitations for specific systems work.
- Meeting schedule for coordination with other work.

- Review of acceptance criteria for substrates to which multiple systems interact and corrective procedure requirements and procedures.
- Acceptable tolerances of the finished all trades.
- Cleaning procedures.
- Protection of completed work.
- Inspection and acceptance procedures and criteria for work in progress.
- Acceptance procedures for completed work.
- Convene a pre-installation meeting between the designer, consultants reviewing construction, the general contractor or construction manager, the trade contractors, and the tradesmen. During this meeting the following should be reviewed: the systems designs, the critical details, acceptable practices, unacceptable practices, and inspection procedures. This meeting helps to establish a strong working relationship that can provide great benefits, especially for systems that are labor intensive on-site.
- Schedule and hold regular progress meetings to review the status of work as it proceeds.
- Review storage conditions periodically to identify and correct deficiencies that may occur.
- Review in-place work regularly to assure conformance to project requirements, including protection as necessary. Notify contractors of required remedial work.
- Review typical detail conditions with contractors and tradesmen as the work is started. This is often the last opportunity to be sure that the work will be performed correctly.
- Identify examples of acceptable work for each system condition and review with contractors and tradesmen. When part of the project, the accepted and approved on-site mock-up should be used as the acceptable level of quality.
- Resolve problems as they are identified; do not procrastinate. Once on-site work is started, delays are not acceptable and may lead to unacceptable construction if not resolved quickly.
- Require a complete set of project documents be available on-site at all times, including submittals, samples, and installation instructions. Refer to these documents regularly.
- Check materials that are being incorporated into the buildings. Read labels, check manufacturer's precautions, and compare to approved submittals and project usage.
- Establish clear lines of communications between the designer, consultants, contractor/construction manager, manufacturers/fabricators, and installing or erection contractors. This is necessary to facilitate identification and resolution of problems with a limited amount of conflict and delay.
- Establish and foster working relationships on individual as well as organizational levels. Good individual working relationships have the potential to provide a very successful QA tool.
- Arrange access for inspection of the work, even and especially at hard to reach locations. If these locations are difficult to access, they may well be difficult to work on also and represent locations more likely to be a source of problems.

- Establish schedule and requirements for in situ progress testing and reporting. Involve all parties and distribute the results of meetings and testing.
- Perform “proof” or compliance testing early in the project to detect potential concerns at the earliest possible time. If you do not, a significant amount of non-conforming work may result before detection.
- Require cleaning test areas to establish acceptable methods, tools, and cleaning materials and end results. Document the results and identify an acceptable example area. Manufacturers of chemicals to be used and finished materials or systems should be represented at these meetings and test cleanings.
- Where possible, perform acceptance review of work as the work is completed by area or system. This will allow corrections to be completed while there is still ready access and help to discover unacceptable conditions that may have been missed. This will also facilitate correction of the problems in the ongoing work.
- Records of job site weather and progress are important and should be rigorously obtained and recorded. These can be valuable in identifying the sources of problems or likely problem areas.
- Require daily reports of work in progress, problems encountered, crew sizes and individuals, and planned areas of work for all exterior envelope work. These records may also prove valuable in problem resolution and can help the exterior wall contractor in planning their work.

Acceptance: Phase 7

The following guidelines apply to all disciplines within specific DER projects.

To assure that the Owner has received systems that meet the requirements of the contract documents, the design team and contractor/construction manager are required to review and confirm the completed product conforms to the contract requirements.

- Require submission of “as-built” records indicating the final construction of the systems. The project contract documents should contain detailed requirements for submission of these documents.
- Review qualifications of the specialized consultants or testing agencies to ensure that they possess the proper knowledge and experience required to perform and evaluate the results of the in situ testing procedures. Perform this review before they proceed.
- Hold a pre-acceptance meeting with the project team to review and confirm the acceptance process and schedule. Document and distribute the results of this meeting to all parties.
- Require that installing contractors, general contractor/construction manager, design team, and manufacturer or fabricator of the systems be represented at testing and/or acceptance reviews. This will facilitate understanding of written review comments and test reports.
- Require system manuals to be submitted and approved.

- For unique systems, require system training to be completed.
- Require written comments documenting the results of review and inspections. Depending on the nature of the project, comments or a legend and elevation drawings and details may also be required. Beware of tagging or marking comments directly on completed construction. Tags tend to blow away or wash off, and markings may damage finishes. If tagging or marking are to be used, review proposed methods with contractors and obtain their permission before proceeding.
- Review results of acceptance observations, inspections, and testing before corrections are implemented. All corrective procedures should be agreed on by the project team before they are started.
- If required, perform on-site testing or observations on examples of proposed corrective procedures before they are implemented. This may be limited to critical conditions where there is a serious flaw and doubts regarding the ability of the proposed corrective procedures to perform.
- Require final sign-off of the work by the designer, consultant, envelope systems sub-contractors, Owner's representative and general contractor, or construction manager.
- In addition to defining the Owner's Project Requirements and criteria through the SOW/OPR process, the methodology for validating both the design and construction will deliver the desired results must be considered and identified within the project contract documents for all parties of the process.
- When used, the third-party agent will assess the acceptance for commissioning.

Post-Occupancy Evaluation: Phase 8

The following guidelines apply to all disciplines within specific DER projects.

QA continues into the occupancy phase of a DER project. Certain post-occupancy tasks are critical for ensuring that DER-related energy performance goals as formalized in the Owner's Project Requirements, Basis of Design, energy simulation models, and construction documents are being met.

- Require the third-party commissioning specialist to return on-site for an inspection and performance data analysis of the DER project within the warranty period and to resolve all outstanding commissioning related issues. Schedule this inspection between 6 and 9 months after beneficial occupancy of the facility to allow for sufficient building operation and enough time for deficiency correction before warranty period expiration.

- Develop an Ongoing Commissioning (OCx) Report targeting certain key factors that best characterize and qualify building performance. Ongoing Commissioning is a continuation of the commissioning process well into the Occupancy/Operations phase to verify that a project continues to meet current and evolving Owner's Project Requirements. Ongoing Commissioning Process Activities occur throughout the life of the facility. Report shall include, at a minimum, evaluation of observed or reported facility use that deviates from design, summary of energy and water usage during the occupancy period, operation of RFP-required energy and water systems, and any other indicators or qualifiers of DER project performance. Highlight any deviation between actual and expected site and source energy intensities.
- Develop and administer occupant comfort surveys. Include survey questions on overall satisfaction with thermal, lighting, and space controllability that capture space areas/times associated with any systemic dissatisfaction. Require online surveys for easy distribution, collection, and analysis of survey content.
- Develop or update the Issues and Resolutions Log. Indicate completed vs. outstanding resolutions and describe whether observed or reported deficiencies are the result of facility use or operation that deviates from design conditions. Quantify the energy and water performance effect calculated or estimated for each resolved and unresolved issue logged. Develop a corrective action plan for any deficiencies that were not resolved.
- Require in the DER SOW/RFP that the contractor correct any issues that are not the result of abnormal facility use or operation relative to design during the warranty period.

Other DER post-occupancy tasks may be considered optional and dependent on Owner budget, QA team preferences, project goals, and post-occupancy facility conditions.

- Perform seasonal testing by repeating endurance performance required in the Construction phase of the DER project. Reuse commissioning trend length and interval requirements for data logging of control systems and evaluate a period no less than 1 operational week (168 consecutive hours).
- Conduct a Lessons Learned Workshop. Attendees shall include representatives from Owner, contractor, designer, maintenance, and commissioning teams. Document all project shortcomings, missed opportunities, process inefficiencies, and any other improvements to be applied to future DER efforts.
- Update relevant project documents based on OCx Report results including equipment inventories, systems manuals, and as-builts.

Annex H-1. Selected Energy-Related Parameters and Their Targets for DER Projects to Be Included in the SOW/OPR

Parameter	Sub-parameter	Criteria	Test procedure	Remarks
Energy targets	Site Energy, EUI	Minimum of _____KWh/m^2 yr [kBtu/ft^2 yr], >50% reduction compared to pre-renovation benchmark, but better than minimum national/agency requirement	For the bid submission, the bidder shall present a review of the energy requirements for the project to include site and source energy targets; energy calculation and modeling methodologies; and any conflicts or questions from the scope. The bidder shall provide a preliminary results of modeling analysis using simulation program allowing for monthly analysis along with an initial list of all energy parameters for the project to include operational runtimes, load peaks/schedules, equipment efficiencies/set points/sizes, insulation values, etc. Each entry/value shall be identified as being supplied from the project SOW, a specific design guide, a specific energy standard. Simulation results of the renovated building concept at 65% design using computer-based program such as DOE-2, Energy Plus, or similar per the ASHRAE Standard 90.1, Section G2.2.1. When economical and feasible compare with the metering data for 1 year on beginning of the building normal occupancy	
	Primary Energy, EUI	better, than minimum national/agency requirement	Building modeling at 100% design using local conversion factors	

Maximum thermal conductivity of the building envelope	External wall assembly above grade	Table 4 or better	Design review and manufacturers' specifications	
	Walls below grade	Table A4	Design review and manufacturers' specifications	
	Roof assembly, attic floor assembly	Table 5 or better	Design review and manufacturers' specifications	
	Slabs above unheated basement	Table A4	Design review and manufacturers' specifications	
	Slab-on-grade floors	Table A4	Design review and manufacturers' specifications	
	Window assembly	Table 13 or better	Design review and manufacturers' specifications	
	Opaque doors	Table A4	Design review and manufacturers' specifications	
	Skylight	Table A4	Design review and manufacturers' specifications	
Thermal bridges		A thermal bridge (or cold bridge) occurs when a thermally conductive material (such as a metal stud, steel frame or concrete beam, slab or column) penetrates or bypasses the exterior insulation system. Design the building envelope to reduce losses from thermal bridging by applying external continuous insulation and aligning all insulating elements, i.e., the continuous wall insulation, insulated glazing, and insulated doors from top of footing to bottom of roof deck. Pay special attention to wall-to-roof, wall-to-wall, and wall-to-window connections. Wrap insulation around roof	Existence of significant thermal bridges shall be verified using thermograph photography on completion of construction before testing air tightness with a pressure difference between inside and outside close to 0 (without building pressurization) and with no direct solar irradiation 3 hours before and at the time of the test	

(continued)

Parameter	Sub-parameter	Criteria	Test procedure	Remarks
		overhangs. Disconnect window and door sills from interior construction. Use thermally broken window and door frames. Provide details to eliminate thermal bridges particularly at floor slabs, roof/wall intersections, steel lintels and relief angles, metal through-wall flashings and at building corners		
Maximum building envelope air tightness		Design and construct the building envelope with a continuous air barrier to control air leakage into, or out of, the conditioned space that shall meet the requirements of Table 20. Design and construct the building envelope with a continuous air barrier to control air leakage in (or out of) the conditioned space. Clearly identify all air barrier components of each envelope assembly on construction documents and detail the joints, interconnections, and penetrations of the air barrier components. Clearly identify the boundary limits of the building air barriers and of the zone or zones to be tested for building air tightness on the drawings	Test the completed building and demonstrate that the air leakage rate of the building envelope does not exceed requirements listed in Table 20 in accordance with ASTM E 779 or ASTM E 1827. Conduct the test in accordance with the Engineer and Construction Bulletin "Building Air Tightness and Air Barrier Continuity Requirements" ECB 2012 -16 and the "US Army Corps of Engineers Air Leakage Test Protocol for Building Envelopes, Version 3, February 21, 2012" (http://www.wbdg.org/references/pa_dod_energy.php)	
		(b) Trace a continuous plane of airtightness throughout the building envelope and make flexible and seal all moving joints. The air barrier material(s) must have an air permeance not to exceed 0.004 CFM/sq ft at 0.3 iwg [0.02 L/s.m^2 @ 75 Pa] when tested in accordance with American Society for Testing and Materials (ASTM) E 2178.		

		Join and seal the air barrier material of each assembly in a flexible manner to the air barrier material of adjacent assemblies, allowing for the relative movement of these assemblies and components		
		(c) Support the air barrier so as to withstand the maximum positive and negative air pressure to be placed on the building without displacement, or damage, and transfer the load to the structure. Seal all penetrations of the air barrier. If any unavoidable penetrations of the air barrier by electrical boxes or conduit, plumbing, and other assemblies are not airtight, make them airtight by sealing the assembly and the interface between the assembly and the air barrier or by extending the air barrier over the assembly. The air barrier must be durable to last the anticipated service life of the assembly. Do not install lighting fixtures with ventilation holes through the air barrier		
		(d) Compartmentalize spaces under negative pressure such as boiler rooms and provide make-up air for combustion		
		e) Provide a mock-up of the exterior wall showing the connections at wall to foundation, wall to window, wall to door, wall to roof, wall to intake/exhaust ducts/pipes/conduits/etc. to ensure continuity of the air barrier		

(continued)

Parameter	Sub-parameter	Criteria	Test procedure	Remarks
Mold and mildew prevention		The DoR shall provide details in the design analysis and design showing steps taken to mitigate the potential growth of mold and mildew in the facility. Perform a wall and/or roof construction moisture analysis to verify appropriate thermal insulation and vapor permeability retardant assemblies to prevent condensation with the wall and/or roof under all foreseeable climate conditions. All gypsum board shall achieve a score of 10, the highest level of performance for mold resistance under the ASTM D 3273 test method. All gypsum board shall be transported, handled, stored, and installed in accordance with the GYPSUM ASSOCIATION – *Guidelines for Prevention of Mold Growth on Gypsum Board* (GA-238-03)	Perform a wall and/or roof construction moisture analysis to verify appropriate thermal insulation and vapor permeability retardant assemblies to prevent condensation with the wall and/or roof under all foreseeable climate conditions. All gypsum board shall achieve a score of 10, the highest level of performance for mold resistance under the ASTM D 3273 test method.	
Maximum duct and plenum air leakage		During construction, ductwork shall be tested using the pressure test [EN 15727:2010-10, DW143, ANSI/SMACNA 016-2012] to meet Class CL 2.1 (USA) or Class C (EU) ductwork air leakage requirement. The air leakage limit (liters/second per square meter of duct surface area) of the tested duct section, ΔQ shall not exceed: ΔQ_(duct system section) ≤ 0.003 x Δp^0.65	ALL supply, return, and exhaust air systems shall be tested for leakage during construction to verify good workmanship and the use of low-leakage components as required to achieve the leakage rate specified by Owner in the contract. Per ASHRAE [2016] as a minimum 25% of the system (including 100% of the ducts to be enclosed in chases and other concealed space and ducts installed outdoor), based on the duct surface area, should be tested during construction and another 25%, if any of initial sections fail. If any of the second 25% fails, the entire	

		where Δp is the maximum for design operating conditions ductwork pressure difference	system should be leakage tested. Sections should be selected randomly by the Owner's representative	
			On the whole system completion, it shall be tested at operating conditions using ASHRAE Standard 215 or EN 12599: 2013-01. Leakage tests should be conducted by an independent party responsible to the Owner's representative	
Minimum duct insulation		Table 12-8	Design review and manufacturers' specifications	
Minimum pipe insulation	Hot water pipe	Table 12-9	Design review and manufacturers' specifications	
	Cold/chilled water pipe	Table 12-9		
	Refrigerant pipe	Table 12-9		
Minimum heat recovery equipment efficiency		Table 12-2	Design review and manufacturers' specifications	
Minimum HVAC systems equipment efficiency		Table 12-1	Design review and manufacturers' specifications	
Lighting systems	Lighting level	Table 22 for specific building type and spaces	Design review	
	LPD	Appendix H for specific building type and spaces	Lighting level measurements on construction completion??	
Electric equipment and appliances	Energy Efficiency	EnergyStar	Confirm submittal equipment complies	
Thermal Comfort	Heating season	ISO Standard 7730, ASHRAE Standard 55 or similar National Thermal Comfort Standard.		
	Cooling season			
Indoor air quality		ISO/TC 146, Determination of volatile organic compounds (VOCs) in indoor air (working group)		

Annex H-2. Example of Accountability and Responsibilities Matrix

PDQA Task		Roles and Responsibilities					
		DOR[1]	CxG	CxD	CxC	COR	O&m[2]
Predesign	DoR/CxG selection	Participate by providing DER experience to O&M	Participate by providing Cx experience to DoR/O&M	~	~	~	Lead DoR/ CxG funding (P&D/ BOS $)
	DER OPR development	Approve	Lead				Participate
	DER energy modeling	Lead (or contract) energy modeling	Participate with energy modeling criteria	~	~	~	Provides operating schedules
	DER model completion	Lead selection of preferred model	Approve DER model systems/ EUIs	~	~	~	Approve DER model systems/ EUIs
	Project data sheet used to state requirements and justifications in support of funding requests	Lead 1391 DER descriptions	~	~	~	~	Review content
	DER cost estimating	Lead 1391 estimating from preferred model	~	~	~	~	~
	PDQA/TBCx criteria development	Review criteria	Lead cost estimating of PDQA and Tot Bldg. Cx	~	~	~	~
	PDQA/TBCx cost estimating	Review CxG PDQA/TBCx prices and add to 1391	Lead cost estimating of PDQA and Tot Bldg. Cx	~	~	~	Participate by contributing to O&M PDQA estimates
	DER RFP generation (D-B)	Lead development of RFP requirements	Review PDQA specific requirements				Review O&M-relevant RFP sections

Design	DER OPR review		Participate	Approve	Lead	~	~	Participate
	DER BoD development		Lead	Approve	Approve	~	~	Review
	Cx Plan development		Review	Approve	Lead	~	~	Participate
	35% energy model review		Review	Approve	Lead	~	~	Review
	35% plans and specs review		Review	Approve	Lead	~	~	Review
	100% energy model review		Review	Approve	Lead	~	~	Review
	100% plans and specs review		~	~	~	~	~	~
	100% TAB, Cx, envelope, HVAC, Utility Monitoring and Control System review		Review	Approve	Lead	~	~	Review
	Construction PDQA submittal register review		Review	Approve	Lead	~	~	Review
	Construction contractor evaluation		Review	Approve	Review	~	Lead	Review
Construction	Construction phase Cx Plan		Participate	Approve	Participate	Lead	Review	Participate
	Weekly Issues and Resolution meetings		~	Approve	~	Lead	~	~
	Submittal review, kickoff, coordination, and change orders for specific facility specifications		Participate	Review	Participate	Lead	Recommends approval	Participate
	Envelope	Air barrier requirement coordination	~	Approve	~	Lead	Participate	Participate
		Window mock-up	~	Approve	Participate	Lead	Participate	~
		Air barrier tightness test	~	Approve	Participate	Lead	Review	~
		Insulation R-value and installation review	~	Approve	~	Lead	~	~
	HVAC	System diagram check of piping connections and appurtenances	~	Approve	Participate	Lead	~	~
		Duct and pipe R-value and installation review	~	Approve	~	Lead	Participate	~

(continued)

PDQA Task			Roles and Responsibilities					
			DOR[1]	CxG	CxD	CxC	COR	O&m[2]
		Plant start-ups: cycling, staging, and resets	~	Review	~	Lead	Participate	~
		HVAC controls: programming logic sequences	~	Approve	~	Lead	~	Participate
		TAB validation: flow deviations, balancers in use, and controls issues	~	Review	~	Lead	Recommends approval	~
		PVT of each plant/terminal equipment	~	Review	~	Lead	Approve	~
		Prefunctional test review	~	Approve	~	Lead	Review	~
		Endurance testing		Approve	Review	Lead	~	
		Cx execution: functional tests, trending, point checks, alarms	Participate	Approve	Participate	Lead	Participate	Review
		Additional TAB/controls work	Participate	Review	Participate	Lead	Recommends approval	Participate~
		Systems Manuals	Participate	Approve	Participate	Lead	Participate	Review
		Summary Cx Report	Participate	Approve	Participate	Lead	Participate	Review
		Training	~	Review	~	Lead		Approve

Annex H-3. Mock-ups

H3.1. Mock-ups: What Are They?

"Mockups are full sized erected assemblies used to ensure understanding and coordination of required construction for testing and observation and for establishing standards by which workmanship will be judged." – CSI Definition. Examples of mock-ups are shown in Fig. H3.1.

Fig. H3.1 Examples of mock-ups

H3.2. Mock-up Construction

The major QAP points the designer must address during this phase of the project include the following procedures:

- Review submittal procedures relative to the mock-up/testing. This process should have been completed in the previous stage and should be confirmed at this point. Any energy performance criteria determined to validate mock-ups are correctly constructed must be agreed to and appropriate tests defined.
- Verify that mock-ups will be erected/constructed by the individuals who will erect/construct the envelope systems on the building.
- Verify location of testing or mock-ups on site (or on the building) have been fully coordinated. In the interest of the other aspects of the project, all ongoing construction activities must be considered.
- Verify the length of time that the mock-ups are required to be retained has been clearly communicated to the construction team. Premature destruction or removal of mock-ups can lead to problems.
- Confirm required levels of performance, and test procedures to be used are clearly verified with the construction team, including those parties performing the testing.
- Review configuration of the envelope mock-ups and the sequence of testing with the construction team. Sequence is particularly important if any of the testing will be destructive.
- Clearly state conditions to be tested and the acceptance criteria. Obtain verification from the construction team.
- Address conditions under which re-testing will be required and who will be responsible for the cost of re-testing. Obtain verification from the construction team.
- Meet with Owner and construction team to review procedural requirements for witnessing of the testing.
- Re-evaluate and confirm schedules for production of test panel or mock-up material, erection of the mock-up, and how access will be provided for on-site testing.
- Confirm the schedule for testing, whether on the building separate from building on-site or at a separate location such as a testing facility, has been properly addressed by the construction team.
- For on-site mock-up or testing, confirm the conditions if any, for which the mock-ups will be acceptable for inclusion in the finished envelope systems.

After these QAP efforts have been completed, the mock-up drawings have shall be developed (see Figs. H3.2 and H3.3), and construction and/or performance testing can proceed.

Additional QAP procedures to be followed for the remainder of mock-up/performance testing include:

- Document each step in the mock-up/performance testing. Include photographic as well as video documentation.

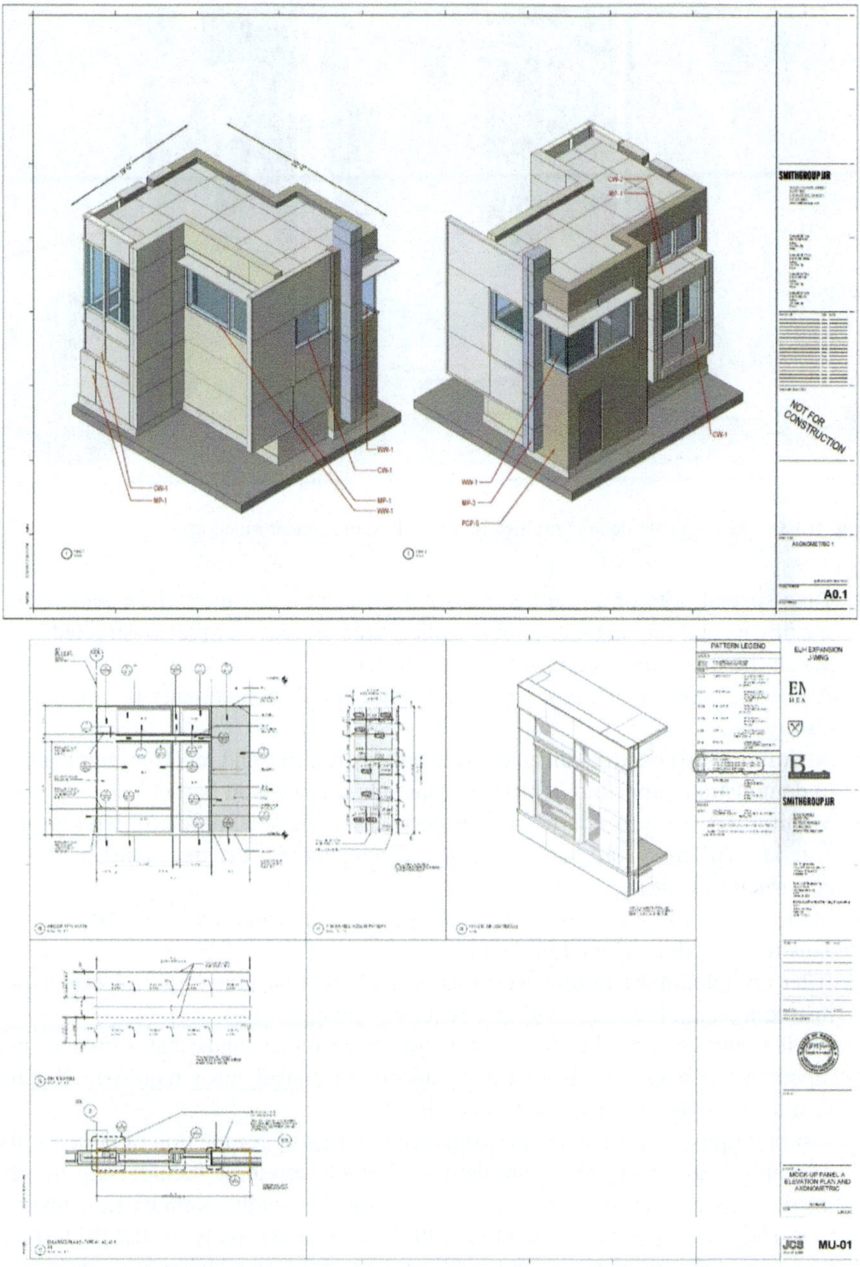

Fig. H3.2 Mock-ups drawings

(a) (b)

Fig. H3.3 Mock-up roof details: (a) historic roof, (b) replacement mock-up

- Take special note of any differences between the mock-up/performance testing specimens and the material to be installed. It is critical that the constructed work match the mock-ups as constructed and tested.
- Allow adequate time for sealants and other "wet" components to cure before testing.
- After mock-up construction and before testing, remove and reglaze sample panels to demonstrate procedures (metal or glass curtain walls in particular).
- Monitor testing procedures and results to confirm compliance with project criteria. Address and confirm any discrepancies or questions with the construction team.
- In the event of failure, modification to system details and retest may be required. This must be documented in detail.
- After completion of testing, verify shop and erection drawings and installation instructions are revised to indicate required changes.
- Review and confirm that the construction team understands that a final written report of test results from the testing agency or consultant is required. Obtain a firm date for receipt of this information.
- List prerequisites for the test performance in terms of completion of systems and assemblies and acceptable completion of other activities. Include step-by-step instructions to exercise the specific systems and assemblies under test. Instructions should include how to configure the system or assembly to start the test and how to restore the system to normal operation at the conclusion of the test.
- List of instrumentation, tools, and supplies required for the test. The list should indicate which of the participants is responsible for each of the items listed. The list should be specific as to make, model, range, capacity, accuracy, calibration, and other pertinent performance requirements.

- List what observations or measurements must be recorded and the range of acceptable results.

Keeping the mock-up and performance testing on the proper schedule track provides a critical link in avoiding performance problems in the completed building envelope.

Annex H-4. Use of Infrared Thermography (IR) to Identify Loss of Thermal Barrier Continuity and Thermal Bridging

H-4.1. Introduction

IR Thermography (Fig. H4.1) can be used to identify thermal bridging through visually accessible portions of the retrofitted building envelope when the IR inspection is performed under the correct environmental conditions:

- The Thermal Bridging Inspector possesses and can competently operate thermal imaging equipment.
- The Thermal Bridging Inspector possesses adequate construction knowledge and experience, familiarity with the drawings and design of the subject structure and locations of potential thermal bridging sites specific to the subject structure, and adequate knowledge to recognize thermal anomalies representing the various building materials components viewed through an infrared camera.
- The Thermal Bridging Inspector possesses the knowledge, experience, and familiarity with the structure design to recognize, understand, and articulate the differences between thermal anomalies caused by thermal bridging and thermal anomalies by air leakage.

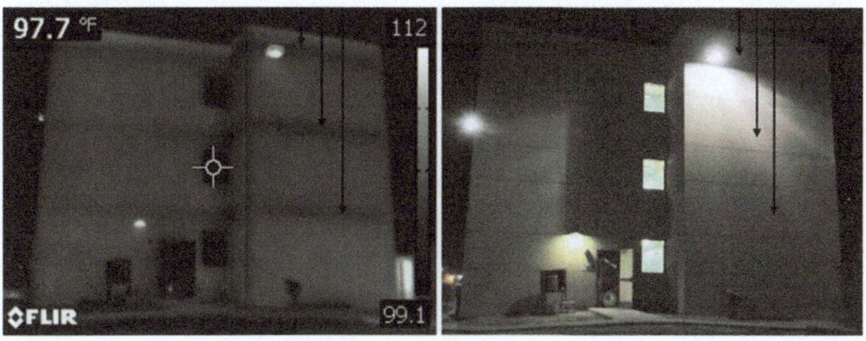

Fig. H4.1 Infrared Thermography image shows loss of thermal barrier continuity due to thermal bridging through exterior wall at floor to wall connections. Note straight anomalies. Source: TMI Air Barrier Testing

H-4.2. Problems and Issues

During the test, it would be ideal, but impracticable to eliminate anomalies caused by air leakage. However, it is impracticable to eliminate air leakage anomalies caused by stack effect. Therefore, for the purpose of real-world building diagnostics and testing under varied locations and weather conditions, it must be assumed that all building envelopes have some air leakage and some stack effect that causes air leakage in opposite directions at the top and bottom of the envelope. Installing a door fan at the top or bottom of the envelope would tend to stratify the envelope pressure and cause all air leakage to occur toward either the outside or the inside of the envelope, depending fan flow direction.

It is impracticable to use door fans to eliminate air leakage anomalies at individual floor heights by adjusting the neutral pressure plane height.

The theoretical neutral pressure plane is the only place in the envelope where stack-induced air leakage would not occur. The location of the neutral plane is not necessarily horizontal and is subject to vacillation due to variables such as stack effect, wind pressure, compartmentalization, or bottlenecks within the building envelope, etc.

Thermal anomalies caused by thermal bridging often occur at the same building envelope location as anomalies caused by air leakage. Anomalies caused by air leakage can overlap, exaggerate, or even obscure anomalies from thermal bridging.

As an example, thermal anomalies at connection joints between insulated steel panels are straight and thin when delta t=10 & delta p=0, but air leaks through these same joints at delta t=10 & delta p=25Pa cause anomalies to become wide and puddle shaped, as the steel surfaces absorb heat from the leaking air (Fig. H4.2).

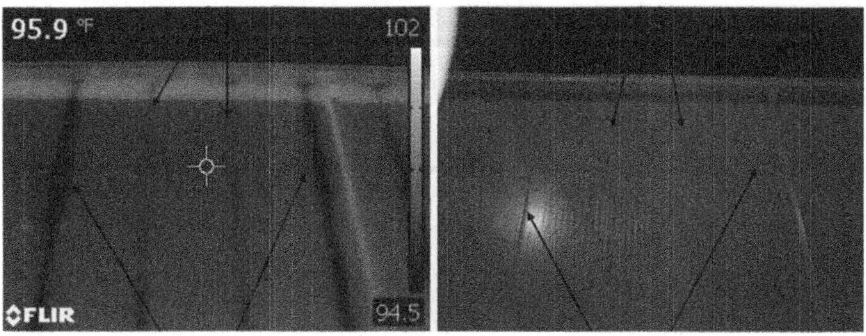

Fig. H4.2 Loss of thermal barrier continuity due to thermal bridging through metal edges of insulated metal panel joints and air leakage under induced pressure differential through insulated metal panel joints. Note curved/puddle shape anomalies. Source: TMI Air Barrier Testing

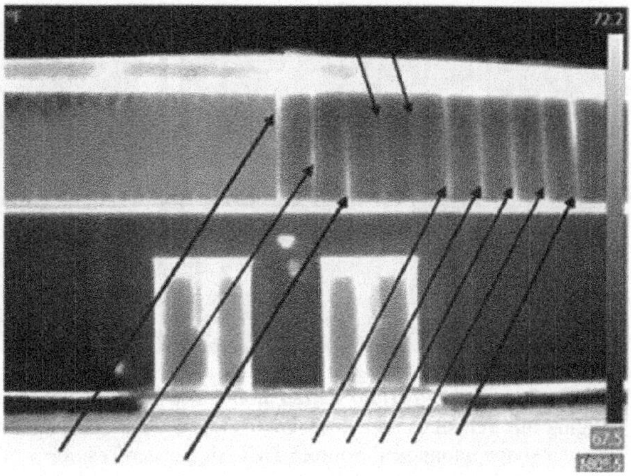

Fig. H4.3 Loss of thermal barrier continuity due to thermal bridging through metal edges of insulated metal panel joints and air leakage under induced pressure differential. Note straight edge anomalies. Source: TMI Air Barrier Testing

Thermal bridging anomalies typically have shapes different from those of air leakage anomalies.

Building components typically have straight edges; therefore anomalies resulting from thermal bridging are likely to consist of straight lines (Fig. H4.3).

Air leakage drawn across smooth surfaces tend to produce anomalies with wispy edges (Fig. H4.4).

Air leakage drawn through complex shapes causes the surfaces to absorb heat from the air and tends to have rounded shapes and smooth edges (Fig. H4.5).

Compare anomalies under natural (stack) condition, pressurization, and depressurization to help differentiate thermal bridging from air leakage.

An analysis of the same envelope anomaly locations under natural (stack) pressure, induced positive pressure, and induced negative pressure affords the best chance to single out and verify thermal bridging locations.

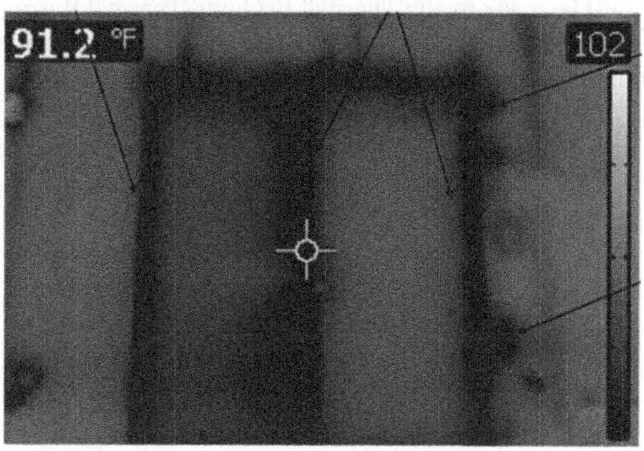

Fig. H4.4 Loss of thermal barrier continuity due to thermal bridging through fenestration frame and air leakage under induced pressure differential. Note straight edge and wispy shape anomalies. Source: TMI Air Barrier Testing

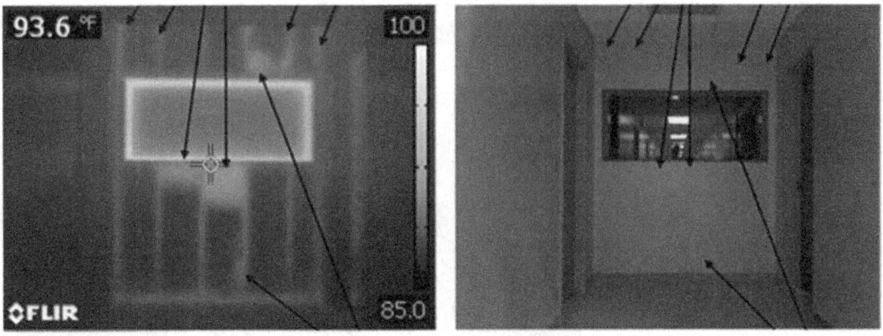

Fig. H4.5 Loss of thermal barrier continuity through voids at missing insulation and thermal bridging through studs and air leakage under induced pressure differential. Note straight edge and irregular shape anomalies. Source: TMI Air Barrier Testing

Anomalies that have straight lines or patterns and do not change in size or shape under all pressure conditions are most likely thermal bridges and not air leakage locations.

Anomalies that are visible during natural conditions and that change shape or size (either larger or smaller) under pressurization or depressurization are most likely air leakage at or adjoining thermal bridging locations.

Anomalies that only appear under pressurization or depressurization are most likely air leakage through or between building components, and less likely to be thermal bridging.

Anomalies that have curved lines or patterns are likely air leakage.

Anomalies that have wispy edges are likely air leakage where air is drawn straight along a smooth surface, such as floor surfaces under a door or gyp-board around a door frame.

These are general examples. However field conditions vary and the Thermal Bridging/Thermal Barrier Inspector must rely on training and experience to accurately analyze the cause of the thermal anomalies observed during the inspection.

Differentiation between anomalies caused by thermal bridging and by air leakage must be made by the Thermal Bridging/Thermal Barrier Inspector who has reviewed and understands the construction drawings and understands the DoR design to minimize thermal bridging.

Compare anomalies under natural (stack) condition, pressurization, and depressurization to help differentiate thermal bridging from air leakage.

An analysis of the same envelope anomaly locations under natural (stack) pressure, induced positive pressure, and induced negative pressure affords the best chance to single out and verify thermal bridging locations.

Anomalies that have straight lines or patterns and do not change in size or shape under all pressure conditions are most likely thermal bridges and not air leakage locations.

Anomalies that are visible during natural conditions and that change shape or size (either larger or smaller) under pressurization or depressurization are most likely air leakage at or adjoining thermal bridging locations.

Anomalies that only appear under pressurization or depressurization are most likely air leakage through or between building components and less likely to be thermal bridging.

Anomalies that have curved lines or patterns are likely air leakage.

Anomalies that have wispy edges are likely air leakage where air is drawn straight along a smooth surface, such as floor surfaces under a door or gyp-board around a door frame.

These are general examples. However field conditions vary and the Thermal Bridging/Thermal Barrier Inspector must rely on training and experience to accurately analyze the cause of the thermal anomalies observed during the inspection.

Thermal Bridging Inspector Qualifications

IR cameras have become quite reasonably priced and commonplace. Numerous venders offer brief, inexpensive training courses and issue documents indicating their trainee is certified to some level of expertise in infrared thermography. Individuals inspecting deep energy retrofit buildings to verify the design provided by the Designer of Record (DoR) was successfully executed in the field need more than a basic IR certification. Additional qualifications include:

- Substantial knowledge and experience in commercial construction, building science, and air/thermal barriers
- Clear understanding of the construction drawings of the subject building
- Clear understanding of the subject building-specific designs to eliminate thermal bridging surveyed
- The knowledge, experience, and ability to place the subject building under the various
- Pressures and conditions needed to execute the IR inspection.

Thermal Bridging Testing Procedures and Protocol

Designer of Record. The DoR shall clearly indicate the thermal boundaries of the structure with a heavy long-short-long dashed line or cross hatched line. Both vertical and horizontal boundaries shall be shown. Thermal boundary lines shall be such that a continuous line of thermal continuity can be traced around the structure and back to the point of beginning without lifting the pencil point. Use a separate plan page if necessary.

The DoR shall clearly detail the thermal boundary across all changes of plane and substrate.

The DoR shall clearly label each component of the thermal boundary along with the R-value of each of those components, including, but not limited to, fasteners and other penetrations through the thermal barrier.

Detail shall be such that a continuous line of thermal continuity can be traced through the assembly without lifting the pencil point.

The DoR shall draw all details at a scale such that installers in the field can clearly see the individual components making up the assembly on the drawings. Use exploded views if necessary.

Design Review. Contractor shall engage services of a qualified Thermal Bridging/Thermal Barrier Inspector. The inspector shall perform a design review of the plans provided by the DoR and provide a written report indicating areas wherein the plan's loss of thermal barrier continuity/thermal bridging occurs. Report shall include both marked-up plans with comments and arrows indicating locations where the loss occurs, a written narrative. Note: a design review for thermal bridging/thermal barrier continuity can be economically performed at the same time as design review for air barrier continuity.

Field Inspections. The Contractor shall engage services of a qualified Thermal Bridging/Thermal Barrier Inspector. Inspector shall conduct inspections at appropriate times and frequency in the construction schedule and shall provide written inspection reports regarding thermal barrier continuity and compliance with the job specifications the plans provided by DoR.

Test Setup and Building Conditions. Setup building for testing in accordance with Air Leakage Testing Protocol/Standard (e.g., USACE Air Leakage Test Protocol for Building Envelopes).

Ensure that the building envelope is free of all solar loads. IR inspection can be conducted on full overcast days. IR inspection of non-masonry buildings on non-overcast days should not start until 3 hours after sunlight is off of the structure. IR inspection of masonry veneer should not start until 6 hours sunlight is off of the structure. Environmental conditions shall comply with current ISO 6781 and ASTM C1060 standards.

Conduct IR thermal bridging/thermal barrier inspection from exterior under natural conditions before inducing any pressure changes on the building.

Take clearly focused thermograms and digital photographs of all visibly accessible surfaces of the structure. Start with large areas. Capture entire elevations in single field-of-view images if possible. Move in and take clearly focused images of wall sections, capturing grade-to-roof in single field-of-view images if possible. Ensure that images overlap each wall section so a complete record of all surfaces is recorded. Move in and take clearly focused images of all anomalies. For each image taken, mark or otherwise indicate on the standing surface the location the inspector stood and the direction the IR camera was pointed when the image was taken. Work in a consistent clockwise or counter-clockwise direction around the entire structure.

Place the building under a negative pressure of 25 Pa (0.0036 psi). Stand in the previously marked locations and take another set of thermograms and digital photographs from the same angles and covering the same fields-of-view. Capture the same elevations, wall sections, and anomalies in the same order. When new anomalies appear, take clearly focused images, mark or otherwise indicate the standing location inspector stood and direction IR camera was pointed when the image of the new anomaly was taken. Record new anomalies order they are observed.

Place the building under a positive pressure of 25 Pa (0.0036 psi). Stand in the previously marked locations and take another set of thermograms and digital photographs from the same angles and covering the same fields-of-view. Capture the same elevations, wall sections, and anomalies. If new anomalies appear, take clearly focused images and mark or otherwise indicate the standing locations inspector stood and direction IR camera was pointed when the image of the new anomaly was recorded. Record new anomalies in the order they are observed.

Compare thermograms of the anomalies taken under natural (stack) pressure, positive pressure, and negative pressure. Review the locations of the anomalies against the design drawings provided by DoR. Provide a report showing all anomalies and indicate the anomalies that were consistent with thermal bridging and those that were consistent with air leakage.

Annex H-5. Air Barrier Testing

One critical aspect of evaluating the performance of the new envelope is to test the new envelope for air leakage by quantifying the leakage and comparing the measured leakage with the design expectations. This critical step of the process will validate the quality of the installation and identify any deficient areas in the construction. The most common method for performing this test is the blower door test, also called a "fan pressurization test," which is a positive and negative pressure test on the envelope, or specifically, of the materials or system performing the air barrier function of the envelope, to determine a leakage rate at a given pressure. As indicated by the data in Table 9.1, specific requirements for a passing or failing leakage rate at a given pressure may vary between countries.

In addition to the blower door test, other diagnostic inspection or leakage location detection methods are performed during pressurization or depressurization to identify leak locations. These may include thermal investigations with a forward-looking infrared cameras, the use of theatrical smoke generation machines to find leak locations, the use of smoke pens, and even the use of simple "find an feel" techniques with one's hands. It is important to periodically perform visual inspections for evdence of leaks, e.g., for dirty insulation at roof/wall intersection, cobwebs, or dirt lines. Any of these methods can be employed to find leaks that need correction before the project is complete.

The time when a test on the envelope can be performed varies based on the contractor's schedule and on the proposed time for testing during the project. The optimal time to perform an envelope test is after the envelope nearing completion, but before any of the interior and exterior finish work has been installed. This provides the maximum access to areas in need of correction without having to remove and replace finishes and finish work.

The contractor constructing the building is ultimately responsible for the testing; the contractor's QC manager should oversee the preparations and process on testing day. The important roles, responsibilities, and functions of the contractor QC staff are:

- The QC should be familiar with the testing process as well as the contract documents related to the particular project to ensure the testing entity is performing the testing and diagnostics properly.
- The QC should review any and all qualifications of testing entities per contract, approve and be familiar with any required testing plans, conduct the prep meeting for the testing, and be present during the testing.
- The QC should be present during the testing of the project and, in particular, during the diagnostic evaluation and thermography. Some issues can be addressed on the spot. However, it is critical for the QC to have a general awareness of the conditions found during testing.
- The QC should ensure that air barrier surface area calculations have been completed and established before the testing.
- The QC should review and approve all final test reports and address any issues or deficiencies found or noted in the test report appropriately.

Once testing has been completed, the final tested performance value should be compared to that of the design expectations. If the tested value exceeds the minimum passing requirements, the contractor is required to find and identify all issues, make corrections, and retest until a passing value is achieved. The contractor is ultimately responsible for the work in place and needs to make the needed corrections. The DoR may need to be involved if design flaws in the air barrier system, beyond field installation quality defects in the system, are found or suspected. Failure to meet the minimum requirements will affect the overall performance and energy conservation design aspects of the building such that it will not perform as intended.

Annex H-6. Duct System Air Leakage Testing

During construction, sections of the air system should be tested using EN 12599 or SMACNA's HVAC Air Duct Leakage Test Manual (2012: Sections 4, 6, and 7). The air leakage should not exceed the air leakage *fraction* calculated using Eq. 12.4 for EU airtightness Class C (0.003 L/s per m^2) or ASHRAE air leakage class 2.1 (cfm per 100 ft^2) at the static pressure listed in Table H6.1. The system air leakage will not exceed 4.5% when the test pressure does not exceed 1500 Pa (6 in. water).

A minimum 25% of the system (including 100% of the ducts to be enclosed in chases and other concealed space and ducts installed outdoor), based on the duct surface area, should be tested and another 25% if any of initial sections fail. If the second 25% fails, the entire system should be air leakage tested. Sections should be selected randomly by the Owner's representative.

ASHRAE Standard 215 will be available late 2017 or 2018 to determine the air leakage airflow and fractional air leakage of operating HVAC air distribution systems. It is recommended that this standard be used when commissioning DER

Table H6.1 Air leakage test pressure for HVAC air systems

Type of system	Minimum test pressure
Small systems; split DX systems usually under 940 L/s (2000 cfm)	250 Pa (1 in. water)
Variable air volume (VAV) and constant air volume (CAV) boxes and associated downstream ductwork	250 Pa (1 in. water)
Single zone, multizone, return ducts, and exhaust duct systems	500 Pa (2 in. water)
Ducts in chases and concealed spaces, main return ducts on VAV and CAV systems, main ducts on general exhaust or outside air systems	750 Pa (3 in. water)
VAV and CAV boxes with upstream boxes	1000 Pa (4 in. water)
High-pressure induction system	1500 Pa (6 in. water)

buildings. Acceptable air leakage fractions should be the same as for pressurization tests during construction.

For most of typical HVAC systems and practical static pressure ranges, air leakage requirements for DER projects can be accomplished using Class $C_L2.1$ (USA) or Class C (EU) ductwork following testing procedures described above. Additionally, casing leakage for the air-handling unit (AHU) of supply air system should not exceed 1% of the design airflow.

Example H6.1 Example of the duct system depicted by Fig. H6.1 with characteristics presented in Tables H6.2a and H6.3 illustrates how to determine the air leakage fraction required for a DER HVAC air system air leakage test and what is the allowable air leakage rates for different test setups.

The system air leakage fraction calculated using Eq. 12.4 shall not exceed 2.7:

$$Q_{L,frac} = \left[\frac{(0.003)\left(750^{0.65}\right)}{\left(\frac{2360}{289.7}\right)}\right] 100 = 2.7\%$$

$$Q_{L,frac} = \left[\frac{(2.1)\left(3^{0.65}\right)}{\left(\frac{5000}{3117}\right)}\right] = 2.7\%$$

Calculations of the allowable tested air leakage rate using Eq. 12.2 for the system described above are summarized in Tables H6.4 and H6.5 for the following test setups:

(1) Each section
(2) Sections 4 and 5 combined
(3) 944 m^2 (500 ft^2) of Section 5
(4) Entire system

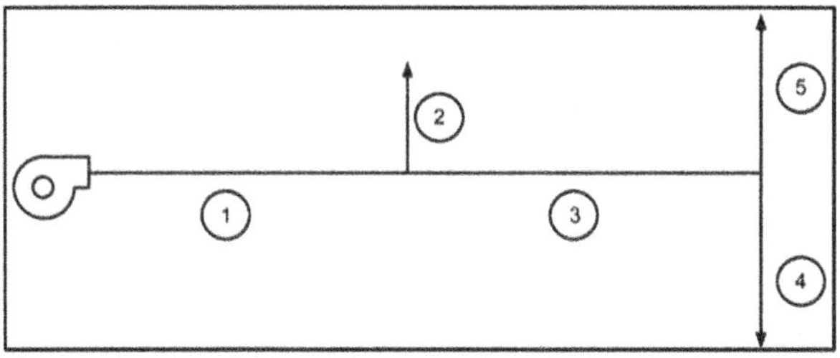

Fig. H6.1 System layout for Example H6.1

Table H6.2 System characteristics (input) (SI units)

Section	Section inlet flow, L/s	$A_{section}$, m^2	Section static pressure, Pa
1	2360	46.5	750
2	944	62.0	250
3	1416	69.7	500
4	472	18.6	250
5	944	92.9	250
Total		289.7	

Table H6.3 System characteristics (input) (I-P units)

Section	Section inlet flow, cfm	$A_{section}$, ft^2	Section static pressure, in. water
1	5000	500	3.0
2	2000	667	1.0
3	3000	750	2.0
4	1000	200	1.0
5	2000	1000	1.0
Total		3117	

Table H6.4 Allowable air leakage for the example system test (SI units)

	Inlet flow, L/s	$A_{section}$, m^2	Section static pressure, Pa	Section airtightness class, L/s per m^2 per $(Pa)^{0.65}$	Section allowable air leakage, L/s
Section	Input			Equation 12-1	Equation 12-2 or 12-3
1	2360	46.5	750	0.003	10.2
2	944	62.0	250	0.006	13.6
3	1416	69.7	500	0.004	15.3
4	472	18.6	250	0.006	4.1
5	944	92.9	250	0.006	20.4
Total		289.7			63.7
4 and 5	1416	111.5	250	0.006	24.5
Partial 5	944	46.45	250	0.006	10.2
—	**2360**	**289.7**	**750**	**0.003**	**63.7**

Table H6.5 Allowable air leakage for the example system test (I-P units)

Section	Inlet flow, cfm	$A_{section}$, ft^2	Section static pressure, in. water	Section air leakage class, cfm per 100 ft^2 per (in. water)$^{0.65}$	Section allowable air leakage, cfm
	Input			Equation 12-1	Equation 12-2 or 12-3
1	5000	500	3.0	2.1	21.7
2	2000	667	1.0	4.3	28.9
3	3000	750	2.0	2.8	32.5
4	1000	200	1.0	4.3	8.7
5	2000	1000	1.0	4.3	43.3
Total		3117			135.0
4 and 5	3000	1200	1.0	4.3	52
Partial 5	2000	500	1.0	4.3	21.7
–	5000	3117	3.0	2.1	135

Appendix I: Economics of DER

Most of major renovation projects include scope of work, which can be either energy- or non-energy-related. Non-energy-related scope of work may include such elements as different construction jobs related to changing floor layouts (e.g., moving/removing internal partitions), adding bathrooms, removing asbestos, adding sprinkler system, etc. An energy-related scope of work for a major renovation project typically includes replacement of existing mechanical and electrical systems, which will meet current minimum standard requirements, electrical work, replacement of some or all windows, replacement of exiting ductwork and plumbing systems. The scope of work with DER can include the same items, but using higher efficiency equipment and systems, as well as additional items that are not often a part of typical major renovation projects (i.e., building envelope insulation, improvement of building airtightness, heat recovery, etc.). Some improvements, such as additional insulation and high-performance windows, increase the cost of renovation; other improvements (e.g., smaller heating and cooling systems, boilers and chillers) allow reduction in cost compared to similar budgeted items. Therefore, the overall budget for DER project is typically higher than one required for a major renovation that seeks only to minimize energy requirements (Fig. I.1).

The budget increase allowance (Δ Budget) compared to the budget allocated for a major renovation project can be calculated using fundamental formula for NPV calculation:

$$\begin{aligned}\Delta \text{Budget}_{max} &= \text{NPV}\left[\Delta \text{Energy}\,(\$)\right] + \text{NPV}\left[\Delta \text{Maintenance}\,(\$)\right] + \\ &\text{NPV}\left[\Delta \text{Replacement Cost}\,(\$)\right] + \text{NPV}\left[\Delta \text{Lease Revenues}\,(\$)\right]\end{aligned} \tag{I.1}$$

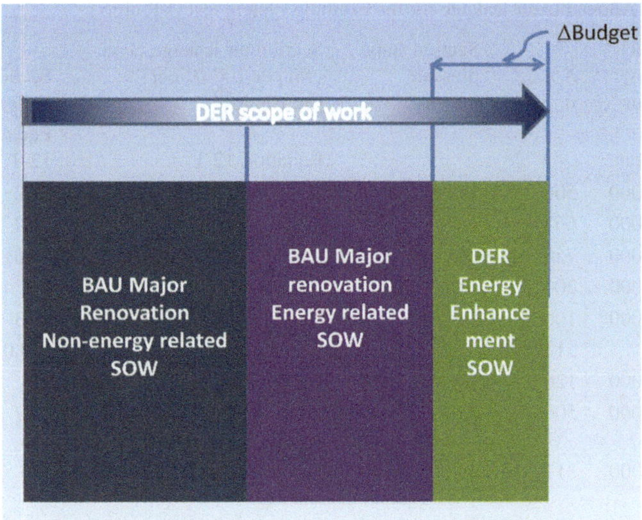

Fig. I.1 Scope of work of DER project

where NPV [Δ Energy ($)] is a present value of future energy cost savings for the project with the project life of N years, due to reduced use of electricity (E), gas (G), and other fuels (OF).

$$\text{NPV}\,[\Delta\,\text{Energy (\$)}] = \text{NPV}\,[\Delta\text{E} \times \text{C}_\text{E}] + \text{NPV}\,[\Delta\text{G} \times \text{C}_\text{G}] + \text{NPV}\,[\Delta\,\text{OF} \times \text{C}_\text{OF}] \tag{I.2}$$

where:

C_E, C_G, C_{OF} are unit fuel prices
ΔE, ΔG, ΔOF – annual electricity, gas, and other fuel saving.

For each fuel type, NPV of energy cost saving NPV can be calculated using the following formula (using gas as an example):

$$\text{NPV}\ [\Delta\text{G} \times \text{C}_\text{G}] = [\Delta\text{G}]_{\text{t}=1} \times \text{C}_{\text{G(t}=1)} \times \sum\nolimits_{t=1}^{t=N} \boldsymbol{It} \times \left[(1+\text{d})^\text{N} - 1\right] / \left[\text{d}(1+\text{d})^\text{N}\right] \tag{I.3}$$

where:

It = projected average fuel price index.
$\mathbf{C_{G(t=1)}}$– **gas unit p**rice in the first year.

To simplify calculations, the energy unit price change from year to year can be assumed to be at a constant rate (or escalation rate) over the study period. The escalation rate can be positive or negative. The formula for finding the present value

(**NPV [ΔG x C_G]**) of an annually recurring cost savings at base-date prices (**$C_{G(t=1)}$**) changing at escalation rate **e** is:

$$NPV\left[\Delta G \times C_G\right] = \left[\Delta G\right]_{t=1} \times C_{G(t=1)} \times (1+e)/d-e) \times \left[1-(1+e)/1+d)\right]^N \tag{I.4}$$

In Eq. I.4:

NPV [ΔMaintenance ($)] – present value of future maintenance cost savings
NPV [ΔReplacement Cost ($)] – present value of future replacement cost reduction
NPV [ΔLease Revenues ($)] – increase in revenues from the space lease.

The formulas for calculating NPV [ΔMaintenance ($)] and NPV [ΔLease Revenues ($)] are based on the discount or inflation rate, d:

$$NPV\left[\Delta Maintenance\ (\$)\right] = \left[\Delta Maintenance\right]_{t=1} \times \left[(1+d)^N-1\right]/\left[d\,(1+d)^N\right] \tag{I.5}$$

where: $[\Delta Maintenance]_{t=1}$ – maintenance costs savings in the first year.

$$NPV\left[\Delta Lease\ Revenues\ (\$)\right] = \left[\Delta Lease\ Revenues\ (\$)\right]_{t=1} \times \left[(1+d)^N-1\right]/\left[d\,(1+d)^N\right] \tag{I.6}$$

where: $[\Delta Lease\ Revenues\ (\$)]_{t=1}$ – lease revenues increase in the first year

$$NPV\left[\Delta Replacement\ Cost\ (\$)\right]_T = \left[\Delta Replacement\ Cost\ (\$)\right]_T \times (1+d)^T \tag{I.7}$$

where: $[\Delta Replacement\ Cost\ (\$)]_T$ – equipment replacement cost saving in the year (T).

This analysis does not address the possibility of financing either any part of the major renovation cost or the cost increase to achieve DER. Therefore, there is no financing cost involved, and there is no need to account for the interest rate of financing.

When the whole scope of the project is financed or DER is financed using a combination of funding sources (see the Annex 61 "Deep Energy Retrofit – Business Guide"), the project budget will include the capital cost financing (see schematic in Fig. I.2).

Figure I.2 shows different scenarios with private funds used to either pay for the total scope of the project or used to extend the capacity of limited public funds. However, this model comes at a cost of capital costs financing. The cost of financing depends upon the study period and the interest for borrowing money.

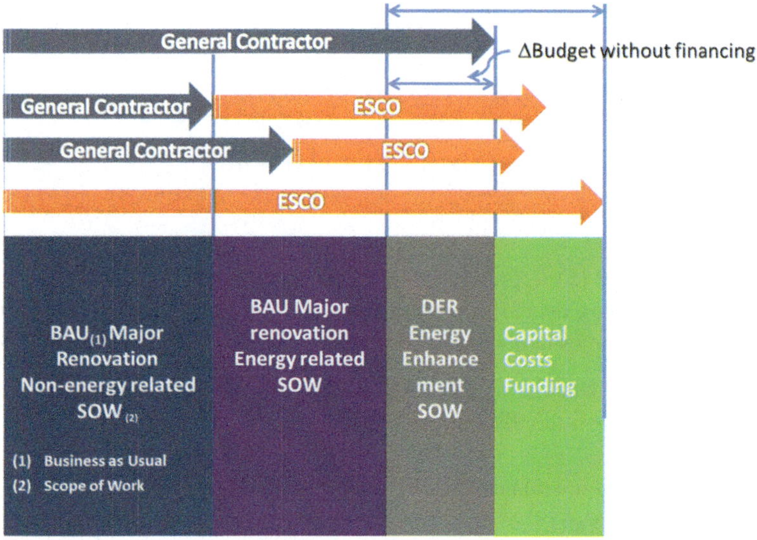

Fig. I.2 Budget required for DER using a combination of public and private funding

Each term in Eq. I.1 can be calculated in terms of net present dollars (€) or constant dollars (€). Instead of calculating the NPV of each term, this can be simplified by using economic scalar ratios (SRs) for energy and scalars (S) for maintenance, lease, and replacement. This simplification avoids the difficulty of selecting all of the individual economic parameters in determining the cost-effectiveness of projects, thus establishing a comparative economic feasibility threshold for analysis.

The SR is a summation of the annual present worth factors over project study life to produce a single present value factor (see McBride 1995 for detailed development). In estimating the NPV of energy savings, the ratio of the discount factor is factored into the fuel cost scalars to form the SR used in the economic analysis. The equation below shows the usage of the SR and S, simplified by neglecting the replacement value:

$$\Delta\,\text{Budget}_{\text{max}} = \text{SR}_{\text{E}}[\Delta\,\text{Energy (\$)}] + \text{S}_{\text{M}}[\Delta\,\text{Maintenance}] + \text{S}_{\text{L}}[\Delta\,\text{Lease Revenues}] \quad \text{(I.8)}$$

where: $\mathbf{S_M}$ and $\mathbf{S_L}$ scalars can be calculated and are the uniform present worth factor series that use the discount rate, the same way as $\mathbf{SR_E}$ with the escalation rate e=0%.

Table I.1 lists examples of SRs and Ss calculated at for a range of discount (Dsc) and escalation (Esc) values. These sample calculations are presented to demonstrate the sensitivity of the scalars with varying input values. The data in Table I.2 indicate

Table I.1 Ranges of escalation and discount rates for Annex 61 participating countries (2015 Data)

Country	Discount rate	Escalation rate
Austria	2.2%	3.9% – electricity
		4.8 – district heating
China	2.25%	2%
Denmark	2%	4%
Estonia	4.0%	3.0%
Germany	2.5%	0, 2, and 4%
Latvia	3.5%	0.9%
UK	3.5%	1.5%
USA (Lavappa and Kneifel 2015)	DOE -3%	2.8 – electricity
	OMB short term (< 7-year projects) – 0.7%	3.8 – gas
	OMB long-term (>30-year projects) – 1.4%	Nominal annual escalation rate averaged across all states at discount rate of 2.1% (for specific location and discount rate, use EERC 2.0-15 tool)
	For US Federal Projects, the discount rates and Escalation Rates are prescribed by the NIST Handbook 135 Annual Supplement	

that with Int = 0%, Dsc = 0%, and Esc = 0%, the SR is equal to the study life. The results of other calculations listed in Table I.2 demonstrate how the interest rate and the cost of money decrease SR, or cost-effectiveness, and fuel escalation increases the SR. Changes in the SR change the budget that can be spent economically on the project. For simplicity, the federal and state taxes are being held constant, but are incorporated in the SR as necessary. Table I.1 also lists (and Fig. I.3 shows) the scalars that are applied to maintenance, lease, and replacement cost delta values and that are just the uniform present worth value summation for study life. Individual values can be applied as necessary for each term depending on the years for the item of interest. If the SR and S are close in value, then these can all be simplified to one SR for clarity. Table I.2 lists ranges of country specific values of discount, escalation, and interest rates.

SRs are not uniform present worth factors, but are used in a similar fashion in that they are factors that are multiplied by the calculated single year energy savings to arrive at the present value cost. In addition, SRs are not equivalent to simple payback (SPB) since they account for the time value of money. Scalars are uniform present worth factors that are used similarly to SRs.

The usage of the SR in the context of the DER is the main focus here. Economic analysis for projects can be calculated by two methods, either absolute or incremental. The first method calculates the total LCC given all of the project costs; the second method calculates the annual savings and uses the incremental costs associated with that annual savings. The steps described in the previous section and in the example below show an incremental method approach.

Table I.2 Examples of SRe calculated for selected values of economic project life, interest, discount, and escalation rates

No.*	Economic life (years)		5	10	15	20	25	30	35	40	45	50
	Discount	**Escalation**										
1	0%	0%	5.0	10.0	15.0	20.0	25.0	30.0	35.0	40.0	45.0	50.0
2	0%	-1%	4.9	9.5	13.9	18.0	22.0	25.8	29.4	32.8	36.0	39.1
3	0%	1%	5.2	10.6	16.3	22.2	28.5	35.1	42.1	49.4	57.0	65.1
4	0%	3%	5.5	11.8	19.2	27.7	37.6	49.0	62.3	77.7	95.5	116.2
5	2%	-1%	4.9	9.5	13.9	18.1	22.2	26.2	30.0	33.6	37.2	40.7
6	2%	1%	5.1	10.5	16.2	22.1	28.2	34.6	41.2	48.1	55.2	62.5
7	2%	3%	5.5	11.8	18.9	27.1	36.4	46.9	58.7	71.9	86.6	103.0
8	4%	-1%	4.9	9.5	14.0	18.3	22.4	26.5	30.5	34.4	38.3	42.2
9	4%	1%	5.1	10.5	16.1	22.0	28.0	34.1	40.5	46.9	53.5	60.2
10	4%	3%	5.5	11.7	18.7	26.6	35.4	45.0	55.4	66.7	78.9	91.8
11	6%	-1%	4.9	9.5	14.0	18.4	22.6	26.9	31.0	35.2	39.3	43.4
12	6%	1%	5.1	10.5	16.1	21.8	27.7	33.7	39.8	45.9	52.1	58.4
13	6%	3%	5.4	11.6	18.6	26.2	34.4	43.2	52.5	62.3	72.5	83.0
*These data (indicated by "No.") relate to the curves in Figure I.3a.												
Scalars for maintenance and leases below, escalation = 0%												
1	0%	0%	5.0	10.0	15.0	20.0	25.0	30.0	35.0	40.0	45.0	50.0
2	1%	0%	4.9	9.5	13.9	18.0	22.0	25.8	29.4	32.8	36.1	39.2
3	2%	0%	4.7	9.0	12.8	16.4	19.5	22.4	25.0	27.4	29.5	31.4
4	3%	0%	4.6	8.5	11.9	14.9	17.4	19.6	21.5	23.1	24.5	25.7
5	4%	0%	4.5	8.1	11.1	13.6	15.6	17.3	18.7	19.8	20.7	21.5
6	5%	0%	4.3	7.7	10.4	12.5	14.1	15.4	16.4	17.2	17.8	18.3
7	6%	0%	4.2	7.4	9.7	11.5	12.8	13.8	14.5	15.0	15.5	15.8
8	7%	0%	4.1	7.0	9.1	10.6	11.7	12.4	12.9	13.3	13.6	13.8

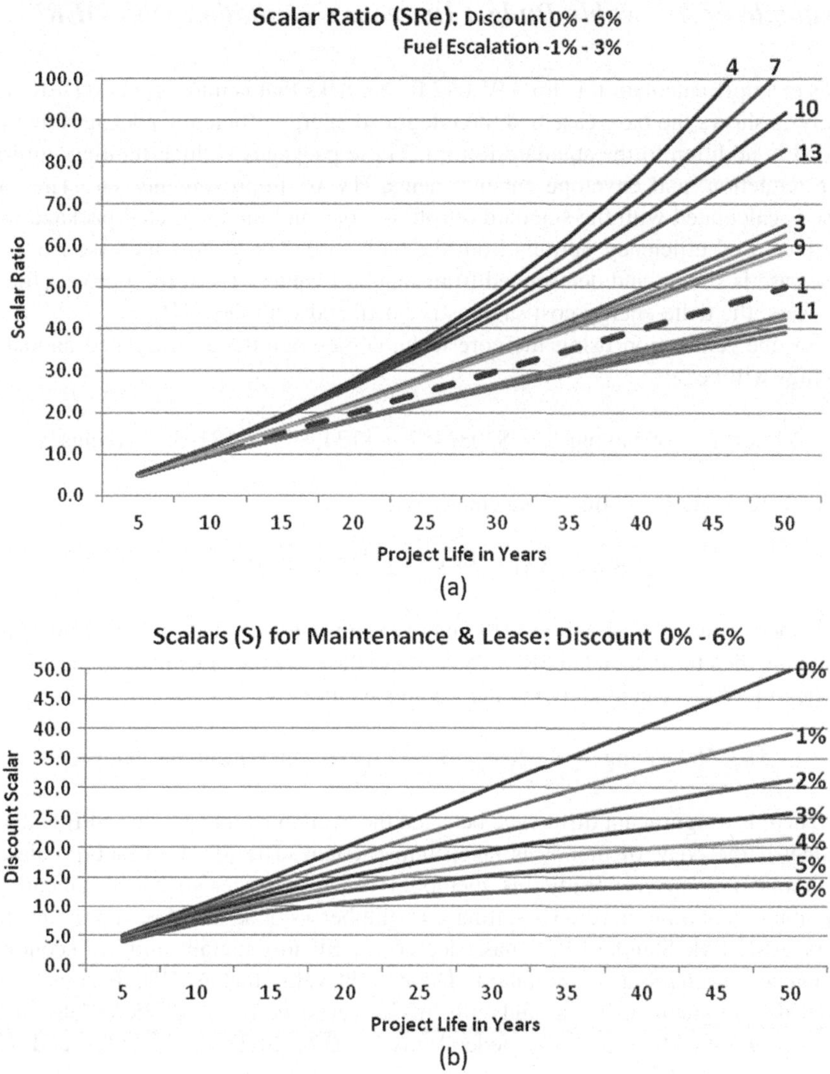

Fig. I.3 Scalar ratio for fuels at varying discount and fuel escalations rates (**a**) (note: curves are identified with the data listed in Table I.2) and scalars for maintenance and leases (**b**)

Example of Allowable Budget Increase Calculation with DER

This example calculation is for a 95,042 ft^2 barracks that is undergoing a DER. The analysis shows the base case and the integrated energy efficiency packages that are added in addition to the standard retrofit. These packages include lighting, infiltration reduction, and envelope enhancements, HVAC improvements, etc. The base case is calculated with the standard retrofit options and the integrated package with the increased efficiency options from the base case. The Δ cost increase from the base case is shown and determined from standard values. Then, for a project life of 25 years, the delta energy cost savings is calculated with the SPB.

For the renovation using the core technologies bundle, accumulated annual Δ savings will be:

$$\Delta \text{Energy Cost Saving \$} = \$208,427 - \$124,845 = \$83,582 \text{ savings/yr}$$

and a SPB available budget is calculated as:

$$\text{SPB for 25 yr} = \$83,582 * 25 = \$2,089,550$$

If there is no cost of money (0% interest) and one uses the same escalation and discount, then the calculated SR = 28.4, where the escalation rate adds to the NPV of money in a LCC calculation and the savings would be:

$$\$83,582 \text{ savings/yr} * 28.4 = \$2,373,729 \text{ (NPV available budget)}$$

There is a significant difference between the available budget using SPB and that budget if the NPV of money is taken into account. The SR does incorporate the economic parameters. With this methodology, it is easier to set a standard or minimum economic hurdle by setting one number vs. a series of economic parameters. ASHRAE Standard 90.1 has adopted the SR to maintain uniform economic stringency throughout the Standard. The current value that ASHRAE is using for 2016 development, using a study life of 25 years, is 14.8. ASHRAE also uses: Dsc = 9.34%, Int = 7.0%, federal tax = 0%, State tax = 0%, and fuel escalation = 2.38%.

Table I.3 lists the results for a 95,042 ft^2 barrack analysis starting with the final integrated package with the base case. The "Delta First Cost" column is the cumulative cost of the efficiency options quoted for the integrated package, $1,639,474. When there is no cost of money (basically using an SPB method), there is more than enough budget to cover the improvements. However, using NPV dollars, the financial package that is needed for this case would be an SR = 20, or a value above than needed to meet the quoted costs for the integrated package to attain 41% savings. With this analysis, you can either work with the financier to bargain for better rates or work with the contractor to establish a quote that fits the available budget. (Note that, for this example, the maintenance costs were neglected for

Table I.3 Economic analysis for a barracks renovation project with a 25-year project study life

	Energy reduction (%)	Electric kWh (kBtu)	Gas kWh (kBtu)	Total kWh (kBtu)	Total Utility Cost ($)	Delta first Cost ($)	Total SPB budget 25 years ($)	SR= 15	SR= 18	SR= 20	SR= 28.4
Base case (minimum standard)		1,431073	2,532018	3,963,091	$208,427						
		4,883,022	8,639,604	13,522,626							
Energy-efficient renovation	41%	878,209	1,456,489	2,334,698	$124,845	$1,639,474	$2,089,550	$1,253,730	$1,504,476	$1,671,640	$2,373,729
		2,996,573	4,969,746	7,966,319							

simplicity.) So for this example, any SR $\geq$ 20 will be cost-effective, so the SR currently used by ASHRAE of 14.8 will not meet the threshold. However, with no need to borrow money or with an interest rate = 0% then the current SR of 28.4 is more than adequate and the initial package could be extended further with the additional available finances.

The SR not only allows for a quick calculation once the analysis is complete, but it makes it easier to calculate and monitor the economic calculation as well. It also allows for quick comparison of one region or country to one another, i.e., a US value compared to a European value. If the USA had an SR = 16 and the EU had a value of SR = 22 for a 25-year study life, it would be easy to compare the economic stringency for each compared to all of the individual economic parameters and their effects.

Abbreviations

AABC	Associated Air Balance Council (AABC)
AAC	Autoclaved aerated concrete
AAMA	American Architectural Manufacturers Association
ABAA	Air Barrier Association of America
AC	Air-conditioning
AE	Architect-Engineer
AEE	Institute for Sustainable Technologies
AG	Army Garrison
AHU	Air handling unit
AMA	Swedish abbreviation for “General Requirements for Material and Workmanship”
ANSI	American National Standards Institute
ASHRAE	American Society of Heating, Refrigerating, and Air-Conditioning Engineers
ASIEPI	Assessment and Improvement of the EPBD Impact (where EPBD = Energy Performance of Buildings Directive)
ASTM	American Society for Testing and Materials
ATTMA	Airtightness Testing and Measurement Association
BC	British Columbia
BEMS	Building Energy Management System
BESA	Building Engineering Services Association
BETEC	Building Enclosure Technology and Environment Council
BS	British Standard
BSRIA	Building Services Research and Information Association (UK)
CAN	Canada
CAV	Constant air volume
CFM	Cubic feet per minute
CIBSE	The Chartered Institution of Building Services Engineers
CMU	Concrete masonry unit

A. Zhivov, R. Lohse, *Deep Energy Retrofit*,
https://doi.org/10.1007/978-3-030-30679-3

COGEN	Cogeneration
COGEN-SIM	The Simulation of Building-Integrated Fuel Cell and Other Cogeneration Systems
COMIS	Conjunction of Multizone Infiltration Specialists
c.z.	Climate zone
DBRI-AAU	Danish Building Research Institute – Aalborg University
DER	Deep energy retrofit
DIN	Deutsches Institut für Normung (the German national standards organization)
DK-DGNB	Danish Green Building Council
DOAS	Dedicated outdoor air system
DOE	US Department of Energy
DS	Dansk (Danish) Standard
DX	Direct expansion
EBC	Energy in Buildings and Communities Programme
EBPD	Energy Performance of Buildings Directive (EU)
ECB	Engineering and Construction Bulletin
ECBCS	Energy Conservation in Buildings and Community Systems
EEM	Energy Efficiency Measure
EFIS	Exterior Finish Insulation System
EnEV	Energieeinsparverordnung ("Energy-Saving Ordinance")
EPBD	Energy Performance of Buildings Directive
EqLA	Equivalent Leakage Area at 10 Pascals
ERDC	US Army Engineer Research and Development Center
ERV	Energy recovery ventilator
ESCO	Energy Service Company
ESPC	Energy Savings Performance Contract
ESTCP	Environmental Security Technology Certification Program
ETICS	External thermal insulating composite systems
EU	European Union
EUI	Energy use intensity
EVS	Estonian Centre for Standardization
FC	Fuel cell
FEBY	Forum för energieffektiva byggnader (Swedish Center for Zero-Energy Housing)
FEMP	Federal Energy Management Program
FIW	Forschungsinstitut für Wärmeschutz (German Research Institute for Thermal Protection)
FVHF	Fachportal vorgehängte hinterlüftete Fassaden (Expert Portal Rear-Ventilated Façades)
FY	Fiscal year
GB	Guobiao Standards (national standards issued by the Standardization Administration of China)
GFA	Gross floor area

HR	Heat recovery
HVAC	Heating, ventilating, and air-conditioning
HYBVENT	Hybrid ventilation
IAQ	Indoor air quality
IBP	Institute for Building Physics
IDEC	Indirect and direct evaporative cooling
IEA	International Energy Agency
IEA-EBC	International Energy Agency - Energy in Buildings and Communities Programme
IECC	International Energy Conservation Code
IESNA	Illuminating Engineering Society of North America
IG	Insulating glass
IMP	Insulated metal panel
IVD	Industrieverband Dichtstoffe E.V. (German Industry Sealant Association)
IP	Inch-pound (British system of units)
ISBN	International Standard Book Number
ISO	International Organization for Standardization
ISSN	International Standard Serial Number
KEA	Klimaschutz und Energieagentur (German Climate and Energy Agency)
LBNL	Lawrence Berkeley National Laboratory
LCC	Life cycle cost
LED	Light-emitting diode
LPD	Lighting power density
NBI	New Buildings Institute
NEN	Nederlands Normalisatie-Instituut (Netherlands Standards Institute)
NFRC	National Fenestration Rating Council
NIST	National Institute of Standards and Technology
NISTIR	National Institute of Standards and Technology Interagency Report
NPV	Net present value
NSAI	National Standards Authority of Ireland
NZEB	Nearly-zero-energy building
O&M	Operations and maintenance
OA	Outside air
OECD	Organization for Economic Cooperation and Development
OIB	Österreichisches Institut für Bautechnik (Austrian Institute for Structural Engineering)
ÖNORM	Österreichisches Normungsinstitu (Austrian Standards Institute)
OPR	Owner's Project Requirements
ORNL	Oak Ridge National Laboratory
OSB	Oriented strand board
PDQA	Product delivery quality assurance process
PHA	Passive House Academy

PHI	Passivhaus Institut (Passive House Institute)
PIER	Public Interest Energy Research
PIR	Polyisocyanurate
PNNL	Pacific Northwest National Laboratory
PU	Polyurethane
PVC	Polyvinyl chloride
QA	Quality assurance
QC	Quality control
R&D	Research and Development
REHVA	Federation of European Heating, Ventilating, and Air-conditioning
RFP	Request for proposal
RH	Relative humidity
RL	Richtlinie (Guideline)
RMI	Rocky Mountain Institute
ROI	Return on investment
SHGC	Solar heat gain coefficient
SI	Systeme Internationale
SMACNA	Sheet Metal and Air-conditioning Contractors' National Association, Inc.
SOW	Statement of Work
SOW/OPR	Statement of Work/Office of Primary Responsibility (SOW/OPR)
STC	Sound Transmission Coefficient
STR	Sound Transmission Rating
TC	Technical Committee
TUT	Tallinn University of Technology (Estonia)
TÜV	Technischer Überwachungsverein ([German] Association for Technical Inspection])
UESC	Utility Energy Service Contract
UIUC	University of Illinois at Urbana-Champaign
UK	United Kingdom
US	United States
USA	United States of America
USACE	US Army Corps of Engineers
USDOE	US Department of Energy
USEPA	US Environmental Protection Agency
UV	Ultraviolet
VAV	Variable air volume
VDI	Verein Deutscher Ingenieure (Association of German Engineers)
WDMA	Window and Door Manufacturers Association
WSVO	Wärmeschutzverordnung (German: Ordinance on Thermal Insulation)
XPS	Extruded polystyrene
ZVEI	Zentralverband Elektrotechnik und Elektronikindus (German Central Association of Electrical and Electronic Industry)

References

American National Standards Institute (ANSI)/Sheet Metal and Air-conditioning Contractors' National Association, Inc. (SMACNA) 016-2012. HVAC air duct leakage test manual. Sheet metal and air-conditioning contractors' National Association, Inc. (2012)

American Society for Testing and Materials (ASTM), Standard practice for building enclosure commissioning. E2813-12. ASTM International (2015).

J. Andersson. "Swedish experience with airtight ductwork," *REHVA J.* (2013, January)

J. Andersson. *AMA and certification of ventilation installers-Two Swedish ways of improving the quality of HVAC systems*. REHVA Journal August: 23-25. https://www.rehva.eu/publications-and-resources/hvacjournal-abstracts/042015-abstracts/amaand-certificationof-ventilation-installers-two-swedish-ways-ofimproving-the-quality-of-hvac-systems.html (2015)

W. Anis, *The impact of airtightness on system design*. ASHRAE J 43, 12. Atlanta, GA: ASHRAE (2001)

Anned 46, Energy and process assessment protocol. International Energy Agency (IEA), Energy Conservation in Buildings and Community Systems (ECBCS). Published by ASHRAE (2009)

ASHRAE, *Investigation of duct leakage*. ASHRAE Research Project (RP) 308 (Final Report) (1985)

ASHRAE. Standard 160-2009, *Criteria for moisture-control design analysis in buildings* (ANSI/ ASHRAE Approved), including Addenda a through e. http://www.techstreet.com/searches/ 13823482 (2009)

ASHRAE, ASHRAE advanced energy design guide for K-12 school buildings (HV14) (2010)

ASHRAE, *Advanced energy design guide for small to medium office buildings. Achieving 50% energy savings toward a net zero-energy building* (2011)

ASHRAE, *2013 ASHRAE Handbook—Fundamentals* (ASHRAE, Atlanta, GA, 2013a)

ASHRAE, ANSI/ASHRAE/IES Standard 90.1-2013, *Energy standard for buildings except low-rise* (2013b)

ASHRAE, *The commissioning process*. Guideline 0-2013 (ASHRAE, Atlanta, GA, 2013c)

ASHRAE, Energy efficiency in existing buildings. ASHRAE Standard 100-2015 (2015)

ASHRAE, ASHRAE Standard 215, *Method of test to determine leakage airflows and fractional leakage of operating air-handling systems*. (2016a)

ASHRAE. Chapter 21, "Duct design," In *ASHRAE handbook-fundamentals*. Atlanta: ASHRAE (2017)

ASHRAE. ANSI/ASHRAE Standard 215-2018, *Method of test to determine leakage airflows and fractional leakage of operating air distribution systems*. Atlanta: ASHRAE. (2018)

A. Zhivov, R. Lohse, *Deep Energy Retrofit*,
https://doi.org/10.1007/978-3-030-30679-3

ASHRAE Handbook systems and equipment: Chapter – Air-handling and distribution.
ASHRAE Handbook, HVAC systems and equipment. Chapter 25 (2008)
ASHRAE. *ASHRAE Handbook*, “HVAC systems and equipment.” ASHRAE. 1791 Tullie Circle, N.E. Atlanta, GA (2016b)
ASHRAE RP-1365. Thermal performance of building envelope details for mid- and high-rise buildings. Report No. 5085243.01. Prepared by Morrison Hershfield. ASHRAE (2011)
ASHRAE Standard 160-2009, *Criteria for moisture-control design analysis in buildings* (ANSI/ASHRAE Approved), including Addenda a through e. http://www.techstreet.com/searches/13823482 (2009)
Associated Air Balance Council (AABC), National Standards for Total System Balance. 1518 K Street, N.W., Suite 503 • Washington, DC, 2000 (2002)
M. Baechler, C. Strecker, J. Shafer. 2011. *A guide to energy audits*. PNNL-20956. (Pacific Northwest National Laboratory [PNNL], Richland, WA), https://www.pnnl.gov/main/publications/external/technical_reports/PNNL-20956.pdf
Z. Bastian, W. Feist, C. Baumgartner, et al., *Planner's guide for renovation with passive house components* (Passivhaus Institut, Darmstadt, 2012)
Bautechnik. IEA energy in buildings and communities programme Annex 61 “Business and Technical Concepts for Deep Energy Retrofit of Public Buildings” www.iea-annex61.org
BC Hydro, *Commercial new construction. Building envelope thermal bridging guide* available at http://www.bchydro.com/powersmart/builders_developers/high_performance_building_program.html?WT.mc_id=rd_construction (2016)
Bekanntmachung der Regeln für Energieverbrauchskennwerte und Vergleichswerte im Nichtwohngebäudebestand vom 30. Juli 2009
M. Bendewald, R. Hutchinson, S. Muldavin, R. Torbert, *How to calculate and present deep Retrofit value* (Rocky Mountain Institute, 2014)
BESA, DW/143. Duct air leakage testing. www.theBESA.com (2013)
K. Bettgenhäuser, R. de Vos, J. Grözinger, T. Boermans, Deep renovation of buildings: An effective way to decrease Europe's energy import dependency. Ecofys project # BUIDE14901by order of Eurima (2014)
E. Brandt et al., *Moisture in buildings (in Danish)* (SBi, Hoersholm, Denmark, 2009)
Brink International. http://www.brinkclimatesystems.nl (2016)
Brussels: RICS Europe.
BS EN 15727:2010-10. Ventilation for buildings – Ducts and ductwork components, leakage classification and testing. London: British Standards Institution (BSI) (2010)
BSRAE, Building Services Research and Information Association (UK) (BSRIA). Innovative M&E data sheets – Pre-gasketed circular ductwork system, Report Act 5/2002 Data sheet 3.1. www.bsria.co.uk (2002)
Building Council, Denmark (DGNB), DK-DGNB. DGNB manual for office buildings. Green Building Council Denmark, http://issuu.com/greenbuildingcouncil/docs/gbc_manual_kontor_1.1_web_singles?e=5618766/10480809 (2014)
Building Engineering Services Association (BESA), Sheet metal Ductwork. DW/144. www.theBESA.com (2013)
Building Envelope Thermal Bridging Guide – Analysis, Applications and Insights. British Columbia (BC) Hydro power smart, https://www.bchydro.com/content/dam/BCHydro/customer-portal/documents/power-smart/builders-developers/final-mh-bc-part-0-introduction.pdf (2014)
Building Regulations (BR10), The Danish Ministry of Economic and Business Affairs Danish Enterprise and Construction Authority Copenhagen 12. December 2010. http://bygningsreglementet.dk/file/155699/BR10_ENGLISH.pdf (2010)
J. Carmody, S. Selkowitz, D. Arasteh, L. Heschong, *Residential windows: A guide to new technologies and energy performance*, 3rd edn. (W.W. Norton & Company, New York, 2007). ISBN-13: 978-0-393-73225-2
R. Carrie, P. Wouters, *Critical steps for wide scale implementation of building and ductwork airtightness* (TightVent Europe, 2011). www.tightvent.eu
R. Carrie, P. Wouters, Critical Steps for wide scale implementation of building and ductwork airtightness. TightVent Europe. www.tightvent.eu

M. Case, A. Zhivov, R. Liesen, M. Zhivov, Building envelope optimization with US Army barracks and office building renovation in 15 DOE climate zones for deep energy retrofit. *ASHRAE Transactions*. Vol. 122, No. 1 (2016a)

M. Case, A. Zhivov, R. Liesen, M. Zhivov, *A parametric study of energy efficiency measures used in deep energy retrofits for two building types and U.S. climate zones*. OR-16-004 (ASHRAE Winter Conference, Orlando, FL, 2016b)

M. Case, A. Zhivov, D. Fisher, R. Liesen, Comparison of approaches to deep energy retrofit of buildings with low and high internal loads and ventilation requirements. ASHRAE Transactions. Vol. 123, Part 1 (2017)

CEN, *Ventilation for buildings –Test procedures and measuring methods for handling over installed ventilation and air conditioning systems* (European Committee for Standardization, 2000)

CEN, *Ventilation for buildings – Ductwork -- Strength and leakage of circular sheet metal ducts, EN 12237-2003* (European Committee for Standardization, 2003)

CEN. EN Standard 14239, *Ventilation for buildings- ductwork-measurement of ductwork surface area*. Brussels, Belgium: European Committee for Standardization. (2004)

CEN. Standard EN 1507, *Ventilation for buildings-sheet metal air ducts with rectangular section-Requirements for strength and leakage*. Brussels, Belgium: European Committee for Standardization. (2006)

CEN. Standard. EN 12599, *Ventilation for buildings-test procedures and measurement methods to hand over air conditioning and ventilation systems*. Brussels, Belgium: European Committee for Standardization (2012)

G. Christensen et al., *Internal additional insulation of a block of flats – building technology, prizes and experience* (in Danish) (SBi, Hoersholm, Denmark, 1978)

CIBSE (The Chartered Institution of Building Services Engineers). *Lighting guide 7: Lighting for offices* (1993) ISBN: 0900953632

Citterio, Marco, Manuela Cocco, Heike Erhorn-Kluttig. 2008. Thermal bridges in the Energy Performance of Buildings Directive [EU] (EBPD) Context: Overview on MS approaches in regulations. ASIEPI Project. Information paper P64 (28-4-2008).

M. Dake, Building simulation modeling report. M.E. GROUP, http://iea-annex61.org/# (2014)

L.D. Danny Harvey, *A handbook on low energy buildings and district energy systems* (Earthscan, Sterling, VA, 2006)

Dean, J., L. Herrmann, E. Kozubal, J. Geiger (National Renewable Energy Laboratory) M. Eastment (Eastment Consulting Inc.) S. Slayzak (Colorado), *Dew-point evaporative comfort cooling*. Energy and water projects demonstration plan SI-0821. Environmental Security Technology Certification Program (ESTCP). TP-7A40-56256-2 November 2012 (2012)

Deep Energy Retrofit – Case Studies, IEA EBC programme. Annex 61, Subtask A, (2017)

Details for Passive Houses Renovation: A Catalogue of Ecologically Rated Construction for Renovation, Ibo Osterreichisches Institut Fur Baubiologie und Okologie (ed). Birkhauser Verlag. ISBN: 9783035609530 (2016)

Deutscher Ingenieure (VDI) 2067, Economic efficiency of building systems - Fundamentals and economic calculation Wirtschaftlichkeit gebäudetechnischer Anlagen - Grundlagen und Kostenberechnung. VDI, Part 1 (2012)

R.C. Diamond, C.P. Wray, D.J. Dickerhoff, N.E. Matson, and D.M. Wang. *Thermal distribution systems in commercial buildings*. Report Number LBNL-51860. Berkeley, CA: Lawrence Berkeley National Laboratory. https://escholarship.org/uc/item/92m8h3vj (2003)

DIN 4108-2 DIN 4108-2:2003-07 *Wärmeschutz und Energieeinsparung in Gebäuden, Teil 2: Mindestanforderungen an den Wärmeschutz* (2003)

DIN V 18599-2: Energetische Bewertung von Gebäuden – Berechnung des Nutz-, End- und Primärenergiebedarfs für Heizung, Kühlung, Lüftung, Trinkwarmwasser und Beleuchtung – Teil 2: Nutzenergiebedarf für Heizen und Kühlen von Gebäudezonen

DOD, Unified Facilities Criteria UFC 1-200-02. High Performance and Sustainable Building Requirments. 1 March 2013. http://citeseerx.ist.psu.edu/viewdoc/download?doi=10.1.1.300.9181&rep=rep1&type=pdf (2013)

DOE, Energy efficiency and sustainable design standards for new federal buildings. www.govpulse.us/entries/2010/05/28/2010-12,677/energy-efficiency-and-sustainable-design-standards-for-new-federal-buildings#idp22682672. Washington, DC: Department of Energy (2010)

Energy in Buildings and Communities Programme (EBC), Deep energy retrofit – A guide to achieving significant energy use reduction with major renovation projects. Subtask A Report (2017b)

DS/EN 12464-1:2011, Light and lighting – Lighting of work places – Part 1: Indoor work places. European Committee for Standardization

W. Emmerich, A. Georgescu, M. Ginter, H. Garrecht, J. Huber, O. Hildebrand, A. Koenig, M. Laidig, E. Gruber, R. Jank, *EnSan Projekt Karlsruhe Görderlerstraße Garrecht* (Volkswohnung, Karlsruhe, Germany, 2011)

EN 12599: 2013-01. Ventilation for buildings – Test procedures and measurement methods to hand over air-conditioning and ventilation systems. London: British Standards Institution (BSI) (2013)

EN ISO 10211-1:1995 Thermal bridges in building construction - Heat flows and surface temperatures - Part 1: General calculation methods (1995)

EN ISO 10211-2:2002 *Thermal bridges in building construction - Calculation of heat flows and surface temperatures - Part 2: Linear thermal bridges* (2002)

Energy and Process Assessment Protocol, IEA ECBCS Annex 46. Published by ASHRAE (2009)

Energy Efficiency in Existing Buildings, ANSI/ASHRAE/IES Standard 100-2015. ASHRAE (2015)

Energy Efficient Technologies and Measures for Building Renovation: Sourcebook, IEA ECBCS Annex 46, http://www.iea-ebc.org/fileadmin/user_upload/docs/Annex/EBC_Annex_46_Technologies_and_Measures_Sourcebook.pdf (2014)

Energy in Business and Communities Programme (ECB). EBC Annex 61 business and technical concepts for deep energy Retrofit of public buildings, http://www.iea-ebc.org/projects/ongoing-projects/ebc-annex-61/ (2015)

Energy.gov, *Types of insulation. Web page* (USDOE, Washington, DC, 2016). http://energy.gov/energysaver/types-insulation

EnEV (Energieeinsparverordnung, "German Energy Saving Ordinance"). http://www.bbsrenergieeinsparung.de/EnEVPortal/DE/EnEV/EnEV2009/Download/EnEV2009,templateId=raw,property=publicationFile.pdf/EnEV2009.pdf (2014)

H. Erhorn, H. Erhorn-Kluttig, M. Citterio, M. Cocco, D. van Orshoven, A. Tilmans, P. Schild, P. Bloem, Kirsten Engelund Thomsen, Jørgen Rose. An effective handling of thermal bridges in the EPBD context. Final Report of the IEE ASIEPI Work on Thermal Bridges. www.asiepi.eu (2010)

H. Erhorn-Kluttig, H. Erhorn, Impact of thermal bridges on the energy performance of buildings. Information paper P148. http://www.buildup.eu/publications/2345 (2009)

EU (European Union). Directive 2010/31/EU of the European Parliament and of the Council of 19 May 2010 on the energy performance of buildings. http://eur-lex.europa.eu/LexUriServ/LexUriServ.do?uri=OJ:L:2010:153:0013:0035:EN:PDF (2010)

European Committee for Standardization (CEN), EN ISO 10211-1:1995 Thermal bridges in building construction – Heat flows and surface temperatures – Part 1: General calculation methods. Standard available via national standardization platforms as printed version or download (1995)

European Committee for Standardization (CEN) 2011. DS/EN 12464-1:2011. Light and lighting – Lighting of work places – Part 1: Indoor work places. European Committee for Standardization (2011)

EuroPhit, The EuroPhit Project. Web page, www.europhit.eu (2016)

W. Feist (editor), Ventilation systems for nonresidential buildings, Workgroup Cost Efficient Passive Houses, Report Book No 44, Darmstadt, (German language), see as well [passipedia.org] (2013)

W. Feist (editor), Energy efficient hot water systems, Workgroup cost efficient passive houses, Report Book No 49, Darmstadt, (German language), see as well [passipedia.org] (2015)

W. Feist (editor), Criteria for the passive house, EnerPHit . . . Building Standard," Passive House Institute, Darmstadt, Available for download at http://passivehouse.com/downloads/03_building_criteria_en.pdf (2016)

W. Feist, B. Kaufmann, R. Pfluger, S. Peper, J. Grove-Smith, Sanierung mit Passivhaus Komponenten– Planungsbegleitende Beratung und Qualitätssicherung, Tevesstrasse Frankfurt. Renovation Project Frankfurt Tevesstrasse, Darmstadt. The full project report is available for download at www.passiv.de (2009)

GreenBuildingAdvisor.com, EnerPHit—The passive house approach to deep retrofit. www.greenbuildingadvisor.com/blogs/dept/green-building-blog/enerphit-passive-house-approach-deep-retrofit (2013)

G. Guyot, F.R. Carrie, "Stimulation of good building and ductwork airtightness through EPBD. Assessment and Improvement of the EPBD Impact (for new buildings and building renovation)," Report April 2010. *Assessment and Improvement of the EPBD Impact* (where EPBD = "Energy Performance of Buildings Directive"). www.asiepi.eu (2010)

S. Herkel, F. Kagerer (editors), Advances in housing Retrofit. Processes, concepts and technologies. A handbook composed in the framework of IEA Task 37, Freiburg, Germany (2011)

A. Hermelink, A. Müller, *Economics of deep renovation—Implications of a set of case studies* (2010)

A.H. Hermelink, A. Müller, Economics of deep energy renovation. Implications of a set of case studies. Ecofys project # PDEMDE101646 by order of Eurima (2011)

IEA Annex 46, *Energy efficient technologies and measures for building renovation: Sourcebook.* IEA ECBCS Programme Annex 46 (2014)

IEA Annex 61, Deep energy Retrofit – Case studies. Energy in buildings and communities programme. Subtask B Report. (In print) (2015)

IEA-EBC (International Energy Agency - Energy in Buildings and Communities Programme). 2017. Deep energy retrofit – Case studies. IEA-EBC Programme. Annex 61, Subtask A

IMCOM-Wide Energy Enterprise Strategy: Energy Efficiency and Conservation. 2011. Draft

Instituto Valenciano de la Edificación (IVE), Catálogo de soluciones constructivas de rehabilitación. Web page. Valencia, Spain: IVE, http://www.five.es/tienda/product_info.php?products_id=108 (2011)

Interest Energy Research (PIER), California Energy Commission (PIER), Lawrence Berkeley National Laboratory (LBNL). Berkeley, CA. (2009, July)

International Energy Agency (IEA) Energy in Buildings and Communities Programme, Deep Energy Retrofit: A Prescriptive Guide to Achieve Significant Energy Use Reduction with Major Renovation Projects. Annex 61, Subtask A (2016, July)

ISO 10211, Thermal bridges in building construction -- Heat flows and surface temperatures (2007)

ISO 13788:2012: Hygro-thermal performance of building components and building elements – Internal surface temperature to avoid critical surface humidity and interstitial condensation – Calculation methods. http://www.techstreet.com/searches/13823594 (2012)

IWU, *End energy for heating: German building's typology* (InstitutWohnen Umwelt, Darmstadt, Germany, 2011). www.iwu.de/forschung/energie/laufend/tabula/

Jorgen Rose, Kalle Kuusk, Kirsten Engelund Thomsen, Targo Kalamees, and Ove Mørck. The Economic Challenges of Deep Energy Renovation: Differences, Similarities, and Possible Solutions in Northern Europe—Estonia and Denmark. OR-16-005. Orlando, FL: ASHRAE Winter Conference

B. Kaufmann, W. Hasper, Examples for detailed design of critical junctions at building envelope, Chapter 3 of [IEA Task 37] original source: [Feist 2009] (2009)

B. Kaufmann, S. Peper, R. Pfluger, W. Fiest, *Sanierung mit Passivhauskomponenten, Planungsbegleitende Berantung und Qualitatssicherung*. Tevesstrasse Frankfurt a.M. Passivhaus Institut. Darmstadt, February 2009. (www.passiv.de) (2009)

K. Kiatreungwattana, *Building energy simulation analysis: Building 619-621-623 & Building 624 Presidio of Monterey* (National Renewable Energy Laboratory (NREL), Monterey, CA/Golden, CO, 2014). http://iea-annex61.org/#

Latvian National Standardization Body Latvian Standard (LVS), *LVS-EN 15193:2009 Energy performance of buildings - Energy requirements for lighting* (LVS, Riga, Latvia, 2009)
P. Lavappa, J. Kneifel, Energy price indices and discount factors for life cycle cost analysis – 2015. Annual Supplement to National Institute of Standards and Technology (formerly National Bureau of Standards) (NIST) Handbook 135. NISTIR 85-3273-30. NIST. US Department of Commerce (2015)
M. Lawton, The impact of thermal bridges on effective thermal resistance and energy use in mid and high rise buildings. https://fpinnovations.ca/media/presentations/Documents/seminars/2012-nrcan/the_impact_of_thermal_bridges_on_effective_thermal_resistance.pdf London: British Standards Institute (BSI)
B. Lawton Mark, Ryan, J. Straube. Solutions and instructional aids for prevention of thermal bridges. W9132 T-10-002 Delivery Order 0024. Prepared by PERTAN Group for US Army ERDC-CERL. August 21, (2014)
M. Lawton, B. Ryan, J. Straube, *Solutions and instructional aid for prevention of thermal bridges* (ERDC-CERL, Champaign, IL, 2014). Work performed by The PERTAN Group under CERL US Army W9132T-10-D-0002, Delivery Order 0024
Lighting Guide 7: Lighting for Offices. The Chartered Institution of Building Services Engineers (CIBSE). ISBN: 0900953632 (1993)
Linab, TÜV test of five duct systems. http://campaign.lindab.com/click/docs/br-tuv-lindab-en.pdf (2009)
U. Lohse, *Evaluation of durability of materials in additional and high insulated houses* (in Danish) (SBi, Hoersholm, Denmark, 1992)
R. Lohse, H. Staller, M. Riel, *The economic challenges of deep energy renovation: Differences, similarities and possible solutions in Central Europe: Austria and Germany. OR-16-006* (ASHRAE Winter Conference, Orlando, FL, 2016)
L.-A. Mattson, Up to standards? A Guide to standards for ventilation duct work. February **19**, 2016 (2016). http://blog.lindab.com/up-to-standards-a-guide-to-standards-for-ventilation-ductwork
M. McBride, Development of economic scalar ratios for ASHRAE Standard 90.1R. Thermal performance of the exterior envelopes of Buildings VI. ASHRAE, DOE, ORNL and Building Enclosure Technology and Environment Council (BETEC) Conference. Clearwater Beach, FL. December 4–8, 1995, pp 663–677 (1995)
A. Miller, C. Higgins, Deep energy savings in existing buildings. CH-15-033 *ASHRAE Transactions*. Vol. 121, No. 1 (2015)
E. Mills, *Building commissioning – A Golden opportunity for reducing energy costs and greenhouse gas emissions*. Lawrence Berkeley National Laboratory (Report prepared for the California Energy Commission Public Interest Energy Researdh [PIER]) (2009)
Ministry of Economics of the Republic of Latvia, Latvian Construction Standard LBN 002-01 of 1 January 2003 "Thermotechnics of Building Envelopes" (2003)
Ministry of Economics of the Republic of Latvia, Cabinet Regulation No. 359 of 28 April 2009 "Labour Protection Requirements in Workplaces" (2009)
Ministry of Economics of the Republic of Latvia. *Regulations regarding the methodology for calculating the energy performance of buildings*, Cabinet Regulation No. 348. (2013, 25 June)
Ministry of Economics of the Republic of Latvia, Cabinet Regulation No. 348 of 25 Jun 2013 "Regulations Regarding the Methodology for Calculating the Energy Performance of Buildings" (2013a)
Ministry of Economics of the Republic of Latvia, Cabinet Regulation No. 383 of 9 July 2013 "Regulations Regarding the Energy Certification of Buildings" (2013b)
Ministry of Economics of the Republic of Latvia, Cabinet Regulation No. 382 of 9 July 2013 "Regulations Regarding the Independent Experts in the Field of Energy Performance" (2013c)
J. Munch-Andersen, *Additional insulation of blocks of flats (in Danish)* (SBi, Hoersholm, Denmark, 2008)
National Institute of Building Sciences (NIBS), *Building Enclosure Commissioning Process BECx. Guideline 3-2012* (NIBS, Washington, DC, 2012)
NBI, *Case studies* (New Buildings Institute, Vancouver, WA, 2014). http://www.newbuildings.org/casestudies

NEN 1068:2001 *Thermische isolatie van gebouwen – rekenmethoden* ('Thermal insulation of buildings – calculation methods') (2001)
L. Nock, C. Wheelock, Energy efficiency retrofits for commercial and public buildings. Energy savings potential, Retrofit business cases, financing structures, policy and regulatory factors demand drivers by segment, and market forecasts. Pike Research (2010)
M. Nørregaard et al., *External additional insulation of a block of flats – building technology, prizes and experience* (in Danish) (SBi, Hoersholm, Denmark, 1981)
M. Nørregaard et al., *Components to external additional insulation of façades* (in Danish) (SBi, Hoersholm, Denmark, 1984)
OIB, *OIB-Richtlinie 6—Energieeinsparung und Wärmeschutz* (Österreichisches Institut Für, Vienna, 2011)
T. Ojanen, H. Viitanen, R. Peuhkuri, K. Lähdesmäki, J. Vinha, K. Salminen, Mold growth modeling of building structures using sensitivity classes of materials. Buildings XI conference. http://www.techstreet.com/searches/13823613 (2010)
L. Olsen, G. Johannesson, *A guidance for the treatment of thermal bridges in building practice, 4th Nordic symposium on Building physics* (Espoo, Finland, 1996)
Österreichisches Institut Für Baubiologie und Okologie (IBO), *Details for passive house renovation: A catalogue of ecologically rated construction for renovation. 2016. Österreichisches Institut Für Baubiologie und Okologie* (Birkhauser Verlag, Basel, Switzerland, 2016). ISBN: 9783035609530
A. Pagan-Vazquez, J. Yu, S. Lux, D. Chu, R. Liesen, N. Alexander. *Thermal bridge mitigation catalogue*. ERDC (2015a)
A. Pagan-Vazquez, D. Chu, J. Straube, R. Ryan, M. Lawton, *Development of Thermal Bridging Factors for Use in Energy Models*. ERDC/CERL TR-15-10 (Engineer Research and Development Center, Construction Engineering Research Laboratory (ERDC-CERL), Champaign, IL, 2015b)
A. Pagan-Vazquez (ERDC, USA), J. Yu (ERDC, USA), S. Lux (ERDC, USA), D. Chu (ERDC, USA), J. Staube (BSC-Canada), M. Lawton (Morrison-Hershfield, Canada), B. Ryan (Passive House Academy, Ireland), Thermal Bridge Mitigation in Buildings on Army Installations. ASHRAE Transactions. Vol 122, Part 1 (2016)
Passipedia, The passive house resource. Web site, http://www.passipedia.org/planning/refurbishment_with_passive_house_components/thermal_envelope/minimising_thermal_bridges (2016)
Project #PDEMDE 101646. Brussels: European Insulation Manufacturers Association: Brussels.
Refurbishment of non-domestic buildings, TM53. CIBSE ISBN 978-1-906846-37-4 (2013)
Residential Buildings. Atlanta: ASHRAE
RICS. 2013. *Sustainable construction: Realizing the opportunities for built environment professionals*
J.M. Riel, R. Lohse (KEA, Germany), H. Staller (Austrian Energy & Environment [AEE], Austria), Building envelope parameters optimization for deep energy retrofit of public buildings in Germany and Austria. ASHRAE Transactions. Vol 122, Part 1 (2016)
J. Rose (DBRI-AAU, Denmark), K.E. Thomsen (DBRI-AAU, Denmark), O.C. Mørck (Cenergia, Denmark), K. Kuusk (TUT, Estonia), T. Kalamees (TUT, Estonia), T. Mauring (University of Tartu, Estonia), Economic challenges of deep energy renovation – differences, similarities and possible solutions for northern Europe – Estonia and Denmark. *ASHRAE Transactions*. Vol. 122, No. 1 (2016)
P. Schild, *Good practice guidance on thermal bridges and construction details – Part 2: Good examples*. Information Paper 189, http://www.buildup.eu/publications/8241 (2010)
P.G. Schild, J. Railio, *Duct system air leakage — How Scandinavia tackled the problem*. ASIEPE Project Technical Report P187, 9 January 2009 (2009)
J. Shonder, C. Nasseri, Achieving deeper savings in federal energy performance contracts.CH-15-032. *ASHRAE Transactions*. Vol. 121, No. 1 (2015)
SMACNA, HVAC duct construction standards – Metal and flexible. 3rd edition. ANSI/SMACNA 006-2006. Sheet Metal and Air-conditioning Contractors National Association, Inc. (2006)
G. Stankevica, A. Kreslins, A. Borodinecs, R. Millers, Achieving deep energy retrofit in Latvian public building – simulation study. Published in P. K. Heiselberg (ed.). 2016. *CLIMA 2016 –*

Proceedings of the 12th REHVA World Congress: Volume 10. Aalborg: Aalborg University, Department of Civil Engineering (2016)

Stanley Mumma. DOAS and building pressurization. ASHRAE Journal (2010, August)

Svensk Byggtjänst. Chapter Q-AMA VVS & Kyl 16: *Allmän material-och arbetsbeskrivning för vvs- och kyltekniska arbeten*. [General material and job description for plumbing and refrigeration works.] Stockholm, Sweden: Svensk Byggtjänst AB. https://byggtjanst.se/bokhandel/kategorier/projektering-upphandling/program-projektering-beskrivning/ama-vvs-kyl-16/ (2015)

M. Tillou, Energy performance path improvements in 90.1-2013. Presented at the ASHRAE Winter Conference, New York, January 18–22 (2014)

TÜV Rheinland, Advisory opinion regarding the comparative test and evaluation of the overall installation costs of various installed air systems under consideration of the functional requirements and seal in accordance with EN 12237 and EN 1507. TÜV order No. 9986501 (2008)

US Army Corps of Engineers (USACE). Air Leakage Test Protocol for Building Envelopes, Version 3, February 21, 2012 (http://www.wbdg.org/references/pa_dod_energy.php) (2012)

USACE, *Lighting design guide for low energy buildings – New and Retrofits* (ERDC, 2013)

USACE, Engineering and construction bulletin. *TechNote: Guidance for indirect evaporative cooling applications in Army buildings*. ECB 2016-15 (2016)

USACE, Engineering and construction bulletin ECB 2012-16, in *Building air tightness and air barrier continuity requirements*, (USACE HQ, Washington, DC)

Verein Deutscher Ingenieure (VDI) 2067, Economic efficiency of building systems - Fundamentals and economic calculation Wirtschaftlichkeit gebäudetechnischer Anlagen - Grundlagen und Kostenberechnung. VDI, Part 1 (2012)

A. Wagner, Wirtschaftlichkeit von Maßnahmen zur Energieeffizienzsteigerung bei der Sanierung der opaken Gebäudehülle; Economic figures of actions [...] thermal insulation of opaque building envelope parts, Master Thesis, Darmstadt (2015)

C.P. Wray, R.C. Diamond, M.H. Sherman. "Rationale for measuring duct leakage flows in large commercial buildings." Proceedings – 26th AIVC Conference, Brussels, Belgium. LBNL-58252, https://escholarship.org/uc/item/38x7347n (2005)

WSVO, Verordnung über einen energiesparenden Wärmeschutz bei Gebäuden. Wärmeschutzverordnung—WärmeschutzV. 11.08.1997, BGBI. I 1554 (1977)

WUFI (Wärme und Feuchte Instationär) – Oak Ridge National Laboratory (ORNL)/Fraunhofer IBP. Holzkirchen Germany

R. Yao (University of Reading, UK), X. Li (Chongqing University, China), B. Li (Chongqing University, China), M. Shahrestani (University of Reading, UK), S. Han (University of Reading, UK), Building envelope parameters optimization for deep energy retrofit of public buildings for different climate zones in UK and China. *ASHRAE Transactions*. Vol 122, No. 1 (2016a)

R. Yao, S. Han, X. Li, M. Shahrestani, and B. Li, Evaluation of building retrofit strategies in different climate zones. *ASHRAE Transactions*. Vol. 122, No. 1 (2016b)

R. Yao, X. Li, S. Han, B. Li, C. Lin, An analysis of energy efficient retrofit strategies for office buildings in China. CIBSE Technical Symposium (2016c)

R. Yao, M. Shahrestani, B. Li, X. Li, S. Han, Evaluation of Building Retrofit Strategies in Different Climate Zones. OR-16-023, Orlando, FL: ASHRAE Winter Conference

A. Zhivov, R. Liesen, D. Herron, G. Zielke, D. Carpio, Towards low energy buildings and building clusters – analysis for US army installations. Proceedings of the 15th international passive house conference. 27–28 May 2011, Innsbruck, pp. 131–138 (2011)

A. Zhivov, D. Herron, J.L. Durston, M. Heron, G. Lea, Air tightness in new and retrofitted US army buildings. *Int. J. Ventilation*. **12**(4) (2014). ISSN 1473 – 3315

A. Zhivov, R. Lohse, J. Shonder, C. Nasseri, H. Staller, O. Moerck, M. Nokkala, Business and technical concepts for deep energy Retrofit of public buildings. *ASHRAE Transactions*. Vol. 121, No. 1 (2015)

ZVEI Manual to DIN EN 12464-1. *Lighting of work places – Part 1: Indoor work places*. ZVEI (Central Association of Electrical and Electronic Industry) (2005)

The manufacturer's authorised representative in the EU is Springer Nature Customer Service Centre GmbH, Europaplatz 3, 69115 Heidelberg, Germany. If you have any concerns regarding our products, please contact ProductSafety@springernature.com

Printed and bound by CPI Group (UK) Ltd, Croydon, CR0 4YY
22/09/2026
02221048-0013